APPLIED ELECTRONIC COMMUNICATION

CIRCUITS, SYSTEMS, TRANSMISSION

APPLIED ELECTRONIC COMMUNICATION

CIRCUITS, SYSTEMS, TRANSMISSION

Robert Kellejian

COLLEGE OF SAN MATEO
CALIFORNIA

SCIENCE RESEARCH ASSOCIATES, INC.
Chicago, Palo Alto, Toronto, Henley-on-Thames, Sydney, Paris

A Subsidiary of IBM

Compositor	Bi-Comp, Incorporated
Illustrator	Christopher Hyde
Text Designer	SRA Staff
Cover Designer	Michael Rogondino
Acquisition Editor	Alan W. Lowe
Project Editor	Ronald Q. Lewton
Technical Editor	Alice Lescalleet

ACKNOWLEDGMENTS

The following items are reproduced or adapted
through the courtesy of the sources shown.

Figures 2–6 and 2–7: J. W. Miller Division, Bell Industries,
    Compton, California
Figure 2–8: Sprague Electric Company, North Adams, Massachusetts
Figure 4–16d: Dynascan Corporation, Chicago, Illinois
Figure 7–10: Marconi Instruments, Northvale, New Jersey, and
    Hewlett-Packard Company, Palo Alto, California
Table 9–1: Reprinted with permission from *Electronic Design,* Vol. 27,
    No. 8, April 12, 1979; copyright Hayden Publishing Co., Inc., 1979
Figure 9–30: J. S. Mayo, Bell Laboratories, Murray Hill, New Jersey
Figure 9–31: Hewlett-Packard Company, Palo Alto, California
Table 10–2 A, B: *Reference Data for Radio Engineers,* Sixth Edition,
    1975, Howard W. Sams & Co., Inc., Indianapolis, Indiana
Figure 12–12: Bunker Ramo Corporation, Amphenol North America
    Division, Oak Brook, Illinois

LIBRARY OF CONGRESS CATALOGING IN PUBLICATION DATA

Kellejian, Robert.
    Applied electronic communication.

    Includes index.
    1. Telecommunication. 2. Electronics. I. Title.
TK5101.K38   621.38   79-18138

ISBN 0-574-21535-2

10 9 8 7 6 5 4 3 2 1

PREFACE

This book is intended for use as a text or reference by students seeking expertise in the *applied* aspects of electronic communication technology. It is a book of engineering practice—the work in which electronic engineers and technicians are engaged. The content is divided into three general topic areas parallel to specializations within the field: Part I—Circuits, Part II—Systems, and Part III—Transmission. The book thus provides a thorough coverage of the communication field, from system concept through circuit execution. Each part may be studied separately, or included in separate courses of study, as required to meet varying needs.

The approach of the text is practical, but quantitative, using mathematics (algebra and trigonometry) to provide both problem-solving experience and verification of theory. Although mathematical derivations and definitions are excluded in the interest of maintaining its practical approach, quantitative *techniques* are emphasized, using current engineering practice, with formulas adapted for convenient calculator solution. Emphasis is also placed on assurance of the students' understanding of both "how" and "why" for each of the topics under study. Engineering practice, after all, entails the transformation of sound ideas into practical hardware. It is thus the purpose of this book to instruct individuals to balance both sides of the "theory-to-practice" equation.

Each chapter has been structured to achieve optimum feedback on the students' learning progress. After an introductory section, specific *learning objectives* are presented. These learning objectives serve as a guide to identify the important topics of the chapter, the sequence of their presentation, and the level of understanding the student is to achieve. The end-of-chapter *problems* are keyed to this list of objectives, and serve to close the "learning loop." Successful solution of these problems will both verify a thorough understanding of the topics within the chapter and provide direct experience with procedures typical of engineering practice. The presentation of the text material within the chapters includes many examples and procedures in *sample problems* to guide the student. Where appropriate, actual laboratory measurement procedures are detailed to further tie theory to practice.

This book is intended to make the complex material of electronic communication both readable and understandable. The formal theoretical material presented in a style appropriate for college and university readership. The book assumes prior knowledge and experience with basic electrical and electronic circuits, as well as a solid foundation in applied mathematics (through trigonometry), facility with a scientific calculator, and some exposure to electronic systems concepts.

The author wishes to acknowledge the encouragement and patience of his colleagues, students, editors, and especially his family members, all of whom contributed significantly to the completion of this work. The author is grateful to Louis Dorren and Wesley R. Muller for their review of the proofs and contributions to the content of the book. Gratitude is also expressed to the principal manuscript reviewers, for their many helpful suggestions:

Richard M. Bernstein	William Rainey Harper College
Jack Blanchard	Arizona State University
Raymond F. Davidson	Texas State Technical Institute—Waco
Edwin Pollock	Cabrillo College
Melvin C. Vye	Community and Technical College, University of Akron

CONTENTS

PART II SYSTEMS

CHAPTER 1

INTRODUCTION

In years past, the study of electronics was concerned primarily with communications—*radio* communication. In fact, the entire field of applied electronics was called "radio," and even the professional organization serving electronics identified itself as the Institute of *Radio* Engineers (IRE). Most textbooks on electronic devices and circuits were likewise concerned with communications, as indicated by their titles: *Radio Physics, Radio Engineering, Elements of Radio,* and so forth. With the increasing growth of electronics over the years, the technology has expanded from just that single specialty—radio—into numerous fields of study and application, including electronic communication, of which radio is a part.

Although modern electronic communication includes radio, it also includes telephone, television, data, microwave, radar, and many other techniques of getting signals from one place to another. The number and diversity of applications in this field are ever increasing and occupying a more significant part in all phases of our daily lives. Your study of electronic communication must necessarily review these many applications and their related fundamentals, including electronic circuits, signal processing functions, and transmission functions.

THE BOOK

This textbook is organized in three parts to accommodate your review of electronic communication applications and their related fundamentals. Part I (Chapters 2 through 5) presents the fundamentals of electrical networks, electronic devices, and circuits. Part II (Chapters 6 through 10) presents signal processing techniques and circuits, and their application to electronic communication systems. Part III (Chapters 11 through 14) presents the principles of signal transmission and their application to transmission lines, antennas, and microwave devices. The topics in each of these three parts generally represent specialization areas within electronic communication

1

technology. Engineers, technicians, and often whole manufacturing companies commonly specialize in one of these areas, where expertise may be
combined and exploited. People who work with components are usually the
most specialized within the structure, whereas those who work with circuits
require a somewhat broader perspective, emphasizing the topics of Parts I
and III of this text. People who work at the systems level are generally the
least specialized, with the emphasis of their expertise directed to the topics
of Parts II and III.

As a student of electronic communication techniques and practices, you
may elect to specialize; but you should do so only after a broader exposure to
the technology as reflected by the three parts of this book. This broad exposure should acquaint you with the differences in the type of work and skills
required within specialty areas. Breadboarding skills and network calculations, for example, are requisites for success with *circuits,* whereas signal
measurement and interpretive skills are necessary for *systems* work. Some
specialty areas entail sedentary ''bench'' work; others require considerable
strenuous ''field'' work in both good and bad weather. Some specialties are
highly analytical; others are more practical. The growth and diversity of
electronic communication technology will no doubt accommodate a wide
range of skills, interests, and capabilities among technologists. Your identification of these personal attributes is thus as important as your learning of
electronic *theory.*

The purpose of this book is to provide a *working* knowledge of electronic
communication—the ability to understand and then apply principles and
techniques to actual hardware. In this sense the book is not heavily theoretical, but it is quantitative. Mathematical derivations are not included here, for
example; but extensive mathematical relationships are employed, thus providing opportunities to solve practical problems of the type encountered in
engineering practice. Much effort has been made to align the mathematics
with those procedures prescribed for modern ''scientific'' calculators. These
''tools of the trade'' incorporate those capabilities most useful to engineering
practice as a result of extensive market analysis by their manufacturers.
Your practice in their use for problem solving throughout this text may thus
closely simulate actual working conditions. This kind of problem solving
yields an added bonus of *judgment* about practical magnitudes, allowing for
impressive ''off-the-cuff'' predictions of ''too large,'' ''too small,'' or ''just
about right,'' when evaluating technical problems.

Each chapter contains a set of specific *learning objectives* which you
should accomplish through study of that chapter's content. The objectives
were selected to define the most important technical aspects of applied electronic communication. Your knowledge of the topics defined by the objectives, and your ability to apply this knowledge to solving technical problems,
will qualify your expertise in this field. The end-of-chapter *problems* were
written to determine your success with the learning objectives of each chap-

ter. Like the *sample problems* within the body of each chapter, these problems are not simple, as they often require a thorough understanding of the topic in question and a structuring of steps to a solution. Considerable effort was expended to assure that these problems provide both a learning experience and exposure to the procedures of engineering practice. You should approach these problems deliberately, taking sufficient time to generalize their structure and procedures to solution, making sure that you have met the learning objectives in the process.

ELECTRONIC DEVICES AND CIRCUITS

A wide variety of electronic circuits are used for communication; but those most often associated with this field require extensive use of passive *LC* networks with active IC and discrete electronic devices. The fundamentals of these *LC* networks are traditionally reviewed in introductory electronics courses. The applied aspects of the networks are anything but fundamental, however, as considerable knowledge is necessary to use them effectively within electronic circuits. The mathematics and techniques for problem solving presented in Chapters 2 and 3 may not be familiar to you, nor identical with the approaches introduced in other textbooks. They are here specifically adapted to enforce conceptual understanding of the topics under study and, at the same time, provide an efficient means of solution using a calculator.

As with the presentations for *LC* networks, the content of Chapters 4 and 5 was most probably studied in more basic courses on electronic devices and circuits. Again, the applied aspects of these topics to electronic communication systems require a more intensive and special treatment than you may have encountered in introductory studies. In Chapter 4 attention is directed to both the electrical and the amplifying properties of active electronic devices (IC's, transistors, FET's, tubes) when they are operated at higher frequencies. This chapter defines *capacitance* as the critical factor limiting the devices' performance in RF circuits; Chapter 5 reviews techniques to compensate for capacitive losses to obtain optimum RF performance with active devices. Chapter 5 also reviews the operation of a variety of popular RF amplifier and oscillator circuits.

Part I of this textbook is thus of greatest importance to individuals who need the skills to determine circuit and component requirements for electronic communication equipment. These skills include:

1. Calculation of component values for given operational requirements
2. Selection of practical, off-the-shelf components (if available) for *LC* networks
3. Specification of coil winding and fabrication in lieu of off-the-shelf acquisition

4. Selection of active IC or discrete electronic devices to obtain given operational requirements
5. Specification of bias and operating conditions for active IC or discrete devices
6. Specification of special layout or fabrication procedures to assure successful circuit performance after assembly
7. Circuit testing and evaluation of performance with regard to operational requirements
8. Circuit device bias or component modification to improve performance as necessary
9. Final calibration and documentation of the project
10. Repair and maintenance of the equipment after installation

As with the other sections of this book, the *how* and *why* are here emphasized to develop skills and understanding of concepts, techniques, and procedures.

ELECTRONIC COMMUNICATION SYSTEMS

Electronic communication involves the transmission and reception of information signals. These information signals may take any number of forms, as required, to be reproduced in aural, visual, or other forms necessary for interpretation and control. In addition to the information signals, the process of electronic communication usually requires other signals to *carry* information within circuits and from one point to another, along wires (cable) or through space. These other signals—which do not impart information but only carry information signals—are aptly called *carrier* signals. The process of applying the information signals to the carrier signals is called *modulation*. The information signals modulate the carrier signals.

Part II of this book reviews, in great detail, the popular forms of modulation applied to electronic communication: AM (amplitude modulation), FM (frequency modulation), pulse modulation, and digital modulation (Chapters 6, 7, and 9). Other chapters in this section cover radio receiver functions (Chapter 8), and the application of electronic systems to the various communications services used for security, business, and entertainment (Chapter 10). The content of Part II differs drastically from that of Part I. Here signal levels and waveforms, and their interaction, are studied in addition to the kind of component and circuit analysis performed in Part I. With a good working knowledge of circuits, however, you should be able, as you proceed through Part II, to shift the emphasis of your study from concern with components to analysis of signal characteristics and processing functions. Although electronic communication requires circuits to achieve signal processing, it is most important in this section to understand why and how circuits are *used,* rather than how they are biased, etc.

The chapters on AM and FM (Chapters 6 and 7) define modulation techniques that are generally used to communicate analog information signals such as audio and video. You are probably more familiar with these forms, as they identify our two broadcast radio services: AM and FM. Both forms and their variations—vestigial sideband (VSB) and double sideband suppressed carrier (DSB-SC)—are used in the broadcast television system for transmission and reception of color video and audio. Another variation of AM— single sideband (SSB) modulation—is used extensively for voice communication where bandwidth economies are necessary. Both AM and FM are used extensively in common-carrier, satellite, and other communications services that you may not *directly* encounter, but that are nonetheless very important to business and government operation.

Chapter 9 defines more recently popular modulation forms that are used primarily to transmit data rather than audio or video. Although the need for data communications has grown with the increased use of computers, these electronic "brains" have also made practical the coding and conversion techniques that now more efficiently process analog information signals in digital form. Data and digital electronic communication is, no doubt, the most dynamic specialty area in electronic communications. Most data signals are converted to frequency-shift keying (FSK), a form of FM, for further modulation and transmission. This technique provides a reliable two-frequency, rather than a two-level, code to pulse and digital signals, so level losses in transmission will not degrade modulation.

Your understanding of electronic communication systems requires a thorough knowledge of modulation techniques as the means by which information signals are processed and combined with carrier signals for transmission. Aside from higher power and frequency considerations, the theoretical concepts that define "sending" or transmitting station operation are based primarily on modulation principles. Communications receivers, however, do employ a variety of unique concepts that make them a topic to be studied separately (Chapter 8), once modulation techniques are understood. Because high-power considerations limit the versatility of transmitting equipment, sophistication in design of receiver functions is exploited to improve the communications process. It is far more efficient to increase receiver sensitivity by 3 dB, for example, than to double the level of an already high-power transmitting station when improved communication capabilities are necessary.

The final chapter (Chapter 10) in Part II of this book defines the many services in which electronic communication systems are used. Like the other chapters in Part II, emphasis is placed on the systems aspects of electronic communication, but the content of Chapter 10 deals almost exclusively with applications rather than theory. The topics include identification of radio frequency allocations, channel bandwidths and information capacities, multiplexing techniques, and the uses for communication systems. Individuals who work with communication systems and services must maintain a broad knowledge of these operational aspects of communications, as well as specific knowledge of the equipment used in such systems and services. Among specific capabilities they may possess the following:

1. Determination of which frequencies are available, and technically appropriate, for a given communication service
2. Determination of which modulation forms are available for a given communication service
3. Specification of the operational limitations with specific modulation forms for utilization of channel capacity
4. Specification of the signal processing functions and signal levels at each stage of the communication system
5. Identification of subsystem components and systems equipment available as off-the-shelf hardware
6. Specification of special equipment and components that must be fabricated in lieu of off-the-shelf acquisitions
7. Specification and performance of installation, calibration, service, and maintenance of equipment
8. Performance of sales and in-plant and customer training functions in support of equipment

TRANSMISSION LINES, MICROWAVE DEVICES, AND ANTENNAS

Part III of this book covers topics that often share the same theoretical basis of function: *distributed constants*. The term refers to shorter wavelength signals along transmission circuits that yield a distribution of voltage and current levels, and hence a distribution of capacitance, inductance, resistance, and conductance along the length of the circuit. In comparison, conventional lower frequency transmission circuits have long wavelengths, so the constants (voltage, current, capacitance, inductance, resistance, and conductance) remain fixed over the length of a circuit. With low frequencies the voltage at one point along the length of a pair of wires is the same as that at any other point. This condition is not true with short wavelength signals of much higher frequency, as the actual voltage at a given instant of time differs from one point to another along a two-wire transmission line. A comparison may be made by referring to water in a bathtub. When the water is still, it has the same level all along the sides of the tub; but when waves are created, the water level varies at different points along the sides of the tub.

Your interest aroused with this short explanation of the distributed constants signal transmission phenomena should spur you on to delve into the theory unraveled in Part III of this book. There, two-conductor and coaxial transmission lines, waveguides, and transmission-circuit components, used extensively in the UHF, SHF, and EHF regions of the spectrum, are introduced and applied. Active microwave devices are also introduced (Chapter 13), as they usually operate with distributed constants within their structure, and have input and output terminals interfacing with transmission circuits. Chapter 14 (the final chapter of the book), deals with antennas and propagation of electromagnetic waves in space and about our atmosphere. The topics of this chapter tie together the concepts of transmission lines and the optical properties of radiation.

Knowledge of applied RF transmission principles is an absolute necessity for everyone who works with electronic communication. Even circuit specialists must consider the characteristic impedance of the conductors that carry higher frequency signals from one point in a circuit to another. Individuals whose specialty is electronic systems must identify transmission lines, cables, and antennas just as they do for other electronic equipment and subsystem components. There are also microwave and antenna specialists whose knowledge of both the theoretical and applied aspects of these topics in Part III is considerable, yet almost every aspect of their work must take into account transmission or radiation phenomena.

The mathematical aspects of transmission and radiation phenomena are perhaps the most difficult to apply. Certainly the mathematics used to define these phenomena in theoretical textbooks is sufficiently complicated to exclude many technical people from understanding the principles under study. Part III of this book attempts to clarify transmission concepts by approaching the topics from a practical viewpoint, using the mathematics as a *tool* for problem solving and reinforcement of theory—not as a definer of theory. As with all specialties in electronics, many of the procedures of engineering practice presented in Part III have been reduced from abstract theory, then converted to basic understandable techniques—techniques that clarify abstract theory through direct experience. Among these techniques is the use of the Smith chart, which is an invaluable tool for both analysis and problem solving with transmission circuit topics. The Smith chart is heavily emphasized in this section of the text to reflect the importance of its use in engineering practice.

The specific abilities required of individuals who successfully direct and execute work in the transmission, microwave, antenna, and propagation aspects of electronic communication include:

1. Determination of impedance transformation networks within higher frequency and microwave circuits for interfacing standard transmission lines with electronic components
2. Selection and specification of cables, connectors, and waveguides based on given requirements for transmitting a signal from one point to another
3. Specification and execution of devices and techniques for impedance matching for maximum power transfer and efficiency

4. Specification of specific microwave and transmission components to achieve given signal processing functions in the UHF, SHF, and EHF regions of the spectrum
5. Specification of signal sources, related power supplies, and operating equipment to generate UHF, SHF, and EHF signals
6. Identification of subsystem components and special equipment available as off-the-shelf hardware
7. Specification and execution of fabrication procedures for antennas and components that must be fabricated in lieu of off-the-shelf acquisitions
8. Determination of antenna characteristics and specification of antenna types and transmitter power to provide given field signal-intensity levels at specified distances from the transmitter
9. Execution of field intensity measurements and plotting of propagation contour maps

You may conclude that this book is intended to help prepare practical technicians and engineers for the field of electronic communication. Although electronic technicians and engineers may be practically oriented, they must nonetheless possess a solid theoretical background for understanding both concepts and relationships. They must have the ability to see general patterns and to predict effects, as well as to employ the mathematical tools at their disposal for effective problem solving. Beyond the numerical results obtained through calculation, these technologists should be able to execute the assembly, calibration, and maintenance of many products they encounter within the field of electronic communication.

PART I CIRCUITS

CHAPTER 2

CIRCUIT ANALYSIS

This chapter begins a review of the detailed technical aspects of RF electronic circuitry. Although this book is concerned with advanced circuits, the fundamentals to be applied are the same as those you probably learned when you began your study of elementary electricity. A thorough understanding of resistance, capacitance, and inductance (R, C, and L) is the key to your success with RF circuits, just as it is with basic electrical circuits. Even though this material may seem "old hat" and unexciting, it is nonetheless important to the successful analysis of the active RF electronic circuits presented in subsequent chapters. As you proceed through this chapter you may encounter approaches to analysis and problem solving that are perhaps new to you. Mastery of these approaches will pay dividends in time and energy when applied to these more advanced topics of RF communications.

The schematic diagram of Fig. 2-1 typifies a modern IC (integrated circuit) AM broadcast radio. As you can see, the external electronics (transistors, diodes, etc.) are minimal, with most of the circuit functions performed within the IC's and thus hidden from schematic detail. This example is presented to indicate the kind of circuitry that increasingly comprises modern radio equipment. The few external components used with such circuits are, for the most part, simply resistors, capacitors, and inductors that work in conjunction with the electronics within the IC chips, just the same as they work with discrete components. Aside from the very basic current and voltage functions of resistors and resistive elements (biasing, etc.), much of the signal processing in RF circuits is achieved with capacitive and inductive elements. These C and L components are used in RF circuits to obtain frequency selectivity, to shift phase, or to transform impedances within circuitry of electronic components.

Working effectively with RF circuits thus requires applying your knowledge of inductance and capacitance, and of Ohm's and Kirchhoff's laws. This chapter provides an applied review of the principles of inductance and capacitance, with specific learning objectives in mind.

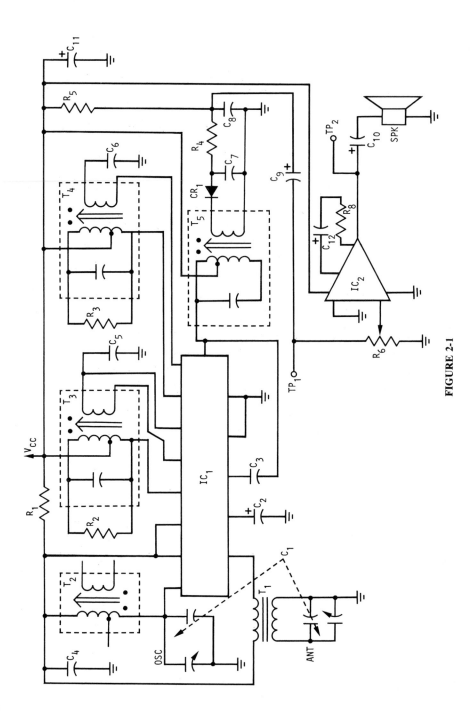

FIGURE 2-1

LEARNING OBJECTIVES

Upon completing this chapter, including the problem set, you should be able to:

1. Understand the principles of inductance and capacitance as they apply to sine-wave signals at radio frequencies.
2. Compute the frequency and phase characteristics of L and C components and circuits.
3. Understand and compute the Q factor of L and C components and circuits.
4. Understand the physical and electrical characteristics of typical coils and capacitors used in RF circuits.
5. Understand the concepts of series- and parallel-resonant circuits.
6. Compute the frequency, phase, Q, impedance, admittance, current, and voltage characteristics of series- and parallel-resonant circuits.
7. Understand and compute the relationship of Q to bandwidth.
8. Understand and compute the relationships of unloaded and loaded Q to circuit loading of series and parallel impedance and admittance.
9. Apply basic network analysis procedures to resonant and nonresonant RLC circuits.
10. Select standard L and C components from catalog listings that are appropriate for particular RF circuit applications.

INDUCTOR AND CAPACITOR BASICS

You may recall from your studies of basic electricity that an *inductor* (coil) stores the effects of current—the electrical form of kinetic energy—in the magnetic field surrounding the coil structure. A *capacitor* also stores the effects of electrical energy—potential energy—in the electric field within the dielectric material separating the capacitor's plates. In RF circuits the electrical properties of inductance (L) and capacitance (C) are found not only in the specific components, but also along the wires transferring signals and along the hardware used to package or house the circuits. Inductance and capacitance are *fixed* properties of electrical/electronic components and hardware in that they usually do not change with frequency. For example, 80 pF is the same capacitance at 20 Hz as it is at 20 MHz, and 80 μH is the same inductance at 20 Hz as it is at 20 MHz.

Reactance

The property of L and C that *does* change with frequency is the *reactance* (X)—the current-limiting factor—that these elements exhibit. For the most part, RF signals are sine waves; so reactance can be defined by using the sine-wave frequency expression $2\pi f$, indicating an alternating current at a given number of hertz (Hz, formerly referred to as cycles per second). The expression $2\pi f$ is so common in RF calculations that it is often abbreviated in engineering practice with the Greek letter ω (omega). For inductors the value of reactance is specified by, and may be computed with, the following formula:

$$X_L = 2\pi fL = \omega L = j\omega L$$

X_L is the inductive reactance in ohms, f is the frequency in Hz, and L is the inductance value in henrys. The j found in some engineering formulas indicates that inductive reactance operates at 90° (¼ cycle) phase shift as compared to the phase of the RF signal voltage applied to its terminals.

The typical values of practical inductors (coils and transformer windings) that are encountered in RF circuits are in the microhenry (μH) range. If you use a 4 μH coil, for example, at a frequency of 8 MHz, the reactance would be computed as follows:

$$X_L = \omega L = 6.28 \times 8 \times 10^6 \times 4 \times 10^{-6} = 201 \ \Omega$$

Once the reactance is known, its value may be employed to determine circuit behavior using Ohm's law, Watt's law ($P = I^2R$, etc.), or other circuit analysis procedures.

For capacitors the value of reactance is specified by, and may be computed with, the following formula:

$$X_C = 1/2\pi fC = 1/\omega C = -j/\omega C$$

X_C is the capacitive reactance, computed (or measured) in ohms, C is the capacitance in farads, and the $-j$ indicates that capacitive reactance operates at minus ($-$) 90° (¼ cycle) phase shift as compared to the phase of the RF signal voltage applied to its terminals.

Practical capacitors encountered in RF circuits typically have values in the picofarad (pF) range. For a 250 pF capacitor, as an example, at a frequency of 8 MHz, the reactance would be computed as follows:

$$X_C = 1/\omega C = 1/(6.28 \times 8 \times 10^6 \times 2.5 \times 10^{-10}) = 79.6 \ \Omega$$

As with inductors, once the capacitor's reactance is known, its value may be employed to determine circuit behavior using conventional circuit analysis procedures.

Estimating Techniques

If you examine the formulas for inductive and capacitive reactance, you will notice the relationship of reactance to frequency, when the same coil or capacitor is connected to an RF signal source, that may operate over a range of frequencies. Inductive reactance changes *directly* with frequency, whereas capacitive reactance changes *inversely* with frequency. In either case a change in frequency will yield a proportional change in reactance—directly with X_L and inversely with X_C. If, for example, you consider the 4 μH coil and the 250 pF capacitor, and round off their reactances at 8 MHz to 200 Ω and 80 Ω, respectively, you can quickly tabulate the reactance at other frequencies by ratio and proportion. This approach provides a quick way to estimate reasonably accurate reactance values after completing one mechanical calculation. With a little practice you can develop a keen sense of judgment about the relationships of reactance, frequency, capacitance, and inductance. This judgment will help you understand the more advanced aspects of RF circuits without getting sidetracked in the tedium of repeated mechanical calculations. Tables 2-1, 2-2, and 2-3 show examples of this "shortcut" approach.

As you may observe, the exact values obtained from circuit calculations are often rounded off for convenience. With experience, you can quickly develop judgment about the accuracy and resolution required of your calculations. In engineering practice the degree of accuracy required depends on the specifications of a circuit or product developed to meet the needs of a customer or market. Since the objectives of this book are to communicate principles and to develop engineering judgment, the author—in applying the judgment of his own experience—has often taken liberties in rounding off or

TABLE 2-1. REACTANCE VERSUS FREQUENCY RATIOS

Frequency (MHz) f_2	f_1	Ratio f_2/f_1	Reactance (Ω) X_L (4 μH)	X_C (250 pF)
8	8	1/1	200	80
4	8	0.5/1	100	160
6	8	0.75/1	150	120
12	8	1.5/1	300	60
16	8	2/1	400	40

TABLE 2-2. REACTANCE VERSUS INDUCTANCE RATIOS

Inductance (μH) L_2	L_1	Ratio L_2/L_1	Reactance (Ω) X_L (8 MHz)
4	4	1/1	200
2	4	0.5/1	100
3	4	0.75/1	150
6	4	1.5/1	300
8	4	2/1	400

TABLE 2-3. REACTANCE VERSUS CAPACITANCE RATIOS

Capacitance (pF) C_2	C_1	Ratio C_2/C_1	Reactance (Ω) X_C (8 MHz)
250	250	1/1	80
125	250	0.5/1	160
188	250	0.75/1	120
375	250	1.5/1	60
500	250	2/1	40

extending the results of calculations. These liberties may serve as a guide until you have accumulated some judgmental experience regarding the results of your own calculations.

Susceptance

When reactances are connected in parallel, their circuit behavior is best described by an *inverse concept*. In parallel circuits the value of current *allowed* to flow is consistent with our accepted concepts of Kirchhoff's laws. Since the term *reactance* defines a current-*limiting* property in a circuit, it is really inappropriate to use this same term to define a current-*allowing* property. The correct term that defines the current-allowing property of inductors and capacitors is *susceptance*. Susceptance (*B*) is simply the inverse of reactance: $B = 1/X$. The principles of inductive and capacitive reactance all function in parallel circuits, but do so inversely. If the reactance of a coil or capacitor is known at a particular frequency, a simple inversion will yield the appropriate susceptance value for the component operating at the same frequency. The 4 μH coil's susceptance at 8 MHz would be calculated as:

$$B_L = 1/X_L = 1/200 \ \Omega = 0.005 \ S = 5 \ mS$$

B_L is the inductive susceptance computed (or measured) in siemens. (Historically, the unit of susceptance was the mho, symbolized by $\mho$, but the present international standard is siemens, abbreviated S.) If the reactance of a coil is not known, the susceptance can be calculated directly using the formula that specifies its function:

$$B_L = 1/2\pi fL = 1/\omega L = -j/\omega L$$

The $-j$ indicates that inductive susceptance operates at $-90°$ ($\frac{1}{4}$ cycle) phase shift as compared to the phase of the RF signal voltage applied to its terminals. Note that the $-j$ is applied to both inductive susceptance and capacitive reactance.

Proceeding with the same approach, using the capacitive reactance of 80 Ω for the 250 pF capacitor:

$$B_C = 1/X_C = 1/80 \ \Omega = 0.0125 \ S = 12.5 \ mS$$

B_C is the capacitive susceptance computed in siemens. The susceptance can also be calculated directly using the formula that specifies its function:

$$B_C = 2\pi fC = \omega C = j\omega C$$

The j indicates that capacitive susceptance operates at $+90°$ ($\frac{1}{4}$ cycle) phase shift as compared to the voltage of the RF signal applied to its terminals. Note that $+j$ is applied to both capacitive susceptance and inductive reactance.

Phase

Now that you have some sense of the *magnitudes* of reactance and susceptance, let us proceed with their second characteristic: *phase*. You may recall
that the application of a sine-wave voltage to an inductor causes a current
which follows the applied sine waveform, but lags the voltage by ¼ of a
cycle. With 360° in one cycle, the current in the inductor lags ¼ × 360°, or
90°, behind the phase of the applied voltage. While the inductor affects
current, a capacitor affects voltage—it has inverse properties. Thus, with the
application of a sine-wave *current* to a capacitor, the *voltage* that develops
across its plates follows the applied waveform, but lags by ¼ of a cycle, or
90°.

By comparing the effects of a voltage signal upon a resistor, where the
current flows through the resistor *in phase* with the applied voltage, as in
Fig. 2-2, you can generalize by stating that current *leads* the voltage in a
capacitive circuit but *lags* the voltage in an inductive circuit. (A gimmick for
remembering this relationship is the mnemonic "ELI the ICE man," where
E—henceforth known as *V*—stands for voltage, *I* for current, *L* for inductance, and *C* for capacitance.) A more appropriate view of this phase
relationship is necessary, however, for a better perspective in analyzing RF
circuit performance.

If a sine-wave signal is applied to three components, *R, C,* and *L* connected in parallel, the voltage is in phase across all three components. The

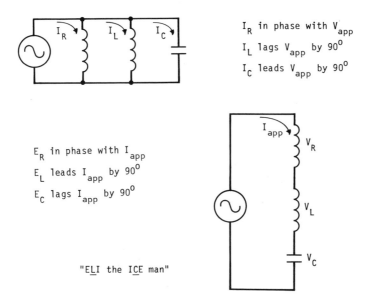

I_R in phase with V_{app}

I_L lags V_{app} by 90°

I_C leads V_{app} by 90°

E_R in phase with I_{app}

E_L leads I_{app} by 90°

E_C lags I_{app} by 90°

"E**L**I the I**C**E man"

FIGURE 2-2

currents that flow through the components, however, are *not* in phase. Only the current flowing through the resistor is in phase with the applied voltage. The current flowing through the coil *lags* the resistor current by 90°, whereas the current flowing through the capacitor *leads* the resistor current by 90°.

If the same sine-wave signal is applied to the same three components connected in series, the current is in phase through all three components. (In a series circuit the current is always the same through each component.) The voltages that develop across the components, however, are not in phase. Only the voltage developed across the resistor is in phase with the applied voltage. The voltage developed across the capacitor *lags* the resistor voltage by 90°, whereas the voltage developed across the coil *leads* the resistor voltage by 90°.

You must be careful not to confuse series and parallel *circuit* concepts with the *phase* concepts of inductors and capacitors. Ohm's and Kirchhoff's laws don't dry up and blow away just because you are working with RF circuits; they are *laws* and they apply universally to *all* circuits at *all* frequencies! When you apply these lead/lag phase relationships, always keep in mind the changing phase of *voltages* across series components and the changing phase of *currents* through parallel components. Once these phase relationships are obtained, you may proceed to compute component and circuit values using Ohm's and Kirchhoff's laws in their conventional form.

NETWORK ANALYSIS

Now that we have reviewed the phase relationships of the individual components $(R, C,$ and $L)$, the next step is to combine these elements into electrical networks and analyze their properties. The approach taken here will allow us to reduce complex networks to one reactive (or susceptive) and one resistive (or conductive) element in order to simplify the mathematics.

Series Networks

Consider a 4 μH coil, a 250 pF capacitor, and a 160 Ω resistor connected in series and operating at 8 MHz. When the circuit is connected to a signal source the voltages that develop across the reactive components are out of phase, while the current through each component is in phase. The voltages are, of course, directly proportional to the ohmic values of resistances and reactances in the series circuit (Kirchhoff's law). Both X_L and the voltage across the coil have a 90° leading phase angle, compared with the resistance R and the voltage across the resistor; whereas X_C and the voltage across the capacitor have a $-90°$ lagging phase angle. Since the phase relationship of X_L to X_C is 180° ($+90°$ to $-90° = 180°$), their effects are opposing and they will

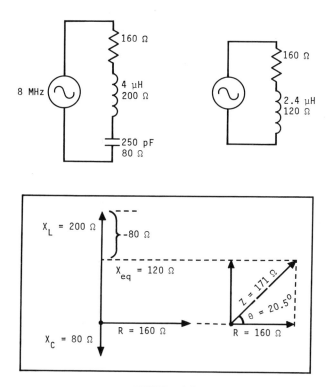

FIGURE 2-3

cancel, yielding a net reactance with the phase characteristics of the larger of the two initial reactance values. In this example the X_L of 200 Ω is canceled (reduced) by the X_C of 80 Ω, leaving 120 Ω of effective inductive reactance (X_L) in series with the 160 Ω of resistance. This circuit, its equivalent network, and the associated phasor diagram are shown in Fig. 2-3.

Simple two-element resistance-reactance networks of this type are called *complex networks;* they are expressed in two forms appropriate for mathematical manipulation: *rectangular* and *polar*. The rectangular-form expression for the complex network in Fig. 2-3 is $Z = R \pm jX$, indicating that Z is a complex impedance made up of 160 Ω of resistance and 120 Ω of inductive reactance; that is, $Z = 160 + j120$ Ω.

The polar-form expression for the same network indicates the net effective impedance of the combined resistance and reactance. It is specified as $Z \angle \theta$, indicating that Z is an effective impedance value with an associated phase angle $(\angle \theta)$, relative to the phase of a purely resistive network. For the network of our example, $Z \angle \theta = 200 \angle 37°$ Ω. This polar form is sometimes called a *phasor,* as it includes both magnitude and angle factors.

In computing network characteristics or circuit response, it is most convenient to use the rectangular form for addition and subtraction of impedance values, and the polar form for multiplication and division of impedance values. Most modern scientific calculators have direct functions for rectangular-to-polar and polar-to-rectangular conversion for convenience in problem solving. If you are obliged to do things "the hard way," however, you can use more basic calculators, tables, or even a slide rule to make the conversions using the following two-step formulas:

Rectangular-to-polar

$$\angle\theta = \arctan X/R \quad \text{and} \quad Z = X/\sin \angle\theta$$

Applying this conversion to our example network,

$$\angle\theta = \arctan 120/160 = 37° \quad \text{and} \quad Z = 120/\sin 37° = 200\angle 37° \ \Omega$$

Polar-to-rectangular

$$R = Z \cos \angle\theta \quad \text{and} \quad X = Z \sin \angle\theta$$

Applying this conversion to our example network,

$$R = 200 \cos \angle 37° \ \Omega = 160 \ \Omega$$
$$X = 200 \sin \angle 37° = 120 \ \Omega$$
$$Z = 160 + j120 \ \Omega$$

If the circuit in our example were operated at 4 MHz instead of 8 MHz, the network analysis would proceed as follows:

i. The frequency ratio f_1/f_2 is $2:1$, so the reactance ratio is also $2:1$.
ii. $X_L = 200/2 = 100 \ \Omega$, $X_C = 2 \times 80 = 160 \ \Omega$, so the net reactance is 60 Ω of X_C in series with 160 Ω of resistance.
iii. $Z = 160 - j60 \ \Omega$, $Z\angle\theta = 60/\sin (\tan^{-1} 60/160) = 171\angle -20.5° \ \Omega$.

From the above steps, note how quickly and simply network analysis can be performed. These results can now be used for further analysis using Ohm's and Kirchhoff's laws, as in the following example.

Sample Problem 1

Given: A series RLC circuit of 160 Ω, 4 μH, and 250 pF, connected to a 2 V RF signal source operating at 4 MHz.

Determine: The total impedance, total current, and the voltages across R, L, and C. (Note that the current through the resistance is *in phase* with that of the *impedance*.*)

$Z = 171\angle -20.5°$ Ω

$i = v/Z = 2/171\angle -20.5° = 0.012\angle 20.5°$ A $= 12\angle 20.5°$ mA

$v_R = iR = {}^*0.012\angle 20.5° \times 160\angle 0° = 1.87\angle 20.5°$ V

$v_L = iX_L = 0.012\angle 20.5° \times 100\angle 90° = 1.2\angle 110.5°$ V

$v_C = iX_C = 0.012\angle 20.5° \times 160\angle -90° = 1.87\angle -69.5°$ V

Parallel Networks

In working with parallel networks of R, L, and C components, it is appropriate to consider the current-*allowing* characteristics of these components. The current-allowing property of resistors is conductance (G): $G = 1/R$ and $R = 1/G$. The current-allowing property of capacitors and inductors is susceptance (B), as detailed earlier. The combination of G and B is admittance (Y), which is the inverse function of impedance (Z): $Y\angle \theta = 1/Z\angle \theta$ and $Z\angle \theta = 1/Y\angle \theta$. The best approach for parallel network analysis is to reduce a complex network to one conductive (G) and one susceptive (B) element; then we can use the same phasor diagrams and mathematics that we used in the analysis of series networks.

If we use the same 4 μH coil, 250 pF capacitor, and 160 Ω resistor, and connect them in parallel to a signal source operating at 8 MHz, the *currents* that pass through the components are out of phase. Since the currents are directly proportional to the G and B values of the components in a parallel circuit, both B_C and the current through the capacitor have a 90° leading phase angle relative to G and the current through the resistor. Likewise, B_L and the current through the coil have a $-90°$ lagging phase angle. Since the phase relationship of B_C to B_L is 180°, their effects are opposing and they will cancel, yielding a net susceptance with the phase characteristics of the larger of the two initial susceptance values. In our example the B_C of 12.5 mS (millisiemens) is canceled (reduced) by the B_L of 5 mS, leaving 7.5 mS of effective capacitive susceptance B_C in parallel with 6.25 mS (1/160 Ω) of conductance. The circuit, its equivalent network, and the associated phasor diagram are shown in Fig. 2-4.

You can analyze parallel networks using the same approaches used for series networks, converting rectangular-to-polar form, and vice versa, where necessary for addition/subtraction or multiplication/division of values. The rectangular-form expression for the parallel network in Fig. 2-4 is $Y = G + jB$, indicating that Y is a complex admittance made up of 6.25 mS of conductance and 7.5 mS of capacitive susceptance: $Y = 6.25 + j7.5$ mS. The polar-form expression for the same network is specified as $Y\angle \theta$, indicating that Y is an effective admittance value with an associated phase angle relative to the phase of a purely conductive network. For the network of our example, $Y\angle \theta = 9.75\angle 50°$ mS.

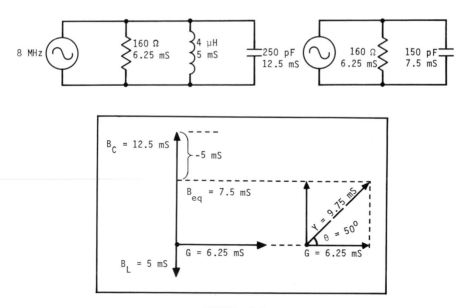

FIGURE 2-4

If, as in an earlier example, the same parallel RLC circuit were operated at 4 MHz instead of 8 MHz, the network analysis would proceed as follows:

i. The frequency ratio f_1/f_2 is $2:1$, so the susceptance ratio is also $2:1$.
ii. $B_C = 6.25$ mS, $B_L = 10$ mS, so the net susceptance is 3.75 mS of B_L in parallel with 6.25 mS of conductance.
iii. $Y = 6.25 - j3.75$ mS, $Y\angle\theta = 7.3\angle -31°$ mS.

These results can now be used for further analysis using modified forms of Ohm's and Kirchhoff's laws in the following example.

Sample Problem 2

Given: A parallel RLC circuit of 160 Ω, 4 μH, and 250 pF, connected to a 2 V RF signal source operating at 4 MHz.

Determine: The total admittance, the total current, and the individual currents through R, C, and L. Find also the total equivalent impedance of the circuit. (Note that the voltage applied across the conductance is *in phase* with that of the *admittance.**)

$$Y = 7.3\angle -31° \text{ mS}$$
$$i = \nu Y = 2 \times 7.3\angle -31° = 14.6\angle -31° \text{ mA}$$
$$i_R = \nu G = *2\angle -31° \times 6.25\angle 0° = 12.5\angle -31° \text{ mA}$$

$$i_C = vB_c = 2\angle -31° \times 6.25\angle 90° = 12.5\angle 59° \text{ mA}$$
$$i_L = vB_L = 2\angle -31° \times 10\angle -90° = 20\angle -121° \text{ mA}$$
$$Z_{eq} = 1/Y\angle\theta = 1/7.3\angle -31° \text{ mS} = 137\angle 31° \ \Omega$$

In rectangular form,

$$Z_{eq} = 137 \cos \angle 31° + j\, 137 \sin \angle 31° = 117 + j70 \ \Omega$$

indicating that this parallel circuit is equivalent to a series circuit with 117 Ω of resistance and 70 Ω of inductive reactance.

Combined Networks

You can analyze more complicated passive *RLC* circuits by using the series and parallel approaches we have just presented, breaking the larger circuit into smaller series and parallel networks and combining the equivalent impedances and admittances. Such analysis tends to become tedious; but if you proceed with the *Z* and *Y* techniques, the analysis will remain uncomplicated. If the same *RLC* components of the previous examples are combined in a series-parallel circuit as shown in Fig. 2-5, we can use the values already calculated to analyze the total circuit.

Sample Problem 3

Given: The series-parallel *RLC* circuit of Fig. 2-5, with both resistors 160 Ω, both coils 4 μH, both capacitors 250 pF, and with the signal source having an output of 2 V operating at 4 MHz.

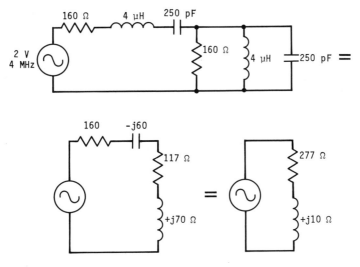

FIGURE 2-5

Determine: The input impedance and output voltage.

(a)   The equivalent series impedance of the parallel network (see Sample Problem 2: Z_{eq} 137∠31° Ω = 117 +j70 Ω) is first added to the series-network impedance (Z = 160 −j60 Ω) to obtain a total equivalent impedance as follows:

$$Z_{tot} = Z_{par} + Z_{ser} = Z_{eq} + Z = (117 + j70) + (160 - j60) = 277 + j10 \ \Omega$$
$$Z_{tot}\angle\theta = X/\sin (\tan^{-1} X/R) = 10/\sin (\tan^{-1} 10/277) = 277.2\angle 2° \ \Omega$$

(b)   The total circuit current is then computed as:

$$i = v/Z\angle\theta = 2/277.2\angle 2° = 7.2\angle -2° \ mA$$

(c)   The output voltage across the equivalent (series) impedance of the parallel network (Z_{eq}) is finally computed as:

$$v_{out} = i_{tot} \ Z\angle\theta_{eq} = 7.2\angle -2° \ mA \times 137\angle 31° = 0.988\angle 29° \ mV$$

PRACTICAL INDUCTORS AND CAPACITORS

Practical RF coils have an inductance rating in henrys and a current rating in amperes (milliamperes). Practical RF capacitors have a capacitance rating in farads and a maximum voltage rating. In choosing inductors and capacitors you need to consider one additional important electrical characteristic: Q. Basically, inductors store energy in the magnetic field surrounding a coil, and capacitors store energy in the electric field between the plates of a capacitor. The ratio of this *stored* energy to any that is *lost* in the respective component is a measure of its *quality*, (Q). Sometimes the inverse term, *dissipation* (D), is used to define the energy lost in an L or C component: $Q = 1/D$ and $D = 1/Q$.

The losses in a simple (air core) coil are due primarily to the series resistance encountered when RF current flows through the wire of the coil's windings. Because of the *skin effect,* which limits the flow of high-frequency currents to the surface of conductors, losses due to this series resistance are significant in RF coils, especially those wound of fine-gauge wire. These losses are often minimized by fabricating coils of heavy-gauge, silver-plated wire or tubing. The Q factor of an RF inductor may be simply calculated as follows:

$$Q = reactance/RF \ resistance = \omega L/R$$

ωL is the inductive reactance and R is the series resistance distributed along the windings of the coil. Quality RF coils can be practically fabricated with Q factors in excess of 100 at lower radio frequencies, and to above 500 for use in the UHF region. High-Q coils are not always necessary for RF circuits. Decisions regarding Q requirements will be discussed later in this chapter.

The losses in a simple capacitor are due primarily to *leakage* through the dielectric when RF voltages develop across the capacitor's plates. Such losses vary depending on the dielectric material and the separation and size of the plates, but they are minimized in small-value capacitors. The Q factor of an RF capacitor may be simply calculated as follows:

$$Q = \text{susceptance}/\text{RF conductance} = \omega C/G$$

ωC is the value of capacitive susceptance and G is the value of leakage conductance through the dielectric between the plates of the capacitor. High-quality RF capacitors have typical Q factors of about 1000.

At the higher radio frequencies (VHF, UHF, and above) the Q factors of coils and capacitors are complicated by a number of electrical factors, including radiation, absorption, lead inductance, package and mounting capacitance, and other environmental effects. The electrical factors tend to reduce the Q of L and C components, with the total circuit Q dependent on an entire circuit *package*.

Sample Problem 4

Given: A 4 μH coil with a specified Q of 70 at 10 MHz.

Determine: The series RF resistance at 10 MHz and the estimated Q at 8 MHz.

$$Q = \omega L/R, \text{ thus } R = \omega L/Q = 250/70 = 3.6 \ \Omega$$

Considering that the RF resistance will not change significantly from 10 MHz to 8 MHz,

$$Q = \omega L/R = 200/3.6 = 56$$

Selection

The practical aspects of inductors and capacitors used in RF circuits relate to mechanical as well as electrical requirements. Among these requirements are physical size, temperature range and stability, heat (power) dissipation, weight, mounting characteristics, range of standard values available, reliability, availability, and, often most important, cost. To select and apply inductors and capacitors to electronic circuits intelligently, you should have access to manufacturers' published materials and be able to interpret specifications based on the requirements of a given circuit. Most manufacturers supply such information in three general forms: *catalog listings,* indicating what is available and pricing/distribution information; *specifications,* describing the electrical and physical properties of their products; and *application notes,* detailing typical ways in which their products may be used in circuits.

To become competent in this important aspect of practical electronics, you should regularly peruse technical magazines and respond to advertisements offering catalogs, specifications, and application notes. The effort will quickly set you up with a library of technical information relating to the "real world" of hardware.

Inductors

Decisions regarding practical RF inductors are generally based on the circuit function in which a coil is employed. Three common categories for coils include their use as RF chokes, tuning elements, and transformer/impedance-matching elements. RF chokes need not be high Q; and because they are noncritical (not precise inductance) components, they are available in a wide range of values at low cost. Inductors used as RF tuning elements, however, must be high Q and critically designed for accuracy and temperature stability. They should also be shielded or specially wound to maintain their immunity to stray magnetic fields. RF transformers and impedance-matching elements are generally considered to be as critical as tuning elements, so they are likewise high Q, accurate, and stable. Inductors are available in both fixed and adjustable types, with an adjustable core

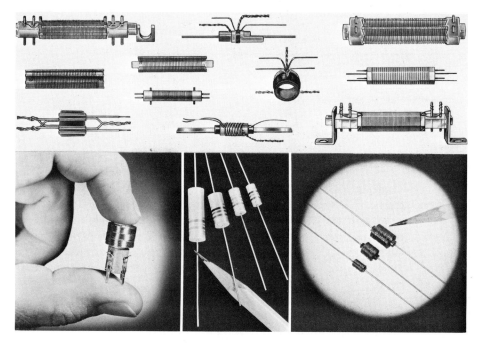

FIGURE 2-6

providing tuning of the most popular adjustable types. Figure 2-6 shows a variety of available coils.

While inductors are available from manufacturers in a wide range of specifications, technicians are often called upon to fabricate their own coils for critical RF circuits or to specify fabrication procedures for producing such coils in quantity. This is where considerable knowledge, judgment, and experience need to be employed. The process may at times be as simple as winding four turns of 16-gauge wire around a cylindrical form; yet it can become quite complicated, with special forms and core materials and with requirements for turns-spacing and special coupling characteristics in transformers. Such prototype RF coils and transformers are generally wound in a laboratory with access to Q-meters, RF bridges, signal generators, and detectors. This equipment incidentally is also used for measuring or characterizing purchased inductors when they are to be used in critical circuits. Figure 2-7 shows various types of winding techniques that are employed in constructing inductors.

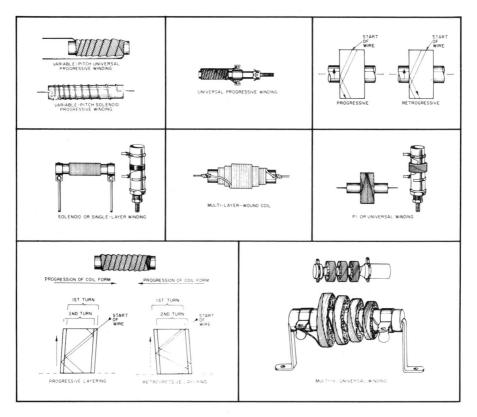

FIGURE 2-7

Capacitors

As with inductors, the decisions made about RF capacitors are also based on the circuit functions in which the capacitors are employed. Capacitors are used for coupling, bypass, tuning, and impedance matching in RF circuits. Popular RF capacitor types are generally specified by their dielectric material; these include ceramic, polystyrene, and mica elements. There are two classes of ceramic capacitors: those used for noncritical circuits (e.g., coupling and bypass elements) and those used for critical circuits. The noncritical class is the most popular because it is inexpensive and available in a wide range of values. Ceramic capacitors for critical circuits are of high Q and accuracy; they are available with special temperature coefficients to compensate for temperature drifts of other components in tuning circuits. Polystyrene and silver-mica (silver plates and mica dielectric) capacitors are stable, accurate, and high Q. Variable or trimmer capacitors are often used for critically tuning RF circuits. They are available with air, polyethylene, glass or mica, and even vacuum dielectric materials, with the air and vacuum dielectric units maintaining the highest Q and stability. Figure 2.8 shows a variety of capacitors typical of those presently on the market.

RESONANT CIRCUITS

One most important property provided by L and C components in electronic circuits is *resonance,* whereby the characteristics of these components allow for the processing of signals at only specific frequencies. In a general sense, resonance refers to a natural vibration activated by a signal. You can hear examples of resonance in musical instruments, where signals from vibrating strings, reeds, or other sources are enhanced by the structure of the instrument, which in turn vibrates and makes the signal louder.

In electronic circuits resonance is commonly achieved through the use of the current and voltage characteristics of inductors and capacitors in combination. If you review the reactance and phase characteristics of L and C in a series circuit, you will note that X_L has a leading phase angle ($+90°$) and increases in magnitude when the frequency of an applied signal increases. X_C has a lagging phase angle ($-90°$) and decreases in magnitude when the frequency of an applied signal increases. In such a series circuit there will be *one* frequency where the inductive reactance and the capacitive reactance are equal; and since their phase characteristics are opposing, the magnitudes of reactance will cancel and yield a net reactance of zero ohms. When this cancellation of reactance occurs in a series circuit, an increased current flow results, since only the remaining resistance of the circuit will act to limit current. This condition is called *resonance,* and the frequency at which this occurs is called the *resonant frequency.* The simple series *RLC* circuit and its associated phasor diagrams are shown in Fig. 2-9.

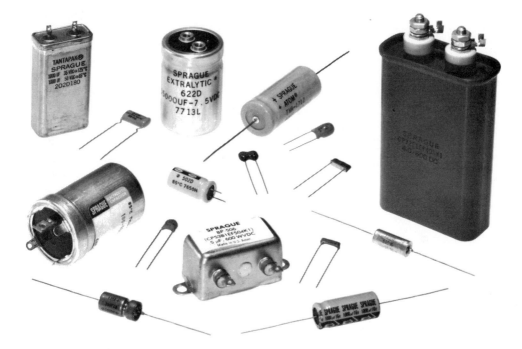

FIGURE 2-8

You may calculate the resonant frequency of a circuit if L and C are known; and since X_L equals X_C at resonance, this frequency may also be calculated if only L or C and X are known. The most popular formula used in computing resonance is

$\omega^2 = 1/LC$, which transposes to: $f = 1/2\pi \sqrt{LC}$

f is the resonant frequency (in hertz), L is the inductance (in henrys), and C is the capacitance (in farads).

You may, for example, compute the resonant frequency of a circuit using the 4 μH coil and 250 pF capacitor:

$$f = 1/2\pi \sqrt{LC} = 1/(6.28 \times \sqrt{4 \times 250 \times 10^{-18}}) = 5.04 \text{ MHz}$$

If the values of C and X were known ($X_L = X_C$ at resonance), the resonant frequency could be computed by extracting f through transposition of the reactance formula:

$$X = 1/\omega C, \quad \omega = 1/CX = 2\pi f; \text{ thus } f = 1/2\pi CX$$

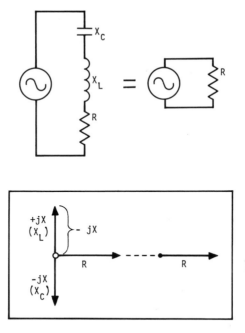

FIGURE 2-9

X is the reactance of the coil or capacitor (in ohms) and C is the capacitance (in farads). Given a reactance of 127 Ω and the 250 pF capacitor, the resonant frequency would be computed as follows:

$$f = 1/2\pi CX = 1/(6.28 \times 250 \times 10^{-12} \times 127) = 5.02 \text{ MHz}$$

If the values of L and X were known, f could be extracted through transposition of the inductive reactance formula:

$$X = \omega L, \; \omega = X/L = 2\pi f; \text{ thus } f = X/2\pi L$$

X is the reactance (in ohms) and L is the inductance (in henrys). Given the reactance of 127 Ω and the 4 μH coil, the resonant frequency would be computed as follows:

$$f = X/2\pi L = 127/(6.28 \times 4 \times 10^{-6}) = 5.06 \text{ MHz}$$

All the formulas for resonance can, of course, be transposed to find f, X, L, and C as necessary, with a most common transposition applied to the resonance formula in order to find L or C for a specific resonant frequency:

$$\omega^2 = 1/LC, \text{ thus } L = 1/\omega^2 C \quad \text{and} \quad C = 1/\omega^2 L$$

If, for example, you were given a 4 μH coil and wanted to compute the value of a capacitor required for resonance at 8 MHz, you could proceed as follows:

$$C = 1/\omega^2 L = 1/[(6.28 \times 8 \times 10^6)^2 \times 4 \times 10^{-6}] = 99 \text{ pF}$$

You may have observed that different values of L and C will yield the same resonant frequency, and you may wonder what effects will result from the use of different L-C combinations in circuits. This most critical aspect of resonant LC circuits is best analyzed at signal frequencies slightly above and below resonance. We will proceed by using the series RLC circuit of our earlier examples (4 μH coil, 250 pF capacitor, and 16 Ω resistor) connected in series to a variable-frequency signal source (Fig. 2-9). The values of L and C will then be changed, but with the frequency of resonance maintained at 5 MHz using a 20 μH coil and a 50 pF capacitor. The network can be quickly tabulated by employing the estimating techniques developed earlier in this chapter and inserting the results in two tables for comparison. Tables 2-4 and 2-5 show network values at resonance and at frequencies 5% and 10% above and below resonance. Figure 2-10 shows the graphical results of the data so tabulated.

Your careful observation of Tables 2-4, 2-5, and Fig. 2-10 should disclose that the frequency-response characteristics of resonant LC circuits are greatly affected by *which* values of L and C are combined for resonance. Higher L/C ratios yield narrowband frequency response; lower L/C ratios

TABLE 2-4. CIRCUIT CHARACTERISTICS VERSUS FREQUENCY DEVIATION

	$L = 4 \mu$H	$C = 250$ pF	$R = 16 \Omega$		
	-10%	-5%	f_o	$+5\%$	$+10\%$
f (MHz)	4.5	4.75	5.0	5.25	5.5
X_L	$+j114$	$+j121$	$+j127$	$+j133$	$+j140$
X_C	$-j140$	$-j133$	$-j127$	$-j121$	$-j114$
X_T	$-j26$	$-j12$	$\pm j0$	$+j12$	$+j26$
R	16	16	16	16	16
Z	$30\angle-58°$	$20\angle-37°$	$16\angle0°$	$20\angle+37°$	$30\angle+58°$

TABLE 2-5. CIRCUIT CHARACTERISTICS VERSUS FREQUENCY DEVIATION

	$L = 20 \mu$H	$C = 50$ pF	$R = 16 \Omega$		
	-10%	-5%	f_o	$+5\%$	$+10\%$
f (MHz)	4.5	4.75	5.0	5.25	5.5
X_L	$+j571$	$+j603$	$+j635$	$+j667$	$+j699$
X_C	$-j699$	$-j667$	$-j635$	$-j603$	$-j571$
X_T	$-j128$	$-j64$	$\pm j0$	$+j64$	$+j128$
R	16	16	16	16	16
Z	$129\angle-83°$	$66\angle-76°$	$16\angle0°$	$66\angle+76°$	$129\angle+83°$

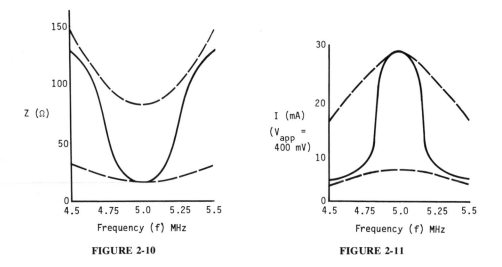

FIGURE 2-10   FIGURE 2-11

TABLE 2-6.   CIRCUIT CHARACTERISTICS VERSUS
FREQUENCY DEVIATION

	$L = 20 \ \mu H$	$C = 50$ pF	$R = 80 \ \Omega$		
	-10%	-5%	f_o	$+5\%$	$+10\%$
f (MHz)	4.5	4.75	5.0	5.25	5.5
X_T	$-j128$	$-j64$	$\pm j0$	$+j64$	$+j128$
R	80	80	80	80	80
Z	$151\angle -58°$	$102\angle -38°$	$80\angle 0°$	$102\angle +38°$	$151\angle +58°$

yield wider, and less discriminating, frequency response. You should note also that both circuits have the same resistance (16 Ω) at resonance, so they will pass an equal resonant current when connected to a signal source, as shown in Fig. 2-11.

Table 2-6 shows the results of changing the value of R in the RLC resonant circuit from 16 Ω to 80 Ω. Note that the frequency-response characteristics with this higher resistance are similar to the 4 μH, 250 pF combination with the 16 Ω resistor, but that the resonant current is reduced by the increased resistance of 80 Ω.

Q (Quality Factor)

Q characteristics are applied to resonant circuits as well as to L and C components individually. In the series-resonant RLC circuit the ohmic value of the resistor is added to any internal resistance of the coil to make up the RF resistance of the circuit. This total resistance is then divided into the circuit reactance (usually X_L) and the result is equal to the circuit Q. If the resistance and reactance values of the components used in tabulating the response noted in Tables 2-4, 2-5, and 2-6 are related, you should see a relationship between Q and L/C ratios in the resonant circuit. A higher L/C ratio yields a higher reactance; and since $Q = X/R$, a higher L/C ratio yields a higher Q. The circuit Q at resonance may be computed using the following formula:

$$Q = X/R = \omega L/R = 1/\omega CR$$

You may quickly compute the Q of the circuit examples noted in Fig. 2-10 using the above formula. For example, if the reactance (X) of the 4 μH coil and the 250 pF capacitor is 127 Ω at the resonant frequency of 5 MHz, then:

$$Q = X/R = 127/16 = 7.9$$

If the L/C ratio were changed, however, to 20 μH and 50 pF, the Q would increase in proportion to the increase in reactance:

$$Q = \omega L/R = 6.28 \times 5 \times 10^6\ 2 \times 10^{-5}/16 = 628/16 = 39.3$$

thus yielding

$$20:4\ \mu H = 5:1 = 39.3:7.9$$

But if the resistance were also changed in proportion to the reactance, the Q would return to its original value. Since $20:4\ \mu H = 5:1$ and $R = 5 \times 16\ \Omega = 80\ \Omega$, then $Q = X/R = 628/80 = 7.9$.

Bandwidth

If you again observe the frequency-response characteristics of the circuits in Fig. 2-10, you should note that the wider response curves (a and c) are identical in form, even though their minimum impedances differ. If you then observe the values of Q calculated for the circuits of Fig. 2-10 (a and c), you should note the relationship of resonant-circuit Q to frequency response. Specifically, there is an inverse relationship of Q and 3 dB bandwidth as follows:

$$Q = f_o/BW \quad \text{and} \quad BW = f_o/Q$$

f_o is the resonant frequency of the tuned circuit, and BW represents the range of frequency, extending above and below f_o, where the response of the circuit varies 3 dB (half power) below the peak response at resonance. If the frequency response is specified or measured as impedance, voltage, or current, this -3 dB level occurs at 0.707 × peak magnitude. If the frequency response is specified or measured as power, this -3 dB level occurs at 0.5 × peak power.

You may compute the bandwidth for the circuit characteristics presented in Tables 2-4, 2-5, and 2-6 as follows:

$$\text{BW} = f_o/Q = \begin{cases} 5 \text{ MHz}/7.9 = 0.63 \text{ MHz} = 630 \text{ kHz} \\ 5 \text{ MHz}/39.3 = 0.13 \text{ MHz} = 130 \text{ kHz} \end{cases}$$

The bandwidths of response curves a and c of Figs. 2-10 and 2-11 are 630 kHz above and 315 kHz below the f_o of 5 MHz; the bandwidth of response curve b is 130 kHz above and 65 kHz below the f_o of 5 MHz.

If you apply network analysis techniques to the circuit of Fig. 2-9 and the curves of Fig. 2-10, you should observe that the net circuit reactance and resistance are equal at the frequencies where they respond at the -3 dB level. Using the values for curve a in Fig. 2-10, for example,

$$f = 5 \text{ MHz} + 315 \text{ kHz} = 5.32 \text{ MHz}$$
$$X_L = \omega L = 6.28 \times 5.32 \times 10^6 \times 4 \times 10^{-6} = 134 \text{ } \Omega$$
$$X_C = 1/\omega C = 1/(6.28 \times 5.32 \times 10^6 \times 250 \times 10^{-12}) = 119 \text{ } \Omega$$
$$X_{\text{tot}} = X_L - X_C = 134 - 119 = 15 \text{ } \Omega$$

Note that the net reactance of 15 Ω is not exactly the 16 Ω value required to match the resistance of the circuit. This variation is due to the liberties taken in rounding off computation values. If this tedious network analysis is repeated for the frequency of 5 MHz less 315 kHz (the -3 dB frequency below resonance), the same net reactance of about 16 Ω will also result. Extending this network analysis for the phase relationship will yield:

$$\angle\theta = \arctan X/R = \arctan 16/16 = \arctan 1 = 45°$$

The phase relationships of current and voltage (remember ELI the ICE man) still apply to resonant circuits. Reactances cancel at resonance, leaving only resistance and a 0° phase angle. Reactances equal resistance at the -3 dB point frequencies, resulting in a 45° leading phase angle above resonance (where the net reactance is inductive, $X_L > X_C$), and a 45° lagging phase angle below resonance (where the net reactance is capacitive, $X_C > X_L$). Beyond the -3 dB point frequencies the circuits become increasingly reactive, with the resulting phase angles approaching 90°.

Series Resonance

The points made in the previous section of this chapter describe properties of resonant circuits. In engineering practice, network analysis is not usually employed to determine circuit characteristics, because it is less tedious to use Q, bandwidth, reactance, and impedance relationships to closely estimate characteristics. The individual reactances or frequencies at the -3 dB points are typically determined by ratio and proportion based on circuit Q as follows:

At the -3 dB points,

$$X_C \text{ or } X_L = X_o \pm X_o/2Q \quad \text{and} \quad f = f_o \pm f_o/2Q$$

X_L is larger above resonance and X_C is larger below resonance, X_o is the reactance value ($X_L = X_C$) at resonance, and f_o is the resonant frequency. You may compute the -3 dB point reactances and frequencies using circuit Q by applying the above formulas to the circuits of Fig. 2-10 as follows:

$f_1 = f_o - f_o/2Q$ = 5 MHz $-$ [5 MHz/(2 $\times$ 7.9)] = 4.68 MHz
$f_2 = f_o + f_o/2Q$ = 5 MHz + [5 MHz/(2 $\times$ 7.9)] = 5.32 MHz
X_L at 4.68 MHz = X_C at 5.32 MHz = 127 Ω; $X_o - [X_o/(2 \times 7.9)]$ = 119 Ω
X_L at 5.32 MHz = X_C at 4.68 MHz = 127 Ω; $X_o + [X_o/(2 \times 7.9)]$ = 135 Ω
X_{tot} at 4.68 and 5.32 MHz = 135 $-$ 119 = 16 Ω

If a sine-wave signal is applied to a series-resonant circuit, only the circuit resistance will limit current at resonance. At the -3 dB point frequencies the current will be limited by an impedance 1.4 times the resistance at resonance and at a phase angle of 45°. At frequencies beyond the -3 dB points the high (reactive) impedance values serve to reduce circuit currents to minimum levels. The circuit current at resonance can be computed using Ohm's law, $I = V/R$, where V is the applied voltage and R is the resistance part of the series RLC circuit. This high resonant current is flowing through all three components of the RLC circuit, with resulting (out-of-phase) voltages developed across the inductive and capacitive reactances. These *reactive* voltages can also be computed using Ohm's law, $V = IX$, where V is the reactive voltage, I is the resonant current, and X is the inductive or capacitive reactance ($X_L = X_C$). If you proceed through the two Ohm's law calculations for the circuits of Fig. 2-10, applying a 2 V signal at 5 MHz, you will find that the reactive voltages exceed the applied voltage from the signal source by the value of Q. This phenomenon is called *Q-rise*.

$I = V_{app}/R$ = 2/16 = 125 mA
$V = IX$ = 0.125 $\times$ 127 = 15.9 V
$V_X = QV_{app}$ = 7.9 $\times$ 2 = 15.9 V

At the -3 dB point frequencies the reactive voltage (V_X) is 0.707 times this peak value which occurs at resonance.

V_X at 4.68 and 5.32 MHz = 0.707 $\times$ 15.9 = 11.2 V

This concept of Q-rise is used extensively in electronic circuits where tuning elements are used to drive voltage amplifiers. This is the basis for achieving selectivity in radio and television receivers.

Sample Problem 5

Given: A series-resonant circuit of 5 μH, 250 pF, and 8 Ω connected to a 400 mV signal source.

Determine: The resonant frequency, Q, -3 dB point frequencies, resonant current, and reactive voltage at resonance.

(a)   First, the resonant frequency is computed as:

$$f_o = 1/2\pi \sqrt{LC} = 1/(6.28 \times \sqrt{5 \times 10^{-6} \times 250 \times 10^{-12}}) = 4.5 \text{ MHz}$$

(b)   The inductive reactance and the network Q are computed as:

$$X_L = \omega L = 6.28 \times 4.5 \times 10^6 \times 5 \times 10^{-6} = 141 \ \Omega$$
$$Q = X/R = 141/8 = 17.7$$

(c)   The bandwidth and -3 dB point frequencies are now computed, based on Q and f_o, as follows:

$$BW = f_o/Q = 4.5 \text{ MHz}/17.7 = 255 \text{ kHz}$$

$$f_{(-3 \text{ dB})} = f_o \pm BW/2 = \begin{cases} 4.5 + (0.255/2) = 4.63 \text{ MHz} \\ 4.5 - (0.255/2) = 4.37 \text{ MHz} \end{cases}$$

(d)   Finally, the resonant current and reactive voltage at resonance are computed as follows:

$$I = V/R = 400 \text{ mV}/8 \ \Omega = 50 \text{ mA}$$

$$V_X = \begin{cases} IX = 50 \text{ mA} \times 141 \ \Omega = 7.06 \text{ V} \\ QV_{\text{app}} = 17.7 \times 400 \text{ mV} = 7.08 \text{ V} \end{cases}$$

(The variation of values 7.06 and 7.08 V is, of course, due to rounding off earlier values of X and Q.)

Parallel Resonance (Antiresonance)

Parallel-resonant circuits are far more popular in RF applications than are series-resonant circuits. One reason for this choice is the fact that parallel-resonant circuits have *high* impedance at resonance, making them appropriate for use in conjunction with voltage amplifiers. As you know, since impedance is the ratio of voltage to current, high-impedance circuits yield higher voltages at given power levels than do low impedance circuits. Practical parallel-resonant circuits are not *purely* parallel, because the resistance of the coil windings is in series with the inductance of the coil, and this series *R-L* combination is placed in parallel with a capacitor to form the parallel-resonant circuit. In most instances this series-parallel arrangement is insignificant to circuit characteristics; but when very low Q components are used, the series R becomes significantly large and must be converted to an equivalent parallel value. An example of this conversion is shown in Fig. 2-12; it is treated in detail later in the text where such conversion procedures are necessary. The approach taken at this point will neglect the required conver-

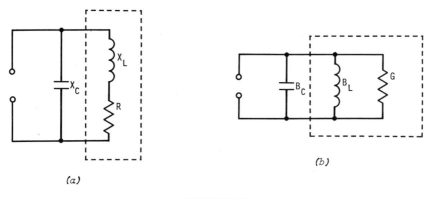

FIGURE 2-12

sion and deal with the concepts of parallel resonance most often used in engineering practice.

Higher Q parallel-resonant circuits are analyzed using two approaches: series analysis of circulating current, and parallel analysis of admittance (or parallel impedance) characteristics. If an instantaneous charge on the capacitor of a parallel-resonant circuit seeks equilibrium by discharging through the current path afforded by the coil, the following sequence of events will occur (refer to Fig. 2-13):

1. The capacitor is charged by momentarily closing switch S_1, then the capacitor discharges through the coil.
2. The increasing (changing) current flowing through the coil builds up a magnetic field around the coil.
3. The capacitor discharges, bringing the voltage across the coil to zero, and current ceases to flow through the coil.
4. The magnetic field collapses into the coil, developing a reverse voltage across the coil and charging the capacitor with a polarity *opposite* to its initial charge in step 1 above.
5. The process now repeats itself, but in reverse, ending back at the polarity of the initial charge in step 1.
6. The process continues to repeat, back and forth, until no energy is left in the circuit; or the process will continue if energy is added to the circuit to replace that which is lost in each cycle (steps 1 through 5).

The series resistance in the coil limits the peak current; thus the amount of energy lost in each cycle is due to this resistance, whereas the amount of energy stored in each cycle is determined by the reactance. As with series circuits, the resistance and reactance dictate the Q of the parallel circuit: Q = energy stored/energy lost = X/R.

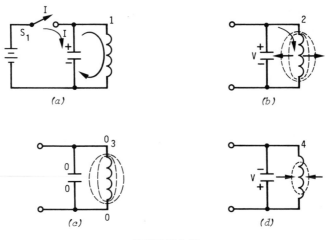

FIGURE 2-13

The time per cycle (360° or 2π radians) is dictated by the size of the coil and capacitor: $t = \sqrt{\omega L C}$. The frequency is computed by the formula: $f_o = 1/2\pi\sqrt{LC}$. Since most parallel-resonant circuits are driven by ac signals, the energy losses in the circuit are replaced by the voltage of the applied signal provided the frequency of the signal is at the resonant frequency of the circuit.

This circulating current approach to analysis considers that the initial charge on the circuit is like the driving voltage of a signal source operating at the resonant frequency of the parallel RLC circuit. Since the circuit reactances (X_L and X_C) cancel at resonance, the circulating current is limited only by the resistance of the circuit, just as with a series-resonant circuit. But the applied voltage is across *each reactive* element rather than both X_L and X_C in series, so the reactive voltage equals the applied voltage, and the voltage across the current-limiting resistance is V_{app}/Q. Thus the circulating current can be computed using Ohm's law: $I_{circ} = V/QR$. The amount of current provided by the signal source now need only replace that amount lost in each cycle, so the ratio of these currents is related to circuit Q:

$$Q = \text{energy stored/energy lost} = I_{circ}/I_{app}$$
$$I_{cir} = QI_{app} = V/QR \quad \text{and} \quad I_{app} = I_{circ}/Q = V/QX$$

Here is an all-important relationship: Although the reactance (X_L or X_C) is driven by the signal source, the current that flows is *not* V_{app}/X because of the *Q-rise* of the circulating current. The current that flows from the driving signal source is $1/Q$ times V_{app}/X. The *effective impedance* of the parallel-resonant circuit has thus increased by the value of Q, thereby reducing the applied current to V/QX. This effective impedance is labeled Z_p. You may use the following formula to compute Z_p based on the circulating current approach to analysis:

$$Z_p = QX = X^2/R = L/CR = Q^2R$$

Sample Problem 6

Given: A parallel-resonant circuit of 20 μH, 50 pF, and 16 Ω, connected in parallel to a 2 V signal source.

Determine: The effective impedance (Z_p), reactance, Q, applied current, and circulating current at resonance.

(a) First, the resonant impedance (Z_p) of the parallel circuit is computed as:

$$Z_p = L/CR = 20 \times 10^{-6}/(50 \times 10^{-12} \times 16) = 25 \text{ k}\Omega$$

(b) The reactance of the L and C components, at resonance, is computed as:

$$X = \sqrt{Z_pR} = \sqrt{25 \times 10^3 \times 16} = 632 \text{ }\Omega$$

(c) Then, the Q of the resonant circuit is computed as:

$$Q = X/R = 632/16 = 39.5$$

(d) The applied current is computed, using Ohm's law, as follows:

$$I_{app} = V/Z_p = 2/25 \text{ k}\Omega = 80 \text{ }\mu\text{A}$$

(e) Finally, the circulating current is computed, using both the Q-rise and Ohm's law approaches, as follows:

$$I_{circ} = \begin{cases} QI_{app} = 39.5 \times 80 \text{ }\mu\text{A} = 3.16 \text{ mA} \\ V/X = 2/632 = 3.16 \text{ mA} \end{cases}$$

The Q, bandwidth, and 3 dB point frequencies apply to parallel-resonant circuits in the same way that they apply to series-resonant circuits. The major differences between series and parallel concepts apply to both resonant and nonresonant RLC circuits: (1) The current is the same through all components of a series circuit, while the voltage across each component depends upon its resistance and/or reactance; and (2) the voltage is the same across all components of a parallel circuit, while the current through each component depends upon its conductance ($G = 1/R$) and/or susceptance ($B = 1/X$). With parallel-resonant circuits the admittance ($Y = 1/Z$) is a minimum (with the impedance thus at a maximum) value and equal to the circuit conductance at resonance. Thus the applied current is at a minimum value at resonance and increases to a level 3 dB *above* minimum at the frequencies $f_o \pm f_o/2Q$. The total susceptance and conductance are equal at the 3 dB

point frequencies, with B_C larger above resonance and B_L larger below resonance. With B_o as the susceptance value ($B_L = B_C$) at resonance, the susceptance ($1/X$) at the 3 dB point frequencies may be computed with the formula:

B at the 3 dB points $= B_o \pm B_o/2Q$

Impedance concepts are often applied to parallel-resonant circuits, in
which case the impedance is maximum at resonance and has no reactance
value (0° phase angle), being purely resistive and equal to QX. At the -3 dB
point frequencies, however, the impedance would be computed using more
awkward network analysis techniques. For this reason it is recommended
that you practice and develop confidence with the parallel concepts of conductance, susceptance, and admittance (G, B, and Y).

Sample Problem 7

Given: The same parallel-resonant circuit of the previous problem (20 μH, 50
pF, and 16 Ω, connected in parallel to a 2 V signal source).

Determine: The admittance at resonance, the susceptance at resonance, the
resonant frequency, the circuit Q, the 3 dB point frequencies, the 3 dB point
susceptance values, the 3 dB point admittance values, and the applied current values at resonance and at the 3 dB points.

(a)   First, the resonant admittance (Y_o) of the parallel circuit is computed,
      based on the 25 kΩ value of Z_p, as follows:

$$Y_o = G_o = 1/Z_p = 1/25 \text{ k}\Omega = 4 \times 10^{-5} \text{ S} = 40 \text{ } \mu\text{S}$$

(b)   The susceptance of the L and C components at resonance is computed
      as:

$$B = 1/X = 1/\sqrt{Z_pR} = 1/632 = 1.58 \text{ mS}$$

(c)   The resonant frequency is computed, now using B and L, as follows:

$$f_o = 1/2\pi LB = 1/(6.28 \times 20 \times 10^{-6} \times 1.58 \times 10^{-3}) = 5.04 \text{ MHz}$$

(d)   The Q of the resonant circuit is computed as:

$$Q = Z_p/X = 25 \text{ k}\Omega/632 = 39.6$$

(e)   The 3 dB point frequencies are computed as:

f_1 and $f_2 = f_o \pm f_o/2Q = 5.04 \pm (5.04/79.2)$
$\qquad = 5.04 \text{ MHz} \pm 63.6 \text{ kHz}$
$\qquad = 4.976 \text{ MHz and } 5.104 \text{ MHz}$

(f)   The 3 dB susceptance values, B_C and B_L above and below resonance,
      are computed, yielding the following net susceptance values:

B_C and $B_L = B_o \pm B_o/2Q = 1.58 \pm (1.58/79.2)$
$= 1.58 \pm 0.02 = 1.6$ mS and 1.56 mS

Above resonance,

$B_C - B_L = 1.6 - 1.56 = +j0.04$ mS

Below resonance,

$B_L - B_C = 1.6 - 1.56 = -j0.04$ mS

(g) The 3 dB point admittance is now computed as:

$Y = G \pm jB = 0.04 \pm j0.04$ mS

In polar form,

$Y\angle\theta = B/\sin(\tan^{-1} B/G) = 0.04/\sin(\tan^{-1} 0.04/0.04)$
$= 0.057\angle\pm45°$ mS

(h) Finally, the applied currents at resonance and the 3 dB point frequencies are computed as follows:

$$I = \begin{cases} V/Z_p = 2/25 \text{ k}\Omega = 0.08 \text{ mA} \\ VG = 2 \times 0.04 \text{ mS} = 0.08 \text{ mA} \end{cases}$$

At the 3 dB points,

$$I = VY\angle\theta = 2 \times 0.057\angle45° = \begin{cases} 0.113\angle45° \text{ mA above resonance} \\ 0.113\angle-45° \text{ mA below resonance} \end{cases}$$

Although the steps of the above problem are fatiguing, you should become thoroughly familiar with this approach as one of a number of ways such a problem may be solved.

Loading

Up to this point in the text the resonant circuits were analyzed without regard to their being connected to a load of any kind. In practical circuits, part of the energy stored in the RLC networks must be used to do some work, e.g., drive an amplifier or an antenna. Whenever a resonant circuit is connected to an electronic device or circuit, the resonant circuit is *loaded* by that device or circuit to which it is connected. Such loading of a parallel-resonant circuit reduces the value of resonant impedance (or increases admittance). This loading effect can be conveniently calculated by employing simple product-over-sum parallel resistance formulas:

$Z_{pL} = Z_{p\,\text{loaded}} = Z_p Z_L/(Z_p + Z_L)$

Z_p is the effective impedance of the parallel-resonant circuit at resonance, and Z_L is the impedance of the circuit *loading* the resonant circuit.

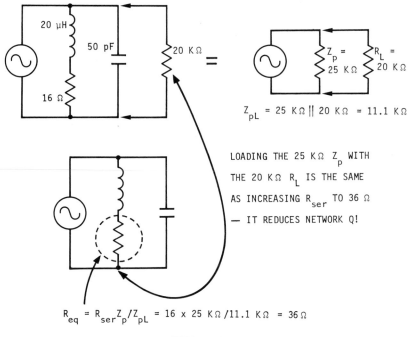

$Z_{pL} = 25 \ K\Omega \ || \ 20 \ K\Omega = 11.1 \ K\Omega$

LOADING THE 25 $K\Omega$ Z_p WITH

THE 20 $K\Omega$ R_L IS THE SAME

AS INCREASING R_{ser} TO 36 Ω

— IT REDUCES NETWORK Q!

$R_{eq} = R_{ser}Z_p/Z_{pL} = 16 \times 25 \ K\Omega \ /11.1 \ K\Omega = 36 \ \Omega$

FIGURE 2-14

If the parallel-resonant circuit of the previous problem (20 μH, 50 pF, and 16 Ω) with a Z_p of 25 kΩ were connected to drive a circuit with a resistive impedance of 20 kΩ, as shown in Fig. 2-14, the resulting loaded parallel resistive impedance would be:

$$Z_{pL} = Z_p R_L/(Z_p + R_L) = 25 \ k\Omega \times 20 \ k\Omega/(25 \ k\Omega + 20 \ k\Omega) = 11.1 \ k\Omega$$

Since Z_p has now been reduced to 11.1 kΩ, a ratio of 11.1 : 25, the remaining parallel-resonant circuit characteristics will be altered by the same ratio: the loaded Q (Q_L) will be 11.1/25 of the original Q value, with an equivalent change in all the circuit factors affected by Q. To properly define the loaded and unloaded circuit conditions, you should first identify all characteristics of the parallel-resonant circuit without loading and label them as unloaded conditions Q_U, Z_{pU}, etc. You should then proceed to derate the unloaded conditions by the ratio of loaded conditions to unloaded conditions. The point is that you need not repeat the tedious calculations, but only alter the results of the calculations by the ratio of loaded-to-unloaded conditions. If you apply the ratio 11.1 : 25 to the previous problem, then the loaded conditions would be computed as follows:

i. Z_{pL} is resistive and 11.1/25 $\times$ 25 kΩ = 11.1 kΩ.
ii. G_o (conductance at resonance) increases as resistance decreases, so $G_{oL} = 25/11.1 \times 40 \ \mu$S = 90.1 μS.

iii. The Q decreases, so $Q_L = 11.1/25 \times 39.6 = 17.6$.
iv. All the conditions affected by Q, bandwidth, and the 3 dB point frequencies, susceptances, admittances, voltages, and currents decrease or increase by the $11.1/25 \times Q_U$ factor, and 17.6 would be used instead of 39.6 in the calculations.

This ratio approach to parallel circuit loading should be used only when loads are predominantly resistive, so that reactive or susceptive factors do not complicate the calculations. Should you encounter complex loads with significant reactance or susceptance values, it is best to employ the more formal approach to analysis, which is the next topic of this chapter.

Admittance Approach

The second analysis technique for parallel-resonant circuits utilizes G, B, and Y (conductance, susceptance, and admittance) concepts, and deals with the parallel-resonant circuit as a network. This approach is more tedious than the circulating current technique, but it can be used effectively for both high and low Q circuits, and it serves as the basis for matching network analysis which will be treated later in this book. The procedure first requires the conversion of all series components, such as the series resistance in the coil, to an equivalent parallel network of conductance (G) and susceptance (B). Since B_C and B_L are equal at resonance, they cancel, leaving only the equivalent G as the current-allowing factor of the circuit. Any resistive loading of the resonant circuit represents an additional conductivity G, with the total conductance simply the sum of G_o and G_L. Reactive or susceptive loading of the resonant circuit will add to the capacitive and inductive elements, thus shifting the resonant frequency.

When a low-Q coil is placed in parallel with a capacitor to form a resonant circuit, the higher resistance distributed within the coil (considered as a series resistance) becomes a significant value. In such a case the resonant frequency is lower than would be obtained by the formula $f_o = 1/2\pi LC$ because the "effective" parallel inductance has been increased by the series resistance. Although there are many ways to compute this frequency shift, the admittance approach is most useful, as the same techniques of analysis can be applied to a wide variety of circuits. The first step in the application of this technique is the conversion of a series X and R to an equivalent parallel B and G. Here you may apply the analysis steps presented earlier in this chapter (see *Network Analysis* and *Parallel Networks*) as follows:

1. Determine the rectangular-form series impedance of the X-R network.
2. Convert rectangular-form impedance to polar-form impedance.
3. Find the polar-form admittance by inverting the polar-form impedance.
4. Convert the polar-form admittance to rectangular-form admittance to obtain the equivalent B and G.

You may also proceed to invert the G and B to show the difference between the parallel X and R and the series X and R that you started with, thus proving that X has been effectively increased for parallel resonance by the series resistance of the coil windings. This conversion is of course unnecessary with the admittance approach, so it need be done once only to prove that the concept is valid. The admittance approach can be applied most conveniently to electronic circuits with the use of an electronic calculator having rectangular-to-polar (and reverse) conversion capability.

If, for example, you want to resonate a low-Q coil with 16 Ω series resistance and 80 Ω reactance at a frequency of 1.5 MHz, as shown in Fig. 2-15, you would first proceed to find the parallel equivalent values of B and G, across which you would place a capacitor with a susceptance (B) equal to that of the coil. (The real inductance of the coil is computed as $L = X/2\pi f_o = 80/(6.28 \times 1.5 \times 10^6) = 8.5 \mu H$.)

i. $\quad Z = R + jX = 16 + j80 = Z\angle\theta = 81.6\angle 78.7° \ \Omega$
$\quad Y\angle\theta = 1/Z\angle\theta = 1/81.6\angle 78.7° = 12.3\angle -78.7° \text{ mS}$
$\quad Y = G - jB = 2.4 - j12 \text{ mS}$

ii. The susceptance of the resonating capacitor should be $+j12$ mS; and the capacitance, C, is:

$$C = B/2\pi f_o = 0.012/(6.28 \times 1.5 \times 10^6) = 1.3 \text{ nF}$$

iii. Now, for comparison only, the effective parallel values of X and R will be computed as:

$$X = 1/B = 1/0.012 = 83.3 \ \Omega$$
$$R = 1/G = 1/0.0024 = 417 \ \Omega$$

and the effective inductance, L, as:

$$L = X/2\pi f_o = 83.3/(6.28 \times 1.5 \times 10^6) = 8.84 \ \mu H$$

Note that when dealing with parallel circuits component values can be conveniently obtained through the admittance approach without reconverting

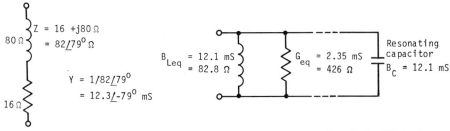

$$Y = 2.35 - j12.1 \text{ mS}$$

FIGURE 2-15

back to resistances and reactances. Note also that the *effective* inductance and resistance are of greater magnitude as equivalent parallel values than they really are as series values.

If additional resistive loading of the circuit in the above example is to be analyzed, the resistance value need only be converted to an equivalent conductance (G) and simply added to the existing conductance. The resulting Q_L would then be computed as B/G_{oL}, where B is the resonant susceptance and G_{oL} is the loaded conductance and the sum of the equivalent G of the resonant circuit and the G of the load: $G_{oL} = G_o + G_L$. If you connected the parallel-resonant circuit of the above example to a 1 kΩ load, the loaded Q and resulting loaded circuit characteristics would be calculated as follows:

i. $Q_U = B/G_o = 12$ mS/2.4 mS $= 5$
    $G_{oL} = G_o + G_L = G_o + 1/R_L = 2.4$ mS $+ 1/1000 = 3.4$ mS
    $Q_L = B/G_{oL} = 12$ mS/3.4 mS $= 3.53$

ii. The ratio which can be applied to all circuit characteristics is
    $Q_L : Q_U = 3.53 : 5$.

Sample Problem 8

Given: A parallel circuit to resonate at 5 MHz, using a 5 μH coil with 26 Ω distributed series resistance and loaded by a 1 kΩ resistance.

Determine: The equivalent parallel admittance of the coil, the required capacitance value, the loaded Q, the 3 dB point frequencies and susceptance values, and the applied current at resonance and the 3 dB points.

(a) First, the reactance of the 5 μH coil and its equivalent rectangular-form impedance are computed as follows:

$X_L = \omega L = 6.28 \times 5 \times 10^6 \times 5 \times 10^{-6} = 157$ Ω
$Z = R + jX = 26 + j157$ Ω

(b) The impedance of the coil is converted to polar form, then inverted to obtain the equivalent (parallel) polar- and rectangular-form admittance of the coil as follows:

$Z\angle\theta = X/\sin(\tan^{-1} X/R) = 157/\sin(\tan^{-1} 157/26) = 159\angle 80.6°$ Ω
$Y\angle\theta = 1/Z\angle\theta = 1/159\angle 80.6° = 6.28\angle -80.6°$ mS
$Y = G - jB = Y\cos\theta - jY\sin\theta = 6.28\cos 80.6° - j6.28\sin 80.6°$
$\quad = 1.03 - j6.2$ mS

(c) The capacitance value required to resonate with the equivalent inductive susceptance (B_L) is now computed as:

$B_C = B_L = 6.2$ mS, so
$C = B/\omega = 6.2$ mS/$(6.28 \times 5 \times 10^6) = 197$ pF

(d) The loaded conductance and the resulting loaded Q are computed as:

$$G_{oL} = G_o + G_L = 1.03 + 1/1 \text{ k}\Omega = 2.03 \text{ mS}$$
$$Q_L = B/G_{oL} = 6.2/2.03 = 3.05$$

(e) The 3 dB point frequencies are computed, based on the loaded Q, as:

f_1 and $f_2 = f_o \pm f_o/2Q_L = 5 \pm (5/6.1) = 5 \text{ MHz} \pm 0.82 \text{ MHz} = 4.18$ MHz and 5.82 MHz

(f) The 3 dB susceptance values (B_C and B_L) are now computed, yielding the following net susceptance at f_1 and f_2:

B_C and $B_L = B_o \pm B_o/2Q = 6.2 \pm (6.2/6.1) = 5.18 \text{ mS}$ and 7.22 mS

Above resonance,

$$B_C - B_L = 7.22 - 5.18 = +j2.04 \text{ mS}$$

Below resonance,

$$B_L - B_C = 7.22 - 5.18 = -j2.04 \text{ mS}$$

(g) Finally, the applied currents at resonance and at the 3 dB point frequencies are computed as follows:

$$I_{(fo)} = VG_{oL} = 2 \times 2.03 \text{ mS} = 4.06 \text{ mA}$$

At the 3 dB points,

$$I_{(3 \text{ dB})} = VY\angle\theta = VB/\sin (\tan^{-1} B/G) = 2 \times 2.04/\sin (\tan^{-1} 2.04/2.03)$$
$$= 5.75 \text{ mA}$$

Calculated more simply,

$$I_{(3 \text{ dB})} = 1.414 \, I_{(f_o)} = 1.414 \times 4.06 = 5.75 \text{ mA}$$

CONCLUSION

After completing this chapter you should have a working knowledge of resonant and nonresonant RLC circuits as applied to RF electronics. You should also be able to employ basic analysis techniques to calculate component values and circuit response. The approaches used in the following chapters apply the terminology and techniques reviewed in this chapter to typical circuit, system, and transmission principles. You should refer to this chapter as necessary to refresh your knowledge of basic terminology and techniques. The problems that follow correlate with the learning objectives given at the beginning of this chapter. Your success in solving these problems will serve as an indication of your knowledge and ability to apply the technical information.

Problems

1. Given: A series *RLC* circuit of 100 Ω, 10 μH, and 150 pF, respectively, driven with a 3 V signal at 8 MHz.

Determine:
(a) The net rectangular- and polar-form impedance
(b) Whether the circuit is operating above or below resonance
(c) The magnitude and phase angle of the circuit current
(d) The magnitude and phase angle of the voltage across the capacitor

Repeat problem steps *a* through *d* for frequencies of 4 MHz and 6 MHz using *estimating* techniques to determine reactance values.

2. Given: A parallel *RLC* circuit of 100 Ω, 10 μH, and 150 pF, respectively, driven with a 3 V signal at 8 MHz.

Determine:
(a) The net rectangular- and polar-form admittance
(b) Whether the circuit is operating above or below resonance
(c) The magnitude and phase angle of the *applied* current
(d) The magnitude and phase angle of the current through the capacitor

Repeat steps *a* through *d* for frequencies of 4 MHz and 6 MHz using estimating techniques to determine susceptance values.

3. Given: The series-parallel *RLC* circuit of Fig. 2-16*a*.

Make: The necessary calculations to convert the circuit to its series equivalent of Fig. 2-16*b*, and to its parallel equivalent of Fig. 2-16*c*, by using impedance/admittance and rectangular/polar conversion techniques.

Determine:
(a) R' in Fig. 2-16*b*   (d) $G \pm jB$ in Fig. 2-16*c*
(b) X'_C in Fig. 2-16*b*   (e) B'_L in Fig. 2-16*c*
(c) $Z\angle\theta_{in}$ in Fig. 2-16*b*

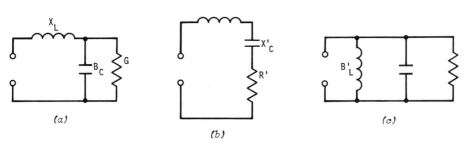

(a)   (b)   (c)

FIGURE 2-16

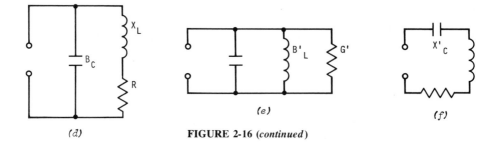

(d)   FIGURE 2-16 (continued)   (e)   (f)

4. Given: The parallel-series RLC circuit of Fig. 2-16d.

Make: The necessary calculations to convert the circuit to its parallel equivalent of Fig. 2-16e, and to its series equivalent of Fig. 2-16f, by using impedance/admittance and rectangular/polar conversion techniques.

Determine:
(a) G′ in Fig. 2-16e   (d) R ±jX in Fig. 2-16f
(b) B'_L in Fig. 2-16e   (e) X'_C in Fig. 2-16f
(c) $Y\angle\theta_{in}$ in Fig. 2-16e

5. Given: A series-resonant RLC circuit with a Q of 20, a capacitor of 40 pF, and driven with a 400 mV signal at its resonant frequency of 25 MHz.

Determine:
(a) Series inductance   (e) Bandwidth
(b) Series resistance   (f) $Z\angle\theta$ at −3 dB point
(c) Current   (g) R ±jX at 22.5 and 27.5 MHz
(d) Voltage across capacitor   (h) R ±jX at 20 and 30 MHz

6. Given: A parallel-resonant RLC circuit (R is in parallel) with a Q of 16, a parallel resistance of 2500 Ω, an inductor of 500 nH, and driven with a 100 mV signal at its resonant frequency.

Using the admittance approach, determine:
(a) Inductive susceptance, B_L (e) Inductor current
(b) Resonant frequency   (f) $Y\angle\theta$ at −3 dB point
(c) Parallel capacitance   (g) R ±jX at −3 dB point
(d) Applied current

7. Given: A parallel-resonant circuit with the series R-L combination of Fig. 2-16d, driven with a 200 mV signal at its resonant frequency and where C = 20 pF, L = 200 nH, and R = 2 Ω.

Using the circulating current approach, determine:
(a) Resonant frequency   (c) Bandwidth
(b) Q   (d) Resonant impedance, Z_p

   (*e*)   Applied current
   (*f*)   Voltage and $\angle\theta$ across R
   (*g*)   Applied current and $\angle\theta$ at -3 dB point

8.   Given: The parallel-resonant circuit with a series R-L combination, as in Problem 7, loaded with a 1000 Ω resistance across the capacitor (the component and signal values are the same as for Problem 7).

   Determine:
   (*a*)   Loaded resonant impedance, Z_{pL}
   (*b*)   Loaded Q
   (*c*) . Bandwidth
   (*d*)   Applied current
   (*e*)   Voltage and $\angle\theta$ across R series
   (*f*)   Equivalent loaded series resistance
   (*g*) $Y\angle\theta$ at -3 dB point

9.   Refer to manufacturers' catalogs and find three fixed and three variable (or trimmer) capacitor types (as designated by their dielectric material) in the range of 5 to 500 pF appropriate for RF circuits.

   Determine the following for each and prepare a table listing them:
   (*a*)   Dielectric material
   (*b*) Q or D factor
   (*c*)   Tolerance (for fixed capacitors) or range (for variable capacitors)
   (*d*)   Temperature coefficient
   (*e*)   Breakdown voltage
   (*f*)   Physical size of representative value
   (*g*)   Cost of representative value per 100
   (*h*)   Manufacturer and type/part number

10. Refer to manufacturers' catalogs and find three fixed and three adjustable coil types (as determined by their physical structure) in the range of 100 nH to 500 μH appropriate for RF circuits.

   Determine the following for each and prepare a table listing them:
   (*a*)   Core material
   (*b*)   Type of winding
   (*c*) Q or series R
   (*d*)   Tolerance (for fixed coils) or range (for adjustable coils)
   (*e*)   Frequency range
   (*f*)   Maximum current
   (*g*)   Physical size of representative value
   (*h*)   Cost of representative value per 100
   (*i*)   Manufacturer and type/part number

CHAPTER 3

APPLIED RF CIRCUITS

Now that you have gained a degree of competence with passive circuit analysis techniques, you may apply these techniques to the kinds of special circuits to be found in RF electronic systems. In particular, electronic devices such as transistors and IC's need to process signals in conjunction with passive *RLC* networks and transformers that provide loading, frequency/phase selectivity, and coupling functions between devices, signal sources, and load elements.

Passive RF circuits fall into two main categories: those that employ special networks and those that employ transformers. A working knowledge of *RLC* network analysis techniques is required for successful application of both these categories. Once you have a thorough understanding of the principles of such techniques, you should be able to understand and analyze even the most complex of circuitry used in electronic systems. The appearance of electronic hardware (components, etc.) may change in different applications, but the *principles* involved do not change. While microwave capacitors and inductors, for example, certainly look different than those used in AM transistor radios, the circuit and network analysis techniques are based on the same principles that are reviewed in these introductory chapters.

LEARNING OBJECTIVES

Upon completing this chapter, including the problem set, you should be able to:

1. Employ advanced network analysis procedures to resonant and nonresonant *RLC* circuits.
2. Understand and compute the characteristics of RF coupling and impedance-matching networks.

3. Understand and compute the characteristics of RF transformers, including mutual inductance, coupling coefficient, and reflected impedance and admittance.
4. Understand and compute the characteristics of resonant circuits employing RF transformers: single-tuned, double-tuned, and stagger-tuned.
5. Understand and compute the relationship of coupling coefficient in RF transformers to loading, bandwidth, Q, and frequency response.
6. Have knowledge of specific applications of passive *RLC* circuits in tuning, filter, phase shift, coupling, and impedance-matching functions.
7. Select standard and special transformers from catalog listings appropriate for particular RF circuit applications.
8. Specify fabrication procedures and materials for the assembly of simple RF coils and transformers.
9. Specify practical RF circuit measurement equipment, techniques, and setups in both the frequency and time domains.

IMPEDANCE-MATCHING NETWORKS

Resonant *RLC* networks are used extensively in matching the impedance of one electronic circuit to another. These networks can be arranged to increase or decrease their impedance—and thus the voltage-to-current ratio—from input to output. Three of the most popular networks used for matching are the L, the T, and the π, which are so designated because of the way they are commonly drawn in schematic form. (See Fig. 3-1.) All three networks are capable of transferring power from their inputs to outputs with a minimum of loss, while simultaneously maintaining frequency selectivity as a resonant circuit. Although the networks are in fact a form of transformer, they are analyzed differently from transformers; and they are considerably more popular in application than transformers in higher-power RF circuits. Additionally, impedance division with matching networks (input-to-output impedance ratio) is achieved as often with capacitors as with tapped coils; whereas impedance division with transformers is achieved exclusively with either tapped or separate coil windings.

Conversion Techniques

The basic approach to analysis of the impedance-matching networks is through conversion of the L, T, or π configuration to a parallel-resonant circuit. The conversion techniques differ for each of the configurations, so each will be approached separately. There are, however, basic conversion techniques which will allow you to quickly convert series R and X values to

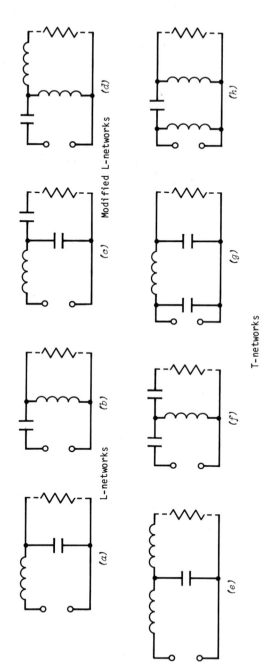

Modified L-networks

(d)

(c)

L-networks

(b)

(a)

T-networks

(h)

(g)

(f)

(e)

FIGURE 3-1

their parallel equivalents, and vice versa. The basis of the conversion techniques is that the phase and Q relationships of the same resistive and reactive components remain the same whether they are connected in series or in parallel, with changes only to voltage or current. If, for example, an R-L circuit had a 45° phase angle of impedance in series and a Q of 1, it would also have a 45° phase angle and a Q of 1 when the same components were connected in parallel, but the impedance value would be lower. This important *phase* relationship determines what value of impedance the components will have when they are connected in parallel if their properties are known when they are connected in series, or vice versa. The general formula for the relationships is specified as follows:

$$Q = \tan\angle\theta = X_s/R_s = R_p/X_p = B_p/G_p = R_s/(X_p - X_s)$$
$$= (R_p - R_s)/X_s = \sqrt{(R_p/R_s)} - 1 = 1/\sqrt{(X_p/X_s)} - 1$$

$\angle\theta$ is the phase angle of the reactive-to-resistive values, X_s and R_s are the series reactance and resistance, X_p and R_p are the parallel reactance and resistance, and B_p and G_p are the parallel susceptance and conductance. Table 3-1 shows these formulas transposed for obtaining any component value.

By applying this conversion approach to the low-Q coil of our earlier example, with 80 Ω reactance and 16 Ω series resistance, you can transpose the above formula to quickly find the equivalent parallel resistance and reactance values as follows: First find $Q = \tan \angle\theta$ by dividing X/R; then transpose $R_s/(X_p - X_s)$ to solve for X_p; then transpose $(R_p - R_s)/X_s$ to solve for R_p.

$$Q = \tan\angle\theta = X_s/R_s = 80/16 = 5$$

TABLE 3-1   TRANSPOSED FORMULAS FOR OBTAINING SERIES-TO-PARALLEL AND PARALLEL-TO-SERIES CONVERSIONS

Solve for:

$Q = \tan \angle\theta = X_s/R_s = R_p/X_p = B_p/G_p = R_s/(X_p - X_s) = (R_p - R_s)/X_s$
$\qquad\qquad\qquad\qquad\qquad\qquad\qquad\qquad\qquad\qquad = \sqrt{(R_p/R_s)} - 1 = 1/\sqrt{(X_p/X_s)} - 1$

$R_s = X_s/Q = Z_s \cos\theta = Q(X_p - X_s) = R_p - QX_s = R_p/(Q^2 + 1)$

$X_s = QR_s = Z_s \sin\theta = X_p - (R_s/Q) = (R_p - R_s)/Q = X_p/[1 + (1/Q^2)] = Q^2X_p/(Q^2 + 1)$

$Z_s = X_s/\sin\theta = R_s/\cos\theta = X_s/\sin(\tan^{-1}Q) = R_s/\cos(\tan^{-1}Q)$

$R_p = QX_p = Z_{par}\sin\theta = QX_s + R_s = R_s(Q^2 + 1) = 1/G_p$

$X_p = R_p/Q = Z_{par}\cos\theta = (R_s/Q) + X_s = (X_s/Q^2) + X_s = 1/B_p$

$Z_{par} = R_p/\sin\theta = X_p/\cos\theta = R_p/\sin(\tan^{-1}Q) = X_p/\cos(\tan^{-1}Q)$

$B_p = QG_p = 1/X_p$

$G_p = B_p/Q = 1/R_p$

$Y_{par} = B_p/\sin\theta = G_p/\cos\theta = B_p/\sin(\tan^{-1}Q) = G_p/\cos(\tan^{-1}Q)$

After transposing,

$$X_p = (R_s/Q) + X_s = (16/5) + 80 = 83.2 \ \Omega$$
$$R_p = (QX_s) + R_s = (5 \times 80) + 16 = 416 \ \Omega$$

This process of conversion can, of course, be reversed to find any of the values of X or R in series or parallel. If values of Z need to be obtained, they can be found by using the following formulas:

$$Q = \tan \angle \theta$$
$$Z_s = X_s/\sin \angle \theta = R_s/\cos \angle \theta$$
$$Z_{par} = R_p/\sin \angle \theta = X_p/\cos \angle \theta$$

(Note: Z_{par} is not to be confused with Z_p, which only refers to the effective impedance of a parallel-resonant circuit, whereas Z_{par} is the impedance of any parallel network.)

The conversion approach can also be executed graphically, just as R-X-Z relationships can be solved by drawing vectors of magnitudes. Referring to Fig. 3-2, the R_s, X_s, and Z_s are plotted, then Z_{par} is drawn perpendicular to Z_s, with the resulting intersect points on the vertical and horizontal axes being the values of R_p and X_p. The process can be reversed or modified to find any single value, as long as Z_{par} and Z_s are perpendicular (90°), and Z_s terminates at the origin of the vertical and horizontal axes. You may use this graphical approach to quickly approximate quantities without resorting to calculations. Such visualization additionally provides a useful point of reference against which you can compare the validity of calculated values.

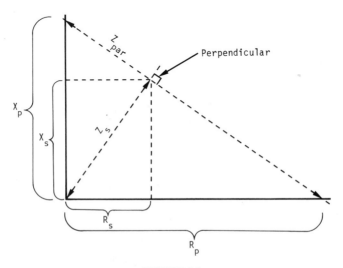

FIGURE 3-2

Practicalities of Impedance Matching

Selection of component values and conditions for impedance matching requires a degree of practical judgment. First, the *loaded* Q will be half of the *unloaded* Q, because a matched impedance will effectively load the network equal to *its* effective impedance, thus yielding a loaded impedance half that of Z_p. Loaded Q values of 20 or less are practical and easy to achieve at even the higher radio frequencies; and they are generally determined, or specified, based on bandwidth requirements. Capacitance values of less than 30 pF present problems due to stray capacitance from wiring, mounting hardware, and chassis enclosures. Inductance values of less than 50 nH also present problems due to short windings and to stray inductance of wiring and mounting hardware. Although some special techniques for fabricating circuits and components exceed the general limitations here recommended, they are of a specialized nature and, as such, are assigned to other portions of this book.

The general approach to practical impedance matching is to remain within the constraints of practicality in Q and component values, which can be achieved by alternative network selection. This requirement gets us back to the popular L, T, and π matching networks, which are selected for use based on the limitations of practical Q and component values. The series components of these networks will raise impedance; the parallel components will load (lower) impedance. Often the capacitors and inductors in the matching network may be interchanged, as shown in Fig. 3-1. This choice is based again on practicality of component values and also on the convenience of blocking dc with capacitors, and of passing dc afforded by inductors.

The L-Network

The L-network is the simplest of matching circuits. It is most appropriate for use where a low resistance is to be matched to a higher resistance of at least Q_U times greater. The circuit may be connected in reverse to match a high resistance to a low resistance. If you refer to the right of the dashed line in the network of Fig. 3-3b, you will note a capacitance in parallel with a

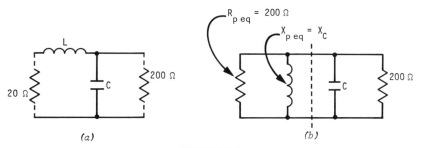

(a)   (b)

FIGURE 3-3

resistance. The relationship of the values of these two components will determine the Q of the RC parallel network. The parallel equivalent RL network to the left of the dashed line of Fig. 3-3b must achieve two functions: (1) The equivalent parallel inductance must resonate with the capacitance to the right of the dashed line; and (2) the equivalent relationship of these two components must match the Q of the output RC network to the right of the dashed line. Since series and parallel equivalent Q values are identical, the series RL network components of Fig. 3-3a can easily be determined, as demonstrated in the following sample problem.

Sample Problem 1

Given: An L-network as in Fig. 3-3a with a loaded Q of 8 at 5 MHz and with an input resistance of 20 Ω to be matched to a 200 Ω load resistance.

Determine: Values of the L and C components.

(a) Q_U is first computed, based on the parallel equivalent circuit of Fig. 3-3b, as follows:

$$Q_U = 2Q_L = 2 \times 8 = 16$$

(b) The reactance and inductance of the series-connected coil are computed as follows:

$$X_s = (R_p - R_s)/Q_U = (200 - 20)/16 = 11.3 \ \Omega$$
$$L = X/\omega = 11.3/(6.28 \times 5 \times 10^6) = 0.36 \ \mu H$$

(c) The equivalent parallel reactance is now computed as:

$$X_p = R_p/Q_U = 200/16 = 12.5 \ \Omega$$

(d) And the parallel capacitance value is computed, based on the equality of X_C and X_p, as:

$$C = 1/\omega X_C = 1/(6.28 \times 5 \times 10^6 \times 12.5) = 2.55 \ nF$$

If you note that the values of L and C in the above problem are somewhat extreme for a 5 MHz RF signal, because of the relatively low 200 Ω load resistance, you will recognize the L-circuit's limitations. Scaling up the load resistance to 2 kΩ would yield an L of 3.94 μH and a C of 255 pF, both of which are most practical values; but if the load resistance were 20 kΩ, C would have to be 25.5 pF—a value too small for general circuit use. This kind of examination should develop your sense of judgment about the limits of matching networks.

To achieve more practical component values for L and C, and yet match lower impedances, the simple L-network may be modified as shown in Fig. 3-4a to raise the *equivalent* parallel resistance, while maintaining a lower load resistance. To analyze the modified circuit of Fig. 3-4 to provide the same

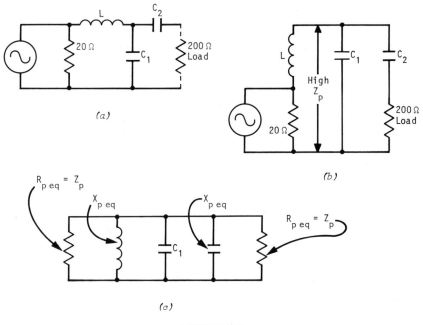

FIGURE 3-4

matching (20 Ω to 200 Ω) as in the previous problem, we would proceed as follows:

(a) After redrawing the circuit as shown in Fig. 3-4b and c, the series reactance (X_s) of the coil is determined as:

$$X_s = QR = 16 \times 20 = 320 \ \Omega$$

(b) Then the equivalent parallel values of R and X are computed as follows:

$$R_p = (QX_s) + R_s = (16 \times 320) + 20 = 5140 \ \Omega$$
$$X_p = (R_s/Q) + X_s = (20/16) + 320 = 321 \ \Omega$$

The function of C_2 is to raise the effective parallel resistance of the 200 Ω load resistor to equal 5140 Ω, with C_1 and the parallel equivalent capacitance of C_2 required to provide a parallel reactance equal to the 321 Ω of the coil's parallel value. To convert the R_L of 200 Ω to 5140 Ω, the Q of the output RC combination will differ from the input RL combination of 16.

(c) The Q of the output section, designated Q', is now computed as follows:

$$Q' = \sqrt{(R_p/R_s')} - 1 = \sqrt{(5140/200)} - 1 = 4.97$$

(d) The capacitive reactance and the equivalent parallel reactance are now computed as:

$$X'_s = Q'R'_s = 4.97 \times 200 = 994 \ \Omega$$
$$X'_p = (R'_s/Q') + X'_s = (200/4.97) + 994 = 1034 \ \Omega$$

(e) Finally, the values of the components are computed as follows:

$$L = X/\omega = 320/(6.28 \times 5 \times 10^6) = 10.2 \ \mu\text{H}$$
$$C_T = 1/\omega X_p = 1/(6.28 \times 5 \times 10^6 \times 321) = 99 \ \text{pF}$$
$$C' = 1/\omega X'_p = 1/(6.28 \times 5 \times 10^6 \times 1034) = 31 \ \text{pF}$$
$$C_1 = C_T - C' = 99 - 31 = 68 \ \text{pF}$$
$$C_2 = 1/\omega X'_s = 1/(6.28 \times 5 \times 10^6 \times 994) = 32 \ \text{pF}$$

Although this procedure is somewhat tedious, you should be able to follow the approach as a sort of two-step L-network solution. You should also note that since the two R_p equivalent resistances are in parallel, the net parallel resistance is $5140/2 = 2570 \ \Omega$; the loaded Q is also half of the unloaded Q, that is, $Q_U = R_p/X_p = 5140/321 = 16$ and $Q_L = Q_U/2 = 8$.

The modified L-network of Fig. 3-4 may be connected in reverse for matching a higher impedance to a lower impedance by interchanging L and C_2. The shape of the network is that of a ''T,'' but the name *T-network* more often applies to two similar series components (two coils or two capacitors), with the single opposite component (capacitor or coil, respectively) making the parallel connection. Figure 3-5 shows the conventional T-network configurations used for impedance matching. Here again the choice of the two-coil or two-capacitor arrangement depends on the practicality of component values and dc path requirements.

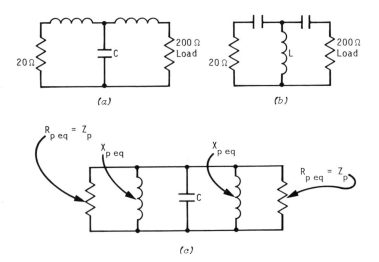

FIGURE 3-5

The T-Network

The analysis techniques used for the conventional T-network are the same as for the modified L-network, where a two-step procedure is required. The approach requires that the parallel equivalent value of R_p for the *input* series RL network be equaled by an equivalent R_p from the *output* series RL network, and that the X of the capacitor (X_C) cancel the total equivalent X_p from both series RL networks. (Of course the approach would be slightly modified for Fig. 3-5b, with series RC input and output networks and a single coil in parallel; but the resistance and reactance values would not change, so the modification is applied only to the calculations for L and C.)

Sample Problem 2

Given: The T-network of Fig. 3-5a with a loaded Q of 8 at 5 MHz and with an input resistance of 20 Ω to be matched to a 200 Ω load resistance.

Determine: Values of the L and C components.

(a) After redrawing the circuit as in Fig. 3-5c, the circuit Q and reactance of L_1 are computed as follows:

$$Q_U = 2Q_L = 2 \times 8 = 16$$
$$X_{L1} = Q_U R = 16 \times 20 = 320 \ \Omega$$

(b) The parallel equivalent values for the input RL network are computed as:

$$R_p = (QX_s) + R_s = (16 \times 320) + 20 = 5140 \ \Omega$$
$$X_p = (R_s/Q) + X_s = (20/16) + 320 = 321 \ \Omega$$

(c) Q' is now computed for the output RL network as:

$$Q' = \sqrt{(R_p/R_s')} - 1 = \sqrt{(5140/200)} - 1 = 4.97$$

(d) Then the reactance of the output coil and its *equivalent* parallel reactance are computed as:

$$X_s' = Q'R_s' = 4.97 \times 200 = 994 \ \Omega$$
$$X_p' = (R_s'/Q') + X_s' = (200/4.97) + 994 = 1034 \ \Omega$$

(e) Then the total equivalent inductive reactance is found by placing X_p in parallel with X_p' as follows:

$$X_L' = (X_p X_p')/(X_p + X_p') = (321 \times 1034)/(321 + 1034) = 245 \ \Omega$$

(f) Finally, the values of the components are computed as follows:

$$L_1 = X/\omega = 320/(6.28 \times 5 \times 10^6) = 10.2 \ \mu H$$
$$L_2 = X/\omega = 994/(6.28 \times 5 \times 10^6) = 31.7 \ \mu H$$
$$C = 1/\omega X = 1/(6.28 \times 5 \times 10^6 \times 245) = 130 \ pF$$

If the arrangement of the T-network were changed to use two capacitors with one coil (Fig. 3-5b), the values of the components would be computed as:

$$C_1 = 1/\omega X = 1/(6.28 \times 5 \times 10^6 \times 320) = 99.5 \text{ pF}$$
$$C_2 = 1/\omega X = 1/(6.28 \times 5 \times 10^6 \times 994) = 32 \text{ pF}$$
$$L = X/\omega = 245/(6.28 \times 5 \times 10^6) = 7.8 \ \mu\text{H}$$

In the case of these above examples, either configuration (Fig. 3-5a or b) yields component values that are practical, and the decision to use one arrangement over the other would probably be based on circuit needs for passing or blocking dc.

The π-Network

The other popular network used for the impedance matching of RF circuits is the π configuration. As with the L and T arrangements, the π-network is used in two forms: two capacitors and one coil, or two coils and one capacitor. The π-network is most appropriate for matching higher input impedances, as the component values are often impractical with input impedances lower than 500 Ω. You can proceed to analyze the network using two approaches: one that is precise (and of course tedious), and another that is convenient and quick. With either approach the first step to analysis requires redrawing the circuit. The precise approach to π-network analysis requires converting the π-network to an equivalent series-resonant circuit, as in Fig. 3-6d, while the quick approach allows for working with the network in the form redrawn as in Fig. 3-6b and f. The first steps employ basic concepts of power transfer which are important to both matching networks and *trans-formers*. (Transformers will be reviewed later in this chapter.) The following sequence of network relationships is based on the principles of power transfer and Watt's law, and is used in the analysis of π-networks, tapping L and C dividers, and transformers.

1. Consider that R_1 and R_2 of Fig. 3-6b and f are the R_{in} and R_{out} of the π-network and that the power is the same in both resistances.

2. Since both resistances are in series, the network current is the same through both R_1 and R_2, so the voltage across the two resistances can be computed with Watt's law: $(E_1)^2 = P_1 R_1$ and $(E_2)^2 = P_2 R_2$.

3. Since the power is the same ($P_1 = P_2$) in both resistances, the voltage across the resistances will vary as the *square* of the resistances: $(E_1/E_2)^2 = R_1/R_2$.

4. And since the voltages (E_1 and E_2) are directly proportional to the reactances (X_1 and X_2), the reactances also vary as the square of the resistances: $(X_1/X_2)^2 = R_1/R_2$.

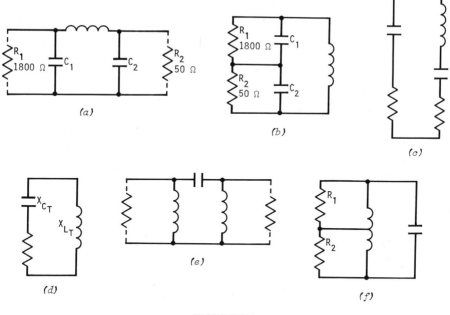

FIGURE 3-6

Sample Problem 3

Given: A network with two capacitors and one coil, with a loaded Q of 8 at 5 MHz, and with an input resistance of 1800 Ω to be matched to a 50 Ω load resistance.

Determine: Values of the C and L components.

(a)   The reactances of C_1 and C_2 are computed as follows:

$$X_1 = R/Q = 1800/8 = 225 \ \Omega$$

and, based on the power transfer relationship, $(X_1/X_2)^2 = R_1/R_2$,

$$X_2 = X_1/\sqrt{R_1/R_2} = 225/\sqrt{1800/50} = 37.5 \ \Omega$$

(b)   The equivalent series reactances of X_1 and X_2 are computed by transposing the formula: $Q = 1/\sqrt{(X_p/X_s) - 1}$ to $X_s = Q^2 X_p/(Q^2 + 1)$, as follows:

$$X_{1s} = (64 \times 225)/(64 + 1) = 222 \ \Omega$$
$$X_{2s} = (64 \times 37.5)/(64 + 1) = 36.9 \ \Omega$$

(c)   The series reactance of the coil is now computed for a value which will cancel the total series capacitive reactance, as follows:

$$X_L = X_{1s} + X_{2s} = 222 + 36.9 = 259 \ \Omega$$

(*d*)   Finally, the component values are computed as follows:

$$C_1 = 1/\omega X_1 = 1/(6.28 \times 5 \times 10^6 \times 225) = 142 \text{ pF}$$
$$C_2 = 1/\omega X_2 = 1/(6.28 \times 5 \times 10^6 \times 37.5) = 849 \text{ pF}$$
$$L = X_L/\omega = 259/(6.28 \times 5 \times 10^6) = 8.25 \text{ } \mu\text{H}$$

The solution above employed the precise approach to analysis of the network. The *quick* approach eliminates the conversion steps that change the parallel reactances of C_1 and C_2 to equivalent series values. The error resulting from the elimination of these steps is only slight, so you can eliminate the extra work involved if you do not need precise answers. The elimination of the conversion to series equivalents would yield the following: C_1 and C_2 would remain the same values, and their parallel reactances would be added for a total, so $X_L \cong 225 + 37.5 = 263\Omega$, and $L = 263/(6.28 \times 5 \times 10^6) = 8.38$ μH. If the two-coil, one-capacitor network of Fig. 3-6*e* were required for use in a circuit, the steps to solution would be the same as above with only the final component calculations changing:

$$L_1 = X/\omega = 225/(6.28 \times 5 \times 10^6) = 7.17 \text{ } \mu\text{H}$$
$$L_2 = 37.5/(6.28 \times 5 \times 10^6) = 1.19 \text{ } \mu\text{H}$$
$$C = 1/\omega X = 1/(6.28 \times 5 \times 10^6 \times 263) = 121 \text{ pF}$$

Tapping

Impedance matching is not always achieved with L, T, or π networks or their variations. These networks are employed when a *maximum* transfer of power is necessary. Many RF circuits and *RLC* networks are used to amplify voltage, however, so they maintain higher impedances as required by the electronic devices they interconnect. *Tapping* is one technique used extensively for higher impedance matching or for coupling signals from one electronic device to another. The concept of tapping is based upon the voltage-dividing/impedance-dividing properties of series circuits. A tapped coil has the same properties of two close-wound (see Chapter 2) coils in series. The voltage division achieved by a tapped coil is in direct proportion to the windings, turns, inductance, and reactance ratio of the tap to the *total* windings, turns, inductance, and reactance of the coil. The impedance division achieved by a tapped coil is also directly proportional to the voltage and reactance, etc., ratios because the series circuit does *not* match input and output power like the matching networks. The characteristics of the dividing function of tapping may be computed (neglecting mutual inductance or loading effects) by simple ratio and proportion as follows:

$$V_{out} = V_{in}[L_2/(L_1 + L_2)] = V_{in}[X_2/(X_1 + X_2)], \text{ etc.}$$

Tapping functions can also be achieved with two capacitors in a series voltage divider, with the same proportional relationships of voltage, reactance, and impedance. The proportionality for capacitance, however, varies

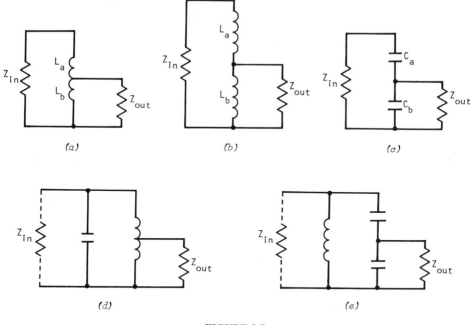

FIGURE 3-7

inversely with voltage, reactance, and impedance. All calculations for tapping with capacitors should be done using reactance values and leaving the capacitance values $(C = 1/\omega X)$ to be calculated as the last steps of the solution procedures. Figure 3-7 shows a variety of tapping connections.

To transfer signals through tapped resonant circuits, the unloaded *effective* impedance (Z_{pU}) must exceed the high output impedance of the element that it is to match. When *part* of the L or C is tapped and loaded, the load on the tap will reduce the effective impedance of the resonant circuit to the extent required to equal the output impedance of the element that it is to match. If, for example, you want to couple the 10 kΩ output of one electronic device to a 1 kΩ input of another, using a tapped resonant circuit, you should first select the *RLC* values that will yield a Z_{pU} greater than 10 kΩ. A good choice for flexibility is to obtain a Z_{pU} of about twice Z_{pL}; thus $Z_{pU} \cong 2Z_{pL}$, which in this case equals 20 kΩ. The 1 kΩ load across the tap would then reduce the 20 kΩ Z_{pU} to the required 10 kΩ Z_{pL}, provided the tap divides the coil or capacitor correctly. The *key* to tapping is in knowing *where* to tap. The techniques used to determine the tap point employ ratio and proportion, as well as parallel-to-series conversion. Although the procedures are tedious, they employ the same techniques used in earlier problems, so you should not find the approach too difficult.

Sample Problem 4

Given: The tapped resonant circuit of Fig. 3-8 with an input resistance of 10 kΩ and an output resistance of 1 kΩ, operating at 5 MHz with a Z_{pU} of 18 kΩ and a loaded Q of 25, and driven with a 2 V input signal.

Determine: The L_{tot}, C, and R_{ser} of the unloaded circuit, the tap inductances, L_a and L_b, the output voltage, and the input-to-output power transfer.

(a)   The unloaded Q is first computed by ratio and proportion as:

$$Q_U = Q_L Z_{pU}/Z_{pL} = 25 \times 18 \text{ k}\Omega/10 \text{ k}\Omega = 45$$

(b)   The series resistance of the resonant circuit (coil resistance) is computed after transposing $Z_p = Q^2R$ to $R = Z_p/Q^2$, as follows:

$$R = Z_p/Q^2 = 18 \text{ k}\Omega/45^2 = 8.89 \text{ }\Omega$$

(c)   Then the reactance and the values of L and C are computed as:

$$X = QR = 45 \times 8.89 = 400 \text{ }\Omega$$
$$L = X/\omega = 400/(6.28 \times 5 \times 10^6) = 12.7 \text{ }\mu\text{H}$$
$$C = 1/\omega C = 1/(6.28 \times 5 \times 10^6 \times 400) = 79.6 \text{ pF}$$

The *effective* series resistance of the 1 kΩ load on the tap will add to the 8.89 Ω *real* series resistance to reduce Z_p from 18 kΩ to 10 kΩ and to reduce Q from 45 to 25 as specified. Then the Q' value (Q' is *not* the circuit Q) of the tap reactance, X_{Lb}, and the 1 kΩ load resistance are found by comparing the parallel resistance to its effective series equivalent value. The parallel inductance is then determined through Q' and from X_{Lb}.

(d)   The total effective series resistance is computed by ratio and proportion as:

$$R_{ser} = R Z_{pU}/Z_{pL} = 8.89 \times 18 \text{ k}\Omega/10 \text{ k}\Omega = 16 \text{ }\Omega$$

(e)   The *added* effective series resistance is now computed as:

$$R_s = R_{ser} - R = 16 - 8.89 = 7.11 \text{ }\Omega$$

(f)   The Q of the tap reactance in combination with the load resistance is computed as:

$$Q' = \sqrt{(R_p/R_s) - 1} = \sqrt{(1000/7.11) - 1} = 11.8$$

(g)   The reactances and inductances, above and below the tap, are now computed as follows:

$$X_{Lb} = X_p = R_p/Q' = 1000/11.8 = 84.7 \text{ }\Omega$$
$$X_{La} = X - X_{Lb} = 400 - 84.7 = 315 \text{ }\Omega$$
$$L_a = L X_a/X = 12.7 \text{ }\mu\text{H} \times 315/400 = 10 \text{ }\mu\text{H}$$
$$L_b = L X_b/X = 12.7 \text{ }\mu\text{H} \times 84.7/400 = 2.7 \text{ }\mu\text{H}$$

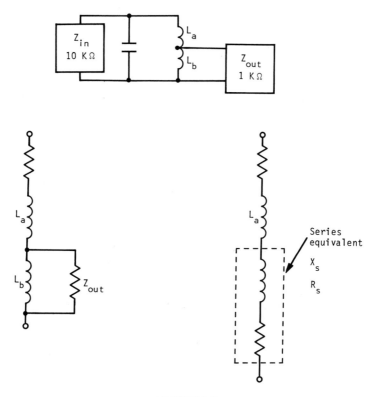

FIGURE 3-8

(*h*)   The series circuit, $L_a - L_b$, is analyzed to determine the tap output voltage as follows:

$$V_{out} \cong V_{in} X_b/X \cong 2 \times 84/400 \cong 0.42 \text{ V}$$

(*i*)   Finally, the input and output power values are computed, using Watt's law, as follows:

$$P_{in} = V^2/R = 2^2/10 \text{ k}\Omega = 0.4 \text{ mW}$$
$$P_{out} = V^2/R = 0.42^2/1 \text{ k}\Omega = 0.176 \text{ mW}$$

so the output power is 176/400, or 44% of P_{in}.

The final results of the calculated solution to the above problem identify the benefits of employing tapped resonant circuits, including relative efficiency in transferring power and voltage while maintaining a high Q for required bandwidth and selectivity. The same procedures of analysis would be used for tapping with series capacitors or for reversing the input and output of the circuits.

RF TRANSFORMERS

RF transformers are used extensively to couple electronic circuits. Transformers perform impedance-matching functions, and also provide isolation from one portion of a circuit to another. The primary, the secondary, or both the primary and secondary windings of a transformer may form a resonant circuit with external capacitors. Additionally, transformers may be tapped for impedance matching and wound with any number of auxiliary windings and taps in combination. RF transformers are among the most versatile of electronic circuit elements.

In your early study of basic transformer principles you most likely learned that electrical energy transfers from primary to secondary windings based on the principles of *electromagnetic induction*. These principles may be summarized as follows:

> *If* the core material of the transformer perfectly concentrates the magnetic field provided by the primary current, and *if* the secondary winding is placed within this magnetic field, then (*a*) the voltage ratio of the primary-to-secondary windings will be *directly* proportional to the turns ratio (number of turns) of the windings; (*b*) the current ratio of the windings will be *inversely* proportional to the turns ratio; (*c*) the impedance ratio will be directly proportional to the *square* of the turns ratio; (*d*) the power in the secondary will equal the power in the primary; and (*e*) the secondary voltage and current will be 180° out of phase with the primary voltage and current.

These relationships are probably familiar to you in application to power transformers and audio frequency (AF) transformers; but as any transformer manufacturer can tell you, the principles are based on two big ''ifs'' that rarely occur in practical transformers. Core materials are *not* perfect, and windings take up space, thus making their placement within an intense homogeneous magnetic field indeed difficult. Additional factors that influence transformer properties become particularly significant at high frequencies. These factors include heat losses, distributed capacitances, stray capacitances and inductances, and the various electrical properties of the core material and the other materials used in the fabrication of transformers. While some undesirable factors can be reduced by careful shielding, by special winding techniques, and by selection of materials used in fabrication, there are many factors in transformer design that make computations most unscientific and inaccurate when they are based simply on the turns ratio of the windings and permeability of the core.

Measurement/Analysis

The approach used for the analysis of RF transformers relies upon accurately measurable electrical properties which may be used to predict the response of such transformers in electronic circuits. These properties are primary inductance (L_{pri}), secondary inductance (L_{sec}), mutual inductance (L_m), and coupling coefficient (K_c); and they are mathematically related as follows:

$$L_m = K_c\sqrt{L_{pri}L_{sec}} \qquad L_{pri} = (L_m/K_c)^2/L_{sec}$$
$$K_c = L_m/\sqrt{L_{pri}L_{sec}} \qquad L_{sec} = (L_m/K_c)^2/L_{pri}$$

As you may observe, three of the factors must be known in order to compute the fourth. In engineering practice L_{pri}, L_{sec}, and L_m of the transformer are generally measured on an RF bridge or Q-meter, with K_c computed with the appropriate formula. The mutual inductance is measured on the bridge by connecting both windings in series-aiding and series-opposing connections, and then computing $L_m = (L_{aid} - L_{opp})/4 = (L_{max} - L_{min})$.

The inductance values obtained from the measurements using the RF bridge may not be accurate, however. Inductance varies with the amount of current passing through the transformer windings, and RF bridges typically do not provide the same RF signals and dc levels as the circuit applies to a transformer. So the in-circuit inductances will most often differ from those measured with an RF bridge. A more accurate approach requires in-circuit measurements (Fig. 3-9), made simply as follows:

1. With the secondary "open," the primary winding of the transformer is connected in series with a resistor to a signal source (and dc power supply if biasing is required) to provide the same signal voltage, frequency, and dc level as expected from the electronic circuit into which the transformer is to be connected.

2. The applied voltage and the voltage across the resistance are measured, and the phase $\angle\theta$ is computed as: $\angle\theta = \cos^{-1} V_R/V_{app}$; then the reactance of the coil is computed as: $X_L = R \tan \angle\theta$; and since the transformer winding is in series with the resistor, the current through both elements is the same and may be computed with Ohm's law: $I_{pri} = V_R/R$.

By employing this procedure the measured values are more accurate, as the conditions simulate those of a real electronic circuit. This same approach can be used to measure any inductor, but the procedure is introduced here to provide a basis for the practical analysis techniques that we will use for transformers.

3. The voltage of the secondary winding of the transformer is measured, and the mutual reactance (the reactance of the mutual inductance, L_m) is computed as: $X_m = V_{sec}/I_{pri}$.

4. Now the transformer windings are reversed and steps 1 and 2 of the procedure are repeated to obtain the secondary-winding reactance.

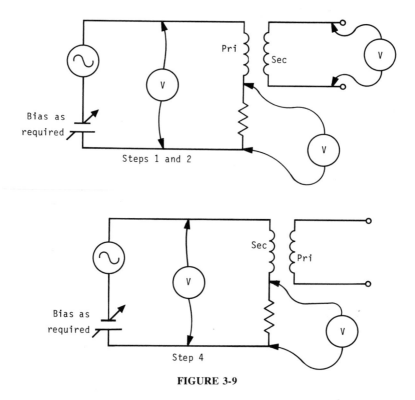

FIGURE 3-9

5. The coupling coefficient, K_c, can now be computed as: $K_c = X_m/\sqrt{X_{pri}X_{sec}}$; and if necessary, L_{pri}, L_{sec}, and L_m can be computed for each value of reactance as: $L = X/\omega$.

This approach to transformer measurements will serve as the basis for the analysis of transformer performance in RF electronic circuits.

Coupling

The coupling coefficient is the ratio of the RF signal out of the secondary winding, as compared to the signal into the primary winding of a transformer. Since signal levels are not a property of transformers, but the reactances of the windings *are,* and since signal levels are proportional to reactances, the coupling coefficient (K_c) also specifies the transfer *reactance.* The transfer reactance is called *mutual reactance* (X_m), while its related inductance is called *mutual inductance* (L_m). If, for example, a transformer with identical primary and secondary inductance of 10 μH and a reactance of 400 Ω transferred 100% of the input signal to its output, the K_c would be 1.0, the L_m would be 10 μH, and the X_m would be 400 Ω. If the same transformer transferred only 20% of the input signal to its output, the K_c would be 0.2, the L_m would be 2 μH, and the X_m would be 80 Ω. For transformers with dissimilar primary and secondary inductances and reactances, L_m and X_m are a percentage of the geometric *mean* of the combined values of the primary and secondary windings: $L_m = K_c\sqrt{L_{pri}L_{sec}}$ and $X_m = K_c\sqrt{X_{pri}X_{sec}}$. A transformer with windings of 10 μH and 40 μH, reactances of 400 Ω and 1600 Ω, and that transferred 20% of its input signal would have an L_m of 4 μH and an X_m of 160 Ω. It is rare to find RF transformers that provide close to 100% signal transfer, with the typical K_c values in the range of 0.01 to 0.5. There are certain advantages to low K_c values, including isolation and reduced circuit loading, so lower K_c values are often desirable. (Maximum power transfer is better achieved with matching networks—not with transformers.) It should also be noted that the phase of current and of voltage shift 180° from primary to secondary of the transformer. Calculations that consider phase should thus assign the $\angle$ 180° to X_m.

Reflected Impedance

To the extent that signals are transferred from the transformer's primary to secondary, they can also transfer in reverse—from secondary to primary. The amount of reverse transfer depends, of course, on secondary current as the source of secondary electromagnetic induction. Because the primary and secondary signals are 180° out of phase, the result of the secondary-to-primary signal transfer is the cancellation of the initial primary signal. Since signal transfer is related to mutual reactance and inductance values, the amount of *reverse* transfer is likewise so related, and it is called *reflected impedance*.

If we use the transformer of our previous example with identical primary and secondary values of 10 μH and 400 Ω, respectively (a K_c of 0.2, yielding an L_m of 2 μH and an X_m of 80 Ω), and connect the primary winding to a 2 V signal source and the secondary to a 20 Ω resistive load, we will find that the following takes place:

An opposing current flowing in the secondary winding and limited by the 20 Ω load resistance causes a reflected resistance back to the primary winding, which in turn causes the primary current to be reduced, with a net loss in output signal. The effects of signal transfer (identified by X_m) are multiplied by encountering X_m twice, so the total effect of the output load resistance on the signal transfer is:

$Z_{ref} = X_m^2/Z_{load} = 80^2/20 = 320$ Ω reflected R (to primary)

The total input impedance of the primary winding is increased by the value of Z_{ref}:

$Z_{in} = Z_{pri} + Z_{ref} = (R_{pri} + jX_{pri}) + R_{ref}$

For this example, R_{pri} is negligible, so

$$Z_{in} = R_{ref} + jX_{pri} = 320 + j400 \ \Omega = 512\angle 51° \ \Omega$$

The primary current as computed with Ohm's law is:

$$I_{pri} = V_{in}/Z_{in} = 2/512\angle 51° = 3.91\angle -51° \ mA$$

and the secondary output voltage is:

$$V_{sec} = X_m \angle 180° \ I_{pri} = 80 \times 3.91\angle -51° = 313\angle 129° \ mV$$

If the output of the same transformer of the previous example were connected to a 20 Ω capacitive reactance as a load, instead of the resistive load, the reflected impedance would still be 320 Ω. In this case the 320 Ω would be a reflected inductive reactance, however, as the transformer shifts the phase of the output capacitive reactance by 180°, making it inductive. This 320 Ω reflected inductive reactance would add to the 400 Ω primary inductive reactance to yield a Z_{in} of $+j720 \ \Omega$. The output voltage under these conditions would be computed as follows:

$$I_{pri} = V_{in}/Z_{in} = 2/720\angle 90° = 2.78\angle -90° \ mA$$
$$V_{sec} = X_m I_{pri} = 80\angle 180° \times 2.78\angle -90° = 222\angle 90° \ mV$$

If the output of the same transformer were connected to a 20 Ω *inductive* reactance as a load, the 320 Ω would be reflected as a *capacitive* reactance, which would cancel the primary inductive reactance and leave a net inductive reactance of 80 Ω as the Z_{in}. The output voltage would then be computed as follows:

$$I_{pri} = V_{in}/Z_{in} = 2/80\angle 90° = 25\angle -90° \ mA$$
$$V_{sec} = X_m I_{pri} = 80\angle 180° \times 25\angle -90° = 2\angle 90° \ V$$

Transformer secondary outputs are most often connected to higher impedance loads, so the effects of reflected impedances are usually negligible. The examples above are unusual in that the load impedances are very low. In such special cases the reflected impedances are significant.

TUNED RF TRANSFORMERS

When the primary or secondary winding of a transformer is connected with a capacitor to make a resonant network, the transformer provides some unique functions which cannot be achieved with a single RLC resonant circuit. Let us proceed using the previous transformer example with identical primary and secondary values: $X_{pri} = X_{sec} = 400 \ \Omega$, $K_c = 0.2$, and $X_m = 80 \ \Omega$. The secondary winding of the transformer is to resonate with a capacitor connected in parallel with its output, as in Fig. 3-10d. In this circuit connection the transformer's secondary winding reactance is canceled by the capacitive reactance, leaving only the series resistance of the coil to limit the circulating current. If the output voltage is taken across the capacitor, the effects will be the same as those achieved with a series-resonant RLC network: there will be a Q-rise in output voltage at resonance. If the transformer's primary winding was connected to the 2 V signal source, as in the previous examples, and the Q of the secondary winding was 40, the output voltage would be computed as follows:

$$I_{pri} = V_{in}/X_{pri} = 2/400 = 5 \text{ mA}$$
$$V_{sec} = X_m I_{pri} = 80 \times 5 = 400 \text{ mV}$$
$$V_{out} = QV_{sec} = 40 \times 0.4 = 16 \text{ V}$$

This example shows one of the benefits of using an RF transformer with a *tuned* secondary for coupling voltage-controlled electronic devices. Of course, the benefit of Q-rise in voltage with this circuit is at the expense of output current. The 16 V output is effectively unloaded, so zero power is obtained from the output. As power is obtained from the output, the voltage, Q_L, and Z_p will be reduced. Maximum output power would be obtained if the load were connected across the capacitor, equal to the Z_p of the parallel-resonant network, thus reducing the Q_U of 40 to a Q_L of 20 and reducing the output voltage to 8 V.

$$Z_{pU} = Q_U X = 40 \times 400 = 16 \text{ k}\Omega$$
$$R_L = Z_p = 16 \text{ k}\Omega$$
$$Z_{pL} = Z_{pU}/2 = 16/2 = 8 \text{ k}\Omega$$
$$Q_L = Q_U/2 = 40/2 = 20$$
$$V_{out} = Q_L V_{sec} = 20 \times 400 \text{ mV} = 8 \text{ V}$$
$$P_{out} = V_{out}^2/R_{out} = 8^2/16 \text{ k}\Omega = 4 \text{ mW}$$
$$P_{in} = V_{in}^2/X_{pri} = 2^2/400 = 10 \text{ mW}$$

Although there is a Q-rise in circulating current when the transformer's secondary winding is tuned with this parallel connection, there is little reflected impedance because the circulating current is 180° out of phase with the applied current, so the load is still at high impedance.

If the primary winding of the same transformer were tuned, and the secondary untuned as in Fig. 3-10b, the Q-rise in effective parallel impedance (Z_p) would reduce the *applied* current by a factor of Q, yet provide the same value of I_{pri} as if the primary were untuned. This circuit connection would yield the same output signal as an untuned transformer, with the advantage of its input impedance being increased for light loading of electronic devices. The characteristics of the circuit would be computed as follows:

$$Z_{in} = Z_p = QX = 40 \times 400 = 16 \text{ k}\Omega$$
$$I_{pri} = Q\, V_{in}/Z_{in} = 40 \times 2/16 \text{ k}\Omega = 5 \text{ mA}$$
$$\phantom{I_{pri}} = V_{in}/X_{pri} = 2/400 = 5 \text{ mA}$$
$$V_{sec} = X_m I_{pri} = 80 \times 5 = 400 \text{ mV}$$
$$P_{in} = V_{in}^2/Z_{in} = 2^2/16 \text{ k}\Omega = 250 \text{ } \mu\text{W (instead of 10 mW)}$$

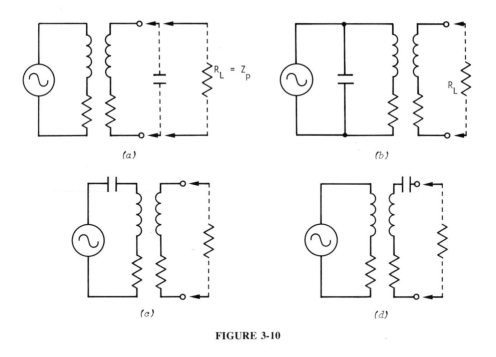

(a)   (b)

(c)   (d)

FIGURE 3-10

The above example assumes a high impedance load on the secondary winding of the transformer. If maximum power transfer was required from the 400
Ω secondary reactance using a 400 Ω load resistance, the reflected impedance would significantly affect the Q of the primary winding and reduce the
input impedance as follows:

$$Z_{ref} = X_m^2/R_L = 80^2/400 = 16\ \Omega$$
$$R_{pri} = X_{pri}/Q_U = 400/40 = 10\ \Omega$$
$$Q_L = X_{pri}/(R_{pri} + R_{ref}) = 400/26 = 15.4$$
$$Z_{in} = Q_L X_{pri} = 15.4 \times 400 = 6.15\ k\Omega$$
$$I_{pri} = Q_L V_{in}/Z_{in} = 15.4 \times 2/6.15\ k\Omega = 5\ mA$$
$$= V_{in}/X_{pri} = 2/400 = 5\ mA$$
$$P_{in} = V_{in}^2/Z_{in} = 2^2/6.15\ k\Omega = 650\ \mu W$$
$$P_{out} = V_{out}^2/R_L = 0.4^2/400 = 400\ \mu W$$

Resonance

The transformer may be connected to resonate in *series* with a capacitor on
either its primary or secondary winding (Fig. 3-10c and d) in order to provide
matching for very low impedances. If the primary winding of the transformer
is series tuned and the secondary is untuned, the reactance of the capacitor
cancels the reactance of the primary winding, leaving only the series coil

resistance as the input impedance. Using the transformer of our previous examples, the input impedance would be:

$$R_{pri} = X_{pri}/Q_U = 400/40 = 10 \ \Omega$$

The primary current, however, would be increased by Q, as compared to the parallel or untuned connections, so the secondary voltage would also be increased by Q.

$$I_{pri} = Q_U V_{in}/X_{pri} = 40 \times 2/400 = 200 \ \text{mA} = V_{in}/R_{pri} = 2/10 = 200 \ \text{mA}$$
$$V_{sec} = X_m I_{pri} = 80 \times 200 = 16 \ \text{V}$$

If maximum power transfer was required, as with the previous example, a 400 Ω load resistance would yield a reflected resistance of 16 Ω, which would in turn add to the R_{pri} of 10 Ω, raising the Z_{in} to 26 Ω and reducing the Q to 15.4. The resulting input and output characteristics would be:

$$I_{pri} = V_{in}/Z_{in} = 2/26 = 77 \ \text{mA}$$
$$V_{sec} = X_m I_{pri} = 80 \times 77 = 6.15 \ \text{V}$$

Then

$$P_{in} = V_{in}^2/Z_{in} = 4/26 = 154 \ \text{mW}$$
$$P_{out} = V_{out}^2/R_L = 6.15^2/400 = 95 \ \text{mW}$$

If the primary winding of the transformer is untuned and the secondary is series tuned (Fig. 3-10d), the reactance of the capacitor cancels the reactance of the secondary winding, leaving the series coil resistance and R_L to limit the circulating current in the secondary. Presuming that R_L is a low value of less that $X/10$, the significant load on the secondary winding is the X_c of 400 Ω in series with R_L. Neither the 180° out-of-phase circulating current nor the $+j16 \ \Omega$ reflected impedance can significantly alter the primary winding reactance of 400 Ω. If the load resistance (R_L) on the secondary series-resonant circuit were 20 Ω, the following circuit characteristics could be obtained:

$$V_{sec} = X_m I_{pri} = 80 \times 5 \ \text{mA} = 400 \ \text{mV}$$
$$I_{sec} = V_{sec}/(R_{sec} + R_L) = 400/(10 + 20) = 13.3 \ \text{mA}$$
$$V_{out} = I_{sec} R_L = 13.3 \times 20 = 267 \ \text{mV}$$
$$P_{in} = V_{in}^2/X_{pri} = 2^2/400 = 10 \ \text{mW}$$
$$P_{out} = V_{out}^2/R_L = 0.267^2/20 = 3.56 \ \text{mW}$$

Double-Tuned Transformers

When *both* the primary and secondary windings of an RF transformer are connected to resonate with a capacitor, the circuit becomes *double tuned.* Although double-tuned transformers are critical, both in adjustment and fabrication, they have a most unusual property which simultaneously provides a broader bandwidth and a higher Q than single-tuned transformers.

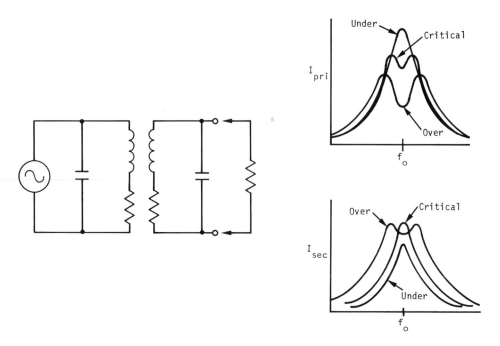

FIGURE 3-11

For this reason, double-tuned RF transformers are a most popular choice for use in electronic circuits requiring the optimum combination of high selectivity, high gain, and broad bandwidth. Using the same basic transformer in this example, you should be able to make some quick comparisons of circuit response against the single-tuned examples used earlier. Referring to Fig. 3-11, you should note that both secondary and primary windings are parallel-resonant networks. A review of general transformer characteristics and those of the parallel single-tuned primary and the parallel single-tuned secondary should disclose the following characteristics of the double-tuned transformer:

Since the primary network Q is a function of the *effective* series resistance, any value of additional resistance reflected from the secondary winding of the transformer will proportionately reduce the Q of the primary network. The maximum efficiency of the primary network occurs when the loaded Q is one-half the unloaded Q, caused by a reflected resistance equal to the value of the primary network series resistance. Since the reflected resistance is a function of the *secondary* network series resistance and the mutual reactance of the transformer ($R_{ref} = X_m^2/R_{ser}$), with both values dependent on Q and K_c, respectively, there is an ideal Q and K_c condition for the double-tuned transformer.

If the transformer of our previous examples was to be used in a double-tuned configuration, the ideal coupling coefficient (K_c), which is called *critical coupling* in engineering practice, could be computed based on the above characteristics as follows:

$$R_{pri} = X_{pri}/Q_{pri} = 400/40 = 10\ \Omega$$
$$R_{ref} = R_{pri} = 10\ \Omega$$
$$R_{sec} = X_{sec}/Q_{sec} = 400/40 = 10\ \Omega$$

Transposing $R_{ref} = X_m^2/R_{sec}$ to solve for X_m,

$$X_m = \sqrt{R_{sec}R_{ref}} = \sqrt{10 \times 10} = 10\ \Omega$$
$$K_{c-crit} = X_m/\sqrt{X_{pri}X_{sec}} = 10/\sqrt{400 \times 400} = 0.025$$
$$= 1/\sqrt{Q_{priU}Q_{secU}} = 1/\sqrt{40 \times 40} = 0.025$$

Observe that the critical-coupling value of 0.025 is considerably smaller than the *real* coupling factor (K_c) of 0.2 for the transformer of our earlier example. When double-tuned transformers have coupling coefficients in excess of the critical-coupling value, they are *overcoupled;* and when the coupling coefficients are less than the critical value, they are *undercoupled.* The transformer of our earlier examples was considerably overcoupled. It would have to be modified in structure to operate efficiently as a double-tuned transformer. The usual practice employed to achieve lower values of coupling coefficient is to wind the primary and secondary coils of the transformer farther apart on the coil form, or to reduce the permeability of the transformer core (if it is not an air core, of course) by selection of alternative core materials.

If we now proceed to analyze the characteristics of the double-tuned transformer at the -3 dB point frequencies, the unique properties of this circuit connection will become apparent. The -3 dB point frequencies are based on the Q characteristics of the transformer's secondary network, as the primary network characteristics have changed owing to the addition of reflected impedance. The -3 dB point frequencies can be computed using the formula: $f_o \pm f_o/2Q$; and the -3 dB point reactances can be computed using the formula: $X_o \pm X_o/2Q$. Recall that the resistance and net reactance of a resonant circuit are equal at the -3 dB point frequencies. With the transformer of our example, the secondary impedance and reflected impedance can be computed at the -3 dB points as follows:

$$Z_{sec} = R \pm jX = 10 \pm j10\ \Omega = 14.14\angle \pm 45°\ \Omega$$
$$Z_{ref} = X_m^2/Z_{sec} = 10^2/14.14\angle \pm 45°\ \Omega = 7.07\angle \pm 45°\ \Omega = 5 \pm j5\ \Omega$$

Now the primary network impedance at the -3 dB points and this reflected impedance will combine. Keep in mind that the signal source is connected across the reactance and resistance of the coil in series, and that the primary (not the applied) current of the coil, limited only by this R-X_L combination, remains relatively constant over the range of the -3 dB point frequencies. If

$X_C = X_L = 400 \ \Omega$ at resonance, while $R_{pri} = 10 \ \Omega$, then the R-X_L portion of the primary winding will have the following characteristics:

At resonance,

$$Z_{pri} = R + jX = 10 + j400 \ \Omega$$

and at the -3 dB points,

$$Z_{pri} = R + j(X_o \pm X_o/2Q) = 10 + j(400 \pm 400/80)$$

$$= \begin{cases} 10 + j395 \ \Omega \text{ (at } -3 \text{ dB point below resonance)} \\ 10 + j405 \ \Omega \text{ (at } -3 \text{ dB point above resonance)} \end{cases}$$

The reflected impedances are now combined with the Z_{pri} of the coil, with the 180° phase reversal of the transformer inverting the character (sign) of the reflected reactance:

At the -3 dB point below resonance,

$$Z_{pri} + Z_{ref} = (10 + j395) + (5 + j5) = 15 + j400 \ \Omega$$

and at the -3 dB point above resonance,

$$Z_{pri} + Z_{ref} = (10 + j405) + (5 - j5) = 15 + j400 \ \Omega$$

Observe that the net reactance of the coil remains at 400 Ω throughout the range of frequencies, and thus the primary current through the coil will be limited by virtually the same net reactance over the range of the -3 dB point frequencies. Since the capacitive reactance of the primary network is $-j405$ Ω at the -3 dB point below resonance, while the coil is $+j400 \ \Omega$, the circuit is not yet exactly resonant. It is closer to resonance due to the reflected reactance of $+j5 \ \Omega$ which effectively increases the reactance of the coil, but an additional amount of reflected reactance of $+j5 \ \Omega$ is required to bring the circuit to resonance.

The reflected impedance $R \pm jX$ is maximum at the -3 dB point frequencies of the secondary resonant network. To make the *primary* network of our example resonant at the -3 dB point frequencies of the *secondary* network, we must increase the reflected reactance to $\pm j10 \ \Omega$ in order to bring the total primary coil reactance to 405 Ω and 395 Ω, and thus cancel the 395 Ω and 405 Ω of the capacitor at these frequencies. The reflected impedance will increase with an increased mutual reactance (X_m), which will in turn be increased by increasing the coupling coefficient (K_c). If we increase the critical coupling coefficient (K_{c-crit}) by a factor of 1.414 (or $\sqrt{2}$), the following circuit characteristics may be computed:

$$K_{c-crit} = 1/\sqrt{Q_{pri}Q_{sec}} = 1/40 = 0.025$$

$$K_c' = 1.414 \ K_{c-crit} = 0.035$$

$$X_m' = K_c'\sqrt{X_{pri}X_{sec}} = 0.035 \times 400 = 14.14 \ \Omega$$

Then, at resonance,

$$Z_{ref} = X_m'^2/R_{sec} = 14.14^2/10 = 20 \ \Omega$$

and at the -3 dB points,

$$Z_{ref} = X_m'^2/Z_{sec} = 14.14^2/14.14\angle \pm 45° = 10 + j10 \ \Omega$$

Now, at resonance,

$$Z_{pri-coil} = R + jX = 20 + j400 \ \Omega$$

and at the -3 dB points,

$$Z_{pri-coil} = (R_{pri} + R_{ref}) + j(X_{pri} \pm X_{ref})$$

$$= \begin{cases} 20 + j405 \ (\text{at } -3 \text{ dB point below resonance}) \\ 20 + j395 \ (\text{at } -3 \text{ dB point above resonance}) \end{cases}$$

Observe that the net reactance of the coil is the same as that of the capacitor at the -3 dB frequencies, and thus a resonant circuit condition occurs over this range of frequencies. This resonant condition was achieved by increasing the critical coupling coefficient by a factor of 1.414 in order to obtain optimum circuit performance. The increased coupling coefficient is properly called *optimum coupling* in engineering practice: $K_{c-opt} = K_{c-crit}\sqrt{2}$.

Under resonant conditions the input impedance of the primary resonant network of the transformer can be computed by employing the formula, $Z_p = QX$, as follows:

$$Q \ (\text{at resonance}) = X_{pri}/(R_{pri} + R_{ref}) = 400/(10 + 20) = 13.3$$
$$Z_p \ (\text{at resonance}) = QX = 13.3 \times 400 = 5.33 \ \text{k}\Omega$$
$$Q \ (\text{at } -3 \text{ dB points}) = (X_{pri} + X_{ref})/(R_{pri} + R_{ref}) = \begin{cases} 405/20 = 20.25 \\ 395/20 = 19.75 \end{cases}$$
$$\text{So } Z_p \ (\text{below resonance}) = QX = 20.25 \times 405 = 8.2 \ \text{k}\Omega$$
$$Z_p \ (\text{above resonance}) = 19.75 \times 395 = 7.8 \ \text{k}\Omega$$

The above calculations for the double-tuned transformer show how the Z_p of the tuned primary rises at frequencies above and below resonance because of the reflected impedance from the tuned secondary winding. The bandwidth of the frequency response has been thus extended to *peak* or "*hump*" (see Fig. 3-11) at the -3 dB frequencies, rather than *fall off*, extending the *real* -3 dB response of the input Z_p without sacrificing circuit Q. The extended bandwidth of the *primary* tuned network drops the Z_p to a -3 dB level when the *net* reactance and resistance are equal. This condition occurs at $f_0 \pm f_0/0.707Q$, where the reactances are $X_0 \pm X_0/0.707Q$, and the primary and reflected impedances combine to yield equal R and net X (including X_C) values. The original bandwidth of the *secondary* tuned network is computed as f_o/Q, or $f_o \pm f_o/2Q$; but when both primary and secondary networks are combined in the single transformer, the bandwidth of the signal transferred through the entire circuit is computed as the product of the two: $2 \times 0.707 = 1.414$. Thus the bandwidth of the entire double-tuned circuit is:

$$\text{BW}_{d-t} = f_o \pm f_o/1.414Q = f_o/0.707Q$$

Sample Problem 5

Given: A double-tuned transformer (Fig. 3-11) with a primary reactance of 400 Ω and a secondary reactance of 600 Ω, and with identical primary and secondary unloaded Q values of 50.

Determine: The optimum coupling and mutual reactance, the Z_p of the primary network at resonance and the hump frequencies, the -3 dB bandwidth of the primary network, and the overall bandwidth of the circuit for a resonant center frequency of 10 MHz.

(a)   First, the critical and optimum coupling values are computed as:

$$K_{c-crit} = 1/\sqrt{Q_{pri}Q_{sec}} = 1/50 = 0.02$$
$$K_{c-opt} = 0.02\sqrt{2} = 0.028$$

(b)   The mutual reactance is computed as:

$$X_m = K_c\sqrt{X_{pri}X_{sec}} = 0.028\sqrt{400 \times 600} = 13.85 \ \Omega$$

(c)   The primary and secondary series resistances are now computed as:

$$R_{pri} = X_{pri}/Q = 400/50 = 8 \ \Omega$$
$$R_{sec} = X_{sec}/Q = 600/50 = 12 \ \Omega$$

(d)   The reflected resistance at resonance is computed and added to the primary resistance as follows:

$$R_{ref} = X_m^2/R_{sec} = 13.85^2/12 = 16 \ \Omega$$

then

$$Z_p = X^2/(R_{pri} + R_{ref}) = 400^2/(8 + 16) = 6.67 \ \text{k}\Omega$$

(e)   The secondary and reflected impedances at the -3 dB point frequencies are computed, in order to proceed with characterizing the hump response, as follows:

$$Z_{sec(-3dB)} = R \pm jX = 12 \pm j12 = 17\angle\pm45° \ \Omega$$
$$Z_{ref(-3dB)} = X_m^2/Z_{sec} = 192/17\angle\pm45° = 8 \pm j8 \ \Omega$$

(f)   The primary-winding coil and the resonating capacitor reactances are computed at the -3 dB point frequencies as follows:

X_L and $X_C = X_o \pm X_o/2Q = 400 \pm (400/100) = 396 \ \Omega$ and $404 \ \Omega$ (X_L above resonance $= X_C$ below resonance $= 404 \ \Omega$; and X_L below resonance and X_C above resonance $= 396 \ \Omega$)

(g)   The Z_p value at the hump frequencies (-3 dB frequencies) is now computed as:

$$Z_p = (X_{pri} + X_{ref})^2/(R_{pri} + R_{ref})$$
$$= \begin{cases} (396 + 8)^2/(8 + 8) = 10.2 \text{ k}\Omega \\ (404 + 8)^2/(8 + 8) = 10.6 \text{ k}\Omega \end{cases}$$

(h)   Finally, the primary and double-tuned bandwidths are computed as:

$$BW_{pri} = 2f_o/0.707Q = 20 \text{ MHz}/(0.707 \times 50) = 566 \text{ kHz}$$
$$BW_{d-t} = BW_{pri}/2 = 566/2 = 283 \text{ kHz}$$

When the double-tuned transformer is undercoupled, the circuit operation is not significantly affected by reflected impedances, so its performance is similar to two single-tuned transformers, with neither double-hump response nor increased bandwidth. The major use for the undercoupled double-tuned transformer is in the coupling of high-impedance electronic devices, providing minimum loading and maximum isolation. While slightly overcoupled double-tuned transformer circuits yield the best double-hump response with high Q and broad bandwidth, increasing the coupling beyond this optimum value will begin to load the circuit and reduce the Q as bandwidth increases. Even though the double-hump response remains, the considerably over-coupled double-tuned transformer is impractical for use with high-impedance devices. A number of alternative connections for double-tuned circuits are found in engineering practice, with their configurations dependent on the bandwidth, Q, and impedance requirements of the devices they interconnect. Figure 3-12 shows four variations of the double-tuned transformer, including the series-resonant connection. The general procedures for analysis of the series-resonant connection are the same as those used for the parallel-resonant connection. Series-resonant networks are, of course, most appropriately used for matching very low impedances.

Stagger Tuning

One alternative approach used to obtain broader bandwidth with high Q response is with *stagger-tuning* techniques, where high-Q tuned transformers or *RLC* networks are connected in series at resonant frequencies above and below the center circuit frequency in order to obtain a required bandwidth. For optimum response each resonant circuit should be separated by a factor of $1.414Q$, as shown in Fig. 3-13. With two such networks, for example, their frequencies could be computed as: $f_2 = f_1 \pm f_1/1.4Q$. The -3 dB bandwidth of

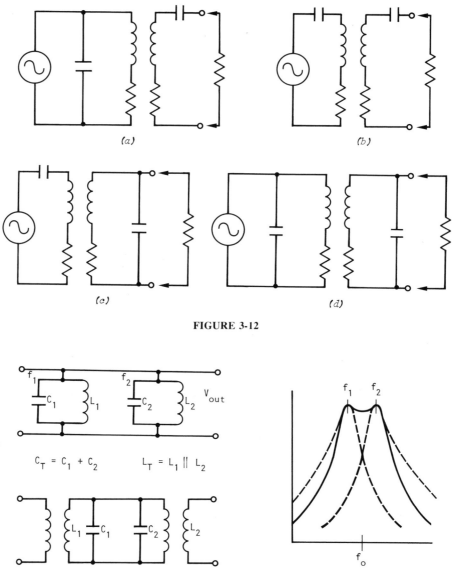

FIGURE 3-12

$$C_T = C_1 + C_2 \qquad L_T = L_1 \parallel L_2$$

FIGURE 3-13

the entire stagger-tuned circuit will extend beyond the resonant frequencies of the individual networks by a factor of Q. If, for example, three stagger-tuned networks were used to extend the bandwidth of a 10 MHz circuit, and yet maintain a Q of 50, the network frequencies and bandwidth would be computed as follows:

$$f_1 = 10 \text{ MHz}$$
$$f_2 = f_1 - f_1/1.4Q = 10 - 10/(1.4 \times 50) = 9.86 \text{ MHz}$$
$$f_3 = f_1 + f_1/1.4Q = 10 + 10/(1.4 \times 50) = 10.14 \text{ MHz}$$
$$\text{BW} = (f_3 - f_2) + f_1/Q = (10.14 - 9.86) + 10/50$$
$$= 486 \text{ kHz} = 0.486 \text{ MHz}$$

APPLICATIONS

RLC networks and transformers are used extensively in electronic circuits where signal processing requires control of frequency, phase, and impedance. The applications of these circuits fall into special categories of coupling, phase shift, tuning, filter, and impedance-matching networks. In many cases *RLC* networks perform multiple functions in electronic circuits, including blocking dc while passing signals, and providing dc paths to power supply connections without loss of signal. If you look closely at the schematic diagram of the frequency converter circuit shown in Fig. 3-14, you should be able to follow the circuit discription as outlined below, categorize the network functions, and even handle some calculations:

L_7 and L_8 make up the tuned secondary-input transformer that provides the benefits of Q-rise in signal voltage across C_{22}, while maintaining a high impedance. The signal is coupled into the gate of Q_3 along with a second signal—coupled through C_{24}—from the drain of Q_4. The two signals are mixed and the resultant signal, at the drain of Q_3, develops across the tapped parallel-resonant network comprised of L_9, C_{26}, and C_{27}. C_{31} locally grounds the $+12$ V dc power supply line feeding the drain of Q_3 through L_9. C_{26} and C_{27} have a voltage/impedance-dividing function, and provide a tapped resonant circuit for impedance matching from the drain of Q_3 to the output load. Q_4 is an oscillator that operates through the impedance matching and phase shift afforded by the parallel-resonant circuit made up of L_{10}, C_{28}, and C_{29}. Part of the output signal at the drain of Q_4 is fed back through the tapped parallel-resonant network, with 180° of phase shift achieved by grounding the tap junction of C_{28} and C_{29}, and feeding the signal at high impedance through C_{30} to the gate of Q_4 for oscillation. RFC_2 is a high-reactance RF *choke* coil that feeds the $+12$ V dc power supply to the drain of Q_4.

Now we'll proceed with some calculations in response to questions about the *RLC* components in the circuit of Fig. 3-14.

1. Given the resonant circuit L_9, C_{26}, and C_{27} (disregarding the output loading across C_{27}), with tuning at 4.0 MHz, and the circuit providing a

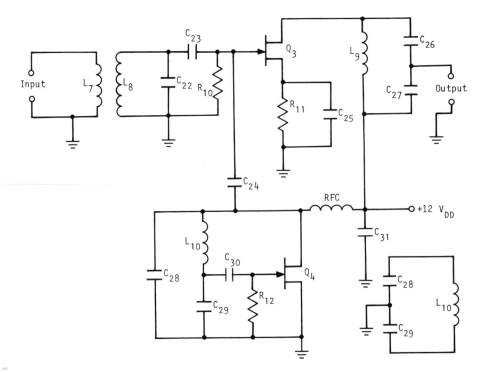

FIGURE 3-14

resonant load impedance of 20 kΩ to the drain of Q_3. Find the reactance of C_{26} and C_{27}, the inductance and coil resistance of L_9, and the network Q and bandwidth.

$$X_{26} = 1/\omega C = 1/(6.28 \times 4 \times 10^6 \times 10^{-10}) = 400 \ \Omega$$
$$X_{27} = X_{26} C_{26}/C_{27} = 400 \times 100/400 = 100 \ \Omega$$
$$X_T = X_{26} + X_{27} = 400 + 100 = 500 \ \Omega$$
$$L = X/\omega = 500/(6.28 \times 4 \times 10^6) = 20 \ \mu H$$
$$R = X^2/Z_p = 500^2/20 \ k\Omega = 12.5 \ \Omega$$
$$Q = X/R = 500/12.5 = 40$$
$$BW = f_0/Q = 4 \ MHz/40 = 100 \ kHz$$

2. Given the single-tuned secondary transformer, L_7, L_8, and C_{22}, with an antenna signal of 100 μV across the primary reactance of 250 Ω, with a resonant impedance of 10 kΩ for the secondary network which is tuned to 27 MHz with a Q of 80, and with the transformer coupling coefficient of 0.2. Find the values of L_7, L_8, and C_{22}, and the signal voltage at the junction of C_{22} and C_{23}.

$$L_7 = X/\omega = 250/(6.28 \times 27 \times 10^6) = 1.5\ \mu\text{H}$$
$$X_8 = Z_p/Q = 10\ \text{k}\Omega/80 = 125\ \Omega$$
So $\quad L_8 = X/\omega = 125/(6.28 \times 27 \times 10^6) = 740\ \text{nH}$
$$C = 1/\omega C = 1/(6.28 \times 27 \times 10^6 \times 125) = 47\ \text{pF}$$

For the signal voltage,

$$X_m = K_c\sqrt{X_{\text{pri}}X_{\text{sec}}} = 0.2\sqrt{250 \times 125} = 35\ \Omega$$
$$I_{\text{pri}} = V_{\text{in}}/X_{\text{in}} = 100\ \mu\text{V}/250 = 400\ \text{nA}$$
So $\quad V_{\text{out}} = QX_mI_{\text{pri}} = 80 \times 35 \times 4 \times 10^{-7} = 1.1\ \text{mV}$

3. Given the resonant circuit L_{10}, C_{28}, and C_{29}, and observing that C_{29} is loaded with R_{12} of 47 kΩ, and C_{28} is loaded with R_{10} of 100 kΩ (neglecting the loading effects of Q_3, Q_4, and other circuit components), with the network resonant at 31 MHz, a resonant impedance of 8 kΩ, and a Q of 80. Find the values of L_{10}, C_{28}, and C_{29}.

$$X_{10} = Z_p/Q = 8\ \text{k}\Omega/80 = 100\ \Omega$$
So $\quad L = X/\omega = 100/(6.28 \times 31 \times 10^6) = 514\ \text{nH}$
$$X_{28} = 2X_{29},\ \text{so}\ X_{29} + 2X_{29} = X_{10}$$
$$X_{29} = X_{10}/3 = 100/3 = 33\ \Omega$$
$$X_{28} = 2X_{29} = 2 \times 33 = 67\ \Omega$$
So $\quad C_{28} = 1/\omega X = 1/(6.28 \times 31 \times 10^6 \times 67) = 76\ \text{pF}$
$$C_{29} = 2C_{28} = 2 \times 76 = 152\ \text{pF}$$

While the above calculations are quite basic, they nonetheless give some indication of how applications of *RLC* networks are analyzed in electronic circuits.

FILTERS

One special, and more popular, application of resonant *RLC* networks is in filtering. *RLC* filters fall into four functional categories: bandpass, band-stop (band-reject), highpass, and lowpass. Although a simple series- or parallel-resonant circuit may be used as a bandpass or band-stop filter, special requirements for broader bandwidths and high Q values often dictate more complex connections for coils and capacitors. Variations of the L, T, and π network connections of *RLC* circuits are most often found in filter applications.

The more popular techniques for analysis of these filter connections fall into two categories, called *constant-k* and *m-derived* in engineering practice. The constant-*k* label comes from the relationship of series-to-shunt (parallel) reactances in the L, T, and π connections, wherein the product of the two reactance values (X_{ser} *times* X_{par}) is constant over a wide range of frequencies. The constant-k networks have moderate frequency cutoff (moderate Q) characteristics, with the advantage of a high rejection factor for unwanted frequencies; while the m-derived networks have sharper cutoff (higher Q) characteristics, but do not provide higher rejection of unwanted frequencies. The circuit for the m-derived filter includes a resonant series or shunt section in the L, T, or π connection. Referring to the networks of Fig. 3-15, observe that the T and π connections of the highpass and lowpass constant-k filters *look* like impedance-matching networks (the component values differ, however), but the m-derived filters have one network element which is resonant. The label *m-derived* comes from a constant value, designated m, which is the ratio of the reactance of the resonant element in the m-derived filter to the same element in the comparable constant-k filter. If this definition is confusing, reread the previous sentence while carefully comparing the filter circuits of Fig. 3-15.

The constant m is also the ratio of the cutoff frequency of a highpass or lowpass filter to the frequency of highest rejection, with typical realistic values of m being from 0.5 to 0.7 in engineering practice. The general formulas for computing the values of L and C in highpass and lowpass filters are based on a cutoff frequency, below and above which (respectively) signals are increasingly rejected. The procedures *first* require calculations for constant-k reactance values, from which m-derived values can be obtained if necessary.

The constant-k reactance values for highpass filters are $X_L = R_L/2$ and $X_C = 2R_L$. The reactance values for lowpass filters are $X_L = 2R_L$ and $X_C = R_L/2$. For T-networks, each series component splits the total reactance, so X_1 and $X_2 = X_T/2$, with the single shunt reactance at full value. For π-networks, the single series reactance is at full value, while each of the shunt reactances are twice the total reactance, so X_1 and $X_2 = 2X_T$.

The relationship of the reactances of these filter networks is no different than with any other circuit: in series, two values of X *add* for a total reactance; while in parallel, two values add reciprocally to yield a total reactance.

Conversion of the constant-k to the m-derived filters requires that the shunt sections become series resonant, or that the series sections become parallel resonant. The resonant frequency of the m-derived section(s) is at maximum rejection (not the cutoff frequency used for the constant-k calculations), so the reactance(s) of the m-derived resonating component(s) (X_L or X_C) is also based upon the constant m. Since the reactance and frequency ratios are directly related, the reactance values can be computed by the factor $4m/(1 - m^2)$. Once the reactances are known, the component values and resonant frequency can be computed.

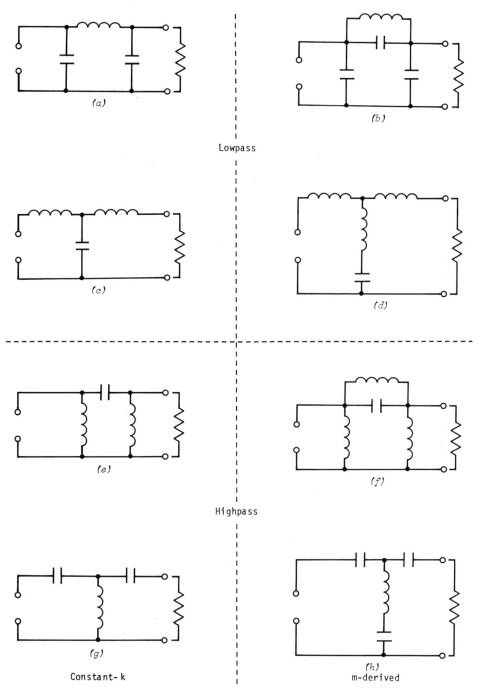

FIGURE 3-15

Sample Problem 6

Given: A constant-k, π-network lowpass filter (Fig. 3-15c).

Determine: The component values for a cutoff frequency of 8 MHz and a load resistance of 100 Ω; then convert the network to an m-derived configuration with $m = 0.5$ (Fig. 3-15d) and compute the component values accordingly.

(a) First, the reactance and inductance of the coil are computed for constant-k operation as:

$$X_L = 2R = 2 \times 100 = 200 \ \Omega$$
$$L = X/\omega = 200/(6.28 \times 8 \times 10^6) = 4 \ \mu H$$

(b) The total reactance and capacitance for each parallel leg of the π-network are computed as:

$$X_{C-tot} = R_L/2 = 100/2 = 50 \ \Omega$$

But each reactance is twice its total value, so:

$$X_{C1} = X_{C2} = 2X_{C-tot} = 2 \times 50 = 100 \ \Omega$$
$$C_1 = C_2 = 1/\omega X = 1/(6.28 \times 8 \times 10^6 \times 100) = 200 \ pF$$

(c) Now the network is converted to an m-derived network, $m = 0.5$, by converting the values of the existing components as follows:

$$L = mL_k = 0.5 \times 4 \ \mu H = 2 \ \mu H$$
$$C_1 = C_2 = mC_k = 0.5 \times 100 \ pF = 50 \ pF$$

(d) Then the resonating capacitance is computed as:

$$C_{res} = C_k(1 - m^2)/4m = 100 \times (1 - 0.5^2)/(4 \times 0.5) = 37.5 \ pF$$

In making the conversion from constant-k to m-derived, the inductance and capacitance values were directly converted, consequently the reactance formulas were inverted for computing capacitance values.

Bandpass/Band-Stop Filters

Bandpass and band-stop filters may have either two or three legs of the T or π network resonant. For the more popular constant-k bandpass networks, the *series* legs are series resonant and the *shunt* legs are parallel resonant; while for the more popular m-derived band-stop networks, the *series* legs are parallel resonant and the *shunt* legs are series resonant. For broader frequency response, such filters are made up of multiple sections with a wide range of L, T, or π configurations, and they may include a combination of highpass, lowpass, bandpass, and band-stop networks. Figure 3-16 includes examples of more basic filters. The component values for the constant-k bandpass filters can be computed using the same general formulas as are

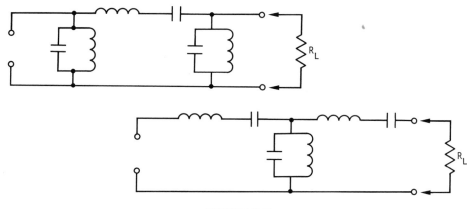

FIGURE 3-16

used for lowpass and highpass filters; but since there are two limit frequencies that indicate the bandwidth of the filters, *both* frequency values are included in the formulas.

For the T and π networks, constant-k bandpass filters, the series-leg, series-resonant network reactance values are $X_L = X_C = 2R_L$; and the shunt-leg, parallel-resonant network reactance values are $X_L = X_C = R_L/2$. As with highpass and lowpass filters, the total series reactance is split in each series leg of the T-network, and the total shunt reactance is doubled in each shunt leg of the π-network. The bandwidth frequency $(f_2 - f_1)$ is used to compute the component values of series inductance and shunt capacitance; while the parallel-equivalent frequency $(f_1 f_2/(f_2 - f_1))$ is used to compute the component values of series capacitance and shunt inductance.

Sample Problem 7

Given: A constant-k, π-network bandpass filter (Fig. 3-16).

Determine: The component values for a passband from 8 to 10 MHz and a load resistance of 100 Ω.

(*a*)   First, the reactances of the series coil and capacitor are computed as follows:

$$X_L = X_C = 2R_L = 2 \times 100 = 200\ \Omega$$

(*b*)   The equivalent frequency and bandwidth are computed as:

$$f_{eq} = f_2 f_1/(f_2 - f_1) = 10 \times 8/(10 - 8)\ \text{(MHz)}$$
$$= 80/2 = 40\ \text{MHz}$$
$$\text{BW} = f_2 - f_1 = 10 - 8 = 2\ \text{MHz}$$

(c)   Then the inductance and capacitance values are computed, using the bandwidth and equivalent frequency, as:

$$L = X/\omega = 200/(6.28 \times 2 \times 10^6) = 16 \ \mu\text{H}$$
$$C = 1/\omega X = 1/(6.28 \times 4 \times 10^7 \times 200) = 20 \ \text{pF}$$

(d)   The reactances of the shunt coil and capacitor are computed as:

$$X_L = X_C = R_L/2 = 100/2 = 50 \ \Omega$$

(e)   The inductance and capacitance values of the shunt components are now computed, using the equivalent frequency and bandwidth, as:

$$L = X/\omega = 50/(6.28 \times 4 \times 10^7) = 200 \ \text{nH}$$
$$C = 1/\omega X = 1/(6.28 \times 2 \times 10^6 \times 50) = 1.6 \ \text{nF}$$

(f)   Finally, the total reactance of *each shunt* leg of the π-network is *twice* the 50 Ω value of step *d* above, so the actual *L* and *C* values are adjusted from those computed in step *e*, as follows:

$$L_{\text{par}} = 2L = 2 \times 200 \ \text{nH} = 400 \ \text{nH}$$
$$C_{\text{par}} = C/2 = 1.6 \ \text{nF}/2 = 0.8 \ \text{nF (or 800 pF)}$$

Tedious filter calculations can yield gigantic headaches if complete ac circuit analysis is required to determine the pass and rejection frequencies and impedances. If you ask an engineer to manually calculate the full characteristics of a four- or five-section filter, you might also prepare yourself for some kind of assault in retaliation. The job of manually calculating the full characteristics of a four- or five-section filter is one that the electronics business has gratefully relegated to the electronic computer.

"Modern" Filters

The constant-*k* and *m*-derived filters are of "classical" design in that they were used extensively before the availability of computers for design-engineering assistance. Modern filter design techniques which are ideally suited to computer analysis provide better performance with fewer components than are required for multisection *classical* filters. Two most popular modern filter designs are the Chebyshev (Tchebysheff) and Butterworth. These filters use a lattice network structure similar to Fig. 3-17, with the frequency-response characteristics noted for each type. Modern filters are seldom fabricated individually for two reasons: (1) They are readily available as a subsystem item from specialty filter manufacturers, and (2) the amount of engineering, design, fabrication, measurement, and trimming experience required for this specialty product makes it uneconomical to fabricate on a low-volume basis. You should nonetheless have an understanding of general

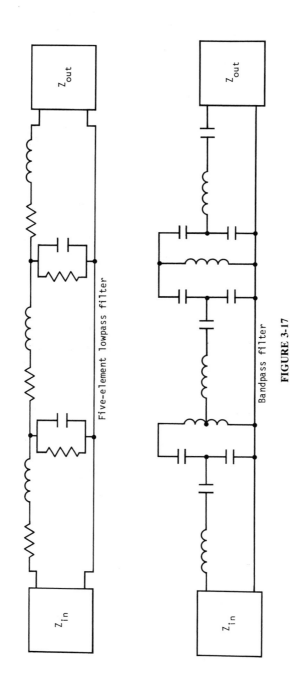

Five-element lowpass filter

Bandpass filter

FIGURE 3-17

filter principles, operation and analysis techniques, along with some familiarity of filter jargon, in order to properly apply even the subsystem units to electronic circuits.

Modern filters are designated by *poles* and *zeros* that identify the number of resonant points over the range of their frequency response. The terminology evolves from mathematical analysis utilizing Laplace transform methods, so there has been some degree of obscurity about the designations. In plain language, however, a *pole* designates a high-impedance resonant frequency, while a *zero* designates a zero-impedance resonant frequency. Parallel *LC* networks as shunt elements in filters, for example, produce maximum impedance at resonance, yielding a pole; while series *LC* networks as shunt elements in filters produce zero impedance at resonance, yielding a zero. Increasing the number of poles in a filter network will flatten the peak of the frequency response, while increasing the number of zeros will steepen the slopes of the frequency response. Figure 3-18 compares the frequency response of representative multipole filters.

COIL FABRICATION

RF coils and transformers are commercially available in great variety and in a wide range of values to supply most circuit needs. Special needs often arise, however, that make the purchase of coils or transformers impractical, and it is under these circumstances that knowledge of coil fabrication can prove invaluable. Here again access to manufacturers' catalogs and applications literature is the primary resource. Besides providing useful data, the

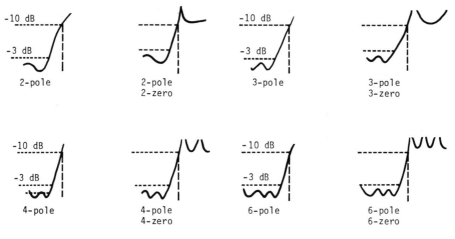

FIGURE 3-18

same manufacturers usually supply coil forms and core materials required for fabrication.

The simplest of coils is the *solenoid* or single-layer-wound type. It is comprised of a series of close-wound turns, and is either self-supporting or wound around a cylindrical form. The inductance of the coil is dependent upon its length, diameter, and number of turns making up the windings. The ratio of the coil diameter to its length determines a value called the *form factor,* which also influences the coil's inductance value. Table 3-2 lists the form factor values (F) for various ratios of coil diameter (d) to length (l). Once the form factor is determined from Table 3-2, the approximate inductance for the single-layer coil can be computed using the following formula:

$$L = Fn^2d$$

L is the inductance in μH, F is the form factor value from Table 3-2, n is the number of turns in the windings, and d is the diameter of the coil. The diameter and length of the coil are calculated or measured from wire centers.

Sample Problem 8 (A Paper Problem)

Given: A 40-turn coil of AWG 30 wire (enamel covered) close-wound on a ¼-inch-diameter form.

Determine: The inductance of the coil.

(*a*)   First, the wire diameter (D) is obtained from a wire-gauge table and multiplied by the number (n) of turns to determine approximate length (l), as follows:

$$l = Dn = 0.011 \text{ in.} \times 39 = 0.430 \text{ in.}$$

TABLE 3-2.   FORM FACTORS FOR
SINGLE-LAYER SOLENOIDS AS A
FUNCTION OF DIAMETER/LENGTH
(d/l)

d/l	F	d/l	F
0.02	0.0005	0.8	0.015
.03	.0008	1.0	.018
.04	.001	1.5	.023
.06	.0015	2.0	.027
0.1	.0025	3	.033
0.15	.0035	4	.036
0.2	.0045	5	.040
0.3	.0065	6	.044
0.4	.0085	10	.051
0.6	.012	20	.062

(b)   Then one wire diameter is added to the form diameter (d_f) to determine the diameter of the wire centers (d) as follows:

$$d = D + d_f = 0.011 + 0.25 \text{ in.} = 0.261 \text{ in.}$$

(c)   The diameter-to-length ratio is computed and referred to Table 3-2, where the corresponding form factor (F) is obtained, as follows:

$$d/l = 0.261/0.430 = 0.6$$

Thus, from Table 3-2,

$$F = 0.012$$

(d)   Finally, the inductance is computed, using F, as follows:

$$L = Fn^2d = 0.012 \times 40^2 \times 0.261 \text{ in.} = 5 \ \mu\text{H}$$

   Because the form factor is reduced as the diameter of the coil increases, the most significant variable in determining coil inductance is its number of turns, with the greater number in the shortest length yielding maximum inductance. In fabricating coils there are practical limits to the amount of inductance that can be obtained through this single-layer approach. First, the greater number of turns in a short coil length requires use of very fine wire, which is both difficult to handle and has higher skin-effect series resistance due to its small diameter, thus yielding low Q values. Second, distributed capacitance in the coil increases with the number of close-wound turns, thereby resulting in the coil becoming self-resonant and making it useless for higher frequency applications. Generally, AWG 36 wire and ¼-inch coil forms are about as small as can be handled for manual coil winding. Given these limits, air-core inductors of up to 50 μH are practical to wind by hand. Of course, coils of less inductance are quite practical and easily wound at inductance values as low as 50 nH. The best overall performance can be achieved with air-core coils that are self-supporting and wound with larger-diameter low-resistance plated wire, and with the turns spaced for minimum distributed capacitance.

Cores

The inductance value for a given coil may be increased or decreased by the use of standard core materials within the coil form. The most popular core types that increase inductance are *ferrite* and *powdered iron*. The ferrite cores are a composition of different metallic *oxides,* mixed in proportions to yield high permeability over a given range of frequencies. Permeability values from 100 to over 2000 times the air-core inductance value are not uncommon with ferrite cores. High permeability does present problems, however, as the ferrite core responds to varying flux levels, temperature, and higher frequencies with poor stability. The powdered iron core material trades off high permeability for stability and is thus more appropriate for use at high frequencies. Powdered iron is also available in a number of mixes, yielding permeability factors from 4 to 80, for higher frequency high-Q applications.

Inductors operating at frequencies above 100 MHz may use core materials of brass or aluminum, which have a permeability of less than 1.0 and so decrease the coil's air-core inductance value, but maintain high Q and stability. All core types are manufactured in a variety of sizes, shapes, and mixtures to meet both the physical and electrical requirements of coil fabrication. Threaded cylindrical cores further provide the advantages of adjustable inductance values for trimming and tuning purposes. Fixed cores are available in rods, toroids, cups, and beads. One popular application of ferrite beads is to thread them over a short length of wire or a component lead. A high-permeability bead, for example, can increase the inductance of a ¼-inch length of wire from 5 nH to perhaps 1 μH or more. Toroidal cores have the advantage of being self-shielding, as most all of the magnetic flux is maintained within the circular core, making toroidal coils almost immune to the effects of stray magnetic fields.

Transformers

The procedures for winding RF transformers are the same as those used for inductors, but there are a few additional factors to consider. First, the primary and secondary coils of the transformer must be wound in the same direction, otherwise the bucking magnetic fields will cancel the mutual inductance and thus prevent the transfer of signals from the primary to secondary windings. Second, the *high side* (sometimes *high end* or *hot side*) of the transformer primary and secondary windings must be connected to the *high side* of the signal source and load, respectively, with the *low side* of the primary and secondary windings connected to an ac ground or neutral point of a circuit. The high sides of transformers are shown with dots on schematic diagrams (see Fig. 2-1 in Chapter 2) to indicate phasing. Fabricating a transformer that does not allow for correct high-side/low-side connections will result in opposing currents which will also cancel the mutual inductance. Third, the proximity of one transformer winding to the other will determine the magnitude of the coupling coefficient (K_c), which in turn will determine the mutual inductance and the degree of signal transfer through the transformer. Small coefficients of coupling are required of double-tuned transformers that dictate separate cores for primary and secondary windings, thus preventing coupling through a common core path. High coupling coefficients are achieved by winding the primary and secondary as adjacent turns, or winding the primary on top of the secondary. The problem with such close winding of coils is an increase in distributed capacitance, causing unwanted resonances. The advantages of toroidal coils include high coupling coefficients without the necessity of close winding, making toroids the ideal choice for high-frequency, *close-coupled* transformers.

MEASUREMENTS

RF coils, capacitors, transformers, *RLC* networks, and resonant circuits are *not* built on paper. As the old saying goes, "Attaching a line cord to even 1000 pages of precise calculations somehow doesn't get us a working product." Although we should not discount the importance of paperwork or computer work, the design and calculation aspects of RF electronics yield only "half a loaf." Electronic measurements are required at all stages of RF circuit work, from initial experimentation to final production testing. The extensive observations are required because so many electrical/mechanical factors influence the total inductance and capacitance of *RLC* circuits. Although the combination of solid theoretical knowledge and accumulated practical experience provides engineers and technicians with guidelines for successfully building and adjusting RF circuits, it is only through the measurements made at each step of the fabrication process that this success can be assured.

The important measurements of individual component characteristics are made initially with an RF bridge or *Q*-meter. Both instruments compare the reactive component under test to their built-in standards, and provide measurements of RF resistance or *Q*, inductance, and capacitance. While such instruments are accurate, they measure only the isolated characteristics of *L*, *C*, and *R*, not how the same components will operate when placed into actual operating circuits. They are used in the first-step identification of individual fabricated or purchased coils, transformers, and capacitors.

After *L*, *C*, and *R* values of reactive components are identified out of circuit, the parts are wired into a working electronic circuit in the form, mounting, and location that they will have in a finished product. At this point *dynamic* measurements are required that provide information about the complete circuit, driven and loaded according to its final specifications. Dynamic measurement techniques employ signal generators, loads, and detectors in conjunction with either meters or oscilloscopes in order to energize the circuits and observe the results of their performance. The more popular dynamic measurement techniques are *swept frequency* and *spectrum analysis*. Both techniques provide a visual oscilloscope display of amplitude vs frequency response of the circuits or output signals under test.

Dynamic Measurement Techniques

Swept-frequency measurements of RF circuits require a sweeping (varying-frequency) signal generator, RF detector, and oscilloscope, connected as shown in Fig. 3-19a. High-frequency oscilloscopes are not required for swept measurements, as the vertical input (Y-axis) of the "scope" display is driven by the *detected* varying *level* of the RF signal—not the RF signal itself. The horizontal input (X-axis) of the scope is externally driven by the relatively slow (typically around 100 Hz) *sweep rate* of the signal generator. Sophisticated sweep-signal generators have built-in features which make the generators a complete test system. These features include: (1) accurate marker signals that mix with the RF output signal and provide *marks* on the scope display in order to calibrate the horizontal axis in frequency; (2) a built-in mixer to combine the RF and marker signals after the RF signals

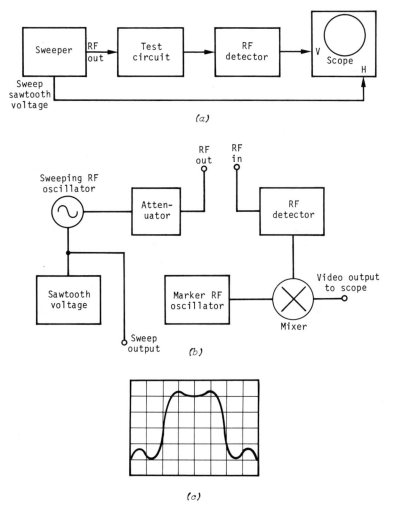

(a)

(b)

(c)

FIGURE 3-19

have passed through the test circuit; (3) accurate RF attenuators, in combinational steps of 1 dB, for making substitution measurements of RF gain and loss; (4) variable sweep width (sweep amplitude) and sweep rate, with a parallel isolated output to drive the horizontal input of an oscilloscope or X-Y plotter; and (5) a built-in broadband RF (sometimes *video*) detector with its output connected to directly drive the vertical input of an oscilloscope or X-Y plotter. Figure 3-19*b* shows the block diagram of a typical *quality* sweep-signal generator; Fig. 3-19*c* shows the kind of *frequency-domain* display obtained from swept-frequency measurement techniques.

Although the swept-frequency technique is most appropriate for measuring the characteristics of RF *circuits,* the approach cannot be used to accurately measure the characteristics of RF *signals*. The dynamic measurement technique that is applied to RF signals is called *spectrum analysis*. Spectrum analysis is considerably more sophisticated than swept-frequency measurements. Instead of applying a sweeping signal to a test circuit, spectrum analysis uses a sweeping *circuit* to which a test *signal* is applied. As such, the complexity (and, of course, cost) of a spectrum analyzer far exceeds that of the sweep-signal generator. Most spectrum analyzers also have a built-in scope display necessitated by calibration requirements that are difficult to achieve with general-purpose oscilloscopes. Figure 3-20 shows the test arrangement, typical block diagram, and the frequency-domain display obtained from spectrum-analysis measurement techniques.

Spectrum analyzers have three major sections: radio frequency (RF), intermediate frequency (IF), and display. The RF section tunes to the center frequency of the signal under test, and includes filters for control of bandwidth, and attenuators for control of signal levels. The frequency-scanning (dispersion) part of the RF section sweep-tunes a range of frequencies across which the test signal is to be measured. The IF section includes an accurately gain-calibrated amplifier with precise attenuators for the measurement of absolute signal power in dBm or mW. The IF section also includes filters in combination with the bandwidth and sweep-tuning rate of the RF section in order to control the resolution of signal-frequency measurements. The display section operates like an accurately calibrated oscilloscope, with logarithmic display of signal amplitude on the vertical axis and calibrated linear display of frequency along the horizontal axis.

The frequency-domain measurement techniques provide the most thorough, accurate, and visual results, but they require both costly equipment and additional setup time. More basic measurement techniques using fixed-frequency signal generators, voltmeters, or time-domain oscilloscopes are far more common for characterizing RF circuits, even though the measured results may be lacking in many respects. Such basic measurement techniques are similar to those used with audio-frequency circuits, but there are two special factors that you should consider when making RF measurements: *loading* and *real-time* response.

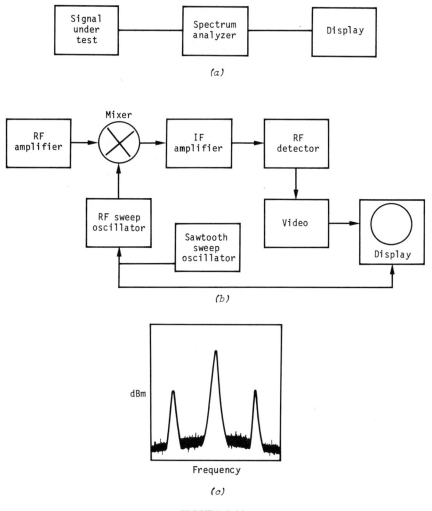

(a)

(b)

(c)

FIGURE 3-20

Isolation

Certain precautions must be taken to provide the proper isolation of circuit elements from the loading effects of electronic instruments. Inexperienced engineers and technicians often have difficulty with RF circuits because of their failure to consider the loading effects of instruments on otherwise perfectly operating circuits.

First, the characteristics of any signal source or measuring instrument must be considered initially in order to obtain the operation of RF networks and circuits. If a high-Q parallel-resonant circuit were driven by a 50 Ω or 600

Ω signal generator, the Z_p of the circuit—which might, for example, be 10 kΩ—would be shunted by the generator impedance, and thus reduce the Q and increase the bandwidth by the ratio Z_{gen}/Z_p. Under such circumstances the circuit would not perform as it should, even though it was electrically perfect. The same kind of loading would take place if a low-impedance detector were connected to measure the RF voltage across any high-impedance element in a circuit.

Second, the capacitance of test leads and inputs to instruments may be as high as 50 pF, which will shift the frequency response and lower the Z_p and Q in proportion to the circuit capacitance. Even with a resonant capacitance of 100 pF (which is a moderately large value for RF circuits), 50 pF of external capacitance will shift the circuit response by 33%.

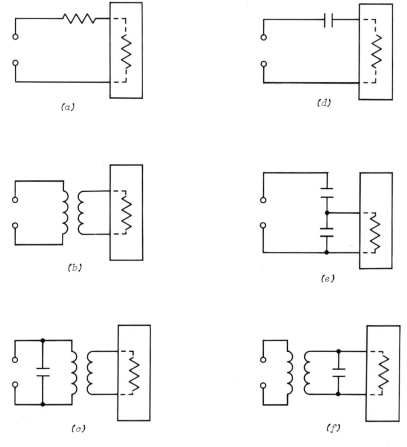

FIGURE 3-21

Low-impedance loading can be minimized where necessary by the use of resistors, capacitors, or transformers in conjunction with electronic instruments. Instrument and test-lead capacitance can either be minimized with low-value, series-connected coupling capacitors, or canceled with small series-connected coils or ferrite beads. Figure 3-21 shows a variety of coupling techniques for minimizing the capacitance and loading effects of instruments. The choice of technique will depend on a number of factors based on the characteristics of the test circuit, test leads, and test equipment, and of course on the convenience of adaptation. The tradeoffs for the benefits of isolation are either a loss of signal level (when using resistors, capacitors, or transformers) or an inconvenience (when using specialized matching networks). If a resistor or single capacitor is used for isolation, the loss in signal voltage will be approximately $Z_L/(R_L + Z_L)$ or $Z_L/(X_C + Z_L)$, where Z_L is the instrument loading impedance and R_L and X_C are the isolating components. If a capacitive voltage divider is used for isolation, and the reactance shunting the instrument loading impedance is at least $5Z_L$, the loss in signal voltage will be approximately $C_1/(C_1 + C_2)$, where C_2 is the shunting capacitor and C_1 is the series capacitor. If a transformer is used for isolation, there will be less voltage loss, as the voltage ratio is the square root of the reactance ratio, so the loss in signal voltage will be approximately $\sqrt{X_{\text{pri}}/X_{\text{sec}}}$, where X_{sec} is driving the instrument loading impedance and X_{pri} is connected to the circuit under test. If more precise measurements with minimum loading are necessary, then resonant matching networks are required. The networks can be attached to the electronic test equipment, or they can be built into the circuit, thus providing convenient access points for tests and measurements.

CONCLUSION

After completing these second and third chapters you should have a good background knowledge of passive *RLC* components, networks, and circuits. With this knowledge you should be adequately prepared to study the electronic circuits that employ active devices, including diodes, transistors, IC's, and specialized vacuum tubes. The content of this chapter reviewed the important tuned passive networks and circuits commonly used to couple active devices, signal sources, and load elements. These network principles also apply to the transmission topics covered in later chapters where lines and associated elements are used to couple power within systems. It is expected that you will use this chapter as a reference for network principles as you proceed through your study of active circuits and trasmission lines. The problems that follow will serve to test your working knowledge and achievement of the learning objectives specified at the beginning of the chapter.

Problems

1. Given: The modified L-network of Fig. 3-4 to match an 8 Ω input resistance to a 72 Ω load resistance, while maintaining a loaded Q of 20 at a resonant frequency of 30 MHz.

Determine:
(a) Inductance of L_1
(b) Equivalent parallel-resonant impedance, R_p
(c) Total resonant capacitance
(d) Capacitance of C_1
(e) Capacitance of C_2

2. Given: A resonant T-network to match a 30 Ω input resistance to a 7.5 Ω load resistance, while maintaining a loaded Q of 10 at a resonant frequency of 50 MHz. Compute the circuit component values for two capacitors and one coil, or two coils and one capacitor, to find which configuration yields the most practical values.

Determine:
(a) Reactance of the input series leg of the "T"
(b) Equivalent parallel-resonant impedance, R_p
(c) Reactance of the output series leg of the "T"
(d) Total equivalent parallel reactance
(e) L_1, L_2, and C values
(f) C_1, C_2, and L values

3. Given: A resonant π-network to match a 600 Ω input resistance to a 50 Ω load resistance, while maintaining a loaded Q of 5 at a resonant frequency of 100 MHz. Compute the circuit component values for two capacitors and one coil, or two coils and one capacitor, to find which configuration yields the most practical values.

Determine:
(a) Reactance of the input parallel leg of the "π"
(b) Reactance of the output parallel leg of the "π"
(c) Reactance of the series leg of the "π"
(d) L_1, L_2, and C values
(e) C_1, C_2, and L values

4. Given: A parallel-resonant circuit (Fig. 3-22a) with a tapped coil to match a 2 kΩ input resistance to a 20 kΩ output resistance, with a loaded Q of 30 and a Z_{pU} of 40 kΩ, and driven with a 100 mV signal at a resonant frequency of 10 MHz.

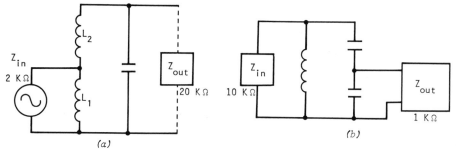

FIGURE 3-22

Determine:
(a)   Series coil resistance (unloaded)
(b)   Total parallel-resonant reactance
(c)   Value of the capacitor
(d)   Loaded Q of the input L-R section (Q')
(e)   Inductance of the loaded section (X_{L1})
(f)   Inductance of the unloaded section (X_{L2})
(g)   Output voltage across the capacitor

5.   Given: A parallel-resonant circuit (Fig. 3-22b) with tapped capacitors to match a 10 kΩ input resistance to a 1 kΩ output resistance, with a loaded Q of 20 and a Z_{pU} of 25 kΩ, and driven with a 100 mV signal at a resonant frequency of 5 MHz.

Determine:
(a)   Series coil resistance (unloaded)
(b)   Total parallel-resonant reactance
(c)   Inductance value of the coil
(d)   Loaded Q of the output C-R section (Q')
(e)   Capacitance of the loaded section (X_{C1})
(f)   Capacitance of the unloaded section (X_{C2})
(g)   Output voltage across C_1

6.   Given: An RF transformer with an untuned 50 μH primary and a resonant 10 μH secondary with a parallel 30 pF capacitor, having identical primary and secondary Q values of 40 and a coupling coefficient of 0.1, and the primary driven with a 500 mV signal at the resonant frequency.

Determine:
(a)   Resonant frequency
(b)   Mutual reactance
(c)   Reflected resistance

(*d*)   Primary current
(*e*)   Output voltage across the capacitor

7.   Given: An RF transformer with an untuned 200 nH primary and a series-resonant 800 nH secondary driving a 20Ω resistive load, with identical primary and secondary *Q* values of 20 and a coupling coefficient of 0.5, and the primary driven with a 50 mV signal at the resonant frequency of 20 MHz.

Determine:
(*a*)   Value of the resonating capacitor
(*b*)   Mutual reactance
(*c*)   Reflected resistance
(*d*)   Primary current
(*e*)   Output voltage across the 20 Ω load resistor

8.   Given: An RF transformer with identical 800 nH, 4 Ω series-resistance primary and secondary windings, with a parallel-tuned primary providing a Z_{pU} of 10 kΩ and driven with a 100 mV signal at a resonant frequency of 40 MHz, and an output voltage of 10 mV across the secondary winding driving a 200Ω resistive load.

Determine:
(*a*)   Primary resonating capacitor
(*b*)   Unloaded *Q* of the primary and secondary
(*c*)   Mutual reactance
(*d*)   Coupling coefficient
(*e*)   Reflected resistance
(*f*)   Input power of the primary
(*g*)   Output power of the secondary

9.   Given: An optimum-coupled double-tuned parallel-resonant transformer with identical primary and secondary windings of 800 nH and 2 Ω series resistance, driven with a 200 mV signal at a resonant frequency of 10 MHz.

Determine:
(*a*)   Value of resonating capacitors
(*b*)   Unloaded *Q* of the primary and secondary
(*c*)   Optimum coupling coefficient
(*d*)   Mutual reactance
(*e*)   Reflected resistance at resonance
(*f*)   Primary Z_p at resonance
(*g*)   Primary Z_p at the hump frequencies
(*h*)   Hump frequencies
(*i*)   Double-tuned bandwidth
(*j*)   Primary (coil) current at resonance
(*k*)   Output voltage across the secondary capacitor at resonance

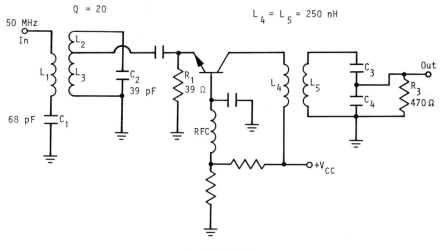

FIGURE 3-23

10. Given: The electronic circuit of Fig. 3-23.

Describe the functions of the following components and circuits:
(*a*) L_1-C_1 and L_2-L_3-C_3
(*b*) L_4-L_5-C_3-C_4
Compute:
(*c*)   Inductance of L_1
(*d*)   Inductance of L_2 and L_3 based on R_1 loading L_3
(*e*)   Capacitance of C_3 and C_4 based on R_2 loading C_4

11. Given: The two-section bandpass filter of Fig. 3-24, for 50 Ω source and load resistances, made up of a constant-*k* π-network lowpass section with a cutoff frequency of 10 MHz, and an *m*-derived T-network highpass section with a cutoff frequency of 6 MHz, where $m = 0.6$.

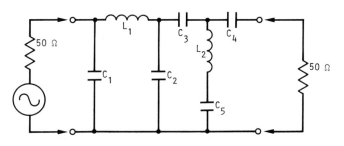

FIGURE 3-24

Determine:

(*a*) Inductance of L_1

(*b*) Capacitance of C_1 and C_2

(*c*) Inductance of L_2

(*d*) Capacitance of C_3 and C_4

(*e*) Capacitance of C_5

12. Given: The circuit of Problem 8 (refer to manufacturers' catalogs).

Specify or prepare:

(*a*) An appropriate resonant trimmer capacitor

(*b*) An appropriate coil form

(*c*) Wire size, number of turns, and procedure for handwinding the transformer

(*d*) A sketch of the transformer showing estimated spacing for K_c and phasing

(*e*) A block diagram showing the setup for measuring the transformer characteristics

(*f*) A step-by-step procedure for measuring the transformer characteristics

(*g*) A block diagram showing the setup for sweep-frequency measurements of the resonant circuit

(*h*) A step-by-step procedure for measuring the characteristics of the resonant circuit using swept-frequency techniques

CHAPTER 4

ACTIVE RF DEVICES

This chapter is concerned with the properties and characteristics of *active* electronic devices used with the passive networks of Chapter 3 for the generation, amplification, and processing of radio-frequency signals. Like the preceding text, this chapter builds upon your understanding of basic principles and relates this information to engineering practice. In particular, this chapter reduces popular electronic devices to their equivalent circuit form, allowing for convenient analysis of their properties. Each device type—bipolar, FET, MOSFET, IC, and vacuum tube—is reviewed in terms of its bias requirements, detailing how variations in bias conditions affect its RF impedance, and gain characteristics. The frequency response, noise figure, and power dissipation characteristics of these active electronic devices, and their control through biasing techniques, are also reviewed. Finally, the power dissipation requirements through the application of heat-sinking techniques are reviewed, as heat sinking, like biasing, is a critical aspect of the operation of solid-state electronic devices. In summary, the material in this chapter should prepare you for success in combining the passive networks of Chapter 3 with active electronic devices in order to achieve the signal processing functions detailed in Chapter 5 and in the systems applications of later chapters.

LEARNING OBJECTIVES

Upon completing this chapter, including the problem set, you should be able to:

1. Understand the electrical parameters of active electronic devices, including transistors, IC's, and vacuum tubes.
2. Understand and relate the characteristics of active electronic devices to a single Y-parameter, equivalent-circuit model.

3. Understand *Y* parameters insofar as they define the input, output, and transfer characteristics of bipolar transistors, JFET's, MOSFET's, VFET's, IC devices, and vacuum tubes intended for RF applications.
4. Understand the relationships of gain and frequency response to the electrical properties of active electronic devices.
5. Understand and apply biasing techniques to achieve specific device operation within its range of electrical characteristics.
6. Apply standard techniques for the measurement of the important characteristics of active electronic devices as required for their operation in combination with external passive networks.
7. Understand and interpret noise characteristics as specified for active electronic devices.
8. Apply noise analysis techniques based on equivalent noise voltage and resistance characteristics of devices.
9. Understand power ratings specified for transistors and IC's and be able to translate these ratings to thermal characteristics.
10. Understand and apply biasing and heat-sinking techniques to assure proper operation of solid-state devices in RF power applications.

ACTIVE DEVICE CHARACTERISTICS

You are probably familiar with the techniques used to characterize electronic devices and specify their *parameters*. The RF characteristics of bipolar and field-effect transistors, integrated circuits, and vacuum tubes may use the same parameters as are applied at lower frequencies. The parameter values are usually *complex,* however, as the small device reactances become significant at the higher frequencies. For this reason the transistor *R* parameters (r'_e, β, etc.) are not useful; and the *h* parameters (h_{ie}, h_{fe}, etc.) are somewhat awkward to apply directly at radio frequencies if shunt capacitances are to be accounted for. The extension of the hybrid-pi transistor model of Fig. 4-1*a* to multitransistor IC's will also result in too many factors to account for. Such mathematical models, while thorough, tend to be impractical, because they include many insignificant factors that have little influence on the operation of active electronic devices.

The emphasis of this chapter is to define the characteristics of transistors, IC's, and vacuum tubes in terms of their *simplified* RF parameters, and to do so without the complications imposed by external circuit components. Once you have the ability to reduce the active RF devices to their equivalent-circuit form, the following task of combining external circuit components becomes a much less complicated process. (The performance of the active electronic devices with external circuit components is covered in Chapter 5.) An example of the equivalent-circuit form to which the active devices will be reduced is shown in Fig. 4-1*b*. Although the schematic appears rather

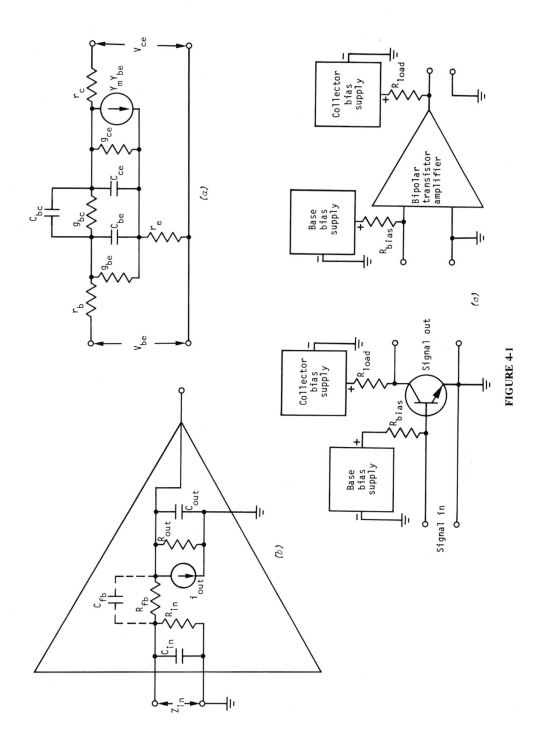

FIGURE 4-1

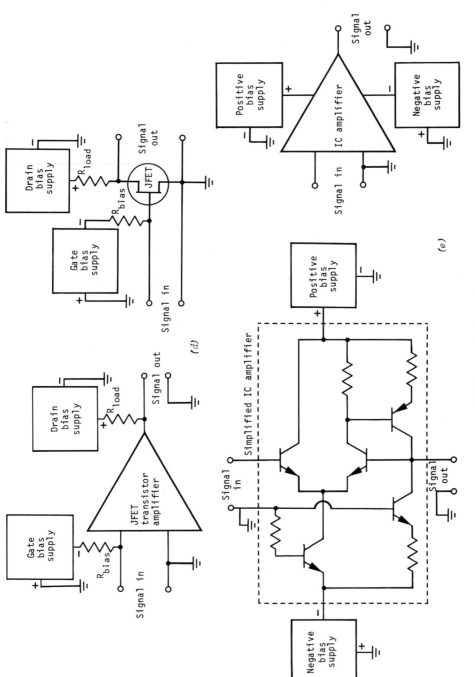

FIGURE 4-1 (*continued*)

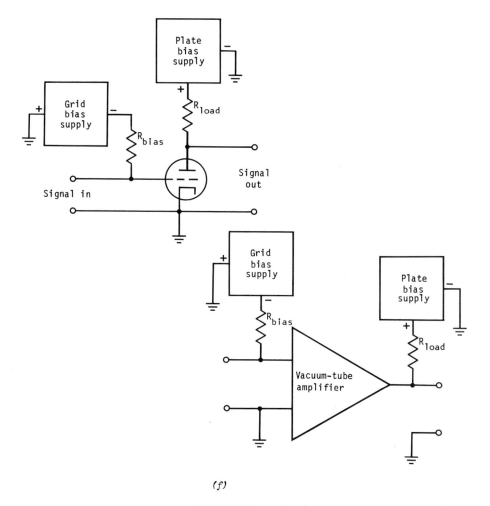

(f)

FIGURE 4-1 (*continued*)

straightforward and easy to work with, two main problems are associated with the process: (1) the interpretation of published or measured data, and (2) the conversion of this data to the equivalent-circuit model. Unfortunately, the terminology and abbreviations that describe the characteristics of active electronic devices are far more confusing than the actual electrical performance of the devices. In this respect you should spend extra time familiarizing yourself with these terms before you apply them to the concepts presented in this chapter.

Transistors, IC's, and vacuum tubes must, obviously, be properly *biased* from a power supply before they can process high-frequency signals. This bias requirement is usually determined prior to ac analysis, and the device parameters are greatly dependent on the bias *operating point*. For this reason there are actually two sets of interrelated factors that determine the operating characteristics of RF devices: *bias* and RF (ac) *parameters*. Your ability to successfully apply device characteristics in order to achieve amplification of RF signals will require a thorough understanding of both factors. Figure 4-1*b* shows the simplified model of an RF amplifying device to be used in this chapter as a basis for amplifier signal analysis. This same model can be used for bipolar and field-effect transistors, linear IC's, and vacuum tubes. The actual representative bias circuits of these devices are shown in Fig. 4-1*c* through *f*. The techniques used to translate bias values and circuit connections to the equivalent-circuit model of Fig. 4-1*b* will be covered in the following paragraphs for each of the device types.

BIPOLAR TRANSISTORS

Bipolar transistors are characterized as current amplifiers, with their output specified as a current source shunted by a conductance, and their input specified as a resistance. The transistors can be connected as single-stage amplifiers in the common-emitter, common-collector, or common-base configurations. Table 4-1 shows typical manufacturer's specifications for one general-purpose, low-power *npn* transistor, with characteristics based on the hybrid-pi model of Fig. 4-1*a*. This table provides all the information required to analyze the transistors' operation in functional circuits. The graphs of the related figure, Fig. 4-2, indicate the typical variation in characteristics for the 2N2102, with variations depending on bias and temperature conditions. You should note that the 25° dc beta (h_{FE}) value of Fig. 4-2*a* reaches a maximum of over 100 at 60 mA I_C and 25°C, but is less than 60 at

TABLE 4-1.   TYPICAL ELECTRICAL CHARACTERISTICS FOR 2N2102
BIPOLAR TRANSISTOR (CASE TEMPERATURE AT 25°C)

Parameter	Specification
Common-emitter, small-signal, forward-current transfer ratio (h_{fe}) at 1 kHz	30 to 150
Magnitude (absolute value, polar form) of h_{fe} ($\|h_{fe}\|$) at 20 MHz	3 (min)
Input resistance (h_{ib}) at 1 kHz	5 to 30 Ω
Reverse voltage transfer (feedback) ratio (h_{rb}) at 1 kHz	3×10^{-4}
Output conductance (h_{ob}) at 1 kHz	0.01 to 1 μS
Output capacitance (C_{ob})	15 pF
Input capacitance (C_{ib})	80 pF
Gain-bandwidth product (f_t)	60 MHz (min)
Noise figure (NF), f_o, and bandwith at 1 kHz	6 dB
Thermal resistance: junction-to-case ($R_{\theta jc}$)	35°C/W
Thermal resistance: junction-to-ambient ($R_{\theta ja}$)	175°C/W

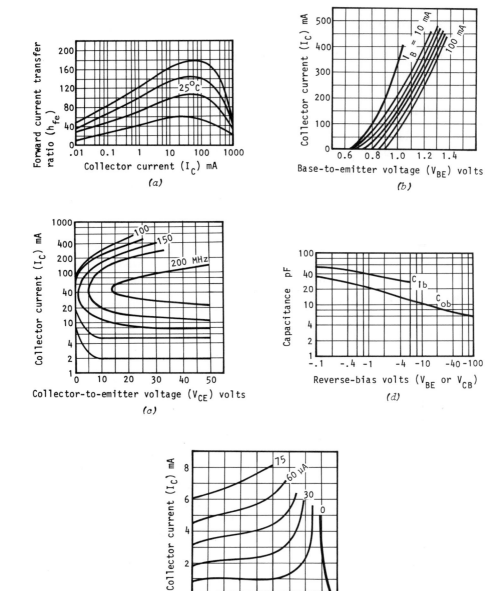

FIGURE 4-2

500 μA. Figure 4-2*b* shows the V_{BE} required to bias the transistor for given base currents and resulting collector currents. Figure 4-2*c* shows the variation in gain-bandwidth product (f_t)* with variations in I_C and V_{CE} bias conditions; and Fig. 4-2*d* shows the related capacitance values that result from bias variations. Finally, Fig. 4-2*e* shows the familiar transfer characteristics for the device.

Bias Conditions

If the 2N2102 is to be operated as an RF amplifier and the bias requirements are selected for optimum high-frequency response, the first reference should be made to Fig. 4-2*c*. The graph shows that the highest f_t will occur with a V_{CE} of 15 V and an I_C of 50 mA. Figure 4-2*a* indicates a dc beta of about 85 at 25°C, and Fig. 4-2*b* indicates a related V_{BE} of 0.65 *V* dc for the operating conditions noted. With this bias condition, Fig. 4-2*d* indicates about 42 pF input capacitance (C_{ib}), and about 10 pF output capacitance (C_{ob}). The operating-point base bias current is obtained from Fig. 4-2*e* as about 60 μA. In practice, engineers and technicians often rely on published data as only a reference, then proceed to make their own characterization measurements if applications or performance are critical.

AC Parameters

Once bias conditions are established for a given RF amplifier transistor, the ac parameters of the device may be obtained for its frequency of operation, then applied to the simplified model of Fig. 4-1*b*. The input resistance can be obtained from combining and expanding Fig. 4-2*b* and *e*, as shown in Fig. 4-3*b* and *c*, respectively. These figures indicate that a hypothetical input signal of 10 mV peak to peak (p-p) yields a change in I_B of about 16 μA, with the collector biased at 15 V dc but at ac ground. The ac current gain can also be obtained from the characteristic curves for the device, as shown in Fig. 4-3*c*, indicating that a change in I_B of 16 μA yields a 1.3 mA change in I_C. Figure 4-3*c* also shows a 10 V change in V_{CE}, yielding a 0.2 mA change in I_C along the 60 μA bias curve. The ac parameters may thus be computed from the graphical data of Fig. 4-3 as follows:

---

* The *gain-bandwidth product* of amplifiers, including transistors and other devices, refers to that frequency where the total gain equals 1.0. This frequency is also designated f_t. The gain-bandwidth product is related to frequency-response characteristics of simple *RC* networks, where the amplitude falls off at 6 dB/octave, thus yielding a 2 : 1 inverse relationship between frequency and output voltage. A transistor with an f_t of 10 MHz, for example, will have a voltage gain of 2.0 at 5 MHz, a gain of 4.0 at 2.5 MHz, and so on. These relationships will be reviewed in greater detail in the *Frequency Response* sections of this chapter.

i. For the common-emitter and common-collector connections (h_{ie} and h_{ic}),

$$r_{in} = \Delta V_{BE}/\Delta I_B = 10 \text{ mV}/16 \text{ } \mu A = 625 \text{ } \Omega$$

and for the common-base connection (h_{ib}),

$$r_{in} = \Delta V_{BE}/\Delta I_E = 10 \text{ mV}/1.3 \text{ mA} = 7.69 \text{ } \Omega$$

ii. For the common-emitter connection (h_{fe}),

$$\beta_{ac} = \Delta I_C/\Delta I_B = 1.3 \text{ mA}/16 \text{ } \mu A = 81.3$$

for the common-collector connection (h_{fc}),

$$\beta_{ac} = \Delta I_E/\Delta I_B = (\Delta I_C + \Delta I_B)/\Delta I_B$$
$$= (1.3 + 0.016 \text{ mA})/0.016 \text{ mA} = 82.3$$

and for the common-base connection (h_{fb}),

$$\alpha_{ac} = \Delta I_C/(\Delta I_C + \Delta I_B) = 1.3 \text{ mA}/(1.3 + 0.016 \text{ mA}) = 0.988$$

iii. For the common-emitter and common-collector connections (h_{oe} and h_{oc}),

$$g_{out} = \Delta I_C/\Delta V_{CE} = 0.2 \text{ mA}/10 \text{ V} = 20 \text{ } \mu S$$

and for the common-base connection (h_{ob}),

$$g_{out} = \Delta I_C/\Delta V_{CB} = \Delta_C/(\Delta V_{CE} - \Delta V_{BE})$$
$$= 0.2 \text{ mA}/(10 - 0.01 \text{ V}) = 20 \text{ } \mu S$$

These calculations, based on graphical data, should identify the close relationship between the h-parameter values. Published specifications thus require only a listing of h_{fe}, h_{ib}, and h_{ob}, as in Table 4-1, to fully characterize the devices. The listing neglects the insignificantly small reverse-voltage transfer ratio h_{rb} (leakage) of the 2N2102 transistor, for example, and most other modern solid-state and vacuum-tube devices.

h_{ie} and $h_{ic} = h_{ib} \times h_{fe} = 7.69 \times 81.3 = 625 \text{ } \Omega$
$h_{fb} = h_{fe}/(1 + h_{fe}) = 81.3/(1 + 81.3) = 0.988$
$h_{fc} \simeq h_{fe}$
h_{oe} and $h_{oc} \simeq h_{ob}$

To establish the same equivalent amplifier approach for transistors, IC's, and tubes, all active devices may be conveniently converted to the simplified model of Fig. 4-1b. The output can be considered as either a current or voltage source, depending on load circuit characteristics; while the *transfer* (input-to-output) function can be either a current or conductance/admittance

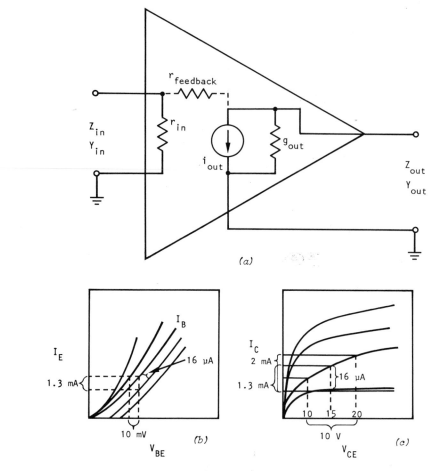

FIGURE 4-3

factor. The transfer function for bipolar transistors and some IC's is specified as a current factor; while FET's, tubes, and some transistors and IC's are specified by *transconductance* or *transadmittance* factors. Given the *h* parameters and capacitance values, the RF performance of a bipolar transistor can be evaluated for a specific operating frequency by deploying the equivalent-circuit analysis technique. The following example proceeds to apply the data of the previous paragraphs for the 2N2102 transistor at 10 MHz.

Sample Problem 1

Given: The 2N2102 bipolar transistor with characteristics as defined in the previous paragraphs and summarized below:

$$h_{fe} = 81.3 \qquad\qquad h_{ie} = 625\ \Omega \qquad\qquad h_{fb} = 0.988$$
$$h_{ib} = 7.69\ \Omega \qquad\quad h_{ic} = 625\ \Omega \qquad\quad h_{oe} = 20\ \mu S$$
$$h_{ob} = 20\ \mu S \qquad\quad h_{fc} = 81.3 \qquad\qquad h_{oc} = 20 \mu S$$

Determine: The equivalent-circuit characteristics for device operation as a common-base amplifier at 10 MHz; record the values on the diagram of Fig. 4-4b.

(a) The susceptance values of the 42 pF C_{ib} and 10 pF C_{ob} are first computed as follows:

$$B_{Cib} = \omega C = 6.28 \times 10^7 \times 42 \times 10^{-12} = 2.64\ mS$$
$$B_{Cob} = \omega C = 6.28 \times 10^7 \times 10^{-11} = 0.62\ mS$$

(b) The input admittance values are computed in rectangular and polar forms, respectively, for the common-base connection as follows:

$$G_{BE} = g_{ib} = 1/h_{ib} = 1/7.69 = 130\ mS$$
$$Y_{BE} = Y_{ib} = G + jB = 130 + j2.64\ mS$$
$$Y\angle\theta = B/\sin(\tan^{-1} B/G)$$
$$= 2.64/\sin(\tan^{-1} 2.64/130) = 130\angle1.16°\ mS$$

(c) The output admittance values are computed as follows:

$$G_{CB} = h_{ob} = 0.02\ mS$$
$$Y_{CB} = Y_{ob} = G + jB = 0.02 + j0.62\ mS$$
$$Y\angle\theta = B/\sin(\tan^{-1} B/G)$$
$$= 0.62/\sin(\tan^{-1} 0.62/0.02) = 0.62\angle88.2°\ mS$$

After reviewing this sample problem you may conclude that device characterization is a rather straightforward process. The analysis of even the most complex device characteristics becomes simple if step-by-step procedures are employed. This approach, of course, requires that you thoroughly understand the introductory aspects of a topic before proceeding to the more complex advanced aspects. This structure will be maintained throughout this and the next chapters, covering first the basic *static* characteristics of the electronic devices, and then proceeding to the more complex *dynamic* characteristics and the effects of external components on device performance.

Common-Emitter Connection

We now proceed to characterize the bipolar transistor operating in the more popular common-emitter (CE) connection as shown in Fig. 4-4c. The static characteristics are obtained with the collector at *ac ground* (bypassed with a large capacitance for biasing), so our approach here to practical analysis

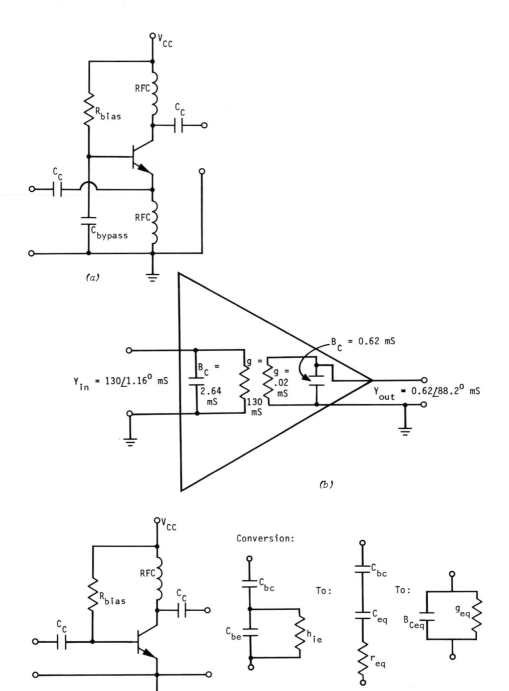

FIGURE 4-4

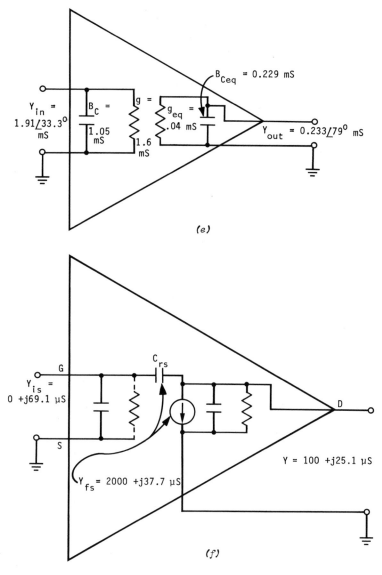

$B_{C_{eq}} = 0.229$ mS

$Y_{in} = 1.91\underline{/33.3^o}$ mS

$B_C = 1.05$ mS

$g = 1.6$ mS

$g_{eq} = .04$ mS

$Y_{out} = 0.233\underline{/79^o}$ mS

(e)

$Y_{is} = 0 +j69.1$ μS

C_{rs}

G

S

D

$Y_{fs} = 2000 +j37.7$ μS

$Y = 100 +j25.1$ μS

(f)

FIGURE 4-4 (*continued*)

may dictate that C_{be} and C_{bc} are actually in parallel with the base input connection. In application as an amplifier, however (see Chapter 5), there is an output load from collector to ground of the CE circuit. This *output* load significantly raises the impedance (reducing the capacitive loading) of C_{bc} on the transistor base input; so for all practical purposes C_{bc} need not be considered in computing the input characteristics of the CE connection of the transistor. This external load *also* affects voltage gain, with a 180° phase shift, from base input to collector output of the transistor. The result of this increased, phase-inverted signal voltage across C_{bc} functionally increases its capacitance (see *Miller Effect* in Chapter 5), thus coupling the input admittance to load the collector output of the CE connection. In addition to the output conductance (h_{oe}) loading the collector, the RC network, including the input admittance, also loads the collector (see Fig. 4-4*d*). The reduction (simplification) of this network requires that a sequence of series-parallel/parallel-series conversions be employed to specify the loading effects of this RC network. Although the series capacitance (C_{bc}) of only 10 pF of this example may appear to be insignificantly small, the Miller effect substantially increases its value in practical loaded circuits, causing increased loading of the collector output in the CE connection.

Sample Problem 2

Given: The same 2N2102 transistor of the previous examples.

Determine: The equivalent circuit for device operation as a common-emitter amplifier at 4 MHz; record the values on the diagram of Fig. 4-4*e*.

(*a*)   The susceptance values of the 42 pF C_{be} and 10 pF C_{bc} are computed as follows:

$$B_{Cbe} = j\omega C = 6.28 \times 4 \times 10^6 \times 42 \times 10^{-12} = 1.05 \text{ mS}$$
$$B_{Cbc} = 6.28 \times 4 \times 10^6 \times 10^{-11} = 0.25 \text{ mS}$$

(*b*)   The input admittance values are computed for the common-emitter connection as follows:

$$G_{BE} = g_{ie} = 1/h_{ie} = 1/625 \ \Omega = 1.6 \text{ mS}$$
$$Y_{BE} = y_{ie} = g + jB = 1.6 + j1.05 \text{ mS}$$
$$Y\angle\theta = B/\sin(\tan^{-1} B/G) = 1.05/\sin(\tan^{-1} 1.05/1.6)$$
$$= 1.91\angle 33.3° \text{ mS}$$

(*c*)   This input admittance for the CE connection must be converted to its series-equivalent value (see Fig. 4-4*d*) in order to combine it with the series feedback reactance of C_{bc}. Then after combination, this series-equivalent network will again be converted back to an admittance in order to determine its loading effect upon the collector output of the 2N2102. The procedure requires parallel-to-series conversion, addition of the reactances, and then conversion back to a parallel-equivalent network, as follows:

$$Z\angle\theta_{eq} = 1/Y\angle\theta_{ie} = 1/1.91\angle 33.3° \text{ mS} = 523\angle -33.3° \ \Omega$$
$$Z_{eq} = R - jX = Z\cos\theta - jZ\sin\theta = 523\cos 33.3°$$
$$-j523\sin 33.3° = 437 - j287 \ \Omega$$

$$X_{Cbc} = 1/B_{Cbc} = 1/0.25 \text{ mS} = 4000 \text{ } \Omega$$
$$Z_{\text{tot}} = R - j(X_{C \text{ eq}} + X_{Cbc}) = 437 - j(287 + 4000)$$
$$= 437 - j4287 \text{ } \Omega$$
$$Z\angle\theta_{\text{tot}} = X/\sin{(\tan^{-1} X/R)}$$
$$= 4287/\sin{(\tan^{-1} 4287/437)} = 4309\angle -84° \text{ } \Omega$$
$$Y\angle\theta_{\text{tot}} = 1/Z\angle\theta = 1/4309\angle -84° = 0.23\angle 84° \text{ mS}$$

In rectangular form:

$$Y_{\text{tot}} = G + jB = Y \cos{\theta} + jY \sin{\theta}$$
$$= 0.23 \cos{84°} + j0.23 \sin{84°}$$
$$= 0.024 + j0.229 \text{ mS}$$

Although step c is extremely tedious, it is included in this example to indicate the relative effects of the input admittance upon the collector output of the transistor connected in the CE configuration. You may conclude by this example that the loading effect of this network is small. Be aware, however, that the additional Miller capacitance, affected by operation of the device in amplifier circuits with external components, will substantially increase the loading of the input network on the collector. The technique reviewed here will apply similarly to those circuits introduced in later sections of this and subsequent chapters; so it is a necessary procedural step in the characterization of electronic devices.

(d) Finally, the output admittance characteristics of the device connected in the CE configuration are computed, combining the equivalent admittance from step c with the h_{oe} value obtained from Table 4-2:

$$Y_{\text{out}} = y_{oe} = (h_{oe} + G_{\text{eq}}) + jB_{\text{tot}}$$
$$= (0.02 + 0.024) + j0.229 \text{ mS} = 0.044 + j0.229 \text{ mS}$$
$$Y\angle\theta_{\text{out}} = 0.229/\sin{(\tan^{-1} 0.229/0.044)} = 0.233\angle 79° \text{ mS}$$

A comparison of the output admittance of this CE example with the simpler CB example of Sample Problem 1 reveals that the equivalent h_{oe}/h_{ob} value of 0.02 mS was effectively increased to 0.044 mS, and that the susceptance value of 0.229 mS only slightly affected a change in output capacitance when compared to the CB example ($+j0.251$ mS for 10 pF at 4 MHz).

Gain and Frequency Response

The procedures of the previous sample problems provide the important equivalent-circuit input and output characteristics necessary to predict and

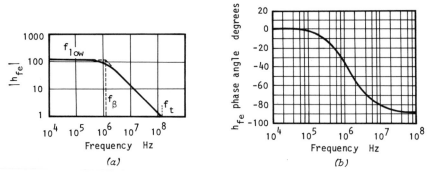

FIGURE 4-5

evaluate transistor RF performance. What remains to be computed are the effects of capacitive losses on the transfer characteristics of the device. Figure 4-5 shows the related magnitude and phase response vs frequency for the 2N2102 of our example. Relating the graphs of Fig. 4-5 to the equivalent input and output networks of Fig. 4-4 reveals that the frequency/phase response of the device is that of an RC network. The slopes of the response curves indicate that an inverse proportionality exists with changes in frequency; for example, a $2:1$ frequency change yields a $1:2$ change in the absolute value of h_{fe}. The frequency at which the magnitude of h_{fe} drops down to -3 dB below its maximum value is shown as f_β on the graphs of Fig. 4-5, indicating a 45° phase shift. A review of the CE input and output admittance values of $1.6 +j1.05$ and $0.044 +j0.229$, respectively, in Sample Problem 2, should reveal that the CE input susceptance, B_C, has the greater effect on the device's high-frequency response.

The transistor f_β can be computed as the frequency where the device input conductance and susceptance are equal:

$$f_\beta = B_C/2\pi C = 1/(h_{ie}2\pi C), \text{ where } B_C = g_{in} = 1/h_{ie}$$

For the 2N2102 of our example:

$$f_\beta = 1/(h_{ie}2\pi C_{be}) = 1/(625 \times 6.28 \times 42 \times 10^{-12}) = 6.07 \text{ MHz}$$

Thus f_β is called the *beta cutoff frequency*. Recall that the operating parameters for this 2N2102 included a low-frequency h_{fe} of about 81 and an f_t above 200 MHz at the bias conditions of Fig. 4-3. This h_{fe} value is reduced by -3 dB at f_β: $0.707 \times 81 = 57$. At frequencies above f_β, h_{fe} drops in inverse proportion to increasing frequency, with its value at 1.0 at f_t. Thus f_t specifies the frequency where h_{fe} is reduced to 1.0; this frequency is called the *gain-bandwidth product*. For this example f_t can be computed, based on f_β and the low-frequency value of h_{fe}, as follows:

$$f_t = f_\beta(0.707 \times h_{fe}) = 6.07 \text{ MHz} \times 57 \cong 346 \text{ MHz}$$

or

$$f_t = 0.707 \ h_{fe}/(h_{ie}2\pi C_{be})$$
$$= 0.707 \times 81/(625 \times 6.28 \times 42 \times 10^{-12}) \cong 346 \ \text{MHz}$$

The comparable cutoff frequency for the common-base connection of bipolar transistors is specified as the f_α of the device. This *alpha cutoff* defines the frequency where the current gain (α in the CB configuration) drops off to 0.707 of its low-frequency value, or where the *currents* through the *input* conductance and susceptance are equal.

Transfer Conductance/Admittance

From the calculations for h_{fe} and h_{fb} the current gain of the bipolar transistor may be considered with two values: the maximum low-frequency value and some decreasing higher frequency value. Since the higher frequency value is decreasing because of the effects of shunt capacitance, these gain characteristics are more appropriately identified in terms of *admittance* values, in order to be consistent with our equivalent-circuit model of Figs. 4-1*b* and 4-4. With the use of admittance parameters the transfer characteristics of any amplifier can be accurately analyzed in terms of *real* and *imaginary* ($\pm j$) parts in order to account for its gain vs frequency.

The real part of a device's transfer admittance is called *transconductance*, often abbreviated g_m. (Historically, g_m was applied to vacuum tubes and called *mutual conductance*.) For this particular bipolar transistor CE connection the transconductance should be specified g_{fe}, for "*forward* transfer conductance ratio in common-*emitter* configuration." This transconductance (g_{fe}) value specifies the ratio of output *current* to input *voltage* in the equivalent circuit of the transistor; and it may be computed from device parameters as the ratio:h_{fe}/h_{ie}. For the 2N2102 of our example, g_{fe} can be computed as:

$$g_{fe} = \Delta I_C/\Delta V_{BE} = 1.3 \ \text{mA}/10 \ \text{mV} = 0.13 \ \text{S}$$
$$g_{fe} = h_{fe}/h_{ie} = 81.3/625 \ \Omega = 0.13 \ \text{S}$$

The high-frequency capacitive losses of the transistor can be computed based on a susceptance value, and combined with the transconductance value, to obtain a complex *transadmittance* (forward transfer admittance), abbreviated Y_{fe} for the CE connection. This value of transadmittance will more consistently define the frequency-response characteristics of any active electronic device without confusion or the necessity of resorting to special-case examples. In the case of our example using the 2N2102 in a CE connection the low-frequency *Y parameters* would thus include a transfer-function value as predominantly a conductance, g_{fe}, of 0.13 S. At higher frequencies, near and above f_β, however, the transfer function becomes predominantly a complex value, Y_{fe}. Since the *Y* parameters have already been applied to the input and output characteristics of the equivalent-circuit model, the values of Sample Problem 2 at 4 MHz can be used to obtain Y_{fe} as follows:

$$Y_{fe} = h_{fe}Y_{ie} = 81.3 \times 1.91\angle 33.3° \text{ mS} = 155\angle 33.3° \text{ mS}$$

Thus the absolute value $|Y_{fe}| = 155$ mS, and the phase shift in the transfer function is 33.3°.

You may relate the value of Y_{fe} at f_β as an absolute admittance 3 dB greater than the low-frequency g_{fe}:

$$|Y_{fe}| \text{ (at 6.07 MHz } f_\beta) = 1.414 \ g_{fe} = 1.414 \times 0.13 \text{ S} = 0.184\angle 45° \text{ S}$$

With this technique you can accurately relate gain and frequency response without relying on the imaginary extensions shown as dashed lines on the frequency vs h_{fe} graph of Fig. 4-5a.

Sample Problem 3

Given: The 2N2102 bipolar transistor with the same characteristics summarized in Sample Problem 1 (see page 115).

Determine: The common-base transfer admittance characteristics at low frequencies, at 10 MHz and 300 MHz, and at the high alpha cutoff frequency, f_α.

(a) The low-frequency transconductance and input conductance are reviewed, generalized as admittance terms, and then compared to obtain a transadmittance factor at 10 MHz and 300 MHz, using data from the previous CB sample problem (Sample Problem 1), as follows:

$$h_{fb} = h_{fe}/(1 + h_{fe}) = 81.3/82.3 = 0.988$$
$$g_{ib} = 1/h_{ib} = 1/7.69 = 0.13 \text{ S}$$

$$B_{Cib} = \omega C = \begin{cases} 0.264 \text{ mS at 1 MHz} \\ 2.64 \text{ mS at 10 MHz} \\ 79.2 \text{ mS at 300 MHz} \end{cases}$$

$$Y_{ib} = g_{ib} + jB_{ib}$$

$$Y_{ib}\angle\theta = B/\sin (\tan^{-1} B/G) = \begin{cases} 130\angle 1.16° \text{ mS at 10 MHz} \\ 152\angle 31.3° \text{ mS at 300 MHz} \end{cases}$$

$$Y_{fb} = h_{fb}Y_{ib} = \begin{cases} 0.988 \times 0.13 = 128.4\angle 0° \text{ mS at low frequencies} \\ 0.988 \times 130\angle 1.16° = 128.5\angle 1.16° \text{ mS at 10 MHz} \\ 0.988 \times 152\angle 31.3° = 150\angle 31.3° \text{ mS at 300 MHz} \end{cases}$$

(b) Then,

$$B_{Cib} = g_{ib} \text{ at } f_\alpha = 130 \text{ mS}$$

(c) So,

$$f_\alpha = B/2\pi C = 0.13/(6.28 \times 42 \times 10^{-12}) = 493 \text{ MHz}$$

Note from the above calculations the relationships of Y and h parameters, and that the gain vs frequency response remains relatively constant over a much higher range of frequencies than the CE connection of the transistor.

You may be encountering difficulties at this point in your progress, because of the many coded and abbreviated terms. It may be of some help here to recall that we have been dealing mainly with two parameter systems: h and Y. The manufacturers' specified h parameters for low-frequency operation have been translated to high-frequency Y parameters to conveniently account for capacitive phase shifts and losses. The Y parameters have been applied to two bipolar transistor connections: the common-emitter and the common-base. The parameters have been abbreviated with two subscripts: the first as i, o, and f, for *input, output,* and *forward-transfer,* respectively, and the second as e and b, for common-*emitter* and common-*base,* respectively. It is recommended that you tabularize and define all the subscripted parameters, including those for R, G, C, X, and B values, in order to become more familiar and comfortable with these abbreviated terms. Incidentally, the common-collector connection has not been reviewed here because it has limited application as an RF amplifier. To analyze its performance, however, the same equivalent-circuit techniques may be applied to obtain Y parameters for the circuit connection.

FIELD-EFFECT TRANSISTORS

Field-effect transistors (FET's) are directly characterized as transconductance amplifiers, because they have a very high input impedance, while having an equivalent output current source shunted by a conductance like the bipolar transistors. Three related types of field-effect transistor are used in RF amplifier circuits: the junction FET (JFET), the insulated-gate/metal-oxide-semiconductor FET (MOSFET), and the dual-gate MOSFET. Figure 4-6 shows the schematic, bias, and equivalent-circuit diagrams of these devices. The three popular connections for FET's are identified as common-source, common-gate, and common-drain. Table 4-2 shows typical manufacturers' specifications for a JFET, a single-gate MOSFET, and a dual-gate MOSFET. These characteristics can be applied directly to the same equivalent-circuit model used for the bipolar transistors in order to obtain Y parameters for the FET devices. Just as with bipolar transistors, the FET electrical characteristics (parameters) also vary with the bias and temperature conditions.

N-channel

P-channel

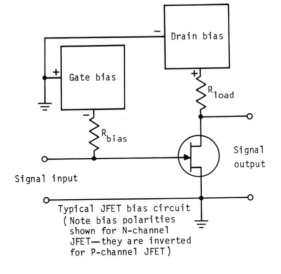

Signal output

Signal input

Typical JFET bias circuit
(Note bias polarities
shown for N-channel
JFET—they are inverted
for P-channel JFET)

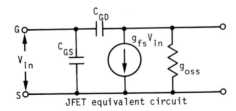

JFET equivalent circuit

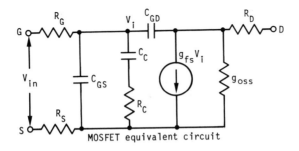

MOSFET equivalent circuit

(a)

FIGURE 4-6

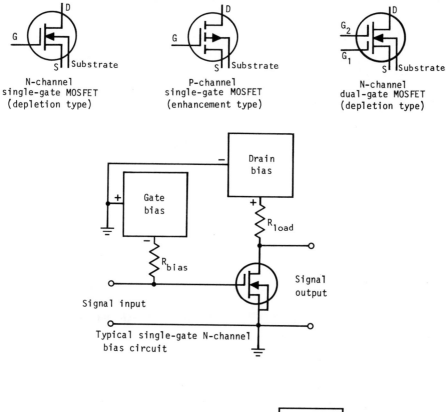

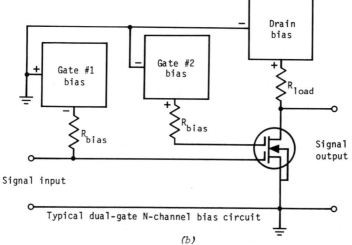

(b)

FIGURE 4-6 (*continued*)

TABLE 4-2. TYPICAL ELECTRICAL CHARACTERISTICS FOR 2N4416 JFET, 3N128 SINGLE-GATE MOSFET, AND 3N140 DUAL-GATE MOSFET (CASE TEMPERATURE AT 25°C)

Characteristic	Device		
	2N4416	3N128	3N140
Forward transconductance (g_{fs}), μS	5000	7500	10,000
Small-signal, short-circuit capacitance, pF			
Input (C_{iss})	4.0	5.5	5.5
Reverse transfer (C_{rss})	0.8	0.25	0.02
Output (C_{oss})	2.0	1.5	2.2
Input admittance (Y_{is}) at 100 MHz, mS	$1 + j2.5$	$0.4 + j0.3$	—
Output admittance (Y_{os}) at 100 MHz, mS	$0.1 + j1.2$	$0.3 + j0.9$	—
Maximum power gain (G_{ps}) at 100 MHz, dB	18	20	20
Noise figure (NF) at 100 MHz, dB	2.0	3.5	3.5

JFET Characteristics

Figure 4-7 graphs the response characteristics of the 2N4416 JFET. Like most field-effect devices this JFET exhibits square-law response for gate-bias voltage-to-drain current, yielding the greatest gain when the device is biased for drain current nearly equal to I_{Dss}. Figures 4-7b and 4-7c show a knee in the capacitance curves for gate-to-drain (C_{gd} or C_{rs}) and gate-to-source (C_{gs}) terminals, respectively. Note that the minimum input capacitance of 1.3 pF is achieved with a V_{GS} bias of -5 V dc, when the device is cut off, then increases to its maximum value near 3 pF at the 0 V dc bias value. Note also that the feedback capacitance increases sharply below 10 volts V_{DS}. Based on these graphs the optimum frequency response of this FET requires a V_{DS} of greater than 10 V dc and a V_{GS} biased for minimum drain current. The tradeoff, of course, is that reduced gain results from biasing the device for minimum capacitance and highest frequency response.

The input of the field-effect transistor is predominantly capacitive, with very low leakage conductivity from gate to source. Since the FET has a voltage-controlled input, however, any small resistive current flow from gate to source will have little effect in the performance of the device as an RF amplifier. The capacitances, C_{gs} and C_{gd}, are essentially series connected across the physical structure of the device. The total device capacitance, C_{ds}, which becomes the output capacitance in the CS connection, is thus the series equivalent of C_{gs} and C_{gd}. With this 2N4416 JFET biased with a V_{DS} of 15 V dc and a V_{GS} at -2.5 V dc, for example, the capacitance values from the graphs of Fig. 4-7 will indicate a C_{gs} of 1.1 pF and a C_{gd} of 0.6 pF. The value of C_{ds} may now be calculated as the series equivalent of the two capacitances, as:

$$C_{ds} = C_{gs}C_{gd}/(C_{gs} + C_{gd}) = 1.1 \times 0.6/(1.1 + 0.6) = 0.4 \text{ pF}$$

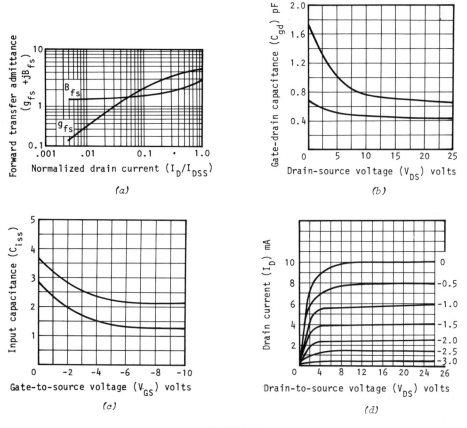

FIGURE 4-7

With the input conductance at an insignificant (leakage) value, the capacitance parameters may be applied to the equivalent-circuit model of Fig. 4-6. For the common-source circuit the input capacitance, C_{is}, is the parallel combination of C_{gs} and C_{gd} when specified with the drain at ac ground; the output capacitance, C_{os}, is C_{ds}; and the reverse feedback capacitance, C_{rs}, is C_{gd}. For the common-gate circuit the input capacitance, C_{ig}, is C_{gs}; the output capacitance, C_{og}, is C_{gd}; and the reverse feedback capacitance, C_{rg}, is C_{ds}. As with the CE connection of the bipolar transistor, the CS connection of the FET with an external load on the drain will reduce the significance of C_{gd} on the input admittance of the device. The resulting Miller-effect capacitance of the externally loaded circuit, however, will significantly affect C_{gd} as a load on the output (drain) of the FET, thereby reducing the frequency response of the device. As mentioned earlier, these effects will be covered in Chapter 5.

Once the bias conditions and capacitance values are determined, the output and transfer conductance values of the JFET can be obtained from the characteristic curves of Fig. 4-7a as follows:

$g_{fs} = \Delta I_D/\Delta V_{GS}$ at $V_{DS} = 15$ V dc $= 2$ mA$/1$ V $= 2$ mS
(This factor is called *transconductance,* or *forward* transfer current (conductance) factor for the common-*s*ource connection of the FET.)

$g_{os} = \Delta I_D/\Delta V_{DS}$ at $V_{GS} = -2.5$ V dc $= 0.1$ mA$/10$ V $= 0.01$ mS
(This factor is the *o*utput conductance for the common-*s*ource connection.)

After the tedium required to fully characterize bipolar transistors, you should appreciate the relative simplicity of the field-effect devices. With the close relationship of the characteristics for the other two (common-gate and common-drain) connections, you now need only *transfer* the common-source values to the other connections. These related characteristics are summarized as follows:

$$g_{fs} = g_{fg} = g_{fd} \quad \text{and} \quad g_{os} = g_{od}$$

Note (see Fig. 4-6) that the FET *channel* conductivity is common to both the output and the input of the common-gate connection, but that it is isolated from ground by C_{gd} and C_{gs}, respectively. The Y parameters for these three connections of the FET may now be defined by combining the conductances with the susceptance values of the capacitances of the previous paragraph:

$$
\begin{aligned}
Y_{is} &= +jB_{is} = +j\omega C_{gs} \\
Y_{os} &= g_{os} +jB_{os} = g_{ds} +j\omega C_{ds} \\
Y_{fs} &= g_{fs} +jB_{rs} = g_{fs} +j\omega C_{gd} \\
Y_{ig} &= +jB_{ig} = +j\omega C_{gs} \\
Y_{og} &= +jB_{og} = +j\omega C_{gd} \\
Y_{fg} &= g_{fg} +jB_{rg} = g_{fs} =j\omega C_{ds} \\
Y_{id} &= +jB_{id} = +j\omega(C_{gs} + C_{gd}) \\
Y_{od} &= g_{od} +jB_{od} = g_{os} +j\omega C_{ds} \\
Y_{fd} &= g_{fd} +jB_{fd} = g_{fs} +j\omega C_{gd}
\end{aligned}
$$

Sample Problem 4

Given: The 2N4416 junction field-effect transistor, operating at 10 MHz ($I_{GSS} = 100$ pA, $V_{GS} = 2.5$ V), with characteristics defined in the previous paragraphs and summarized as follows:

$g_{fs} = g_{fg} = g_{fd} = 2000$ μS	$C_{ds} = C_{os} = C_{od} = 0.4$ pF
$g_{os} = g_{og} = g_{od} = 100$ μS	$C_{gd} = C_{rs} = C_{og} = 0.6$ pF
$C_{gs} = C_{is} = C_{ig} = 1.1$ pF	

Determine: The equivalent-circuit characteristics and Y parameters for operation as a common-source amplifier at 10 MHz; record the values on the diagram of Fig. 4-4f.

(a) The susceptance values of C_{gs}, C_{gd}, and C_{ds} are first computed as follows:

$$B_{Cgs} = \omega C_{gs} = 6.28 \times 10^7 \times 1.1 \times 10^{-12} = 69.1 \ \mu S$$
$$B_{Cgd} = \omega C_{gd} = 6.28 \times 10^7 \times 0.6 \times 10^{-12} = 37.7 \ \mu S$$
$$B_{Cds} = \omega C_{ds} = 6.28 \times 10^7 \times 0.4 \times 10^{-12} = 25.1 \ \mu S$$

(b) The input admittance is identified as follows:

$$Y_{is} = +jB_{is} = +jB_{gs} = +j69.1 \ \mu S$$

(c) The output admittance is identified as:

$$Y_{os} = g_{os} + jB_{os} = g_{os} + jB_{ds} = 100 + j25.1 \ \mu S$$

(d) The transfer admittance is now identified as:

$$Y_{fs} = g_{fs} + jB_{rs} = g_{fs} + jB_{gd} = 2000 + j37.7 \ \mu S$$

Evaluating the results of this sample problem and comparing them with those of the earlier bipolar transistor problems yields some conclusions about the FET devices that make them useful as RF circuit components. First, the Y_{fs} (gain) characteristics of the FET are much lower than those of the bipolar devices; but the FET has reduced capacitance, making it a better choice for many RF applications. Second, the high input and output impedances of the FET device are of value in resonant LC circuits which require light loading—an impossibility with bipolar devices. And third, the square-law response of the FET devices reduces their intermodulation distortion products (see Chapter 6), making FET's the best choice for processing modulated RF signals.

MOSFET's

One of the problems with the RF operation of the JFET is its variation in capacitance with bias, requiring an additional sacrifice in gain to achieve lower capacitance operation. The insulated-gate structure of the MOSFET eliminates this necessity of higher V_{GS} values, thus allowing the device to be operated at optimum gain while maintaining low capacitance. Figure 4-8 graphs the typical response characteristics of the 3N128 MOSFET. You should carefully review these graphs, which are already specified in Y parameters. Optimum square-law response of the 3N128 is indicated at negative gate-bias values, even though the device will operate at zero gate bias. Rather than showing capacitance characteristics, the admittance graphs of Fig. 4-8 plot values of susceptance (B), as an indication of $relative$ capacitance, against V_{DS} and I_D. In general, biasing for a V_{DS} of 15 V dc and an I_D

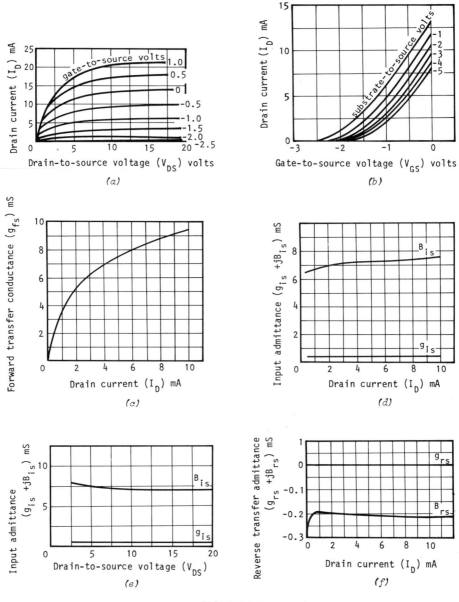

FIGURE 4-8

between 0.5 and 3 mA should assure minimum capacitance and highest frequency response for this device. The higher 3 mA value of I_D will yield the greatest gain (g_{fs}), however, so the gate bias (V_{GS}) should be about -1.5 V dc, as in Fig. 4-8a and b.

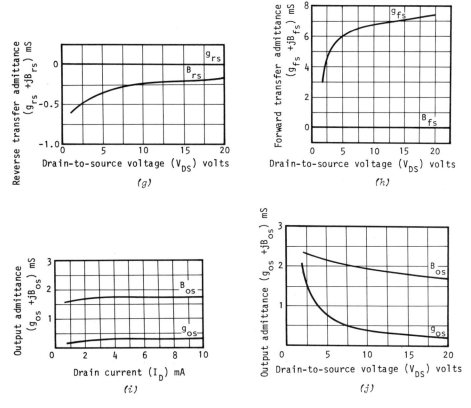

FIGURE 4-8 (*continued*)

With the bias conditions as specified in the previous paragraph, the typical operating characteristics can be obtained for the 3N128 and applied directly to the Y parameters of the equivalent-circuit model of Fig. 4-6.

$$V_{GS} = -1.5 \text{ V dc} \qquad I_D = 3 \text{ mA dc}$$
$$V_{DS} = 15 \text{ V dc} \qquad f_o = 200 \text{ MHz}$$

therefore,

$$Y_{is} = +jB_{is} = +j7 \text{ mS}$$
$$Y_{os} = g_{os} + jB_{os} = 0.3 + j1.8 \text{ mS}$$
$$Y_{fs} = g_{fs} + jB_{rs} = 6 + j0.2 \text{ mS}$$

These graphs save considerable time in the calculation of Y parameter values when applying this circuit at 200 MHz. For operation at other frequencies or with other circuit configurations (common-gate (CG) or common drain (CD)), the actual capacitance values can be extracted, as:

$C = B/\omega = 1.8 \times 10^{-3}/(6.28 \times 2 \times 10^8) = 1.43$ pF for C_{ds}
$B/\omega = 2 \times 10^{-4}/(6.28 \times 2 \times 10^8) = 0.16$ pF for C_{rs}
$B/\omega = 7 \times 10^{-3}/(6.28 \times 2 \times 10^8) = 5.57$ pF for $(C_{is} + C_{rs})$

Thus

$C_{is} = 5.57 - 0.16 = 5.41$ pF

While operation of the MOSFET is specified for a common-source connection at 200 MHz, the conductance (g) values do not change with other frequencies of operation, and they may be easily translated to other configurations (CG or CD) as detailed on page 128. Only the susceptance (b) values of admittances (Y) so specified in the graphs will change with frequency. Since the susceptance varies in direct proportion with frequency, however, it is a rather simple mathematical operation to obtain susceptance values at other frequencies by using *ratio* and *proportion*. If, for example, you wish to compute the performance of this 3N128 to that of the 2N4416 JFET at 10 MHz, the following procedure will yield the Y parameters for the 3N128 at that frequency.

$B'_{is} = B_{is} f_2/f_1 = 7$ mS $\times$ 10 MHz/200 MHz $= 350$ μS
$B'_{os} = B_{os} f_2/f_1 = 0.3$ mS $\times$ 10/200 $= 15$ μS
$B'_{fs} = B_{fs} f_2/f_1 = 0.2$ mS $\times$ 10/200 $= 10$ μS

The resulting Y parameter values at 10 MHz are:

$Y_{is} = g_{is} + jB'_{is} = 0 + j350$ μS
$Y_{os} = g_{os} + jB'_{os} = 300 + j15$ μS
$Y_{fs} = g_{fs} = jB'_{rs} = 6000 + j10$ μS

A comparison of the Y parameters at 10 MHz will thus indicate that the 3N128 has higher gain and lower transfer and output capacitances, but the 2N4416 has a lower input capacitance.

Dual-Gate MOSFET's

The dual-gate MOSFET is useful for a wide variety of RF circuit applications with two active input gates. The graphs of Fig. 4-9 show the varied device performance characteristics available with the 3N140. A careful review of these graphs will quickly disclose the relative complexity, but versatility, of the device. As you can see, the drain characteristics indicate optimum square-law response about the zero-bias level of gate 1, the signal input gate for most RF amplifier applications. As with the 3N128, these admittance graphs of Fig. 4-9 for the 3N140 plot values of susceptance against bias conditions. You should note that there is gate conductance (g_{is}) of a significant value here. After careful examination of the graphs you may conclude that a V_{DS} of greater than 7 V dc and an I_D of 2 to 5 mA will provide lower susceptance and higher frequency response for the device. The optimum transconductance (gain) of the device occurs at higher bias levels for gate 2,

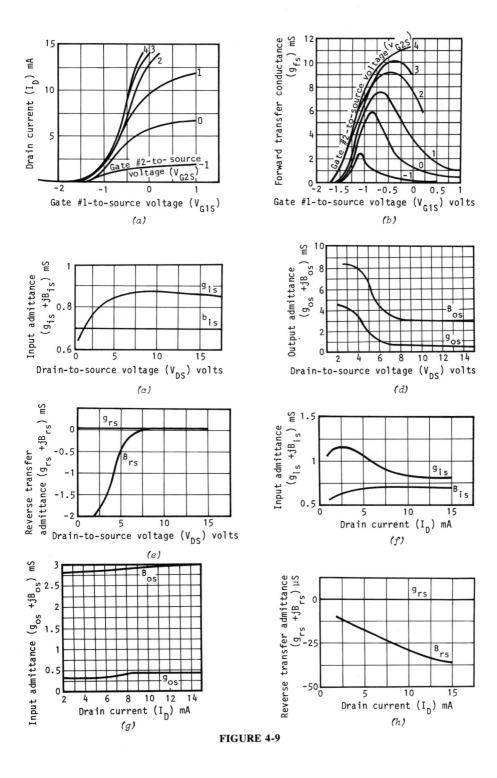

FIGURE 4-9

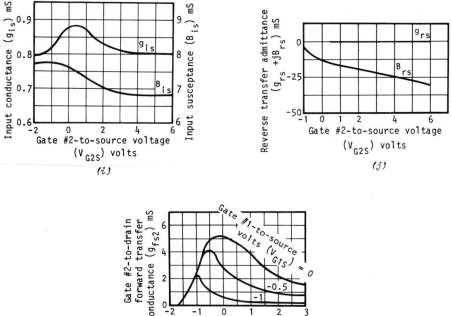

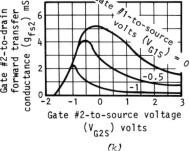

FIGURE 4-9 (*continued*)

as shown in Fig. 4-9*b*, so a compromise for the above I_D value may be a gate 1 V_{GS} of -0.8 to -1.0 V dc, and a gate 2 V_{GS} of $+3$ V dc, yielding a g_{fs} of about 8 mS.

With the bias conditions specified in the previous paragraph, the operating characteristics for the 3N140 can be obtained from the graphs of Fig. 4-9, yielding the following Y parameters at 200 MHz:

$$Y_{is}(\text{gate 1}) = g_{is} + jb_{is} = 1.1 + j7 \text{ mS}$$
$$*\,Y_{is}(\text{gate 2}) = g_{is} + jB_{is} = 0.8 + j7 \text{ mS}$$
$$Y_{os} = g_{os} + jB_{os} = 0.3 + j2.6 \text{ mS}$$
$$Y_{fs}(\text{gate 1}) = g_{fs} + jB_{rs} = 8 + j0.05 \text{ mS}$$
$$*\,Y_{fs}(\text{gate 2}) = g_{fs} + jB_{rs} = 0.4 + j0.2 \text{ mS}$$

---

* With the signal input to gate 1 in this configuration, gate 2 is used for bias only; so the Y parameters do not apply for this RF amplifier application.

A comparison of the performance of this 3N140 with that of the 3N128 points up some tradeoff characteristics for the two devices, including a highly significant input conductance, but an almost *in*significant feedback capacitance with this dual-gate MOSFET.

The frequency-response characteristics of the field-effect transistors, JFET's, and MOSFET's extend beyond those of bipolar transistors, primarily because of reduced capacitance values. Resorting to the common-base connection in order to achieve broad frequency response with the bipolar device requires driving an extremely low input impedance. If a resonant network is required for tuning, etc., the low input impedance of the CB connection will load the circuit and destroy its Q. The CS input impedance of the FET, however, has low conductivity, even though it may have substantial capacitance (7 pF in our examples). The C_{is} of the device can be included as a part of the LC network capacitance. With very small (almost zero) conductance, the FET input will maintain the high-Q factor of the resonant network. The output characteristics of FET's and bipolar transistors are similar, but it is the small value of internal reverse capacitance (C_{gd} or C_{rs}) that makes FET's distinctly superior in the amplification of low-level RF signals.

Since the internal feedback capacitance (C_{gd} or C_{rs}) of FET's is small, the frequency where Y_{fe} drops by -3 dB is very high. Under these circumstances the frequency limitations of the device may be more greatly influenced by the output capacitance of the FET. In the case of the 2N4416, for example, the *cutoff frequencies* for the transfer admittance (f_g) and output admittance (f_{go}) can be computed as:

i. $f_g = B_C/2\pi C = g_{fs}/2\pi C_{gd}$

    where $B_C = g_{fs}$, $g_{fs} = 2$ mS, $C_{gd} = 0.6$ pF

    $f_g = 2 \times 10^{-3}/(6.28 \times 6 \times 10^{-13}) = 531$ MHz

ii. $f_{go} = B_C/2\pi C = g_{os}/2\pi C_{ds}$

    where $B_C = g_{os}$, $g_{os} = 0.1$ mS, $C_{ds} = 0.4$ pF

    $f_{go} = 10^{-4}/(6.28 \times 4 \times 10^{-13}) = 39.8$ MHz

Obviously the high-frequency capability of the transfer characteristics cannot be used with an output cutoff frequency of only 39.8 MHz. Since the output terminals of the device are accessible, however, it is but a simple matter to either *load* the output or utilize its capacitance as a part of an LC network. Resistive loading of the output by a factor of 20 will extend the frequency response in the same proportion, for example, and revert the frequency-response characteristics back to that of the transfer-conductance cutoff frequency (f_g). Loading, matching, and other external circuit techniques are fully reviewed in Chapter 5.

VACUUM TUBES

It wasn't too many years ago that the vacuum tube, with all its limitations, was the only available amplifying device for radio-frequency signals. While such electron tubes are obsolete devices for new designs, many are still in reliable service in electronic communications systems. For very high power applications in radio transmitters and similar equipment, tubes are still being applied in new designs. Also, a wide variety of special-purpose vacuum tubes have as yet no solid-state counterparts. Included in this list of special devices are the display and printing cathode ray tubes; X-ray and other radiation tubes; photomultiplier, infrared, and other detector tubes; and a wide range of microwave tubes, some of which are reviewed in Chapter 13.

Tubes designed for RF amplification are categorized by their internal grid structure as triodes, tetrodes, and pentodes for three-, four- and five-element devices, respectively. The tubes have gain characteristics similar to those of FET's; and they are generally biased like depletion-mode, N-channel FET's, but with higher supply voltages. They can be connected in the same three basic configurations as the other electronic amplifying devices: common-grid, common-plate, and the most popular common-cathode connection. Table 4-3 lists the characteristics of a popular triode, tetrode, and pentode tube; and Fig. 4-10 graphs the characteristic curves for each of the devices. Observe that the response curves for the tetrode and pentode devices are similar to those of the FET's, whereas the response curves for the triode are unique. With the temperature of tubes, of course, stabilized by their heaters, their electrical characteristics will vary with bias conditions, which should be selected for optimum RF performance.

Figure 4-10d shows the equivalent-circuit model of the vacuum tube as being similar to the FET. The input, transfer, and output characteristics of the electron tubes are similar to those of the dual-gate MOSFET's. In fact, the multigrid tetrode and pentode tubes use grid 2 bias to vary the signal processing characteristics of the device, much like gate 2 bias varies the

TABLE 4-3. TYPICAL ELECTRICAL CHARACTERISTICS FOR 6CW4 NUVISTOR TRIODE, 6CY5 TETRODE, AND 6CB6 PENTODE VACUUM TUBES

| | | Device | |
Characteristic	6CW4	6CY5	6CB6
Transconductance (G_m), μS	10,000	8000	8000
Plate resistance (r_p), Ω	6000	100,000	280,000
Capacitance, pF			
Grid-to-plate (C_{gp})	0.92	0.03	0.02
Grid-to-cathode (C_{gk})	4.3	4.5	6.5
Plate-to-cathode (C_{pk})	0.18	3.0	3.0

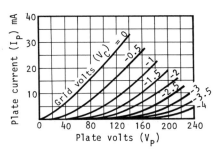

(a)

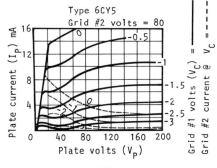

(b)

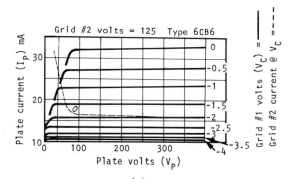

(c)

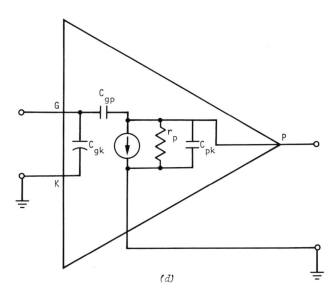

(d)

FIGURE 4-10

response of the dual-gate MOSFET. Like other electronic devices, the interelectrode capacitance values for the tubes vary with bias conditions. The more significant (grid 1) grid-to-cathode capacitance (C_{gk}) increases, almost linearly, in *direct* proportion to a tube's transconductance. The reverse (feedback) grid-to-plate capacitance (C_{gp}) decreases, almost linearly, in *inverse* proportion to a tube's plate current. As may be observed from the data of Table 4-3, the reverse capacitance in tubes is typically very small, like the dual-gate MOSFET. The plate-to-cathode (C_{pk}) capacitance varies inversely with plate current from about 40 to 70% of C_{gk} for a given transconductance value. The response curves that graph these capacitance vs bias characteristics are not generally available from manufacturers' published data, as they are for newer solid-state devices. In application, therefore, you should interpolate capacitance values from data specified in tabular form.

Triode Characteristics

The simplest of tube designs is, of course, the triode; its characteristics are discussed in the following paragraphs. The small, low-noise, metal package type 6CW4 "nuvistor" is used as our example. Referring to Table 4-3 and Fig. 4-10a, consider the following operating conditions for this device based on a compromise for gain (transconductance, g_m or g_{fk}) vs input capacitance. As you can see from the curves of Fig. 4-10a, the transconductance is greatest at zero grid-bias volts and higher plate voltages. The maximum C_{gk} value of about 4.5 pF will thus most likely occur at a V_P of 140 V dc and a V_{GK} of 0 V dc. The transconductance at this bias point can be computed as:

$$g_{fk} = \Delta I_P/\Delta V_{GK} \text{ with } V_P \text{ constant}$$
$$g_{fk} = 12 \text{ mA}/0.5 \text{ V} = 24 \text{ mS}$$

Operating at this bias point is not only a poor choice for capacitance, but it is more wasteful of power, which is a concern with high plate voltages. A compromise in favor of lower capacitance and dc power dissipation is the operation of the tube with negative grid bias, while maintaining reasonable transconductance and plate current for lower power dissipation. One practical alternative with low-level RF input signals is for operation with a plate voltage of 60 V dc and a grid bias (V_{GK}) of -0.5 V dc. This operating point will yield a relative transconductance and approximate capacitance which may be computed as follows:

$$g'_{fk} = \Delta I_D/\Delta V_{GK} = 4 \text{ mA}/0.5 \text{ V} = 8 \text{ mS}$$
$$C'_{gk} = C_{gk(max)} \, g'_{fk}/g_{fk(max)} = 4.5 \text{ pF} \times 8 \text{ mS}/24 \text{ mS} = 1.5 \text{ pF}$$

The grid-to-cathode conductance of the 6CW4 triode results from those electrons that are intercepted by the grid structure as they pass from cathode to plate. This low conductance factor varies in direct proportion with plate current, and is insignificantly low for RF signals. The plate-to-cathode conductance for the 6CW4 can be obtained from the characteristic curves of Fig. 4-10a about the bias point as follows:

$g_{pk} = \Delta I_C / \Delta V_P$, with V_{gk} constant
$g_{pk} = 3 \text{ mA}/10 \text{ V} = 0.3 \text{ mS}$

This value is comparable to the output conductance of both FET's and bipolar transistors. An estimate of the plate-to-cathode capacitance of the 6CW4, at the specified bias point, should be about 60 to 70% of C'_{gk}, since the plate current is relatively low. A quick estimate of two-thirds (67%) of C'_{gk} yields a C'_{pk} of about 1 pF. The grid-to-plate capacitance (C_{gp}) of the 6CW4 is at maximum when the plate current is minimum; and it is reduced to nearly zero at maximum plate current. The specified maximum value of 0.92 pF for the 6CW4 (from Table 4-3) must then be derated in proportion to the increased conductivity of the tube. An estimate of this C_{gp} value can be computed as approximately:

$$C'_{gp} = C_{gp(max)}(I_{Pmax} - I_P)/I_{Pmax} = 0.92 \times (32 - 4 \text{ mA})/32 \text{ mA} = 0.8 \text{ pF}$$

Y parameters can be applied to electron-tube devices for operation as RF amplifiers, in the same manner as the Y parameters are applied to transistors. The Y parameters for the 6CW4 of our example, operating as a 10 MHz common-cathode amplifier and biased as specified ($V_P = 60$ V dc and $V_{GK} = -0.5$ V dc), are thus as follows:

$$
\begin{aligned}
Y_{ik} &= Y_{gk} = +j\omega C_{gk} \\
&= +j(6.28 \times 10^7 \times 1.5 \times 10^{-12}) = 0 + j0.094 \text{ mS} \\
Y_{ok} &= Y_{pk} = g_{pk} + j\omega C_{pk} \\
&= 0.3 + j(6.28 \times 10^7 \times 10^{-12}) = 0.3 + j0.063 \text{ mS} \\
Y_{fk} &= Y_m = g_{fk} + j\omega C_{gp} \\
&= 8 + j(6.28 \times 10^7 \times 8 \times 10^{-13}) = 8 + j0.05 \text{ mS}
\end{aligned}
$$

The Y parameters for the same 6CW4, operating as above but in the *common-grid* configuration, can be specified as follows:

$$
\begin{aligned}
Y_{ig} &= Y_{gk} = +j\omega C_{gk} = +j(6.28 \times 10^7 \times 1.5 \times 10^{-12}) = +j0.094 \text{ mS} \\
Y_{og} &= Y_{gp} = +j(6.28 \times 10^7 \times 0.8 \times 10^{-12}) = +j0.05 \text{ mS}
\end{aligned}
$$

The Y parameters for this 6CW4, biased as above but in common-plate (cathode-follower) configuration, can be specified as follows:

$$
\begin{aligned}
Y_{ip} &= Y'_{gp} = +j0.05 \text{ mS} \\
Y_{op} &= Y_{pk} = 0.3 + j0.063 \text{ mS} \\
Y_{fp} &= Y_{fk} = Y_m \text{ (mutual admittance)} = 8 + j0.05 \text{ mS}
\end{aligned}
$$

Thus the vacuum tube, like other electronic devices, has the same electrical Y-parameter equivalent circuit for the common-cathode *and* common-plate configurations.

Other Tubes

The procedures applied to reduction of the triode tube to the equivalent-circuit model of Fig. 4-1*b,* and its related *Y* parameters under certain bias conditions, apply similarly to the tetrode and pentode tubes. Although some liberties were taken in the establishment of capacitance values under given bias conditions, the techniques, again, reflect the procedures of engineering practice, where thorough graphical information is not as easily obtained for tubes as it is for modern solid-state devices. Also, this section of the text covered only the small-signal RF characteristics of the tubes in order to show their similarity to bipolar and field-effect transistors. Of course, conventional electron tubes are now found in few small-signal RF applications, because the performance, reliability, size, power consumption, cost, and mechanical stability of modern solid-state devices are so far superior to the tubes. For high-power applications, where tubes may be more suitable than solid-state devices, the control grid (grid 1) *does* conduct current, thus considerably changing the *Y* parameters of the tubes. These high-power applications will be thoroughly reviewed in Chapter 5.

INTEGRATED CIRCUITS

Modern solid-state integrated circuits combine the characteristics of multistage transistors with passive components to achieve special signal processing functions. In addition to the obvious size, weight, and cost advantages that IC's offer, their ready availability has reduced the complexity of electronic circuits and systems for easier understanding. Rather than facing pages of large schematics with hundreds of resistors, capacitors, coils, and discrete active devices, many modern systems using IC's can be viewed on a single-page schematic comprising IC blocks with only a few external *R*, *C*, and *L* components. While knowledge of discrete devices is important, applying discrete-device circuit analysis to multistage IC's is inefficient. The approach used in engineering practice with IC's is, in fact, the same as that used with discrete devices: reduction of the device to an equivalent-circuit model with its related Y parameters, concurrent with the selection of bias conditions to achieve desired operating characteristics. We will thus proceed to apply the same approach to the analysis of IC devices, after an introductory review of their circuit structure as used in the preceding sections of this chapter for discrete transistors and tubes.

"Building Block" RF Circuits

Integrated circuits are used for a wide variety of RF circuit functions in modern communications applications. The basic internal circuit structure of most of these RF IC's falls into two categories: the *differential-amplifier* connection and the *cascode-amplifier* connection. Figure 4-11 shows these connections in a variety of typical circuit arrangements. The differential amplifier is basically a pair of matched, balanced transistors connected to a common current source. The circuit can be connected as a common-collector/common-base unit, as shown in Fig. 4-11b. The connection provides high forward transconductance with very little reverse conductance or capacitance, and is used as a basic building block for more complex IC functions. The cascode amplifier is also a two-transistor connection (without the common current source) arranged for series operation. This circuit is generally connected as a common-emitter/common-base unit, as shown in Fig. 4-11d. The cascode connection provides higher gain than the differential connection, but at the expense of increased reverse admittance.

A wide variety of RF IC devices employ the basic building blocks of the differential and cascode circuit arrangements. Reviewing a large number of these devices would certainly be beyond the scope and purpose of this book, so only representative examples will be used here to provide an approach to IC analysis and application. It is strongly recommended, however, that you obtain data books and application notes from manufacturers of transistors and IC's, and that you review their contents in order to relate the material in this text to the real world. Table 4-4 and the graphs of Fig. 4-12 compare the characteristics of the NE510 IC, which may be connected as either a differential or a cascode amplifier. As you can see from the graphs, the transconductance of the NE510 "diff-amp" (differential amplifier) can extend to over 30 mS, whereas the same NE510 in cascode connection can provide a transconductance in excess of 60 mS. The Y parameters for the diff-amp connection, biased as given in Table 4-4, with a V^+ of 12 V dc can be summarized in reference to four-terminal designations, as follows:

$$Y_{in} = Y_{11} = g_{11} + j\omega C_{11} = 0.4 + j(\omega 4.5 \text{ pF}) \text{ mS}$$
$$Y_{out} = Y_{22} = g_{22} + j\omega C_{22} = 0.01 + j(\omega 2.5 \text{ pF}) \text{ mS}$$
$$Y_f = Y_{21} = g_{21} + j\omega C_{21} = 21 + j(\omega 0.1 \text{ pF}) \text{ mS}$$

The Y parameters for the cascode connection, biased as above, can be summarized as follows:

$$Y_{in} = Y_{11} = g_{11} + j\omega C_{11} = 2 + j(\omega 10 \text{ pF}) \text{ mS}$$
$$Y_{out} = Y_{22} = g_{22} + j\omega C_{22} = 0.01 + j(\omega 2.5 \text{ pF}) \text{ mS}$$
$$Y_f = Y_{21} = g_{21} + j\omega C_{21} = 75 + j(\omega 0.1 \text{ pF}) \text{ mS}$$

The significant differences between the two connections are a higher input conductance (g_{11}) and capacitance (C_{11}) in the diff-amp connection, with the tradeoff a high transconductance (g_{21}) value in the cascode connection.

Another popular IC used in RF communications applications is the CA3028/3053, which may also be connected in either differential or cascode

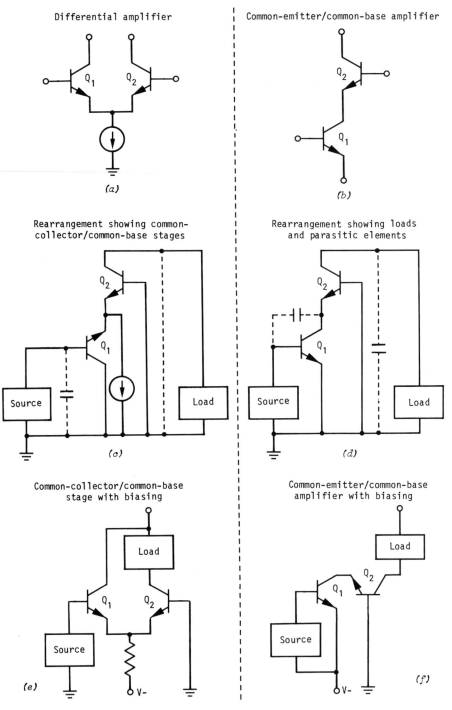

Differential amplifier

(a)

Common-emitter/common-base amplifier

(b)

Rearrangement showing common-collector/common-base stages

(c)

Rearrangement showing loads and parasitic elements

(d)

Common-collector/common-base stage with biasing

(e)

Common-emitter/common-base amplifier with biasing

(f)

FIGURE 4-11

TABLE 4-4. TYPICAL ELECTRICAL CHARACTERISTICS FOR NE510 INTEGRATED CIRCUIT*

Characteristic	CC-CB		CE-CB	
	+6 V dc	+12 V dc	+6 V dc	+12 V dc
Bias network current, mA	1.5	3.2	1.5	3.2
Quiescent input current, μA	25	50	50	100
Quiescent output current, mA	0.6	1.4	1.2	2.8
Input conductance (Y_{11} or g_{in}), mS	0.25	0.4	1.0	2.0
Output conductance (Y_{22} or g_{out}), mS	0.01	0.01	0.01	0.01
Input capacitance (C_{in}), pF	4.0	4.5	8.0	10.0
Output capacitance (C_{out}), pF	3.0	2.5	3.0	2.5
Feedback capacitance (C_f), pF	0.1	0.1	0.1	0.1
Forward transconductance (Y_{21} or g_f), mS	11	21	45	75

* IC is connected as a common-collector/common-base (CC-CB) and a common-emitter/common-base (CE-CB) amplifier (differential and cascode connections, respectively).

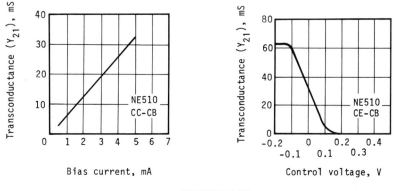

FIGURE 4-12

configurations. The basic circuit configuration for the CA3028 and its connection as a diff-amp and cascode amplifier are shown, along with the complete sets of response graphs, in Fig. 4-13. A careful review of these graphs should disclose the optimum operating parameters for the device. Graphs (d) through (g) plot the four Y parameters for the diff-amp connection; graphs (h) through (k) plot the four Y parameters for the cascode connection; and graphs (l) and (m) plot the frequency response of the circuits with recommended power supply voltages of 9 and 12 V dc. The review of these graphs indicates that such IC's with internal resistors do not provide the wide range of bias options available with discrete devices; this is an advantage in some

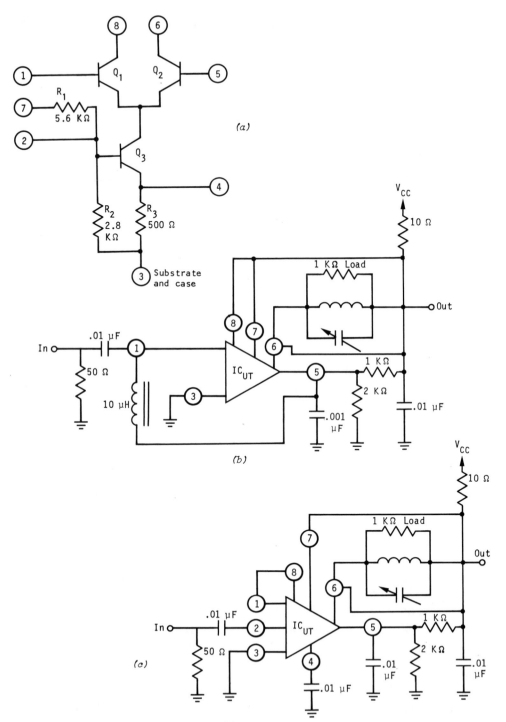

FIGURE 4-13

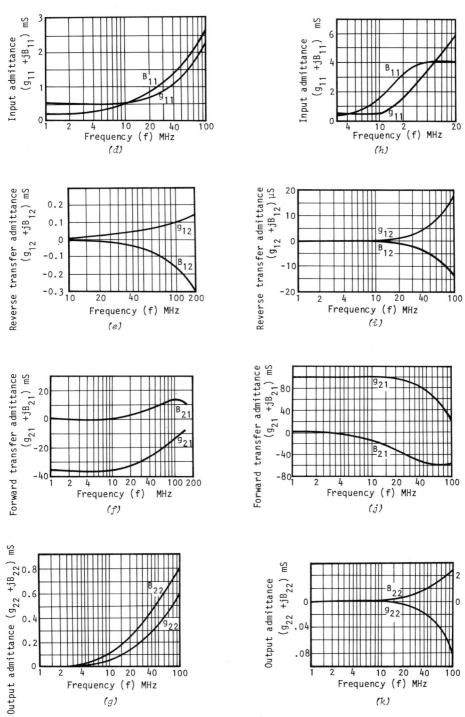

FIGURE 4-13 (*continued*)

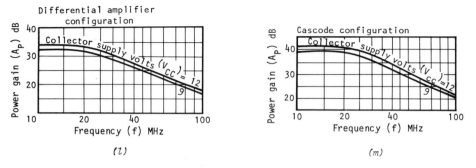

FIGURE 4-13 (*continued*)

cases, but a disadvantage in others. The high-production-volume IC's intended for specific applications (FM-IF amplifiers, for example) have bias and operating parameters closely specified. Other high-production-volume IC's intended for a wide variety of applications, of which the NE510 is an example, will operate over a range of bias and operating conditions, thus making them more versatile.

MEASUREMENTS

The preceding sections on transistors, tubes, and IC's demonstrated that the characterization of electronic devices is a most important function. In the following chapter, device characteristics will be used to obtain specific circuit performance, and it will be shown that successful application of electronic devices in RF circuits requires detailed information about device characteristics. Manufacturers of electronic devices expend considerable effort in thorough device characterization, with related publication of data and techniques for use by engineers and technicians. In some instances the cost of device characterization exceeds the cost of manufacture, but the devices would often be worthless without characterization and support documentation for their application into circuits.

Measurements technology has become more sophisticated and automated in recent years, and thus serves to provide extensive information about electronic devices which was impractical to obtain in earlier times. As a consequence, detailed graphical and tabular data, like those of Figs. 4-2, 4-7, 4-9, and 4-13, were just not available for the comparably low cost vacuum-tube devices of a few decades past. With the availability of sophisticated, automated test and measurement equipment, manufacturers can also inexpensively sort and select devices within limits specified by user needs. The result of these advances in measurements technology is an increasing utilization of manufacturers' data by technicians and engineers, with an emphasis on the external circuit components required for specific applications. This emphasis will be reflected in Chapter 5, which reviews electronic circuits in detail. A number of nonautomated measurement techniques are also useful as an alternative to characterizing electronic devices when manufacturers' data is not available. Use of the curve tracer and the impedance bridge are two of these techniques.

Curve Tracer

The transistor curve tracer has become one of the more popular fixtures of the electronics laboratory. The curve tracer will provide input, output, and transfer characteristic curves from which the *real* (conductance, or *g*) parts of the low-frequency) Y parameters may be obtained. Figure 4-14a shows the popular common-emitter "family of curves," collector characteristics vs base input current steps, for a typical bipolar transistor. The dashed lines and projections from this graph provide the information to compute h_{fe}. As you can see in this figure, an operating point of 2.1 mA I_C and 12 V_{CE} indicates a base current of 25 μA. A variation in this base current (I_B) of 10 μA shows a related variation in collector current (I_C) of 1.5 mA, with V_{CE} fixed at 12 V dc, yielding an h_{fe} of 150.

Figure 4-14b shows the same 2.1 mA I_C, 12 V_{CE} operating point with the vertical gain of the display increased to obtain increased slope of the base-step response curves. The dashed lines and projections from this graph provide the information to compute h_{oe} (see formula for h_{oe} earlier in this chapter). Figure 4-14b also shows a variation in V_{CE} of 8 V and a related variation in I_C of 0.14 mA, yielding an h_{oe} of 17.5 μS.

Figure 4-14c shows a family of common-emitter curves that compare V_{CE} and V_{BE} against base-input current steps. The dashed lines and projections from this graph indicate that the operating point of 12 V_{CE} and 25 μA I_B yields a V_{BE} of 35 mV (above the 600 mV knee of the V_{BE} diode conductivity level). A variation in this base current of 10 μA shows a related variation in base-emitter voltage (V_{BE}) of 12 mV, with V_{CE} fixed at 12 V dc, and yielding an h_{ie} of 1200 Ω.

Figure 4-14d shows the same 25 μA I_B, 12 V_{CE} operating point, with the vertical gain of the display increased to obtain increased slope of the base-step response curves. The dashed lines and projections from this graph indicate a slope of 1 mV for a 10 V change in V_{CE}, along the 25 μA I_B curve, yielding an h_{re} of 10^{-4}, or one part in 10,000—an insignificantly small value.

Figure 4-14e and f shows comparable curve-tracer plots for the common-base connection of the bipolar transistor. In practice, only two such plots for a single connection (CE or CB) are made, with this data used to compute the parameters for the other connections.

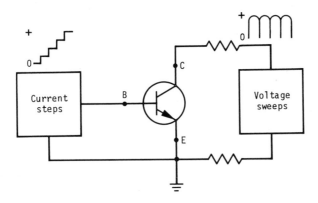

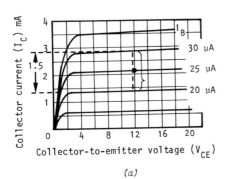

(a)

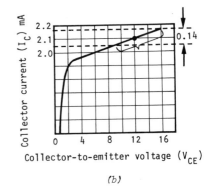

(b)

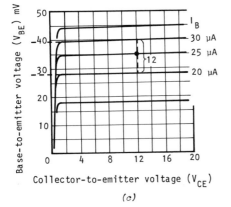

(c)

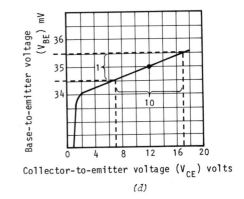

(d)

FIGURE 4-14

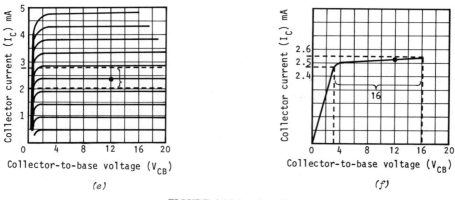

(e)   (f)

FIGURE 4-14 (*continued*)

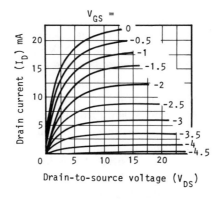

FIGURE 4-15

Figure 4-15 shows the curve-tracer plots for the field-effect transistor. As you can see, the transfer characteristics for the FET are similar to those of the bipolar transistor, with the exception of the base steps, which are in increments of reverse voltage rather than forward current. Both the forward transfer conductance (transconductance, g_{fs}) and the output conductance (g_{os}) can be obtained from this plot for the common-source connection. Since the input conductance and reverse-voltage ratio factors are insignificant with FET devices, other plots are not generally obtained for characterizing these factors. Moreover, as with the bipolar devices, the characteristics for the other connections (CG and CD) can be translated from the single (CS) plot of curves.

Discrete device characteristics, such as those shown in Figs. 4-14 and 4-15, can be applied to IC's at the development level, where discrete "chips" are used in order to characterize the performance of an IC prior to its reduction to monolithic form. For those IC's typically found in RF applications, curve-tracer characteristics cannot be practically obtained, thereby making operational measurements more practical. In earlier times, curve tracers were also used to characterize vacuum tubes. With the impracticality of specialized high-voltage power requirements and infrequent use, these tracers are not usually found in the RF electronics laboratory. Consequently, with IC's, tubes, and often even transistors, other specialized test equipment and measurement techniques are employed to obtain operating parameters. RF bridge-type instruments are among the more popular of these alternatives which provide Y parameters for externally biased devices.

Impedance Bridge

Figure 4-16a and b shows the equivalent circuits for the measurement of input/output and transfer admittances of active electronic devices using the conventional impedance (admittance) bridge. Most bridges of this type have accessory test mounts that hold the devices under test and provide facilities for bias access to their terminals. The mounts have known electrical characteristics that can be calibrated out of the measurement results. The input and output admittance values are measured directly in $g + jB$ units; whereas the transfer characteristics generally require a comparison measurement after connecting the output to the input of the amplifying device, as shown in Fig. 4-16c. Figure 4-16d shows an example of an inexpensive, direct-reading transconductance meter with a built-in signal source and detector. With "flat" or compensated frequency response in the signal source and detector, the instrument basically measures the gain of the device under test, in either the common-emitter or common-source connection, with the frequency increased until the output drops by -3 dB. The obvious advantage of bridge measurements over those from the curve tracer is the indication of capacitance values in addition to conductance values. The tradeoff disadvantage of the bridge is its inability to indicate biasing information. Both the curve tracer *and* bridge are necessary for thorough device characterization.

NOISE CHARACTERISTICS

Noise is among the most important characteristics of RF devices and systems. (The systems aspects of noise are covered in Chapter 8.) The noise characteristics of a transistor, IC, or tube determine both the lowest level of signal that can be amplified and the quality of that signal. Noise is defined as random electrical signals that result from the movement and exchange of

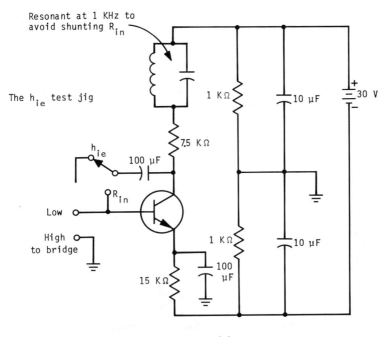

(a)

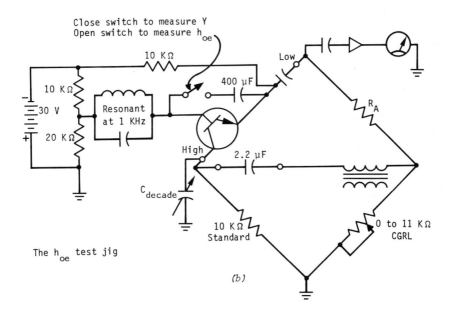

(b)

FIGURE 4-16

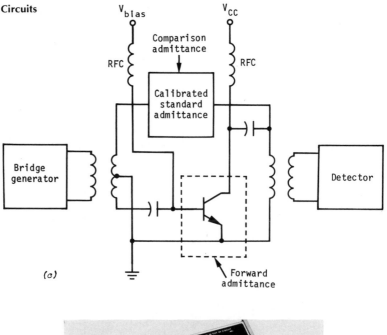

(c)

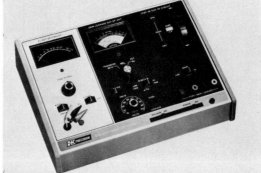

FIGURE 4-16 (*continued*)   (d)

electrons within the atomic structure of active and passive components. When the noise is completely random and its amplitude is uniform over a given range of frequencies, it is called *white noise,* based on the relationship to "white light" as resulting from the equal distribution of energy across the visual spectrum. When the noise is less intense at the higher frequencies within a given range—across the audio spectrum, for example, where the noise amplitude tapers off above 5 kHz—it is called *pink noise,* like the light that is difficient in its higher frequency (blue) energy. Figure 4-17 shows some typical oscilloscope and spectrum analyzer displays of noise energy.

Noise Figure

The frequency range and amplitude variations of the noise signals produced by electronic components vary considerably, depending on the structure and

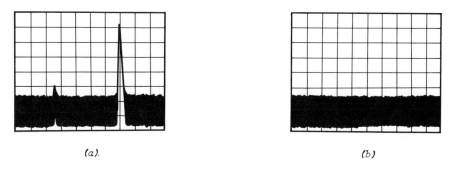

(a). (b)

FIGURE 4-17

function of such active devices as they operate in a particular circuit. Such electronic devices are most often specified in terms of *noise figure,* a quantity which defines the *amount* of noise that components and devices *add* to the signals they amplify. Once the noise figure of a device is known, the signal-to-noise ratio can be computed for a wide variety of applications. Since noise figure is logarithmic, it is most often expressed in dB; it is defined as the *ratio* of the input signal-to-noise ratio to the output signal-to-noise ratio of an electronic device, circuit, or system. Thus the noise figure for a device that increases a 30 : 1 input signal-to-noise ratio by a factor of 2, for example, can be computed as:

$$NF = 20 \log (S/N_{in})/(S/N_{out}) = 20 \log (30:1)/(15:1) = 20 \log 2 = 6 \text{ dB}$$

The noise figure specifications for the same electronic devices that were used as examples earlier in this chapter are listed in Table 4-5; and some representative graphs of noise figure vs bias conditions are shown in Fig. 4-18. The plots of these graphs indicate that there is an obvious tradeoff between noise figure and transfer admittance (Y_f) for all the devices, with higher gain factors yielding higher noise figures. Also, they show that noise figure generally increases with increasing bias current and device operating temperature. The range of practical noise figure values for single-stage RF amplifier devices (not IC's) typically falls between an excellent 2 dB and a poor 8 dB.

Noise Power

In application the noise figure of an electronic device is computed, along with its other parameters, into the total circuit (including external components) in which it operates. These circuit effects of noise are detailed in Chapter 5, and the important systems effects are reviewed in Chapter 8. In these applications the device noise figure may also be expressed as *equivalent noise resistance* or *equivalent noise temperature.* The terms relate to the theoretical basis of noise generation, derived from gas-law physics, which specifies the

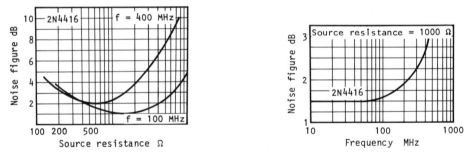

FIGURE 4-18

TABLE 4-5.   TYPICAL TRANSISTOR NOISE FIGURE
SPECIFICATIONS

Device	Noise figure, dB	Device	Noise figure, dB
2N2102	8	3N128	5
2N4416	4	3N140	4
2N3053	10	CA3028	8
2N6660	10	NE510	6

amount of *noise power* developed by the movement and exchange of electrons within the atomic structure of devices and components. This noise
power is called *thermal agitation noise,* since electron activity increases with
temperature. (Noise power is also often called *Johnson noise* after its discoverer.) Noise power may thus be computed by applying the gas-law
relationship:

$$P_n = kTB$$

P_n is the noise power transferred to a load with an input impedance equal to
the internal impedance of a theoretical noise generator (matched impedances
yield maximum power transfer). k is Boltzmann's constant of the statistical
distribution of energy due to electron motion, and is expressed as 1.38 ×
10^{-23} joule/kelvin (or Celsius) (1 joule/second = 1 W). T is the temperature
of a given device or component generating noise, in kelvin (0°C = +273 K). B
is the −3 dB bandwidth of a given device or circuit, in Hz.

  The noise characteristics of active electronic devices are based on the *kTB*
relationship, an equivalent noise resistance (R_n), and a theoretical, open-
circuit noise generator (V_n). The open-circuit noise voltage can be computed
as:

$$V_n = 2\sqrt{PR} = 2\sqrt{kTBR_n} = \sqrt{4kTBR_n}$$

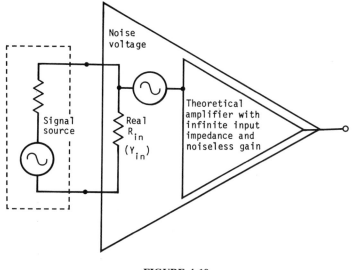

FIGURE 4-19

If, for example, the equivalent-circuit model of Fig. 4-19 is operating at 20°C (293 K), with a bandwidth of 1 MHz and an equivalent noise resistance of 5 KΩ, the open-circuit noise voltage can be computed as:

$$V_n = \sqrt{4kTBR_n} = \sqrt{4 \times 1.38 \times 10^{-23} \times 293 \times 10^6 \times 5 \times 10^3}$$
$$= 9 \ \mu V$$

Once this theoretical open-circuit voltage is known, it can be applied to the transfer characteristics (Y_f) of a given active device, in series with its input, but after its real input impedance or admittance (Y_i), as shown in Fig. 4-19.

Noise Resistance

To fully utilize the noise characteristics of an electronic device when it is applied in a circuit function, the manufacturers' noise figure specifications must be translated to equivalent noise resistance (R_n). Once the value of the equivalent noise resistance is known, it can then be used to obtain signal-to-noise ratios for a variety of circuit applications for the device. The conversion process requires reflecting the output noise of a device back to its input as a noise resistance value proportional to the device's input admittance (Y_i). This equivalent noise resistance value can then be used to compute the open-circuit voltage of a theoretical noise generator, as detailed in the previous paragraph and Fig. 4-19. The noise resistance can be computed as:

$$R_n = [((\log^{-1} NF/20)^2 - 1)/Y_i] - (1/Y_i)$$

R_n is the equivalent noise resistance, proportional to the input admittance and noise figure of a given electronic device; NF is the noise figure, in dB, specified for the device; and Y_i is the input admittance for the particular active device as obtained with the techniques detailed earlier in this chapter.

From Table 4-5 the noise figure specified for the 2N4416 JFET, for example, is 4 dB at an f_{co} of 400 MHz. Now recall (from calculations made earlier in this chapter) that the input admittance (Y_{is}) computed for the device is 69.1 μS ($\simeq$ 70 μS). Based on these specifications the equivalent noise resistance can be computed as follows:

$$R_n = [((\log^{-1} NF/20)^2 - 1)/Y_{is}] - (1/Y_{is})$$
$$= [((\log^{-1} 4/20)^2 - 1)/(70 \times 10^{-6})] - (1/70 \times 10^{-6})$$
$$= 7313 \ \Omega$$

Once the equivalent noise resistance is known, the theoretical open-circuit noise voltage of the 2N4416 can be computed as follows:

$$V_n = \sqrt{4kTBR_n} = \sqrt{4 \times 1.38 \times 10^{-23} \times 293 \times 4 \times 10^8 \times 7313}$$
$$= 218 \ \mu V$$

The results of these two calculations indicate that a theoretical open-circuit voltage of 218 μV, placed in series with the device input (Fig. 4-19), will account for the output noise of the 2N4416. This noise voltage value provides the data necessary to determine required input *signal* levels for a required signal-to-noise ratio. These relationships will be applied to actual circuits in Chapter 5. The procedures require circuit analysis using the real input signal, theoretical noise signal, source and circuit resistances, and the device's input admittance. This section should be reviewed, as necessary, when the related portions of Chapters 5 and 8 are completed.

POWER DISSIPATION

The ideal electronic device is one that will process many watts of power while remaining cool. The laws of physics contradict this ideal as ever being achieved, however. Unfortunately, there is no "free lunch" in electronics. Transistors, IC's, and tubes, because of their inefficiency, do convert some of the signals they process into heat. Even though the efficiency of their operation can be optimized in RF power applications, electronic devices seldom achieve efficiencies above 80%, with 50% being the more typical operating value. As a result, power dissipation techniques must be employed to remove heat from the structure of the devices. In addition to the optimization of bias and electrical characteristics, thermal optimization techniques are also applied to keep RF power devices cool; these techniques include package selection and mounting, heat sinking, and forced air and liquid circulation.

Bipolar transistors are particularly susceptible to thermal problems, because their semiconductor materials have negative temperature coefficients. There is thus a critical temperature/power rating which, if exceeded, will result in *thermal runaway*. This condition results from higher currents heating the transistor, reducing its collector-to-emitter resistance, and allowing yet higher currents to flow, thereby further increasing its temperature and reducing its resistance, until the device destroys itself. To assure that thermal runaway does not occur with bipolar transistors, bias compensation and signal derating techniques must be applied, along with heat sinking, based on the *ambient temperature* (temperature of the surrounding space in the vicinity of the transistor) in which the device is to operate. Heat sinking is required in circumstances where the transistor package structure cannot dissipate heat faster than it can build up within the semiconductor material of the device.

Figure 4-20 shows derating curves and *operating area* curves for the 2N2102 transistor used as an example earlier in this chapter. Note from Fig. 4-20a that the $I_{C(max)}$ value of 1 A at 25°C can be maintained to 5 V_{CE} for conventional operation of the device as an RF amplifier. For pulse operation the $I_{C(max)}$ can be maintained at the 1 A value, at increasing V_{CE} values to V_{CBO}, but breaks at levels that are inversely proportional to the duration of the pulse signal. Figure 4-20b shows the derating curve from the 5 W dissipation, at a 25° case temperature, to the 125° case temperature maximum dissipation of 1 W. This 1 W rating limit is shown as the shaded portion of Fig. 4-20c, superimposed over the operating area curves of Fig. 4-20a. Similar derating techniques can be applied to other bipolar devices, since their general thermal characteristics are similar.

Bias Compensation

There are a variety of techniques for bias compensation of bipolar transistors. Two of the more popular bias arrangements are shown in Fig. 4-20d and e, with each reducing V_{CE}, as I_C (and I_E) increases. You should be able to select appropriate bias resistors to reduce V_{CE} with increasing values of I_C to assure device operation within the "safe" region. Figure 4-20f shows the addition of a thermistor (diodes may also be used) into the bias circuit. This thermal element may be placed in contact with the case of a given transistor or its heat sink, and serves to limit device dissipation. Many IC devices that deploy bipolar transistors in their structure use similar diode temperature compensation to prevent drift, stabilize operation, and protect the higher power-dissipation segments from thermal runaway.

Sample Problem 5

Given: The 2N2102 bipolar transistor, with power ratings as specified in the preceding paragraphs and Fig. 4-20.

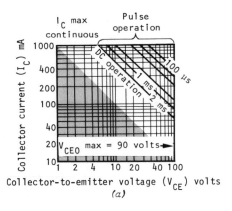

(a)

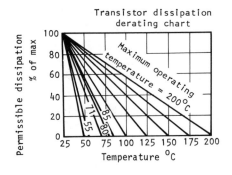

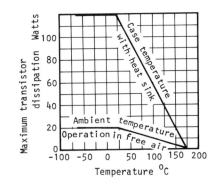

(b)

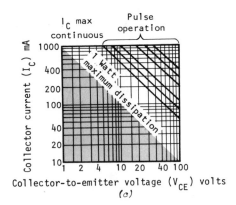

(c)

FIGURE 4-20

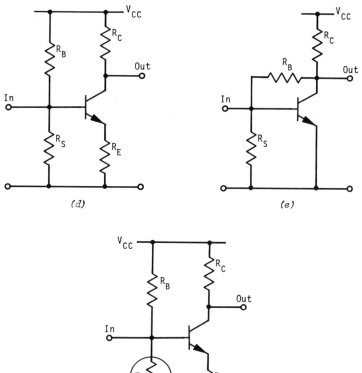

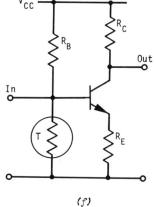

(f)

FIGURE 4-20 (*continued*)

Determine: The appropriate dc bias conditions; select thermistor characteristics that will reduce the power dissipation along the derating curves of Fig. 4-20*b* for a V_{CC} of 25 V dc.

(*a*) The combined values of R_C and R_E are first computed, based on the slope of the "dc operation" curve of Fig. 4-20*a*, as follows:

$$I_C = P_{max}/(V_{CC}/2) = 5 \text{ W}/12.5 \text{ V} = 0.4 \text{ A}$$
$$(R_C + R_E) = (V_{CC}/2)/I_C = 12.5/0.4 = 31.25 \text{ }\Omega$$

With V_{RE} selected for operation at approximately $0.2 \text{ } V_{RC}$,

$$R_C = 5R_E$$

Thus, by substitution:

$$R_T = (5R_E + R_E) = 31.25 \ \Omega$$
$$R_E = 31.25/6 = 5.2 \ \Omega$$
$$R_C = 5R_E = 5 \times 5.2 = 26 \ \Omega$$

(b)   The values of R_B and R_S are selected based on a specified h_{FE} of 80 and a V_{BE} of 1.12 V (from Fig. 4-2) as follows:

$$I_{RS} = I_C/\sqrt{h_{FE}} = 400/\sqrt{80} = 45 \text{ mA}$$
$$V_{RS} = V_{RE} + V_{BE} = (I_E R_E) + V_{BE} = (0.4 \times 5.2) + 1.12 = 3.2 \text{ V}$$
$$R_S = V_{RS}/I_{RS} = 3.2/0.045 = 71 \ \Omega$$
$$I_{RB} = I_{RS} + I_B = I_{RS} + (I_C/h_{FE}) = 45 + (400/80) = 50 \text{ mA}$$
$$V_{RB} = V_{CC} - [(R_C I_C) + V_{RS}] = 25 - [(10.4) + 3.2] = 11.4 \text{ V}$$
$$R_B = V_{RB}/I_{RB} = 11.4/0.05 = 228 \ \Omega$$

(c)   The reduction in collector and base currents required to achieve 1 W power dissipation may be computed, based on the 31.25 Ω dc load resistance, as follows:

$$V_{CC} - I_C(R_C + R_E) = V_{CE} = 25 - 0.4(31.25)$$
$$= 12.5 \text{ V for maximum power transfer, where } V_{CE} = V_{CC}/2$$
$$P_d = V_{CE} I_C = 12.5 \times 0.4 = 5 \text{ W}$$

Substituting for V_{CE} in terms of I_C, the first formula becomes:

$$V_{CC} - I_C(R_C + R_E) = P_d/I_C, \text{ where } V_{CE} = P_d/I_C$$

Transposing the formula to obtain I_C for the 1 W derated value yields the quadratic formula:

$$I_C{}^2(R_C + R_E) - I_C V_{CC} + P_d = 0$$

Thus

$$I_C = \frac{V_{CC} \pm \sqrt{V_{CC}{}^2 - 4(R_C + R_E)P_d}}{2(R_C + R_E)} = \frac{25 \pm \sqrt{25^2 - 4 \times 31.25 \times 1}}{2 \times 31.25}$$
$$= 0.758 \text{ A or } 0.042 \text{ A}$$

(d)   The characteristics of a thermistor, replacing R_S, required to reduce the power dissipation along the derating curves may now be computed as follows, based on a linear reduction of current from 400 mA to 42 mA, as case temperatures vary from 25° to 125°C: (V'_{BE} at $I_C = 42$ mA, as obtained from Fig. 4-2, is 0.7 V.)

$$V'_{RS} = V'_{RE} + V'_{BE} = (I_E R_E) + V_{BE} = (0.042 \times 5.2) + 0.7 = 0.92 \text{ V}$$
$$V'_{RB} = V_{CC} - [(R_C I_C) + V'_{RS}] = 25 - [(26 \times 0.042) + 0.92] = 23 \text{ V}$$
$$R'_S = R_B V'_{RS}/V'_{RB} = 228 \times 0.92/23 = 9 \ \Omega$$

The thermistor (R_S) varies from 71 Ω at 25°C to 9 Ω at 125°C to derate the power dissipation from 5 W to 1 W; so its temperature coefficient is:

$$T_{co} = \Delta R/\Delta T = (71 - 9)/(125 - 25) = 0.62 \text{ Ω/°C}$$

It is evident from the procedures of Sample Problem 5 that, since the bias resistors are fixed, V_{CE} increases from 12.5 to 23.7 V as I_C decreases from 400 to 42 mA, thus obtaining the derated power of 1 W. While this technique obviously decreases device efficiency, it is a worthy tradeoff for protection against thermal runaway.

One of the major advantages of field-effect transistors, including the modern vertical metal-oxide-silicon (VMOS) power devices, is their *positive* temperature coefficient drain-to-source resistance. This property limits drain current as the temperature of the FET increases, thus providing self-protection against thermal runaway. The initial biasing procedures, as outlined in parts (*a*) and (*b*) of Sample Problem 5, are applicable to FET's as well as bipolar devices. The bias compensation techniques of parts (*c*) and (*d*) of the problem, however, are not required for power FET's, thus reducing circuit complexity and providing increased reliability and efficiency over the bipolar transistor. Figure 4-21 shows the operating area curves and bias circuit for the 2N6660 power FET for comparison with the 2N2102 characteristics shown in Fig. 4-20. The thermal properties of vacuum tubes require special treatment, with compensation generally applied to the tube's structure in manufacture. Since tubes are thus specialized, analysis of their power characteristics will not be here reviewed beyond the generalization that their electrical properties are similar to those of the power FET devices.

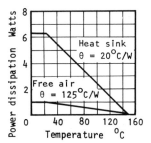

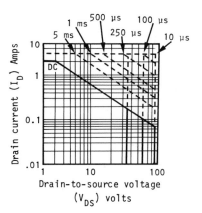

FIGURE 4-21

Thermal Fundamentals

Obviously the highest power and efficiency of active electronic devices are achieved at lower temperatures, so thermal techniques must be applied to the device packages to obtain lower temperature operation. While it is not the purpose of this text to review the physics of thermodynamics, some basic concepts must first be identified if we are to achieve reduced operating temperatures with electronic devices. The basic requirements for heat transfer include a temperature difference (*gradient*) and a path of conduction or convection. The terms *conduction* and *convection* relate to the transfer of heat in solids (and stationary liquids) and in moving air (moving gases or liquids), respectively. For simplicity, basic thermodynamic concepts can be related to their electrical equivalents, as follows:

1. Just as electrical current flows in a conductor from a higher to lower voltage point, so does heat transfer in a conductor (of heat) flow from a higher to a lower temperature point.

2. Just as electrical resistance opposes the flow of current in a conductor, so does *thermal resistance* oppose the flow (transfer) of heat in a conductor (of heat).

3. Just as electrical networks can be analyzed in terms of their series and parallel branches, so too can thermal networks be analyzed as equivalent series and parallel thermal resistance values.

As with electromagnetic radiation, heat can be transmitted and received in the form of radiant energy. Now recall the concepts of radiant heat reflecting off light-colored and optically reflective surfaces while being absorbed by dark, dull, and optically absorptive surfaces. These properties of material surfaces are also important for the temperature reduction of electronic devices. To generalize, then: heat can be removed from an electronic device by conduction, convection (natural or forced), and radiation. The first condition that must be met for such heat removal, however, is that there must be a temperature gradient, high to low, between the electronic device and its surroundings. A transistor obviously cannot be cooled, for example, if its surrounding, *ambient,* temperature is greater than its case temperature.

Thermodynamics is a complex study like other fields of applied science including electricity and electronics. Fortunately, the demand for standards in application of materials and practice to temperature reduction of electronic devices makes information available in a useful form. All that is needed to apply temperature reduction techniques are a few basic relationships and procedures, keeping in mind the basic concepts advanced in the two preceding paragraphs. The first piece of information required is the *thermal resistance* (θ) of a given electronic device in its manufactured pack-

TABLE 4-6
A.   TYPICAL CASE-TO-AMBIENT THERMAL RESISTANCE
FOR POPULAR JEDEC SEMICONDUCTOR PACKAGES

Package/ case	Thermal resistance (θ_{ca})	Package/ case	Thermal resistance (θ_{ca})
TO-18	300	TO-60	70
TO-46	300	TO-66	60
TO-5	150	TO-3	30
TO-39	150	TO-36	25
TO-8	75		

B.   TYPICAL THERMAL RESISTANCE FOR 2N2102 AND
2N6660 TRANSISTORS

Type	Thermal resistance, °C/W	
	Junction-to-case	Junction-to-ambient
2N2102	35	175
2N6660	25	125

age. Table 4-6 provides this information for general bipolar transistors in popular semiconductor packages. The table also lists the specific thermal resistance values, as obtained from manufacturers' specification sheets, for the 2N2102 and 2N6660 devices used in our preceding examples.

Once the specific thermal resistance for a device is known, its temperature rise above ambient can be directly computed. If, for example, the thermal resistance (junction-to-ambient) of the 2N2102 TO-39 package is specified at 220°C/W, the temperature above ambient would be excessive at high power levels. Even with the ambient temperature (T_A) at only 20°C (68°F), the power level (P_{max}) at which 125°C (T_{max}) operation results is high; it can be computed as:

$$P_{max} = (T_{max} - T_A)/\theta = (125 - 20)/220 = 0.48 \text{ W}$$

This example shows that the rated power of an electronic device cannot be obtained with high package thermal resistance values. Even though the 2N2102 is rated at 5 W, it cannot be operated at even 10% of that value without heat sinking.

Heat Sinks

Figure 4-22 and Table 4-7 list some popular heat sink types and their respective thermal resistance values. In applying these heat sinks the device *junction-to-case,* rather than *junction-to-ambient,* thermal resistance specification is required. In the 2N2102 example the junction-to-case thermal resistance is 58°C/W. If this thermal resistance value is placed in series with that

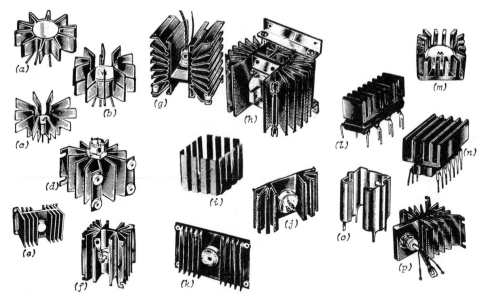

FIGURE 4-22

TABLE 4-7. TYPICAL THERMAL CHARACTERISTICS OF
REPRESENTATIVE SEMICONDUCTOR HEAT SINK TYPES
(Fig. 4-22)

Heat sink type	Thermal resistance, °C/W	Heat sink type	Thermal resistance, °C/W
a	15	I	6
c	10	L	5
e	4	N	7
g	2	P	4

of a large 2°C/W heat sink, as shown in Fig. 4-22g, the total thermal resistance for the device/sink combination would be 60°C/W. With reference to Fig. 4-20, and noting that the 5 W rating for the 2N2102 is specified for a case temperature of 25°C, you should conclude that a most conservative posture must be maintained when reading published device specifications. For this example the junction temperature of the 2N2102 can be computed as:

$T_j = P\theta = 5 \times 58 = 290°C$ above a case T of 25°C

This total junction temperature of 315°C (600°F) is just about hot enough to start a major fire of at least five alarms! Applying this same relationship to the sink, however, yields more realistic values:

$T_c = P\theta = 5 \times 2 = 10°C$ above ambient

For a 25°C case temperature the ambient temperature for the heat sink must be 15°C (59°F), a more realistic value than for the junction. You may react to the implications of this paragraph in favor of the manufacturers' specifications for a given electronic device. However, actual power limitation of a particular device generally requires considerable temperature analysis before the power ratings are accepted at face value.

Sample Problem 6

Given: A 2N3053 bipolar transistor with specifications of 5 W dissipation at a case temperature of 25°C and thermal resistance values of 35°C/W θ_{jc} and 175°C/W θ_{ja}, for junction-to-case and junction-to-ambient, respectively.

Determine: The junction temperature for the specified 5 W dissipation, the case-to-ambient thermal resistance, and the thermal resistance of a heat sink required for a maximum power dissipation of 2 W at a maximum ambient temperature of 140°F.

(a) The maximum junction temperature is first computed based on the specifications of 5 W at 25°C case temperature and a junction-to-case thermal resistance of 35°C/W, as follows:

$$T_{j(max)} = T_{case} + (\theta_{jc}P_{max}) = 25 + (35 \times 5) = 200°C$$

(b) The case-to-ambient thermal resistance is now computed based on the specified values of θ_{jc} and θ_{ja} as follows:

$$\theta_{ca} = \theta_{ja} - \theta_{jc} = 175 - 35 = 140°C/W$$

(c) The case temperature for 2 W dissipation is computed based on the maximum junction temperature as follows:

$$T_{case} = T_{j(max)} - (\theta_{jc}P_d) = 200 - (35 \times 2) = 130°C$$

(d) The ambient temperature is converted to °C and the required thermal resistance of the case and heat sink are computed as follows:

$$T_A(°C) = [T_a(°F) - 32]5/9 = (140 - 32) \times 5/9 = 60°C$$
$$\theta_{ca+sink} = (T_{case} - T_A)/P_d = (130 - 60)/2 = 35°C/W = \theta_{tot}$$

(e) Since the case and heat sink represent a parallel combination of thermal resistance, the heat sink resistance may be computed the same as with parallel resistors, where the total resistance and one resistance value are known, as follows:

$$\theta_{sink} = \theta_{tot}\theta_{ca}/(\theta_{case} - \theta_{tot}) = 35 \times 140/(140 - 35) = 46.7°C/W$$

Figure 4-22 and Table 4-7 identify those heat sinks, appropriate for the TO-5 package of the 2N3053 and with thermal resistance values of 56°C or less, which may serve the application needs of the above sample problem. For this example, requiring only minimum heat sinking, a beryllium-oxide washer mounted below the TO-5 package and smeared with zinc-oxide or silicon paste will serve adequately.

CONCLUSION

This chapter reviewed those properties and characteristics of active electronic devices which are important for their successful application in RF signal processing circuits. Combined with your knowledge of RF components and networks, your working knowledge of the material in this chapter will have prepared you to successfully pursue learning the active RF electronic circuits of Chapter 5 which follows. The emphasis here has been on the reduction of those electrical characteristics of various active devices to a *single* equivalent-circuit model defined with Y parameters. Once this reduction is achieved, the active device can be treated as a simple electrical network terminating or driving a circuit. In addition to simplifying the input and output admittances, the gain of a device is translated to an equivalent admittance value (Y_f), from which frequency and phase response can be easily determined. The noise characteristics, which are so important in the processing of low-level RF signals, were defined and applied to the active devices. These noise relationships are further applied in Chapters 5 and 8. Finally, the power dissipation limitations of the active devices were reviewed and related to the techniques of biasing and heat sinking to obtain optimum higher power operation.

As you continue your study of RF circuits that deploy transistors, FET's, IC's, and tubes, you should refer to this chapter to clarify any electrical equivalent specified for the active devices in those circuits. The following problems will test your understanding of device characteristics, and your ability to interpret them.

Problems

1. Given: A 2N2102 bipolar transistor (see Fig. 4-2 for specifications) biased to operate at a V_{CC} of 12 V dc and an I_C of 25 mA.

Determine:

(a) Dc beta	(d) C_{be}	(g) h_{ib}
(b) V_{BE}	(e) C_{bc}	(h) h_{ob}
(c) f_t	(f) h_{fe}	(i) h_{ie}

2. Given: The measured specifications for a bipolar transistor as follows:

$h_{fe} = 80$	$h_{ob} = 0.02$ mS	$C_{be} = 30$ pF
$h_{ib} = 12\ \Omega$	$h_{rb} = 2 \times 10^{-4}$	$C_{bc} = 8$ pF

Determine: The following common-base equivalent-circuit characteristics (computed in both rectangular and polar forms), as shown in Fig. 4-1b, at an operating frequency of 100 MHz:

(a) Input admittance (Y_{ib})　　(c) Transfer admittance (Y_{fb})
(b) Output admittance (Y_{ob})　　(d) Cutoff frequency (f_{α})

3. Given: The measured specifications for the transistor of Problem 2.

Determine: The following common-emitter equivalent-circuit characteristics (computed in both rectangular and polar forms) as shown in Fig. 4-1b at an operating frequency of 30 MHz:

(a) Input admittance (Y_{ie})　　(c) Transfer admittance (Y_{fe})
(b) Output admittance (Y_{oe})　　(d) Cutoff frequency (f_{β})

4. Given: The specifications for the 2N4416 JFET (Fig. 4-7), biased for a V_{DS} of 15 V dc and a V_{GS} of -3 V dc.

Determine:

(a) The characteristics g_{fs}, g_{os}, C_{gs}, C_{ds}, and C_{gd} as listed in Table 4-2. Then:
(b) Draw and label a common-source (CS) equivalent-circuit diagram (see Fig. 4-1b).
(c) Draw and label a frequency response graph for the FET (similar to Fig. 4-5a), plotting Y_{fs} vs frequency and showing the cutoff frequency (f_g) and gain-bandwidth product (f_t). (Use semilog graph paper for your plot.)

5. The specifications for the 3N140 dual-gate MOSFET (Fig. 4-9).

Determine:

(a) The bias values of gate 2-to-source voltage and gate 1-to-source voltage, which will yield an I_D of 10 mA.
(b) The characteristics g_{fs}, g_{os}, C_{gs}, C_{ds}, and C_{gd} for the dual-gate device. Then:
(c) Draw and label a common-gate (CG) equivalent-circuit diagram (Fig. 4-1b), showing both the rectangular- and polar-form admittance values.

6. Given: The 6CW4 nuvistor triode vacuum tube with general specifications shown in Table 4-3 and Fig. 4-10.

Determine:

(a) The grid-bias voltage (V_{GK}) required to obtain a plate current (I_P) of 10 mA, with a plate voltage (V_P) of 180 V dc. Then:
(b) Graphically determine the transconductance (g_m) of the tube about the 180 V, 10 mA bias point of part a

(c)   Compute the approximate grid-to-cathode capacitance (C_{gk}) at this bias point

(d)   Graphically determine the plate-to-cathode (output) conductance (g_{pk}) about the above bias point

(e)   Estimate the approximate grid-to-plate capacitance (C_{gp}) at this bias point

(f)   Look up and specify the heater voltage and current requirements for the 6CW4 in any convenient reference on vacuum tubes at your disposal.

7.   Given: The CA3053 IC, as connected and graphically characterized in Fig. 4-13.

Determine: Input, output, and transfer admittance characteristics (rectangular and polar form) for operation of this device at 30 MHz and a +9 V dc supply voltage as:

(a)   A differential amplifier

(b)   A cascode amplifier

8.   Given: The requirement to assemble a test jig for the quick (and repetitive) testing of transistor input admittance, using the impedance bridge technique shown in Fig. 4-16.

Prepare:

(a)   A schematic diagram of the test circuit with the required bias connections for a V_{CC} of 6 V dc and an I_B of 10 μA

(b)   A detailed step-by-step procedure to be followed by a less-qualified "test person" in obtaining the necessary data

9.   Given: A transistor specified as having an input admittance of 0.1 mS, with a noise figure of 5 dB.

Determine:

(a)   The equivalent noise resistance of the device

(b)   The theoretical open-circuit noise voltage in series with the signal at the device input

(c)   The required level of input signal from a 600 Ω signal source to obtain a 40 dB (100 : 1 voltage ratio) signal-to-noise ratio at the device output

10. Given: The 2N2102 bipolar transistor (see Fig. 4-2) operating with a V_{CC} of 20 V dc, and h_{FE} (dc β) of 36, and limited to a 2 W (25°C case temperature) maximum power dissipation. (The transistor is connected as in Fig. 4-20.)

Determine:
- (*a*)   Collector current (I_C)
- (*b*)   Value of collector resistor (R_C) for $V_{RC} = 0.5 \ V_{CC}$
- (*c*)   Value of emitter resistor (R_E) for $V_{RE} = 0.1 \ V_{RC}$
- (*d*)   Value of the swamping resistor (R_S)
- (*e*)   Value of the base resistor (R_B)
- (*f*)   The maximum power dissipation with an ambient temperature of 40°C
- (*g*)   The reduction in collector and base currents to meet the limits of 40°C operation (see *f* above)
- (*h*)   The new value of R_S required to obtain reduced power operation (see *f* and *g* above)

11. Given: The 2N4416 transistor with specifications including 2 W dissipation at a case temperature of 100°C, a junction-to-case thermal resistance (θ_{jc}) of 50°C/W, and a TO-39 package with a thermal resistance (θ_{ca}) of 150°C/W.

Determine, for a maximum ambient temperature of 150°F:
- (*a*)   The junction temperature for the specified 2 W rating
- (*b*)   The case temperature at 2 W dissipation (if the device doesn't melt)
- (*c*)   The required thermal resistance rating of a heat sink as required to reduce the case temperature of the transistor to 100°C

CHAPTER 5

ACTIVE RF CIRCUITS

This chapter concludes Part I of this book; it combines passive networks with active electronic devices to achieve amplification, generation, and processing of RF signals. The major emphasis of the chapter is RF amplification, since most other electronic functions—in particular, circuit configurations—are based on the amplifying properties of electronic devices. Emphasis is also placed on the variety of oscillator circuits used to generate fixed- and variable-frequency RF signals. Many special variations of amplifier and oscillator circuits are widely used in communications systems to obtain necessary signal processing functions. The more popular of these special circuits will be reviewed in this chapter, together with techniques that should allow you to apply successful analysis to other circuits not reviewed in the text.

RF amplification circuits typically fall into two categories: narrowband *tuned amplifiers* and broadband *untuned amplifiers*. In a concise sense, untuned amplifiers use electrical networks in combination with transistors, IC's, and vacuum tubes to *extend* the frequency response of a circuit; tuned amplifiers use the networks with solid-state and vacuum-tube devices to *limit* the frequency response (bandpass) of a circuit. There are many reciprocal tradeoffs with the two functions, including *gain* for *bandwidth, noise* for *bandwidth,* and *power efficiency* for *bandwidth*. One significant power efficiency vs bandwidth tradeoff in favor of tuned amplifiers allows for nonlinear operation of the active amplifying device. This *class* of operation is specified "B" or "C," compared to the class "A" operation specified for linear amplifier response. Under these classes of operation the active device needs to conduct (be pulsed) for only a portion of each cycle of the applied RF signal, with the *LC* tuned network maintaining a sinusoidal waveform, thus achieving higher efficiency than the linear amplifier circuits. The content of this chapter will include tuned and untuned linear amplifiers, as well as the special classes of tuned nonlinear amplifiers.

RF oscillator circuits are, by definition, tuned circuits because of their single-frequency operation. The oscillator circuits also fall into two categories, like the RF amplifiers, and can be configured for linear (class A) or nonlinear (class B or C) operation. The two *categories* of oscillator operation are defined by the circuit characteristics that provide feedback to achieve oscillation: *RC* and *LC*. Resistance-capacitance (*RC*) oscillator circuits are usually limited to operation at frequencies below 20 MHz, and are analogous to the linear video amplifiers in function. Inductance-capacitance (*LC*) oscillator circuits provide lower harmonic distortion (more precise sine waveform); and, while they are sometimes practical for operation below 1 MHz, they are best suited for generating higher power and/or higher frequency RF signals. Power oscillator circuits operating in class B or C nonlinear conduction require tuned *LC* networks to maintain sinusoidal waveforms at their outputs. These circuits, along with the *LC* linear oscillators, are analogous in function to the *tuned* amplifier circuits. Although multivibrators and other switching circuits may be categorized as a type of oscillator, they provide nonlinear output waveforms. The operation and application of such switching-circuit oscillators are more appropriately treated in textbooks on pulse circuits—not communications circuits—so those circuits are not covered in this text.

Nonlinear operation is required of communications circuits used to create harmonic or intermodulation distortion products. Such circuits are applied in systems as mixers, modulators, demodulators, harmonic generators, frequency multipliers, and other specialized functions. In communications applications such nonlinear RF circuits often use *LC* tuned networks at their inputs and outputs to restore a sinusoidal waveform to the non-sine-wave distortion signals. These circuits are somewhat related to the class B and C tuned amplifier configurations used for improving the power efficiency of RF power amplifier circuits, because the electronic devices in the circuits do not conduct equally during both the positive and negative halves of the applied RF signal waveforms. While it is impossible to review and identify all the varied specialized linear and nonlinear, *RC* and *LC,* RF circuits in this chapter, the concepts and approach to analysis presented here should equip you to understand most of the popular circuits found in RF systems.

Many of the calculations employed in this chapter are based on techniques for computation developed in Chapters 2, 3, and 4. It is therefore recommended that you look back at those chapters, as required, to review the procedures for network and device analysis and characterization.

LEARNING OBJECTIVES

Upon completing this chapter, including the problem set, you should be able to:

1. Apply the impedance-matching techniques of Chapter 3 to the inputs and outputs of the active electronic devices of Chapter 4 to obtain optimum conditions for amplification.

2. Deploy active devices with passive networks, and compute the circuit characteristics necessary to achieve wideband *video* and linear RF amplifier circuits.

3. Compute the circuit characteristics necessary to achieve *tuned,* narrowband linear RF amplifier circuits.

4. Combine the characteristics of active/passive circuits to achieve high-gain, multistage amplifier circuits, and compute the interrelated effects of multistage operation.

5. Compute the circuit characteristics necessary to achieve tuned class B and C (nonlinear) RF power amplifier circuits.

6. Understand and apply circuit conditions required to achieve stable oscillation using active devices and passive networks.

7. Compute the circuit characteristics necessary to achieve popular *RC* oscillator circuits with given frequency and output specifications.

8. Compute the circuit characteristics necessary to achieve popular *LC* oscillator circuits with given frequency and output specifications.

9. Understand the operation of piezoelectric crystals and elements and their application to RF oscillators, filters, and frequency-selective networks.

IMPEDANCE MATCHING

Based on those skills gained in Chapter 3, you should be able to apply electrical networks to the inputs and outputs of active electronic devices in order to achieve a desired transfer of power into and out of the devices. With a thorough understanding of Chapter 4 and with access to published or measured device characteristics, you should also be able to define the specific input and output admittance of active devices. Impedance matching of active devices thus simply requires determining the values of passive input and output networks. The character of these networks depends, of course, on the device characteristics; but it depends also on the kind of amplification required. For example, highest input-to-output voltage gain of a device is achieved when the input impedance is low and the output impedance is high. Highest current gain, conversely, is achieved when the input impedance is high and the output impedance is low. Highest power gain is achieved when input and output impedances match those of the source and load of a circuit. The choice of circuit configuration and impedance match, therefore, depends on the function to be served by a given amplifier circuit.

Amplifier Application

The steps that must precede impedance matching, then, include (1) definition of the function to be served by an electronic device, (2) selection of a specific device and specification of its circuit configuration, and (3) specification of its Y parameters. Figure 5-1a and b shows the equivalent-circuit parameters and circuit configuration of a 2N2102 bipolar transistor operated as a common-emitter *linear* (not class B or C) power amplifier. Optimum performance for this device will result when its input is driven with a generator admittance equal to G_{ie}, and its output drives a load admittance equal to G_{oe}. Since Y_{ie} and Y_{oe} for this device are complex, the input and output networks must be the conjugate values of these device parameters. For this example, then, the driving admittance must be 1.6 $-j1.31$ mS, and the load admittance must be 0.04 $-j0.231$ mS for operation at 4 MHz.

Figure 5-1c and d shows the equivalent-circuit parameters and circuit configuration of the 2N2102 operated as a common-base linear voltage amplifier. With the emitter and collector currents almost identical, the voltage gain of this configuration is approximately proportional to the input and output admittances. You should recall, however, that this ideal can only be achieved with a zero-impedance source and infinite impedance load, both of which are impractical and would render the circuit useless. So, in practice, the "ideal" voltage gain is actually reduced by generator limitations and load requirements. For this example the input admittance at 4 MHz is 0.13 $+j0.0016$ S, indicating an insignificantly small capacitance value across an equivalent input resistance of only 7.7 Ω. The output admittance at 4 MHz is 0.02 $+j0.25$ mS, however, indicating a significant capacitance value.

There would be little value in deploying this CB circuit for high voltage gain if, in the process, its low input impedance drops the voltage level of its driving signal. Driving the 7.7 Ω input directly with a 10 mV, 50 Ω signal generator, for example, would yield a loss in signal voltage with much of the signal power dissipated in the generator impedance. The application of a simple transformer, in this example, will yield a lower driving voltage, but without power loss due to dissipation in the generator impedance. On the output side of the CB circuit the capacitance may be canceled with an inductive susceptance of $-j0.25$ mS at 4 MHz; but any load admittance will reduce the potential gain of the circuit. At this point in the circuit a high-Q resonant network with transformer output coupling will provide the required low load admittance to assure optimum·voltage gain. Careful selection of component values may include the internal C_{ob} as a part of the output resonant network. Figure 5-1e shows the driving and output circuitry for this CB example with the 2N2102.

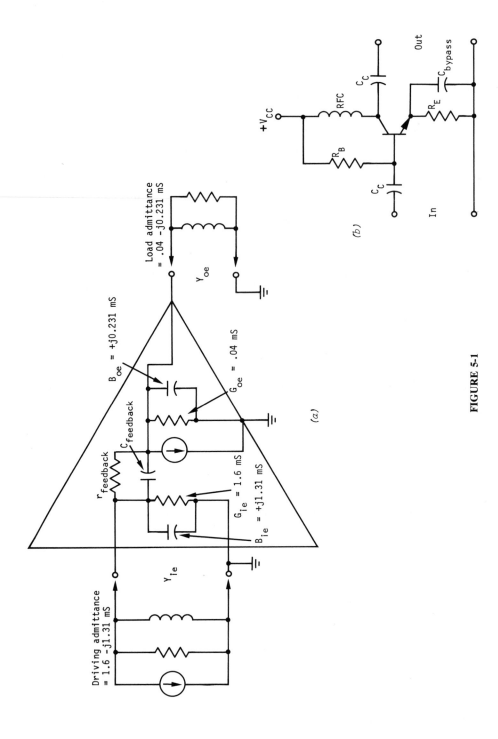

FIGURE 5-1

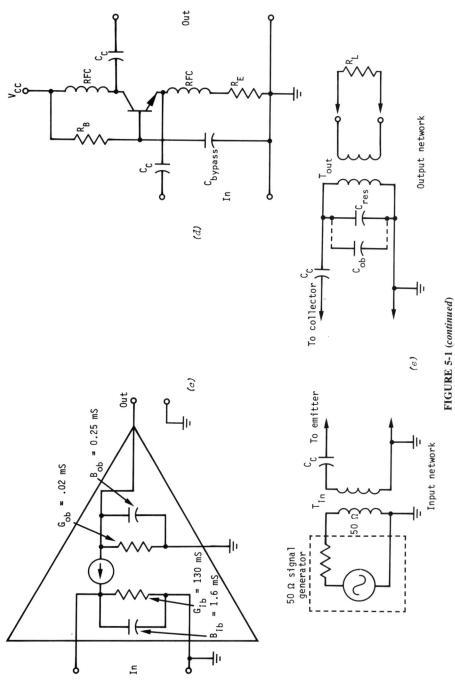

(d)

(c)

(e)

FIGURE 5-1 (continued)

Systems Requirements

There are many choices, variables, and tradeoffs in the selection of components external to an electronic device. The procedures of engineering practice generally begin with a given *system* requirement, with devices and circuits selected, if available, to meet that requirement. With the availability of modern discrete and IC devices, this procedure is not too much of a chore. In fact, most semiconductor manufacturers maintain ongoing programs for the preparation of applications notes on their products, with circuits and components identified to meet specific systems needs. It is not implied that engineering practice merely involves "copying," but rather that it involves "exploiting" all the information resources available to the technologist— there is no point in reinventing the wheel every time you need to achieve an electronic circuit function. Your capability with these devices and circuits will allow you to *adapt* designs and applications information to specific systems needs, which, in turn, will yield reliable electronic products. There is certainly a place for the research aspects of electronics in the technology, but it is the purpose of this book to emphasize the *practical applications* aspects of the technology.

Sample Problem 1

Given: A systems requirement for the single-stage tuned RF voltage amplifier of Fig. 5-2, with the 2N4416 JFET from Sample Problem 4 (Chapter 4) specified as the active device.

Determine: The input and output resonant-circuit requirements for operation at 10 MHz.

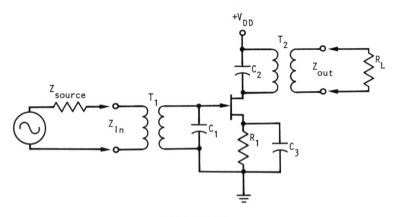

FIGURE 5-2

(a) After examination of the input parameters for the 2N4416, the minimum value of resonant-circuit capacitance is determined to be 25 pF, based on stray and device shunting (parallel) capacitances of about 2.5 pF. The capacitor selected for the resonant circuit is appropriately a 5 to 30 pF ceramic trimmer with a high Q of 500+. Its value will be adjusted to provide a total of 25 pF capacitance, including stray and device capacitance values.

(b) The inductance value of the resonant secondary coil is now computed, based on the 25 pF capacitance value, as follows:

$$L = 1/\omega^2 C = 1/[(6.28 \times 10^7)^2 \times 2.5 \times 10^{-11}] = 10 \ \mu H$$

(c) Assuming a practical Q value for the coil to be at least 40, the approximate minimum parallel tank impedance and admittance for the input transformer secondary can be computed as:

$$X = \omega L = 6.28 \times 10^7 \times 10^{-5} = 628 \ \Omega$$
$$Z_{par} = QX = 40 \times 628 = 25 \ k\Omega$$
$$Y = 1/Z_{par} = 1/(25 \times 10^3) = 40 \ \mu S$$

(d) The output tank impedance is computed based on optimum gain being achieved with minimum loading with highest Q (not considering bandwidth), and with a transformer deployed to achieve minimum loss in reducing the output impedance to a lower value to drive a following circuit. As with the input tank, the minimum capacitance value is determined to obtain the highest load impedance (minimum load admittance). Based on the output and stray capacitance values of less than 1 pF, a 10 pF capacitance value is selected in the form of a 3 to 15 pF ceramic trimmer with a high Q of 500+.

(e) The inductance value of the resonant primary coil and the parallel tank impedance and admittance are now computed, based on the 10 pF capacitance value and practical coil Q of 40, as:

$$L = 1/\omega^2 C = 1/[(6.28 \times 10^7)^2 \times 10^{-11}] = 25 \ \mu H$$
$$X = \omega L = 6.28 \times 10^7 \times 25 \times 10^{-6} = 1570 \ \Omega$$
$$Z_{par} = QX = 40 \times 1570 = 62.8 \ k\Omega$$
$$Y = 1/Z_{par} = 1/62.8 \times 10^3 = 15.9 \ \mu S$$

You may conclude from the results of these calculations that the low input and output admittances of an active device will allow for the highest practical voltage to be developed across a resonant ("tank") network without loading, thus achieving optimum conditions for voltage amplification. In practice, the device's input admittance is usually smaller than that of the resonant input network; whereas the device's output admittance usually exceeds that of the resonant network. In this example, impedance matching of the signal generator and load is achieved with transformer coupling. Other matching techniques are possible, however, including tapping, or application of the L, T, and π networks described in Chapter 3. To provide some indication of voltage-gain performance, Sample Problem 2 extends the above example to a second phase of calculations.

Sample Problem 2

Given: The above 2N4416 RF voltage amplifier, with its input transformer coupled to a 70 Ω antenna/transmission line as a signal source, and its output coupled to a 1 kΩ load.

Determine: The output voltage across the load, and the overall gain if the antenna signal is 100 μV and the transformer coupling coefficients are each 0.5.

(a) The reactance values of the transformer input and output windings are selected to match the signal source and load impedances, respectively, and the transformer characteristics are computed as follows:

$$X_{\text{pri}}T_1 = 70 \ \Omega; \ X_{\text{sec}}T_1 = 628 \ \Omega$$
$$X_m T_1 = k\sqrt{X_{\text{pri}}X_{\text{sec}}} = 0.5 \sqrt{70 \times 628} = 105 \ \Omega$$
$$X_{\text{pri}}T_2 = 1590 \ \Omega; \ X_{\text{sec}}T_2 = 1000 \ \Omega$$
$$X_m T_2 = k \sqrt{X_{\text{pri}}X_{\text{sec}}} = 0.5 \sqrt{1590 \times 1000} = 630 \ \Omega$$

(b) The input voltage to the gate of the 2N4416 is computed as the voltage across the resonant network, based on the mutual reactance of T_1 and the network Q, as follows:

$$V_{\text{gate}} = QX_m i_{\text{pri}} = QX_m V_{\text{pri}}/X_{\text{pri}} = 40 \times 105 \times 10^{-4}/70 = 6 \ \text{mV}$$

(c) The output admittance of the 2N4416 in parallel with the output resonant network is now computed based on a Y_{os} of 100 $-j25 \ \mu$S, with $-jB_{os}$ included in the capacitance of the resonant network; then the voltage gain of the JFET is computed based on this admittance value:

$$Y_{\text{out}} = Y_{\text{network}} + Y_{os} = 15.7 + 100 \ \mu\text{S} = 115.7 \ \mu\text{S}$$
$$A_v = Y_{fs}/Y_{\text{out}} = 2000/115.7 = 17.3$$
$$V_{\text{drain}} = V_{\text{out}} = A_v V_{\text{in}} = A_v V_{\text{gate}} = 17.3 \times 6 \ \text{mV} = 104 \ \text{mV}$$

(d) The load voltage is computed, based on the secondary reactance of T_2 being equal to R_L, yielding a -3 dB (45° phase shift) signal across the output with respect to the drain signal:

$$V_{\text{sec}} = X_m i_{\text{pri}} = X_m V_{\text{drain}}/X_{\text{pri}} = 630 \times 0.104/1590 = 41 \ \text{mV}$$
$$V_{\text{load}} = 0.707 \times 41 = 29 \ \text{mV}$$

(e) Finally, the overall gain is computed, based on the input and output impedances, as:

$$A_v = V_{load}/V_{ant} = 41/0.1 \text{ mV} = 410$$
$$A_p = (V_{load}^2/Z_{load})/(V_{ant}^2/Z_{ant}) = (0.041^2/1000)/(0.0001^2/70) = 11{,}800$$
$$= 10 \log 11{,}800 = 40.7 \text{ dB}$$

Your review of these calculations should disclose the value of impedance-matching and resonant-network techniques in achieving optimum active circuit amplification. The optimum gain of the JFET alone is only:

$$A_{v(opt)} = Y_{fs}/Y_{os} = 2000/100 = 20$$

yet the overall circuit voltage gain is 410 with the appropriate drive and load impedance matching and with device capacitance losses eliminated. In addition to conveying the value of impedance-matching and resonant-network applications, the above problems will serve as a procedural guide for performance analysis of tuned RF voltage amplifier circuits.

FEEDBACK IMPEDANCES AND SIGNALS

In Chapter 4 you studied the capacitance values between terminals of active electronic devices. You may recall that in the common-emitter, common-source, and common-cathode configurations of bipolars, FET's, and tubes, respectively, the feedback capacitance (C_{bc}, C_{gd}, and C_{gp}) significantly influenced the input admittance of the devices. There is, unfortunately, an additional factor that further increases this feedback capacitance; it is called *Miller effect* after its discoverer. You may also recall from your earlier studies that Miller's effect yields a net increase in the value of a capacitance if the difference in signal voltage across its plates is increased due to amplification. This Miller effect is significant in bipolars, FET's, and tubes, but is insignificant in most IC's because of their greater output-to-input isolation.

Miller Effect

Using the 2N4416 JFET of our preceding sample problems as an example, the device value of C_{gd} is specified as 0.6 pF. The device is operated in an amplifier circuit with a (device) voltage gain (A_v) of 17.3; so the signal voltage appears at the drain, out of phase (by 180°) with the gate signal and 17.3 times greater in amplitude. The Miller effect thus defines the feedback through the capacitor as though it were 17.3 times its C_{gd} value, because the extent of feedback voltage increases by that gain factor. The total *effective* feedback capacitance is the sum of C_{gd} and A_v times C_{gd}. This effective feedback capacitance and the resulting total input capacitance of the device can now be recomputed as:

$$C_{\text{feedback}} = C_{gd} + A_v C_{gd} = 0.6 + (17.3 \times 0.6) = 11 \text{ pF}$$
$$C_{is} = C_{\text{feedback}} + C_{gs} = 11 + 1.1 = 12.1 \text{ pF}$$

Referring to the preceding sample problems, you should have observed that the choice of the 5 to 30 pF ceramic trimmer provides the "bailout" from the effects of increased Miller capacitance. The trimmer can be adjusted to include the Miller effect in the total 25 pF of the resonant network—about 12 pF in the device input and 13 pF in the trimmer.

Miller capacitance is basically an unwanted, "nuisance" feedback effect which must be considered in determining circuit values and performance in RF amplifiers. Other impedance and signal feedback effects, however, are desirable in order to improve performance of RF amplifiers and other circuit applications. One of the more important of these, called *neutralization,* relates to the Miller effect *and* device feedback capacitances.

The calculation for Miller capacitance reviewed in the preceding paragraphs assumed the drain (output) circuit to be at resonance, and thus resistive. Since it is an LC network, however, there are frequencies where the network is capacitive or inductive as well. At lower frequencies, below resonance, where the LC output network is capacitive, it will reflect back an out-of-phase resistive component (X/Q) of the output signal that will load the input admittance of the circuit, thus favorably reducing its off-frequency gain. At higher frequencies where the LC network is inductive, however, it will reflect back an in-phase resistive component of the output signal that will enhance the off-frequency gain of the circuit. This equivalent capacitive feedback may be of sufficient magnitude in high-gain amplifiers to cause them to oscillate. Since the Miller capacitance is a function of gain, such unwanted oscillation will surely result with devices with high values of C_{bc}, C_{gd}, or C_{gp} for bipolars, FET's, and tubes, respectively.

A modification of the Miller formula that may be used to determine the total feedback capacitance, taking into account varied load and gain characteristics, is here applicable for bipolars, FET's, and tubes as:

$$C_{\text{feedback}} = C_{bc} + A_v C_{bc} \cos \theta = C_{gd} + A_v C_{gd} \cos \theta$$
$$= C_{gp} + A_v C_{gp}$$

θ is the phase angle of the output load impedance. Examination of the formula indicates that maximum feedback capcitance will occur when θ is $0°$, but that oscillation will result when θ approaches $+90°$ (given that A_v has $180°$ phase shift, often expressed with a minus sign).

Neutralization

Neutralization techniques are deployed to prevent oscillation and optimize circuit performance by cancellation of the feedback signals identified in the preceding paragraph. This cancellation is achieved most often by applying an additional feedback signal that is equal in magnitude but out of phase with the unwanted feedback signal. Figure 5-3 shows various techniques used to obtain neutralization feedback in a number of popular RF amplifier circuits. The figure shows bipolar transistor circuits, as these devices have the greatest capacitance (C_{bc}) requiring neutralization. Figure 5-3a and b represents the more popular of the neutralization circuits, with out-of-phase feedback signals obtained from the center-tapped primary or high side of the RF output transformer, respectively. The value of the neutralizing capacitor for the examples of Fig. 5-3a should be approximately that of the feedback capacitance; those capacitor values for the other circuits of Fig. 5-3 will vary in inverse proportion to the magnitude of the out-of-phase signals at the feedback point.

You are probably familiar with the feedback techniques used to stabilize lower frequency *RC* amplifier gain through your earlier studies of electronic circuits. Similar feedback techniques are applied to both discrete and IC, *RC* and *LC*, amplifier circuits to extend or alter their frequency response in RF applications. The concepts of device *cutoff* frequency resulting from the shunting effects of device capacitances were reviewed in Chapter 4. When electronic devices are placed into circuits, these shunting effects are combined with those of the circuit components and mounting structure (package) to further reduce this cutoff frequency. The shunt capacitances can be effectively canceled if they are included as a part of the resonant *LC* tank networks in narrowband amplifiers, as was done in the previous example of the 2N4416 sample problems. The shunt capacitances cannot be so easily canceled for wide frequency-response amplifiers, however; so rather than attempt cancellation, gain compensation is applied to the circuits to reduce capacitive losses. This gain compensation generally requires the application of an inductor which will raise the load impedance of an amplifier circuit at frequencies above cutoff. Figure 5-4 shows some popular frequency compensation networks as applied to *RC* and *LC* amplifiers to extend their high frequency response. The related graphs in the figure show how series, shunt, and combination networks can extend higher frequency circuit gain almost to the devices' f_t frequencies.

WIDEBAND AND VIDEO AMPLIFIERS

Through your study of the procedures of the previous sample problems, related to Fig. 5-2, you should understand the basic techniques required for the analysis of transformer-coupled, *LC* tuned RF amplifier circuits. In this portion of the chapter we will review the analysis for wideband *RC* and *LC* impedance-coupled amplifier circuits. The term *video* amplifier originated from the early development of radar and television circuitry. This type of amplifier usually operates from dc to some high-frequency limit required by a

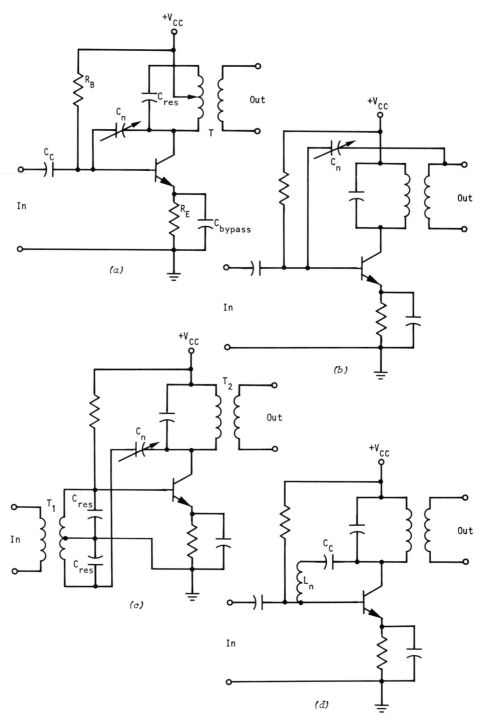

FIGURE 5-3

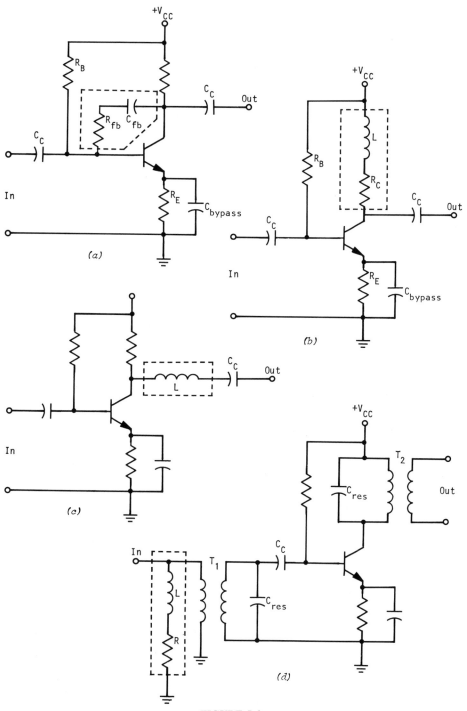

FIGURE 5-4

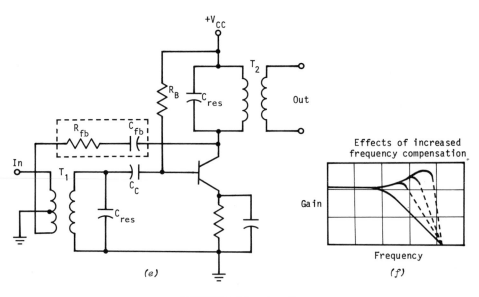

FIGURE 5-4 (*continued*)

given system; and it thus employs direct-coupled *RC* circuitry with frequency compensation. Excellent examples of wideband video amplifier circuitry can be found in oscilloscope *vertical* amplifiers, some of which provide uniform gain from dc to 50 MHz and above. Wideband *RF* amplifiers are typically low-frequency limited, with their high-frequency operation extending far beyond the comparable video amplifiers. Typical quality broadband RF amplifiers may operate over two or three decades (100 kHz to 100 MHz, or 1 MHz to 1 GHz, for example) with uniform gain. As such, these amplifier circuits are composed of *R*, *L*, and *C* components, are usually impedance rather than transformer coupled, and employ extensive frequency compensation elements.

The integrated circuit is perhaps the most appropriate choice for deployment in video and wideband RF amplifier circuits. The IC has increased input-to-output isolation, small stray capacitance and inductance, and stable gain—properties that are most impractical to obtain with multistage discrete device circuits. Figure 5-5a and b shows diagrams of discrete and IC video amplifier circuits, respectively. You should observe that both circuits have resistive loads as their operation must extend down to dc. The function of the inductors in series with the load resistances, and resistance-shunting capacitors, is to provide the frequency compensation described in the previous section of this chapter. The tradeoff in video amplifier performance is simply that gain varies directly with the value of R_L; whereas the -3 dB cutoff frequency varies inversely with the value of R_L for a given device capacitance. The practical extent of frequency compensation will thus ultimately influence the choice of R_L and the resulting circuit gain.

We may consider the bipolar transistor of Fig. 5-5a as an example, using the specifications from Sample Problem 1 in Chapter 4 for device characteristics. With an exceptionally high Y_{fe} of 0.13 S, a Y_{oe} of 0.02 mS for the 2N2102, and an estimated total device, stray, and mounting capacitance of 30 pF, the cutoff frequency and low-frequency gain of the circuit can be easily computed for a 5 kΩ load resistor as follows:

$$f_{co} = 1/2\pi CX = 1/2\pi CR = 1/(6.28 \times 3 \times 10^{-11} \times 5 \times 10^3) = 1.06 \text{ MHz}$$

At low frequencies,

$$Y_{out} = Y_{oe} + 1/R_L = 0.02 + 1/(5 \times 10^3) = 0.22 \text{ mS}$$
$$A_v = Y_{fe}/Y_{out} = 130 \text{ mS}/0.22 \text{ mS} = 591$$

With a cutoff frequency of only 1.06 MHz the circuit can hardly qualify as a wideband video amplifier. One alternative, of course, is the extension of frequency response by reducing the value of R_L. In this example, reducing the value of R_L by a factor of 20, to 250 Ω, will still yield a respectable gain of nearly 30 ($591/20 = 29.5$), with an extended cutoff frequency of over 21 MHz ($1.06 \times 20 = 21.2$ MHz). While the application of frequency compensation to both circuits will increase their response, the widest frequency response will result from lower values of R_L, as shown in the response curves of Fig. 5-5c.

Frequency Compensation

Figure 5-5d graphs the frequency-compensation response of the circuit for various values of inductance. The "flat" response results from the *RLC* combination that yields what is called *optimum damping,* where the circuit is neither resonant nor degenerative. This flat-response inductance value may be computed for given load resistance and shunt capacitance values as:

$$L = 0.36R^2C$$

L is the value of compensation inductance for flat response, C is the total shunt capacitance of the circuit, and R is the output load resistance of the circuit. This value of L is based on a resonant-circuit Q of 0.6 for the R, C, and L network, as required for optimum damping, with its derivation obtained from the resonance formula, as follows:

$$Q = 0.6 = X/R; \qquad R = X/0.6 = 1/(0.6\omega C)$$

Transposing,

$$\omega = 1/(0.6RC); \text{ also } \omega = 1/\sqrt{LC}$$

By substitution

$$\omega^2 = 1/LC = 1/(0.6RC)$$

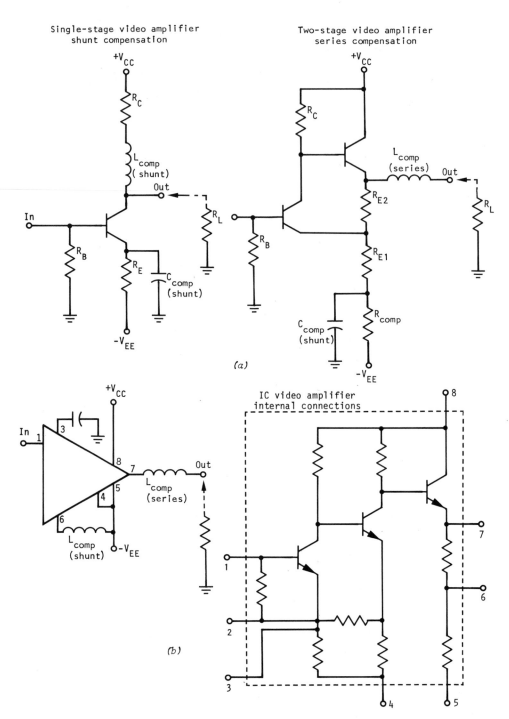

FIGURE 5-5

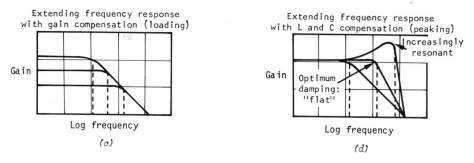

FIGURE 5-5 (*continued*)

Thus

$$LC = 0.6^2R^2C^2 \quad \text{and} \quad L = 0.36R^2C$$

If we apply this relationship to the examples of the 2N2102 of the preceding paragraph, we can compute the values of inductance for flat-response frequency compensation as follows:

For the R_L of 5 kΩ,

$$L = 0.36R^2C = 0.36(1/Y_{\text{out}})^2C = 0.36(1/G_{\text{out}})^2C$$
$$= 0.36 \times (1/2.2 \times 10^{-4})^2 \times 3 \times 10^{-11} = 223 \ \mu\text{H}$$

For the R_L of 250 Ω,

$$L = 0.36R^2C = 0.36 \times (250)^2 \times 3 \times 10^{-11} = 675 \text{ nH}$$

Some implied conditions in the procedures of the above problems may require clarification if you are confused by the use of Y_{out}, G_{out}, and R_L. If you recall from the previous chapter that the 2N2102 device output admittance is specified as $Y_{oe} = G_{oe} + jB_{oe}$. The susceptance part of this output has been combined with the stray and mounting capacitances to make up the total of 30 pF in this problem example, with the device output capacitance accounting for about 10 pF of the total. The gain and frequency-compensation calculations must take this total capacitance into account in order to determine f_{co}; but only the *real* part of Y_{out} is required to determine the low-frequency gain, where the capacitance is, by definition, insignificant. For this reason the Y_{out} and Y_{fe} values in the general formulas considered only the real part of the admittances, not the capacitive parts. Finally, the frequency-compensation formulas considered $Y_{oe} + 1/R_L$ for Y_{out} when the real part of the Y_{oe} was significant with respect to the load resistor. In the case of the 250 Ω load, however, the device admittance was insignificant, so the R_L value of the circuit was used directly. As you gain experience with the calculations of engineering practice, you will gain the sense of magnitudes and significance that this book has attempted to convey.

Now to proceed with the frequency-compensation problem, there yet remains to determine the newly compensated frequency response and its -3 dB cutoff frequency. As you have already concluded, the frequency-compensation inductor forms a resonant network ($Q = 0.6$) at the output of the amplifier circuit. The resonant frequency can thus be computed based on the L and C values of this output network, then on its -3 dB cutoff frequency obtained from the resonant value. You may recall from Chapter 2 that the upper -3 dB frequency of an LC network is $f_o + f_o/2Q$. Combining these relationships, the resonant frequency affected by frequency compensation and the resulting upper cutoff frequency can be determined as:

$$f_R = 1/2\pi \sqrt{LC} = \begin{cases} 1/(6.28 \sqrt{2.23 \times 10^{-4} \times 3 \times 10^{-11}}) = 1.92 \text{ MHz} \\ 1/(6.28 \sqrt{6.75 \times 10^{-7} \times 3 \times 10^{-11}}) = 35.4 \text{ MHz} \end{cases}$$

$$f'_{co} = f_R + f_R/2Q = \begin{cases} 1.92 + 1.92/(2 \times 0.6) = 3.51 \text{ MHz} & \text{for 5 k}\Omega \text{ load} \\ 35.4 + 35.4/(2 \times 0.6) = 64.9 \text{ MHz} & \text{for 250 } \Omega \text{ load} \end{cases}$$

To summarize your examination of the calculations of the preceding paragraphs, you should have concluded that the benefits of frequency compensation include the extension of the -3 dB f_{co} by a factor of at least $3:1$. This extension in frequency, of course, depends upon the operation of an active electronic device below its gain-bandwidth product (f_t).

Wideband Amplifiers

The concepts of frequency compensation of gain, as applied to video amplifiers, can also be applied to wideband transformer-coupled RF amplifier circuits. As noted earlier, these ac coupled wideband circuits may operate over two decades of RF, often with bandwidths in the range of 100 to 500 MHz. Figure 5-6 shows series, shunt, and feedback frequency-compensation networks as used with transformer-coupled RF amplifier circuits. Like the video amplifiers, the wideband RF amplifier circuits must maintain a low Q in their reactive networks to eliminate the possibility of resonant peaks. The low-Q values are assured with the use of lossy ferrite transformer and inductor cores, and by resistive loading of the input and output networks.

The obvious tradeoff in comparing broadband vs tuned amplifiers is that of bandwidth for gain. You may recall that the tuned, transformer-coupled RF amplifier circuit of Fig. 5-2 yields an overall voltage gain of 410, with an optimum device gain of only 20. This high gain value of 410 is thus a function of the high Q (minimum loading) in the tuned networks of the circuit. If the same circuit were hypothetically deployed with network Q's of 0.6, as is

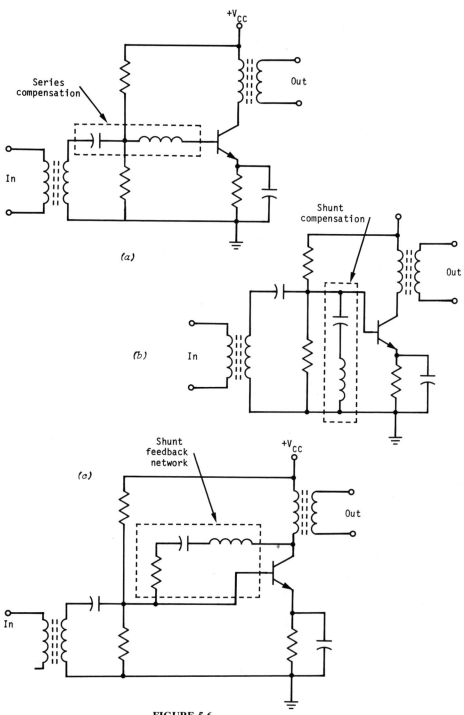

FIGURE 5-6

optimum for the video amplifier, for example, the resulting wideband circuit gain would be *less* than that of the device alone. While this is an over-simplified relationship, you should appreciate that the extent to which bandwidth is increased, for a given device and circuit, is a function of the extent to which gain is reduced.

Transformer Coupling

The broadband coupling transformer is perhaps the most important component affecting successful wideband RF amplifier circuits. These transformers are generally wound on ferrite cores of varying shapes and sizes, assuring close magnetic coupling, high permeability, and low Q. A special category of these transformers is called *broadband transmission-line transformers*. The transmission-line transformer has special properties that make it most appropriate for frequency compensation of the wideband RF amplifier: it acts like a conventional transformer at low radio frequencies, but like a transmission line at the higher frequencies. By proper selection of transformer characteristics, its changing properties with frequency can cancel the capacitive high-frequency losses in the amplifier circuit in order to achieve uniform gain well into the hundreds of megahertz.

The electrical properties of the quarter-wave transformer are detailed in Chapters 10 and 11. In the RF amplifier application referenced here, you need know only that a quarter-wave section of transmission line transforms impedances, just as is achieved with transformers. At high frequencies, quarter-wave transformers may be more efficient, however, since winding capacitance and core resonance effects of conventional RF transformers are eliminated. In the RF amplifier applications described here, the transmission line function is achieved simply by twisting two or more wire pairs and winding them around a ferrite core. The high dielectric constant and permeability of the core will reduce the effective wavelength of the high-frequency signals in the "line" to achieve quarter-wave transformation. At low frequencies, of course, the quarter-wave effect is eliminated because of longer low-frequency wavelengths; thus the core and windings function as a conventional transformer at these lower frequencies. Figure 5-7 shows a number of ferrite core structures and a variety of winding techniques used to obtain a given frequency response in RF coupling transformers. Among the examples in Fig. 5-7, you should note the designation for *bifilar* windings. Bifilar winding of transformers is obtained when the primary and secondary windings are wound in the same direction (often parallel wires), with a resulting cancellation of inductance due to mutual coupling. This cancellation of inductance is desirable, often necessary, at high radio frequencies where impedances must be maintained at low values. The dots on the transformer schematics of Fig. 5-7 indicate the starting and ending of the windings, which are the electrical equivalent of phase indications.

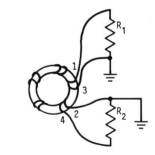

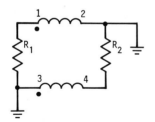

(a)

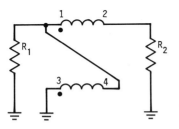

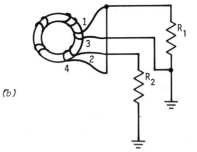

(b)

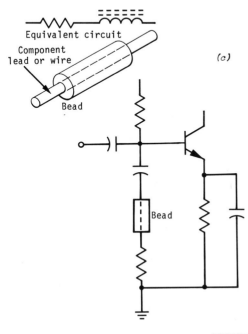

Equivalent circuit

Component
lead or wire

Bead

(c)

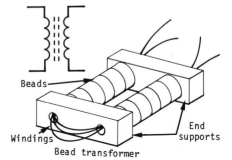

Beads

Windings

End
supports

Bead transformer

(d)

Bead

Bead

FIGURE 5-7

The ferrite beads shown in Fig. 5-7 are often all that is required for high-frequency compensation of transformer-coupled and video wideband RF amplifier circuits. These beads are conveniently slipped over component leads, most often adjacent to the active devices; and being electrical insulators, they do not require insulation or isolation from conductors. The ferrite beads reach their maximum inductance value at the lower radio frequencies in the range of 1 to 10 MHz, and they become lossy attenuators at the higher frequencies. If you recall that a significant part of the capacitive loading of circuits originates from stray and mounting sources, such loading can be isolated from the device with the ferrite bead mounted on the device lead, adjacent to its case, thus increasing circuit high-frequency gain. The ferrite beads are selected for particular applications by size and material *mix* from manufacturers' catalogs which detail their inductance and attenuation properties over a given range of operating frequencies.

Integrated circuit wideband amplifiers deploy degenerative feedback in their monolithic structure to obtain frequency compensation. While this technique can be used with discrete components, the additional capacitance of lead and mounting structures usually defeats any advantages of degenerative feedback. Since the interconnections within IC structures are short, affecting minimum stray inductance and capacitance, they are ideally suited to this technique. Figure 5-8a and b shows simplified two-stage amplifier circuits that employ a combination of voltage and current feedback. The function served by such degenerative feedback is to achieve a low collector (output) capacitance and a low base resistance using transistor elements that have high f_t values and that can drive low-impedance loads. These conditions will achieve optimum wideband frequency response. The circuit of Fig. 5-8a provides emitter degeneration in the second transistor through R_E; and the *signal* across R_E is fed back through R_F to the first transistor to effectively lower its base resistance. This circuit extends its frequency response to about 0.5 f_t at the expense of current gain, which is approximately equal to the ratio of R_F/R_E. The circuit of Fig. 5-8b extends its frequency response to only about 0.1f_t, but with less sacrifice in gain, thus obtaining feedback from its output load, and with emitter degeneration in the first transistor only.

MULTISTAGE CIRCUITS

Electronic systems most often combine the functions of single discrete devices and their related components into *stages*. An IF (intermediate frequency) amplifier with three distinct amplifier circuits in sequence, for example, is called a *three-stage IF strip*. With the entry of IC devices the terminology for multistage circuits becomes a little confusing in the sense that two, three, or often more, stage circuits have been replaced by a single IC. The single IC device is still identified as a single-stage circuit, however,

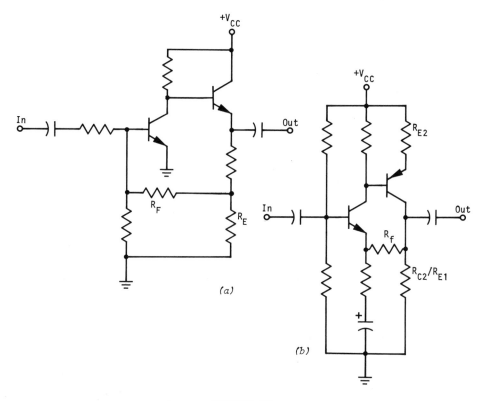

FIGURE 5-8

even though it may provide the *equivalent* of multistage discrete circuit performance. In terms of your knowledge you need to know the effects of combining circuits, whether they utilize IC or discrete devices. In this sense your reference back to the equivalent-circuit model of Chapter 4, and to the passive components of Chapters 2 and 3, as combined in the previous sections of this chapter, should serve to define single-stage circuit performance. The effects of combining such single-stage circuits will be reviewed here as multistage circuits.

Cascaded Stages

When amplifier stages are combined, so that the output of a preceding stage drives the input of a following stage, the center-frequency signal is amplified directly by the gain of the combined stages. The gain of the combined stages does not, however, apply to the off-frequency signals in such a direct relationship. The *off-frequency* signals of tuned, video, and wideband amplifier circuits refer to those frequencies where there is a significant reactive (or susceptive) phase shift in either the parameters of the electronic device, its associated passive components, or any stray and mounting-structure effects. As you may conclude, most of these phase shifts will be capacitive in origin; but with inversion through the active devices or transformers, the effects may be either phase leading or phase lagging.

If two circuits are connected for multistage operation, the resulting bandwidths will decrease, as only the *real* part of the complex gain value can result in signal amplification. With a 45° phase shift, for example, the real part of the gain is 0.707 that of the in-phase gain, resulting in the -3 dB point in the circuit frequency response. If two such circuits are connected, and operated at the -3 dB frequency, the resulting gain will be:

$$A_{v(\text{overall})} = A_{v1} \times A_{v2} = (0.707 \times 0.707)A_v = 0.5\, A_v$$

A simple approach for quick reference, based on the relationships of the above formula, can be generalized as: The -3 dB bandwidth of a given stage will decrease by 3 dB for each identical stage of a multistage amplifier. A multistage RF amplifier comprising three identical 1 MHz bandwidth stages, for example, would have an overall bandwidth of less than 1 MHz, with the frequency response (gain) down to -9 dB at 1 MHz. The -3 dB overall bandwidth of the three-stage amplifier may then be determined by the combined gain values of the individual circuits, as:

$$\text{BW}_n = \text{BW}_1\sqrt{2^{1/n} - 1} = 1\,\text{MHz}\ \sqrt{2^{1/3} - 1} = 1\,\text{MHz}\ \sqrt{0.26} = 0.51\,\text{MHz}$$

The above formula will apply to all multistage amplifiers composed of n stages of identical-response circuits. If the circuits have varying responses, the calculations, unfortunately, have to be "brute force" efforts, taking the complex gain of each stage into account at a given frequency.

Figure 5-9a–c graphs the response of multistage single-tuned, double-tuned, and stagger-tuned amplifier circuits, respectively. You may recall that the response of the optimum-coupled, double-tuned RF transformer maintained the broad frequency-response curve shown in Fig. 5-9b. Since this network already indicates a modified frequency response, the addition of the active electronic device to multistage, double-tuned circuits will provide less total bandwidth reduction. The formula of the preceding paragraph may be modified, however, to simplify the double-tuned bandwidth relationships, using the fourth root rather than the square root of the quantity:

$$\text{BW}_n = \text{BW}_1\ \sqrt[4]{2^{2/n-1}}$$

Stagger Tuning

The frequency vs gain responses of multistage, stagger-tuned amplifiers shown in Fig. 5-9c and d represent a compromise between the narrowband high-gain and wideband low-gain RF amplifier circuits. With careful tuning

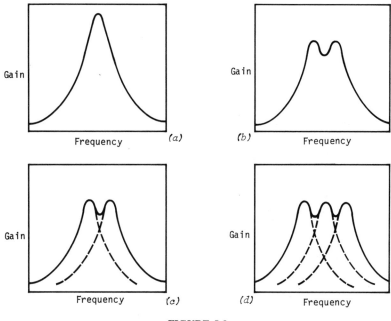

FIGURE 5-9

the isolation provided by each electronic device enhances the combination of the equally leading and lagging phase response of the adjacent tuned *LC* networks. The net result of this technique yields a bandwidth that depends both on the number of tuned networks in the multistage circuit and on their *Q* values. If two networks are tuned to intersect at their adjacent -3 dB frequencies, for example, each will have an opposing 45° phase shift; and the relative amplitude at this frequency may be computed as the square root of the real part of their impedance: $\sqrt{0.707} = 0.84$. On this basis the maximum spacing for 3 dB peak-to-trough variation in the response should occur where the adjacent networks are tuned to intersect at their -6 dB frequencies; and for a 1 dB variation, at about their -2 dB frequencies. As you may conclude, nonuniform frequency response, as required for television and radar IF amplifiers, may easily be achieved with stagger tuning of multistage amplifier circuits by varying the spacing and *Q* of the adjacent tuned networks.

RF POWER AMPLIFIER CIRCUITS

Power amplification at radio frequencies can be achieved with both linear and nonlinear operation of electronic devices, with consequent tradeoffs in circuit function and performance. As mentioned earlier in this chapter, the major requirement for power amplification is matched impedances. At first glance you may conclude that impedance matching of electrical networks to electronic devices is a rather straightforward though often tedious process. You may expect the tedium to increase here, however, because there are some factors relating to higher power RF circuits that have not yet been mentioned. Among these factors are high voltages in vacuum-tube circuitry, high currents in transistor circuitry, and high temperatures in both. These factors complicate the physical, mechanical, and resulting electrical properties of these circuits, making successful RF power amplification somewhat difficult to achieve.

At present both bipolar and VFET transistors are practical for moderate power amplification of RF signals well into the UHF region, with high-power vacuum tubes used for *very* high power applications. Although almost all low-power RF amplifier circuits use solid-state devices, a significant proportion of high-power circuits still use vacuum tubes. These vacuum-tube circuits are found primarily in the final stages of radio transmitters and similar equipment types that require RF power levels in excess of 100 W. References to tube and transistor circuits in this chapter are limited to those applications at frequencies below about 500 MHz; the higher frequency UHF and microwave power devices and circuits are reviewed in Chapters 12 and 13.

Linear Amplifiers

Linear RF power amplifier circuits are often categorized as *class A* amplifiers, since operation of the active devices requires that current be flowing and *controllable* during the entire 360° period of each cycle of the applied RF signal. In addition to providing full control of signal amplitude with low output distortion, the linear power amplifiers have higher gain than their nonlinear counterparts. The tradeoff liabilities of these class A-operated circuits include both electrical and thermal inefficiency, with electrical efficiencies of 25% or less being typical. Figure 5-10*a* and *b* shows popular single- and two-stage RF power amplifier circuits. You should observe that their input networks are biased for full conductivity for the maximum swing of the input signal.

Linear operation of RF power amplifier circuits can also be achieved by using two active devices, operated in a *class B* push-pull arrangement, as shown in Fig. 5-10*c* and *d*. Such *linear* class B operation requires that each electronic device be conducting for alternate 180° periods of each cycle of the applied RF signal. While this alternate conductivity may be achieved with complementary transistors in audio-frequency circuits, the practice is impractical for RF circuits, where transformer phase-inversion techniques provide both correctly phased tuning and impedance matching. The obvious tradeoff advantages of the class B circuits include improved electrical and thermal efficiency and a distribution of thermal dissipation requirements to two devices rather than one.

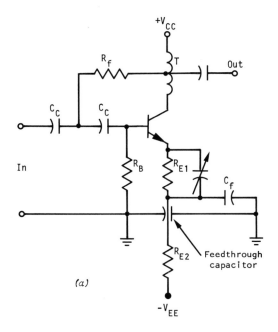

(a)

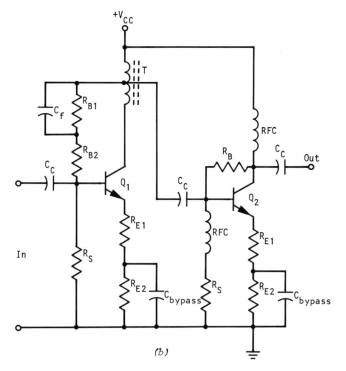

(b)

FIGURE 5-10

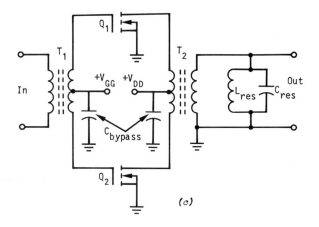

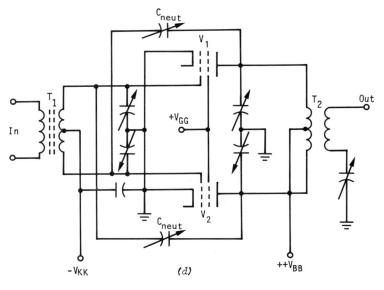

FIGURE 5-10 (*continued*)

Class C Amplifiers

The popular class C RF power circuits operate as nonlinear amplifiers, with active device current flowing during only a *portion* (less than half) of the 360° period of the applied RF signal. You may question the distortion that results from "cutting off" the signal through the device. This distortion is virtually eliminated, however, since high-Q tuned *LC* output networks resonate to reproduce the sine-wave character of the driving signal even though the device is cut *off* during a portion of each cycle. Figure 5-11a comparatively graphs the currents and voltages at various points in the class C amplifier circuit for your reference. The obvious advantage of class C amplifiers is *efficiency*. Since they conduct for only a portion of each cycle, their power dissipation factors are far less than class A amplifiers with similar electrical and physical characteristics.

The major tradeoff disadvantage of class C amplifiers is *control,* or *modulation capability,* or both. With the active devices driven to obtain cutoff and saturation in order to fully load their output *LC* tank networks, a variation in the amplitude of the driving signal will not affect a change in the circuit output. If the driving signal is greatly reduced, however, class B or AB operation of the device occurs, resulting in reduced efficiency and defeating the advantages of class C operation. The consequence of these tradeoff limitations requires that a fixed-amplitude, continuous-wave (CW) or low-deviation FM signal drive the class C amplifier, with any amplitude control limited to varying the power supply or the character of the output network of the active device. Figure 5-11*b* shows the results of varying the amplitude of the driving signal; Fig. 5-11*c* and *d* shows how output control can be achieved by varying the power supply and electrical characteristics of the output section of the class C circuit.

With all its apparent problems you might conclude that the class C RF amplifier circuit is impractical. This is *not* the case, however, because when you're dealing with high-power levels, efficiency and thermal dissipation are

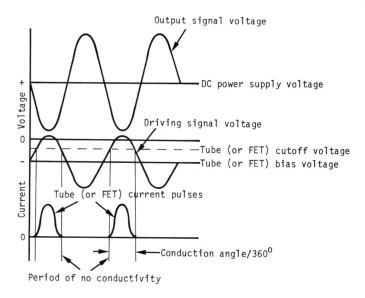

(a)

FIGURE 5-11

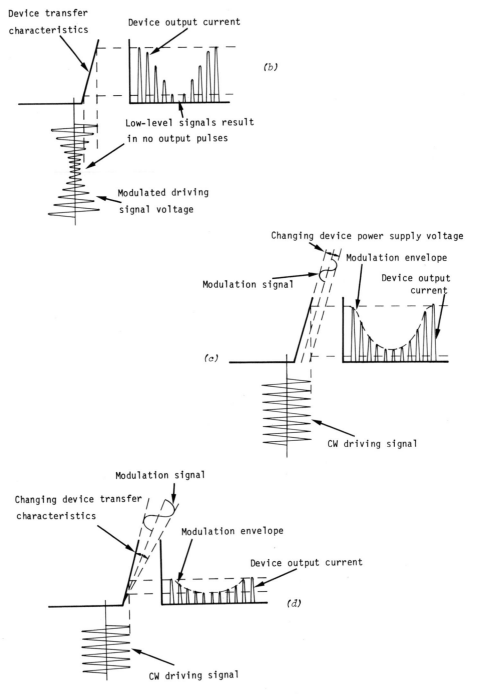

Device transfer characteristics

Device output current

(b)

Low-level signals result in no output pulses

Modulated driving signal voltage

Changing device power supply voltage

Modulation envelope

Modulation signal

Device output current

(c)

CW driving signal

Modulation signal

Changing device transfer characteristics

Modulation envelope

Device output current

(d)

CW driving signal

FIGURE 5-11 (*continued*)

of primary importance. An optimum 80% efficiency class C circuit, for example, will provide 80 W of RF output, with only 20 W required to be dissipated in heat for 100 W of dc power consumed. The same circuit operating in class A with 20% efficiency will provide 80 W of RF output, but will require the dissipation of 320 W in heat and will consume 400 W of dc power in the process. With your understanding of the thermal characteristics and limitations of the electronic devices, as detailed in Chapter 4, you should quickly conclude that class C operation is often the *only* practical means of obtaining high RF power—unless, of course, you attach a large refrigeration plant to each circuit.

The characterization of class C RF amplifier circuits begins with an analysis of their operating efficiency, usually at higher power levels. There is actually no point, of course, in using a class C circuit unless higher power and efficiency are required, because the gain and linearity tradeoffs will not otherwise justify the choice of this circuit. From this *power* viewpoint, then, there are a few terms and concepts that require definition. First, the term *input power* refers to the average dc values from the circuit's power supply—not the power of the driving signal. Don't be confused: the *signal* input is *not* the *dc* input to the circuit. Second, the ac signal voltage and current of the resonant *LC* output tank network are sinusoidal, and thus rise above and fall below the average dc values from the power supply. Third, the device transfer voltage and current are *not* sinusoidal. The voltage waveform of the input driving signal *is* sinusoidal; but since it exceeds the limits of the device conductivity that it affects, it can be considered as being clipped. The current waveform is the most complex since it is *unipolar* and has an unusual waveshape, with characteristics that depend on both the device linearity and the conductivity angle of the driving signal voltage. Figure 5-11*a* shows the typical waveform obtained from class C operation of a vacuum tube device. As you may observe, the true average value of the device current must be obtained on paper by integration, or preferably in operating circuits by direct measurement.

Much of the confusion about class C amplifier circuits appears to originate from the varied symbols and abbreviations used to describe their operation. As such, it is recommended that you become comfortably familiar with the symbols and terms used to describe the driving signals, device signals, bias and power supply values, and output signals. These factors and their abbreviations are tabulated in Table 5-1 for tubes and transistors, with a cross reference to the waveforms of Fig. 5-12*a* and *b*.

Output Requirements

The graphical techniques of Fig. 5-12 are most appropriately suited for visualizing the bias, operating conditions, and limits of signal swing for the class C amplifier circuit. Examining Fig. 5-12, you should observe the following:

TABLE 5-1 CLASS C AMPLIFIER ABBREVIATIONS AND DEFINITIONS

Abbreviation	Definition
v_{rms}, i_{rms} v_{peak}, i_{peak} v_{p-p}, i_{p-p}	Refers to the characteristics of the signal alone, generalized in values of voltage or current. (Note the use of lowercase v and i.)
V_{rms}, I_{rms} V_{peak}, I_{peak} V_{p-p}, I_{p-p}	Applied to general conversion formulas for obtaining different forms of the same signal. (Note the use of caps for V and I.)
V_{max}, I_{max} V_{min}, I_{min}	Refers to the maximum or minimum instantaneous voltage or current at any point in a circuit, referenced to common or ground. (Usually requires superposition of signal and dc values.)
V_{dc}, I_{dc}, P_{dc}	Refers to the power supply or dc bias values, often called the "power input" values of the class C amplifier. (Not to be confused with the "signal input" of the amplifier.)
v_{in}, v_{out} i_{in}, i_{out} P_{in}, P_{out}	Refers to the input and output *signal* voltage, current, and power. (Not to be confused with the dc inputs, above.) Note also that a distinction is made here between uppercase and lowercase abbreviations.
v_{driv}, i_{driv} P_{driv}	Refers to the driving signal voltage at the input of the device.
V_{co}	Refers to the voltage, with reference to common or ground, at which the device ceases to conduct current (cutoff).
V_{cond}	Refers to the voltage range of the driving signal during which the device conducts; with reference to V_{co}.

1. The maximum signal swing is limited by the breakdown voltage of the active device such that the peak instantaneous signal voltage above the power supply level at the output cannot exceed the interelectrode device breakdown voltage.

2. The maximum signal swing is also limited by the intersection of the instantaneous driving signal voltage (grid, base, or gate) and the instantaneous *minimum* output voltage (plate, collector, or drain).

3. The optimum power supply voltage (plate, collector, or drain) should be the sum of the instantaneous peak driving signal and output signal voltages, but less than the difference between the breakdown and instantaneous peak output signal voltages.

If, for example, the driving signal swing on a class C, CE bipolar transistor circuit yields an instantaneous V_{BE} of 1.5 V, and the device breakdown is 32 V, the maximum collector signal swing and supply voltage can be computed as:

$$v_{out(p-p\ max)} = V_{CEO} - v_{BE(max)} = 32 - 1.5 = 30.5 \text{ v(p-p)}$$
$$V_{CE} = V_{CEO} - v_{out(peak)} = 32 - (30.5/2) = 16.75 \text{ V dc}$$

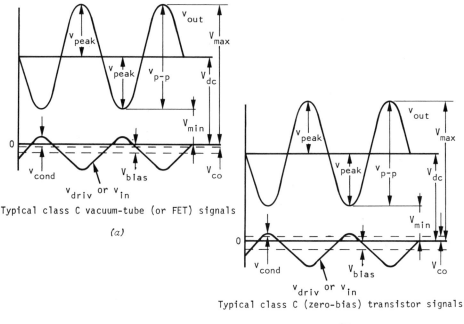

Typical class C vacuum-tube (or FET) signals

(a)

Typical class C (zero-bias) transistor signals

(b)

FIGURE 5-12

(Note: *V* refers to device parameters, but *v* refers to signal values.) These computed values are obviously the maximum limits of the device capability. In practice, marginal design/performance is generally eliminated by the application of a safety factor which reduces the signal swing. In this example a 20% safety factor will reduce the signal swing to about 24 v(p-p), while maintaining the collector supply voltage at about 16.75 V dc.

Once the power supply and output signal swing values for the class C amplifier are obtained, you can proceed to determine the power and load requirements, then follow with driving signal requirements for the circuit. The Watt's law relationship is applied to the device output to determine the load impedance that will yield optimum power transfer; since we are dealing with power amplifiers, maximum power transfer requires impedance matching. If we use the example of the previous paragraph to obtain 10 W RF power output, the resistive part of the load impedance can be computed as:

$$R_L = (v_{out(rms)})^2/P_{out} = (v_{out(p-p)})^2/8P_{out}$$
$$= (24)^2/(8 \times 10) = 7.2 \ \Omega$$

The load impedance on the collector of this bipolar transistor must be a 7.2 Ω, zero phase angle (resistive), resonant *LC* network. In practice, this network must also provide impedance transformation to a transmission line or other circuit, using the π, L, or T configurations described in Chapter 3.

Drive Requirements

Thus far in this section we have defined and characterized the output requirements for the class C amplifier, and applied the information to a practical example. We will now proceed to the device *signal* input in order to define and characterize the drive requirements for the circuit. You may recall that the transfer characteristics were defined for the equivalent-circuit model described in Chapter 4. Most of the transfer characteristics (Y_f) there defined were based on linear amplification. Unfortunately, the transfer characteristics are *not* the same for linear and nonlinear operation of a given active electronic device. Thus, the gain of a class C amplifier is significantly less than its class A counterpart. Additionally, the average dc power supply current for class C circuits depends directly on the signal input to the device—not on the biasing. Without an input signal the device doesn't conduct. Integration techniques can be applied to analysis of the class C operation of a device, however, and factored against its linear characteristics in order to determine its nonlinear values.

In the class C bipolar transistor of our previous example the output signal was specified at 24 v(p-p), for 10 W power output, and with a 7.2 Ω resistive load. The dc input of the power supply was specified at 16.75 V dc. Given these conditions, the output *signal* current is computed, based on Ohm's law, as:

I(p-p) $= v$(p-p)$/R = 24/7.2 = 3.33$ A (p-p)
I(rms) $= 0.707$ I(p-p)$/2 = 0.707 \times 3.33/2 = 1.18$ A (rms)

Even though the *actual* instantaneous device current is in the form of a (nonlinear) pulse, the resonant LC circuit fortunately integrates this pulse to a sine wave, from which the equivalent rms value of 1.18 A is computed. The next step in the analysis procedure requires obtaining the *true* instantaneous signal pulse and dc power supply current values, based on a given input signal conduction angle for the active device. Figure 5-13a graphs the integrated rms output signal current (i_{out}) to dc power supply current (I_{dc}) vs device conduction angles, and shows the related waveforms for driving and output signals. Your review of Fig. 5-13a should disclose that the signal current to dc current, and thus the efficiency of the circuit, increases with decreasing conduction angles. Incidentally, a 180° conduction angle is considered *class B* operation; whereas practical class C operation extends from less than 180° down to about 60°. High-voltage vacuum tubes are typically operated at smaller conduction angles than lower voltage tubes and transistor devices used in class C amplifier circuits.

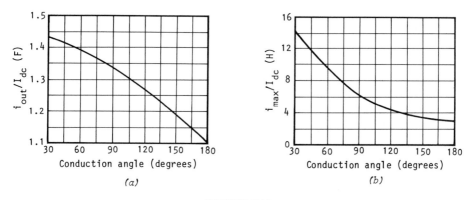

FIGURE 5-13

Proceeding with the analysis of the bipolar transistor example of the previous paragraphs, the dc power supply current may be computed for a typical 150° conduction angle as follows:

$$I_{dc} = i_{out}/F = 1.18/1.2 = 0.98 \text{ A}$$

where F is the i_{out}/I_{dc} ratio for a given conduction angle (from Fig. 5-13a).

The efficiency of this class C amplifier circuit can now be computed as the ratio of the RF power output (P_{out}) to the power supply dc input power (P_{dc}):

$$\eta = P_{out}/P_{dc} = (V_{out}i_{out})/(V_{dc}I_{dc})$$
$$= (8.48 \times 1.18)/(16.75 \times 0.98) = 0.61, \text{ or } 61\%$$

where v_{out} of 8.48 volts in the above equation is the rms value obtained from the 24 v(p-p) signal:

$$v(rms) = 0.707 \text{ v(p-p)}/2 = 0.707 \times 24/2 = 8.48 \text{ v}$$

With a conduction angle of 150°, the greater portion of the input driving signal voltage of our example is below the transistor's 600 mV cutoff value. Figure 5-14 shows the basic circuit diagram and the related waveforms for this example. Since the base circuit is at a zero-bias level, the applied signal voltage must reach 600 mV in 15° of its positive rise, as shown in Fig. 5-14b. The resulting peak and rms signal voltage values can thus be computed for this example as:

$$V_{peak} = V_i/\sin \theta = 0.6/\sin \angle 15° = 2.32 \text{ v}$$
$$v(rms) = 0.707 \ V_{peak} = 0.707 \times 2.32 = 1.64 \text{ v}$$

Most important, the conduction voltage is

$$V_{cond} = V_{peak} - V_{co} = 2.32 - 0.6 = 1.72 \text{ V}$$

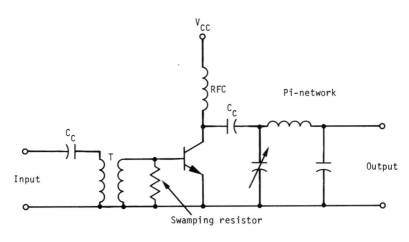

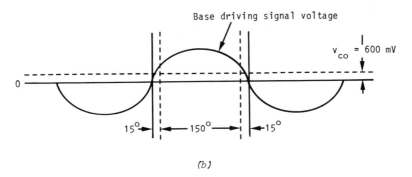

(b)

FIGURE 5-14

V_{peak} and v(rms) refer to the peak and rms values of the input driving signal; V_i is the instantaneous voltage at a given angle of conductivity, and V_{co} is the current cutoff of the active device (0.6 V for the transistor for this example). V_{co} and V_i are the same values because of the zero-bias operation of the active device: the input driving signal voltage swings about zero.

The instantaneous output current pulse results from the conduction voltage of the signal driving the input of the active device. Figure 5-13b graphs the maximum instantaneous output current (I_{max})-to-dc power supply current (I_{dc}) vs device conduction angles. For the 150° conduction angle of our example, the maximum instantaneous output current may be computed as:

$$I_{max} = HI_{dc} = 3.75 \times 0.98 = 3.68 \text{ A}$$

where H is the I_{max}/I_{dc} ratio for a given conduction angle (from Fig. 5-13b).

Finally, the device characteristics and input driving power can be determined as follows:

$$Y_f = Y_{fe} = i_{out}/v_{in} = I_{max}/V_{cond} = 3.68/1.72 = 2.14 \text{ S}$$

Since $Y_{fe} = h_{fe}/h_{ie} = h_{fe}Y_{ie}$, an appropriate power transistor may be selected with a ratio of these two characteristics, yielding a factor of 1.63. The 2N3733 transistor could be selected as a most appropriate device for operation approximately as specified in the procedures of the preceding paragraphs.

With the 2N3733, a typical base spreading resistance of about 5 Ω can be expected for a nonlinear-region h_{fe} of about 8; so the maximum instantaneous input driving current can be computed as:

$$I_{max(in)} = I_{max(out)}/h_{fe} = 3.68/8 = 0.46 \text{ A}$$

This maximum instantaneous driving current value may be converted to an equivalent rms value, in reference to the graphs of Fig. 5-13, as (for the 150° conduction angle):

$$FI_{max}/H = i(rms) = 1.2 \times 0.46/3.75 = 0.147 \text{ A}$$

Now, finally, the input driving power may be computed as:

$$P_{driv} = v(rms)i(rms) = 1.64 \times 0.147 = 0.241 \text{ W}$$

The power gain of the circuit is:

$$A_p = P_{out}/P_{driv} = 10/0.241 = 41.9 = 16.2 \text{ dB}$$

If you have been able to follow the procedures of the last few pages, you should feel confident about the general operating requirements of class C amplifier circuits. Do not be reluctant to review this section, however, as this topic is among the more complex of those covered in this chapter. You should now appreciate the difficulty in applying such RF power amplifier circuits, because, in addition to this circuit analysis, (1) the procedures of impedance matching/transformation outlined in Chapter 3 must *also* be applied to completely specify the passive components; (2) the procedures of device characterization and thermal dissipation outlined in Chapter 4 must *also* be applied to specify the active electronic device and its mounting; and (3) the procedures of engineering practice, including art and experience, must be applied to make the whole circuit practical to build and operate.

RADIO FREQUENCY OSCILLATORS

Oscillation involves an amplifier circuit that feeds back a portion of its output signal to its input for regeneration (reamplification) in order to sustain that output signal without an input from any other signal source. This type of circuit thus generates electrical signals at the amplitudes and frequencies required in electronic communication systems. The oscillator circuit sub-types are classified by the tuning/feedback networks used in their structure as *RC*, for resistance-capacitance networks; *LC*, for inductance-capacitance networks; and *crystal*, for piezoelectric resonant networks. *LC* oscillators are the predominant type used for general RF circuit applications, with the crystal oscillators used where high-frequency stability is required. *RC* oscillators are often used at the lower radio frequencies, below 10 MHz, where frequency stability is not critical or where a wide tuning range may be re-quired. *RC* oscillators are most practical at low frequencies, however, where the required size of inductors makes *LC* networks less practical. In addition to the three *basic* subtype oscillator circuits, a number of specialized IC devices have been developed to provide oscillator functions. The *phase-locked loop* IC (the circuit is achieved with discrete components at the higher radio frequencies) provides stable and accurate signals in its *synthesizer* func-tion; and the *waveform generator* IC provides a less stable, but most practical source of lower frequency signals with a variety of waveforms.

Conditions for Oscillation

The criteria for circuit oscillation are based on Barkhausen's principle, which goes something like:

> The overall gain of an oscillating amplifier circuit, including feedback and output loading, must be slightly greater than 1.0.

The two related criteria for oscillation are thus summarized as:

1. The gain of an amplifier circuit must exceed the losses in the feedback and output networks; and
2. The feedback signal must be in phase with the input requirements of the amplifier; e.g., there must be *positive* feedback for oscillation.

The Barkhausen relationship is expressed in the following formula:

$$A_{vf} = G/(1 - GF)$$

where A_{vf} is the overall circuit voltage gain, *with* feedback, required for oscillation; G is the amplification circuit gain (transistor, tube, IC, etc.), without feedback (this factor is often called *open loop gain*); and F is the feedback network *loss* factor (a gain of less than 1.0). (See Sample Problem 3.)

RC Oscillators

Figure 5-15 shows basic *RC* oscillator circuits. You may be familiar with the operation, performance, and analysis of these circuits through your experience with prior general circuits courses. They are identified as *phase-shift, twin-T,* and *Wien-bridge* in Fig. 5-15a, b, and c, respectively. Since these circuits are not widely used in communications or RF applications, extensive coverage of their characteristics will not be provided beyond a general description of their performance.

Phase-Shift Oscillator

The *phase-shift* oscillator circuit is simple and low in cost. It is often used in textbooks as a direct example of Barkhausen's principle, with its gain, feedback, and operating frequency easily computed. The FET is the most practical choice as an active device for circuit amplification, since its high-impedance input won't load and complicate the three 60° *RC* phase-shift networks required to obtain in-phase feedback to the device input. The frequency of oscillation for the circuit can be computed as:

$$f_o = 1/(2\pi \sqrt{6} \, RC)$$

with equal *R* and *C* values in the feedback network.

Since the losses of the three phase-shift networks each reduce the feedback signal by a factor of 3, the total feedback loss (*F*) of the circuit is 27, requiring a recommended open-loop gain (*G*) in the active device of about 29 to assure stable oscillation. The operation of the phase-shift oscillator is generally limited to frequencies below 1 MHz where the active device capacitive losses are insignificant.

Twin-T Oscillator

Like the phase-shift oscillator of Fig. 5-15a, the *twin-T* oscillator circuit of Fig. 5-15b provides 180° phase shift from the active device output to its input. Unlike the phase-shift oscillator, the twin-T circuit is more stable and has less distortion in its output waveform. The twin-T feedback network is most suitable for use with operational amplifier IC devices where it drives their inverting inputs. The twin-T is composed of two networks: a resistive-T, with a capacitor in its shunt leg; and a capacitive-T, with a resistor in its shunt leg. With both T-networks balanced, the effects of the shunt-leg *R* and *C* components cancel at the frequency where *R* and X_C are equal. The resulting phase shift in the resistive path is 0° with respect to the output of the op amp; while that in the capacitive path is 180°. The effect thus introduces

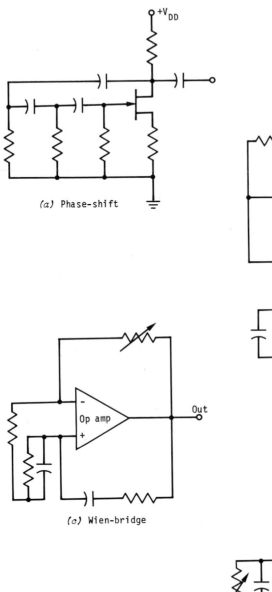

(a) Phase-shift

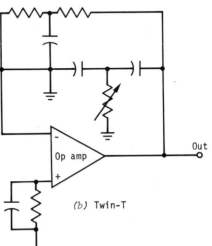

(b) Twin-T

(c) Wien-bridge

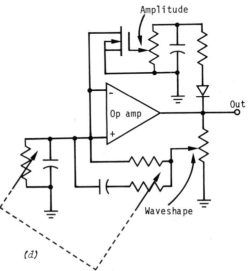

(d)

FIGURE 5-15

cancellation of both feedback signals. With a slight imbalance in the networks, however, caused by the charging of the shunt-leg capacitor, the 180° signal amplitude exceeds that of the signal fed back at 0° through the resistive T, yielding positive feedback with optimum op amp gain, thus meeting the conditions for oscillation. The frequency of oscillation for the twin-T circuit can be computed as:

$$f_o = 1/(2\pi RC)$$

with the series components equal in value to R and C, respectively, and with the shunt resistor equal to $R/2$ and the shunt capacitor equal to $2C$.

If the values of R are low with respect to the input impedance of the op amp, the open- and closed-loop gain factors are directly related to the losses in the feedback twin-T network, thus automatically achieving Barkhausen's criteria for oscillation with the op amp device. The frequency range of operation for the twin-T oscillator is thus limited by that of the op amp devices, which at the date of this writing are limited to frequencies below 1 MHz.

Wien-Bridge Oscillator

The *Wien-bridge* oscillator of Fig. 5-15c is the most popular circuit choice for wide-range variable-frequency signal generators. The advantage of this circuit is that its frequency of oscillation can be varied by adjusting either both R or both C components in the bridge circuit. At its operating frequency the junction of the series and parallel RC sections of the bridge have a 0° phase shift with respect to the output signal driving the bridge. Feeding this point back to the input of a 360° phase-shift amplifier will result in oscillation. The voltage loss in the feedback network is 0.33, so a minimum overall amplifier section voltage gain of only 3 is required to sustain oscillation. This circuit is also most suitable for use with IC operational amplifier devices, with the output signal fed back to the noninverting input of the amplifier. When these circuits are used in signal generators with variable R or C ganged controls, the loading and resulting losses in the feedback network vary with the control setting. This variation is compensated through the gain of the op amp with an automatic gain control element, most often a voltage-variable-resistance FET, between the output and inverting input of the op amp as shown in Fig. 5-15d. The frequency of oscillation for the Wien-bridge circuit can be computed, based on the value of identical R and C elements in the series/parallel feedback bridge leg, as:

$$f_o = 1/(2\pi RC)$$

With careful selection of components and high-frequency discrete or IC amplifier devices, the Wien-bridge circuit is most practical with frequencies up to and above 10 MHz in applications where frequency stability is not critical.

LC Oscillators: Harley and Colpitts

Figure 5-16a and b shows two of the more popular tuned LC oscillator circuits: the *Hartley* and *Colpitts* oscillators, respectively. Both circuits function similarly, with their feedback network coupled back to the amplifier input through a tap on the inductor, or through a capacitive voltage divider, respectively. The required 180° phase shift from the active device output to its input is obtained from the LC tank network *circulating current*. As you recall, the circulating current in a parallel-resonant LC tank network is 180° out of phase with its applied current, so the *tap,* or divider feedback voltage on the tank circuit, arises from the voltage drop: $v = i_{circ}X$. Among the advantages of tuned LC oscillators are greater frequency stability than their RC counterparts, and, of course, stable higher radio frequency operation through 1000 MHz. Additionally, these circuits can operate in class C for higher power, efficient performance; and they may be tunable through varying either the capacitor or the inductor in the resonant LC network. Their frequency of operation can be computed as:

$$f_o = 1/(2\pi \sqrt{LC})$$

Unlike the RC oscillators, which vary frequency directly with a change in R or C values, the frequency of LC oscillators varies with the *square root* of any change in L or C values, limiting their practical tuning range to about a 3 : 1 ratio.

The forward gain of Hartley and Colpitts oscillator circuits depends primarily on the Q and parallel impedance of their LC tank networks, just as with RF tuned voltage amplifier circuits. The feedback tap or voltage divider ratio is thus selected as a function of the device gain and input loading of the tap/divider point. In fact, a practical procedure often employed in determining the tap/divider ratio in these circuits involves first specifying the circuit as a tuned RF amplifier, operating in either class A, B, or C as power and distortion requirements demand. Once the forward *gain* factor for stable amplification is obtained, you need simply feed back *that* portion of the input signal to satisfy Barkhausen's criteria. Connecting the feedback tap/divider signal to the active device input will load the LC tank network, however, thereby reducing its Q, parallel network impedance, and thus the gain of the amplifier. The tap/divider signal must be increased by this loading factor, compensating for the reduced gain in order to assure stable oscillation.

LC Oscillators: Armstrong and Clapp

Figure 5-16c and d shows the popular Armstrong and Clapp oscillator circuits, respectively. The Armstrong circuit simply employs a separate transformer winding rather than a tap in order to provide feedback to the active device input. The Clapp circuit employs the series-resonant network and a

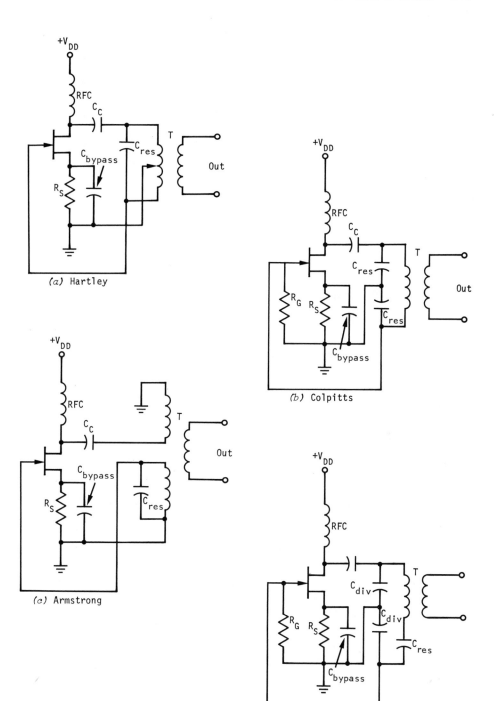

(a) Hartley

(b) Colpitts

(c) Armstrong

(d) Clapp

FIGURE 5-16

capacitive divider to feed back a portion of the output signal to the active device input. This technique is used to reduce the active device output loading on the resonant tank network in order to obtain high tank impedance and higher stable forward gain. You may encounter many possible variations of *LC* oscillator circuits, some of which indicate no apparent feedback path. At high frequencies, of course, there is enough interelectrode capacitance in the active electronic device to couple a portion of an output signal back to the device's input to achieve oscillation. This factor is the bane of RF amplifier design, where it is often more difficult to neutralize a high-gain RF amplifier than to build a stable oscillator. You may recall that Murphy's 32d law states that "all high-gain amplifier circuits oscillate, except when they are *designed* as oscillators."

Sample Problem 3

Given: The 2N4416 JFET to be used in the amplifier portion of the Colpitts oscillator circuit of Fig. 5-16*b,* and operating as specified in Table 4-2.

Determine: The component values for the circuit.

(*a*) The output tank impedance is computed, as for the RF amplifier example (Sample Problem 1), based on optimum gain with minimum loading of the transformer output and very low admittance at the FET's input. The minimum total resonant (tank) capacitance is first determined to obtain the highest load impedance for optimum forward gain. Using the calculations from page 177 for the voltage amplifier output tank network, a 10 pF capacitance and a 25 μH inductance with a Q of 40 yield a practical Z_{par} of 64 kΩ.

(*b*) The output admittance of this resonant *LC* network, placed in parallel with the output admittance of the 2N4416, is now obtained from the data for Sample Problem 2, yielding a Y_{out} of 115.7 μS and a resulting voltage gain of 17.3.

(*c*) The next step in the procedure is the determination of the loss factor for the feedback circuit and the ratio of the voltage divider C_1 and C_2. Since the input loading of the 2N4416 is small (insignificant conductance and about 1 pF capacitance), the capacitive loading of the circuit will result primarily from stray and mounting effects, which can be controlled. This insignificant loading simplifies the circuit requirements, so the feedback ratio can be computed to comply with Barkhausen's criteria, where $A_{vf} = -1.0$, as follows:

$$F = (1 - (G/A_{vf}))/G = (1 + G)/G = (1 + 17.3)/17.3 = 1.058$$

and the feedback ratio is computed as:

$$v_{out}/v_{in} = 1/(F - 1) = 1/(1.058 - 1) = 17.3:1$$

(d) The most convenient way to obtain the divider ratio is to deal with the *reactances* of C_1 and C_2 as a reactive divider:

$$v_{out}/v_{in} = (X_1 + X_2)/X_1 = 17.3$$

thus $X_2 = X_1(17.3 - 1) = 16.3X_1$

(e) The *total* and individual reactances of the 10 pF output tank network capacitance is now computed as:

$$X_T = 1/(\omega C_T) = 1/(6.28 \times 10^7 \times 10^{-11}) = 1592 \ \Omega$$
$$X_1 = X_T/G = 1592/17.3 = 92 \ \Omega$$
$$X_2 = 16.3X_1 = 16.3 \times 92 = 1500 \ \Omega$$

(f) The capacitance values for C_1 and C_2 are now computed as:

$$C_1 = 1/(\omega X_1) = 1/(6.28 \times 10^7 \times 92) = 173 \ pF$$
$$C_2 = 1/(\omega X_2) = 1/(6.28 \times 10^7 \times 1500) = 10.6 \ pF$$

As you may observe, the 173 pF input feedback capacitance is large enough to make any stray, mounting, or device input capacitances insignificant; and a small change in *its* value will not significantly affect the total tank capacitance (0.058, or 17.3 : 1). In practice, the value of C_1 is increased by about 10% over these computed values in order to assure stable operation of the oscillator; and C_2 is often a variable, precision trimmer for critical frequency adjustment. For this example a 180 pF or 200 pF fixed capacitor would be an appropriate choice for C_1; and a 3 to 15 pF glass piston or ceramic trimmer would be appropriate for C_2.

Crystal Oscillators

Piezoelectric "crystal" oscillator circuits are used in applications that require critical stability and accuracy in operating frequency. The piezoelectric element is a quartz crystal that mechanically vibrates with electrical excitation and exhibits properties similar to series- and parallel-resonant *LC* networks. Figure 5-17 shows the equivalent circuit and resonance characteristics of the crystal element. The series-resonant mode occurs at a lower frequency for these devices than the parallel-resonant mode. A variety of manufacturing techniques are employed in the fabrication of crystals for oscillator applications. These techniques determine the general circuit requirements for proper operation of the crystal, including frequency range, operating mode, load capacitance, and drive levels. Since such crystals are usually purchased from specialized manufacturers, you should review the published literature for determination of popular crystal characteristics and operating recommendations.

The crystal oscillator circuits are configured like their *LC* counterparts, the most popular of which are the *Pierce* and *Miller* circuits of Fig. 5-18a and

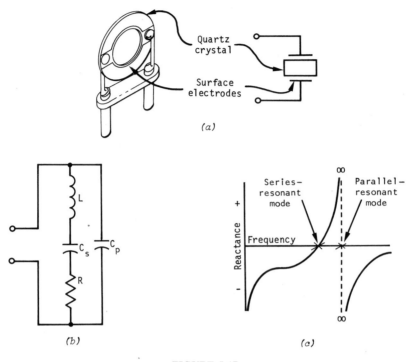

Quartz
crystal

Surface
electrodes

(a)

L

C_s C_p

R

(b)

Series–
resonant
mode

Parallel–
resonant
mode

Reactance

+

Frequency

–

∞

∞

(c)

FIGURE 5-17

b, respectively. The Pierce oscillator is basically a variation of the Colpitts circuit, with the capacitive divider providing feedback in proportion to the gain of the amplifying device, and with the crystal, operating in parallel mode, providing the effective LC resonant network. For operation in this mode the crystal manufacturer will specify both the total series capacitance value of the divider and the drive value, which will determine the device gain and power supply requirements. The Miller oscillator is a variation of an oscillator circuit with tuned input and output networks, with feedback coupled through the interelectrode (Miller) capacitance of the active device. (When used with vacuum tubes this circuit was identified as the "TPTG," for *tuned-plate, tuned-grid* oscillator.) With high LC tank impedances affecting high forward circuit gain, this circuit can deploy low-drive-signal crystals and yet obtain high output power. This circuit application also requires the operation of the crystal in its parallel-resonant mode.

As with LC oscillator circuits, there are many variations of crystal oscillator circuits, each utilizing a particular property of the piezoelectric element to its advantage. Such circuits are often so specialized, for example, that different manufacturers' circuits may require specialized crystal "cuts" for operation, making the crystals or circuits noninterchangeable. In general,

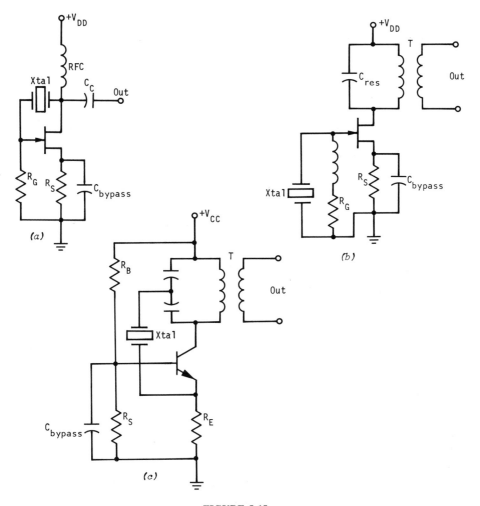

FIGURE 5-18

fundamental frequency crystal oscillator circuits operate at frequencies below 10 MHz, with maximum drive levels below 10 mW. Odd-harmonic *overtone* crystal oscillator circuits operate at the higher frequencies, with even some UHF oscillators stabilized with high-harmonic-number resonant crystals. Since the crystal elements are made thinner with increasing frequency, the drive levels must be proportionately reduced with higher frequency operation. The 160 MHz, 9th-harmonic overtone crystal oscillator circuit of Fig. 5-18c, for example, uses the crystal in its series-resonant mode, driving the low-impedance input of the common-base transistor amplifier, with its drive power limited to below 2 mW.

Because of their frequency stability and accuracy, crystal oscillators are used as secondary frequency standards, with the elements placed in miniature heating chambers, called *crystal ovens,* in order to maintain frequency accuracy. Crystal oscillators are also used in digital frequency synthesizers that use IC's in a counter/divider chain, in combination with a phase-locked-loop IC, to obtain a large number of discrete crystal-stabilized frequencies for use in systems. These IC applications of oscillators and other electronic circuits will be described in the following chapters, where circuit functions meet specific systems needs.

CONCLUSION

The purpose of this chapter was to describe the popular RF circuit applications that are used as the basis for communications systems. This chapter concludes Part I of this text, which is devoted to the circuit aspects of electronic communications. Each chapter in this section (Chapters 2 through 5) reviewed the important aspects of circuits, including passive devices and networks, and active devices and basic circuit functions. While this chapter brought together some of the concepts of the previous three chapters, its problems and examples were somewhat simplified in order to emphasize the circuit function. To add the impedance-matching aspects required for circuit operation could result in your "losing sight of the forest for the trees." Rest assured, however, that real world circuit development requires the application of extensive impedance-matching/loading techniques, as well as a broad knowledge of the availability and performance of discrete solid-state, IC, and vacuum-tube devices. If you specialize in *circuits* in electronics, you will, no doubt, develop your ability to deal with the combined aspects of the topics in Chapters 2 through 5, as well as those outlined in the systems and microwave sections of this book. At this point in your progress, however, your understanding of the basic concepts of this chapter should prove satisfactory. The problems that follow will test your knowledge and understanding of basic electronic circuits as applied to RF communications. You should refer back to the content of this and previous chapters, as necessary, should you encounter difficulty with the problem solutions.

Problems

1. Given: A 3N128 MOSFET to be used as a 100 MHz RF voltage amplifier in a circuit similar to that of Fig. 5-2, with characteristics for the 3N128 as listed in Table 4-2.

Determine:
(a)   The device input, output, and transfer admittance values and the device capacitance values (from Table 4-2)
(b)   A selected minimum value of *input resonant network* capacitance (based on about 10 times the total of stray, mounting, and device capacitances)
(c)   The inductance value of the resonant secondary coil
(d)   The minimum parallel input tank impedance and admittance (based on a practical loaded Q of 50)
(e)   The capacitance, inductance, impedance, and admittance for the *output resonant network* (repeat steps b, c, and d for the output circuit)
(f)   The voltage gain of the device (based on Y_{fs} and Y_{os} in parallel with the output resonant network)
(g)   The value of Miller capacitance reflected back to the input of the device
(h)   The selection of a practical input resonant network trimmer capacitor to provide a range of tuning necessary to accommodate the stray, input, and Miller capacitance computed for this circuit

2.   Given: The MOSFET circuit of Problem 1 used as a "front-end" RF amplifier in an FM broadcast-band receiver. The following specifications apply to the circuit:

Input transformer primary-to-secondary impedance ($X_{pri}:X_{sec}$)
   $= 300: 12$ kΩ

Coupling coefficient (k) = 0.2
Resonant secondary network $Q = 30$

Output transformer primary-to-secondary impedance $= 8$ k$\Omega : 2$ kΩ

Coupling coefficient $= 0.4$
Resonant secondary network $Q = 40$

Device voltage gain (A_v) = 80
Input signal (across the 300 Ω primary) = 20 μV

Determine:
(a)   The mutual reactance of the input transformer
(b)   The input signal voltage at the gate of the 3N128
(c)   The signal voltage at the drain of the 3N128
(d)   The mutual reactance of the output transformer
(e)   The signal voltage across the secondary of the output transformer
(f)   The overall voltage gain of the circuit

3.   Given: The single-stage tuned linear RF voltage amplifier circuit, shown in Fig. 5-19, using a 2N2405 bipolar transistor as the active device. The circuit is to be used in a 30 MHz (10 meter) linear amplifier, and the device voltage gain is specified to be 100. The typical input, output, and transfer characteristics for the 2N2405 at 30 MHz are specified as follows:

$$Y_{in} = Y_{ie} = 5 + j9.4 \text{ mS}$$
$$Y_{out} = Y_{oe} = 0.001 + j1.9 \text{ mS}$$
$$Y_{transfer} = Y_{fe} = 100 + j2.4 \text{ mS}$$

Determine:

(a)   The input, output, and reverse (transfer) capacitance values at 30 MHz

(b)   The admittance and equivalent Z_p value of the output resonant network (transformer primary) to obtain the specified voltage gain

(c)   The reactance of the output transformer based on a practical Q of 20

(d)   The value of capacitance required to resonate the primary of the output transformer (including the device output capacitance)

(e)   The value of Miller effect capacitance and the resulting increased input admittance to the base of the device

(f)   The required inductance at the input transformer secondary to resonate with the total (including Miller effect) input capacitance

(g)   The value of C_c required to increase the transformer secondary inductance (by *tapping* with C_c and C_{in}) to a more practical value of 100 nH

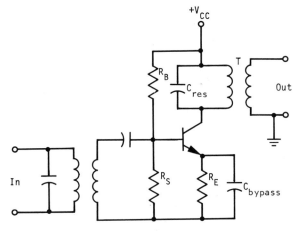

FIGURE 5-19

4. Given: The 2N2405 bipolar transistor of the previous problem used in an *RC*-coupled video amplifier, with a relatively uniform voltage gain of 100 from dc through at least 10 MHz. The circuit configuration for the amplifier is similar to that of Fig. 5-5*a*.

Determine:
  (*a*) The low frequency output admittance and the resulting value of load resistance (R_L) for a device voltage gain of 100
  (*b*) The cutoff frequency (f_{co}) that will result with a minimum of 40 pF total device and stray capacitance
  (*c*) The value of inductance required to achieve "optimum damping" flat response for the output network
  (*d*) The resonant frequency affected by the addition of the series inductance from item *c* above
  (*e*) The new "compensated" cutoff frequency
  (*f*) The value of R_L, and resulting reduced gain, required to extend the compensated cutoff frequency to 50 MHz

5. Given: Three RF amplifier stages combined to form a high-gain, 60 MHz "strip amplifier." The input stage uses a MOSFET in a circuit similar to that of Problem 2; the two following stages use bipolar transistors in circuits similar to that of Fig. 5-19. The functional block diagram equivalent of the strip amplifier is shown in Fig. 5-20. The following specifications apply to the circuit:

  Input transformer: $X_{pri} : X_{sec} = 300 : 12 \text{ k}\Omega$

  $k = 0.2$ and $Q_{pri} = Q_{sec} = 30$

  Interstage transformers (two): $X_{pri} : X_{sec} = 8 \text{ k}\Omega : 1 \text{ k}\Omega$

  $k = 0.4$ and $Q_{pri} = Q_{sec} = 30$

  Output transformer: $X_{pri} : X_{sec} = 8 \text{ k}\Omega : 2 \text{ k}\Omega$

  $k = 0.3$ and $Q_{pri} = Q_{sec} = 30$

  MOSFET device voltage gain $= 60$
  Bipolar device voltage gains $= 80$ (each device)
  $C_c : C_{in}$ (tap) capacitive voltage divider ratio at base inputs to both bipolar transistors $= 4 : 1$ (20%)
  Input signal (across the 300 Ω primary) $= 10 \ \mu\text{V}$

Determine:
  (*a*) The signal voltage at the resonant primary of the first interstage transformer

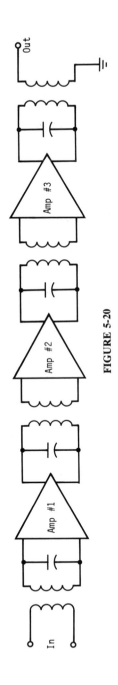

FIGURE 5-20

(*b*)   The signal voltage at the base of the first bipolar transistor

(*c*)   The signal voltage at the base of the second bipolar transistor (all resonant networks peak tuned)

(*d*)   The signal voltage across the secondary winding of the output transformer (all resonant networks peak tuned)

(*e*)   The overall circuit gain (all stages: input to output) with all resonant networks peak tuned at 60 MHz

(*f*)   The bandwidth of *each* of the four individual resonant networks

(*g*)   The total bandwidth with all resonant networks peak tuned at 60 MHz

(*h*)   The overall circuit gain with the resonant networks stagger tuned to intersect at their individual −6 dB frequencies (each of the four resonant networks tuned to a different frequency)

(*i*)   The total bandwidth with the resonant networks stagger tuned as in item *h*

6.   Given: The class B push-pull RF amplifier circuit of Fig. 5-10*d* using a pair of VMP-4 VMOS power FET's. Since these are enhancement-mode devices, they are biased to conduct during the positive 180° swing of the applied driving (input) signal. The specifications for the circuit include the following:

> VFET device transconductance (Y_{fs}) = 80 mS
> VFET device efficiency (η) = 75%
> Power (supply) input voltage (V_{DD}) = 24 Vdc
> Transformers' coupling coefficient (k) = 0.9
> Load resistance (R_L) = 20 Ω

Determine:

(*a*)   The "ideal" maximum $(V^2/2R)$ power output if η and k were 1.0; and the true power output, taking η and k into account

(*b*)   The peak (positive output voltage swing at the drain of each VFET device)

(*c*)   The peak output current swing through each device's drain, and the transformer and load resistor

(*d*)   The peak driving (input) signal voltage to each VFET's gate

7.   Given: A class C amplifier circuit similar to that of Fig. 5-14 operated to obtain 25 W RF power output with a 40 V dc collector supply voltage. The transistor is driven with a base input signal which results in a 120° conduction angle of current flow through the device.

Determine:

(*a*)   The required dc input (supply) power if the device is operating at 65% efficiency (η)

(b) The dc input (supply) current
(c) The rms value of output signal current and voltage (see Fig. 5-13)
(d) The value of load resistance
(e) The peak and rms values of the driving (input) signal voltage and the "conduction" voltage
(f) The maximum instantaneous output current
(g) The required minimum transconductance (Y_{fe}) and current gain (h_{fe}) characteristics for the device
(h) The required instantaneous driving current and driving signal power

8. Given: Three RC oscillator circuits: the phase-shift, twin-T, and Wien-bridge as shown in Fig. 5-15. Each circuit is to operate at 2 MHz and provide an output of 1 V (rms) across a 600 Ω resistive load. This problem requires that you compute the electrical values of the components and determine their cost. Actual part numbers and item costs should be obtained from catalogs and other sources.

Determine: The component values and costs for the following circuits:
(a) A phase-shift oscillator with a source-follower as a buffer, using two 3N128 MOSFET's
(b) A twin-T oscillator using an LM118 operational amplifier
(c) A Wien-bridge oscillator using a "diff-amp"-connected CA3053 IC

9. Given: A cascode-connected CA3028 IC deployed as the active device in the 100 MHz Hartley oscillator circuit of Fig. 5-21. The following specifications apply to the circuit:

Output resonant network:
Total capacitance (including the output capacitance of the device) = 20 pF
Loaded Q (Q_L) = 80
CA3028 IC cascode amplifier:
Transconductance (Y_f) = 60 +j0.1 mS
Input admittance (Y_{in}) = 1.5 +j5 mS
Output admittance (Y_{out}) = 0.01 +j2 mS

Determine:
(a) The Z_p and equivalent admittance of the output tank network
(b) The value of the tank inductance
(c) The voltage gain of the amplifier circuit

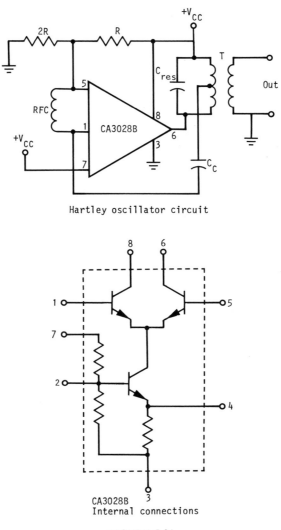

Hartley oscillator circuit

CA3028B
Internal connections

FIGURE 5-21

(d) The loss factor for the feedback circuit: the ratio of reactances above and below the tap point on the coil
(e) The actual physical characteristics (number of turns, spacing, tap point, etc.) if the coil is wrapped around a ¼-inch form using 16-gauge wire (see Chapter 2)

PART II SYSTEMS

CHAPTER 6

AMPLITUDE MODULATION AND DETECTION

Part II of this text deals with radio signals and how they are processed by electronic systems to perform communications functions. RF communications signals are typically *modulated* with intelligent information in the form of sound, video, or other lower frequency waveforms.* This chapter reviews the modulation techniques that are based on controlling the *amplitude* of the RF communications signal. These techniques are categorized, with their respective code designations, as:

1. Double sideband amplitude modulation (DSB-AM) (A3)
2. Vestigial sideband amplitude modulation (VSB-AM) (A5C)
3. Double sideband suppressed carrier (DSB-SC) (A3B)
4. Single sideband (SSB) (A3J)

This chapter also reviews the techniques for *detecting* (or *demodulating*) these modulated RF signals and extracting the intelligent information they carry.

Amplitude modulation was historically the first technique used to transmit sound on radio signals. Interestingly, this technique of modulation and detection preceded the invention of electronic devices. Conversely, it was through the use of electronic devices that AM radio became practical; and it was through concurrent progress with AM radio and electronic techniques that our present advanced technology evolved. Before getting involved with advanced concepts, we owe some recognition to the simplicity of the first AM modulator—a carbon microphone—and to the first detector—a piece of galena crystal. Although neither of these elements were *electronic* devices, their functions of modulation and detection of RF

---

* *Modulation* is defined as the process by which the characteristics of *one* signal are varied by a *second* signal. In communications practice the higher frequency signal, which is modulat*ed,* is called the *carrier;* while a lower frequency signal, which varies the carrier, is called the modulat*ing* signal.

signals were little different from what we presently achieve with the most sophisticated electronic circuits. Since our initial concern is to understand signals and signal processing, we will use these simple elements—the carbon microphone and the galena crystal—to introduce the concepts of amplitude modulation and detection.

The carbon microphone is an audio-frequency (AF), voice (or sound)-controlled variable resistor. In their principal use in telephone systems, carbon "mikes" control AF current by their variable resistance. When they were used in RF systems, carbon mikes were connected in series with the RF signal source to control the RF current to the antenna at an AF rate. Figure 6-1a and b compares the telephone system with the AF modulation of early radio systems. With no sound input, the carbon mike has a moderate resistance value, so in the radio application there was a continuous RF antenna signal with or without the AF signal. This level of RF signal with zero AF modulation was called the *resting* level. With sound input to the microphone, its resistance would vary, which would, in turn, vary the amplitude of the RF signal. A loud sound would vary the RF signal to its extremes: to twice the resting amplitude and to zero amplitude. A quiet sound would only slightly vary the amplitude of the RF signal. A high-frequency sound would yield a high *rate* (not level) of amplitude variation, and a low-frequency sound would yield a low rate of amplitude variation. So the RF signals would be transmitted from early radio antennas with waveforms as shown in Fig. 6-1c, d, and e. These waveforms are incidentally the same as those transmitted from modern AM radio systems.

The second half of the early radio systems required reception and detection of the modulated RF signals. Early radio reception was simply accomplished with a tuned *LC* network connected to an antenna (called *aerial* in those times), and the most critical requirement was the detection process. Here a mineral crystal was used to extract the original sound-signal waveform from the modulated RF signal. The crystal had properties similar to our modern diodes, such that when the polarity of the applied RF signal was + to − the element had low electrical resistance, and when the polarity was reversed it had high resistance. The crystal allowed only positive current of the modulated RF signal to pass, yielding a rectified RF waveform with its level varying as the original sound signal that modulated the RF. Figure 6-2 shows the reception/detection process using the tuned antenna, crystal detector, lowpass filter, and headphone load. This entire process of radio reception is accomplished without electronic components or any external source of power. Although much progress has been made in the application of electronic circuits to this process, the signal functions of this crude "crystal-set receiver" are identical to those used in modern AM radio receivers.

This historical review is presented for your conceptual understanding of signals and functions before electronics is applied to the achievement of

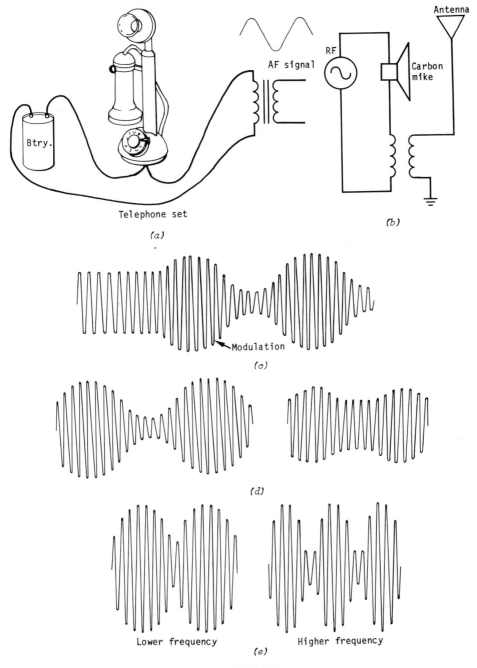

AF signal

RF

Antenna

Carbon mike

Btry.

Telephone set

(a)

(b)

Modulation

(c)

(d)

Lower frequency

Higher frequency

(e)

FIGURE 6-1

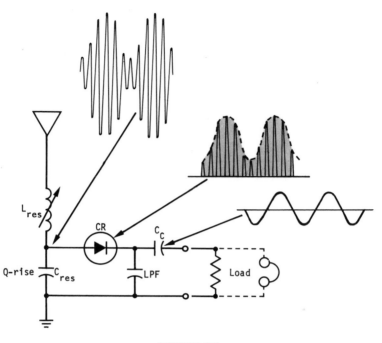

FIGURE 6-2

these processes. Although the electronic processes and hardware are in a continual state of change, functional principles and signal characteristics are stable phenomena; new discoveries and new techniques serve only to expand communications technology. This presentation of amplitude modulation and detection will thus follow the emphasis on signals and functions, and then support this emphasis with examples of signal processing circuits. Your working knowledge of the circuit principles presented in Part I is assumed, so your attention is now directed to learning about signals, functions, and processes.

LEARNING OBJECTIVES

Upon completing this chapter, including the problem set, you should be able to:

1. Understand the concepts of signal mixing products and compute their values.
2. Understand the principles of amplitude (envelope) modulation and compute modulation index.

3. Understand and compute sideband signal frequencies and power distribution in amplitude-modulated RF signals.
4. Understand the concepts of time-domain and frequency-domain measurements as applied to AM RF signals.
5. Calculate critical signal and circuit values for AM modulators, based on your knowledge of specific modulation circuit techniques.
6. Understand the principles, computations, and measurements of other AM-related modulation techniques, including VSB, DSB-SC, and SSB.
7. Understand the special circuit and function requirements for obtaining efficient VSB, DSB-SC, and SSB operation.
8. Understand the principles of envelope detection of amplitude-modulated RF signals, and compute the special circuit values for envelope detectors.
9. Understand the principles of product (synchronous) detectors of AM and sideband modulated RF signals, and compute the special circuit values for product detectors.
10. Understand how integrated circuit devices are used in modulation and detection circuit functions, and adapt these devices to specific requirements.

MIXING AND MODULATION

Electronic signals can be *mixed* in two ways: linear and nonlinear. A microphone or studio audio mixer is an example of a linear mixer, where distinct signals are combined with little distortion. An RF mixer, however, is a *nonlinear* mixer because the amplitude function of the circuit transferring the RF signal is changing. Consider the simple diode mixer response of Fig. 6-3a. With the AF voltage merely changing the Q point of the RF signal along the linear portion of the diode response curve, the RF signal amplitude will remain constant, producing linear mixing, with the associated output waveform as shown in the figure. If the two *sine-wave* signals were shifted to the nonlinear region of the diode's response curve, however, the AF voltage would, in effect, control the *amplitude* of the RF output voltage as the Q point shifted along the nonlinear portion of the curve, as shown in Fig. 6-3b.

Distortion Products

Nonlinear circuits produce *harmonic distortion* frequencies as they transfer a signal from input to output; they produce *intermodulation distortion* frequencies when two or more signals are combined and processed from their input to output. The intermodulation frequencies are called *mixing products,* and they are mathematically related by Fourier-transform methods as the sum

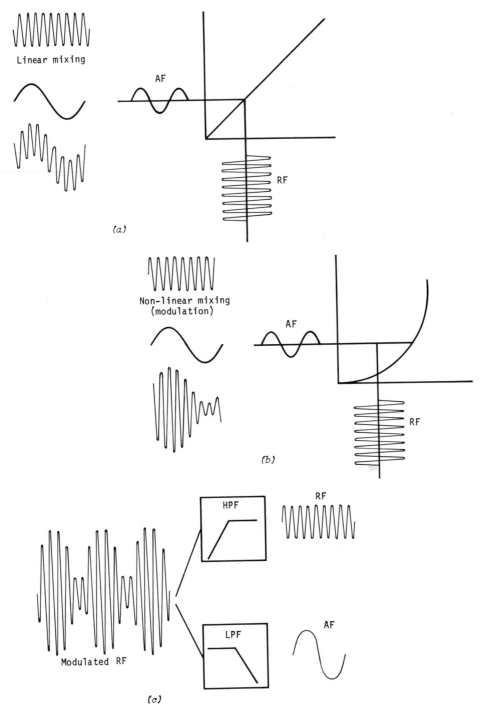

Linear mixing

AF

RF

(a)

Non-linear mixing
(modulation)

AF

RF

(b)

HPF

RF

Modulated RF

LPF

AF

(c)

FIGURE 6-3

and difference of the original signal frequencies that are combined, plus the less intense harmonic distortion products that may be produced by the mixing process.

If, for example, two sine-wave signals of 6 and 8 MHz are combined in a nonlinear mixer, there will be four predominant signals (excluding *harmonic distortion products*) at its output: 6 MHz, 8 MHz, 2 MHz, and 14 MHz. If three sine-wave signals are combined in the nonlinear mixer, there will be nine predominant signals at its output, with the number of predominant output signals increasing as the *square* of the sine-wave input signals. The basic frequencies of such signals may be simply computed on a sum and difference basis as follows:

$$f_3 = f_1 - f_2 \qquad f_4 = f_1 + f_2 \qquad \text{with two signals } (f_1 \text{ and } f_2)$$

$$\left. \begin{array}{ll} f_4 = f_1 - f_2 & f_5 = f_1 + f_2 \\ f_6 = f_1 - f_3 & f_7 = f_1 + f_3 \\ f_8 = f_2 - f_3 & f_9 = f_2 + f_3 \end{array} \right\} \quad \text{with three signals } (f_1, f_2, \text{ and } f_3)$$

and so on

If the two signals driving the nonlinear mixer were not sine waves, there would be a great number of harmonic mixing products to be accounted for—a task just a bit too painful to pursue with the above approach. Fortunately, most examples used to describe the modulation process and mixing products are confined to just two sine-wave signals in order to maintain some sanity in describing the process.

A general description of how the sum and difference frequencies are obtained is here presented in lieu of resorting to the application of Fourier's abstract mathematics. The source of these sum and difference frequencies is simply the changing phase relationships of the original signals at the mixer input. You can probably develop a visual understanding of the mixing process by reviewing our earlier example, that of mixing 6 and 8 MHz sine-wave signals and graphically displaying them for a 1 μs period of 6 and 8 cycles, respectively; with a dual-trace scope set at a sweep rate of 0.1 μs/cm (see Fig. 6-4a). The corresponding time periods of the signals would be 0.167 μs and 0.125 μs, respectively. Combining the two signals in a linear mixer would result in the waveform of Fig. 6-3a, with the weaker 8-cycle signal "riding on" the waveform of the stronger 6-cycle signal. Combining these same two signals in a nonlinear mixer, however, will now result in the unusual waveform of Fig. 6-3b. This waveform not only changes in amplitude, but the time periods of the individual half-cycles vary corresponding to the original 0.167 μs and 0.125 μs, along with 0.5 μs and 0.07 μs. Converting these time periods to frequency ($f = 1/t$) gives us the original frequencies, plus the sum and difference of the two. If a tunable resonant circuit were placed at the output of the mixer, it could resonate and transfer power at

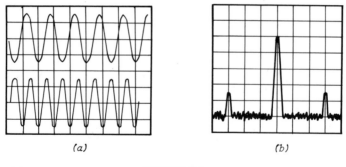

(a)   (b)

FIGURE 6-4

frequencies of 2 MHz, 6 MHz, 8 MHz, and 14 MHz. This is the technique, incidentally, that is used to measure intermodulation distortion, with the least distortion yielding minimum signal levels at the sum and difference frequencies.

Amplitude Modulation

The process of amplitude modulation requires that the intensity of an RF signal be controlled with an AF (or lower frequency) signal. If such modulation were attempted with a linear electronic circuit, the resulting waveform would be similar to that of Fig. 6-3a, with the RF signal merely riding on the AF signal. A nonlinear electronic circuit is required to achieve amplitude modulation in order to obtain the required variation in RF signal intensity as controlled by the AF signal. As was observed with the nonlinear mixer, the results of combining two signals in a (nonlinear) modulator circuit will yield the original signal frequencies, plus the sum and difference of the two, at its output. Let us take an example from the typical AM broadcast band, with a 1.0 MHz RF signal amplitude modulated by a 1 kHz AF signal, which will yield 1 kHz, 1 MHz, 0.999 MHz, and 1.001 MHz at the output of the modulator. If the modulator circuit has a tuned RF output network, the 1 kHz signal will be shorted, leaving only three signals of 0.999, 1.00, and 1.001 MHz across the tuned network. These three signals are identified as the *lower sideband signal,* the *RF carrier signal,* and the *upper sideband signal,* respectively. The signals are shown on the frequency-domain display of Fig. 6-4b.

In the modulation process both the frequency and average power of the *RF carrier signal* remain fixed; whereas both the frequency and level of the *AF modulating signal* vary with the voice, music, or code, etc., that constitutes the intelligence information superimposed on the carrier signal. What changes in the RF output is, of course, the variation in the carrier signal voltage and current, along with the voltage, current, power, and frequency variation of the sideband signals. The significant frequencies of the sideband signals can be computed as:

$$f_{LSB} = f_C - f_M \quad \text{and} \quad f_{USB} = f_C + f_M$$

f_{LSB} is the lower sideband frequency, f_{USB} is the upper sideband frequency, f_C is the RF carrier frequency, and f_M is the modulation frequency. Since the average power of the RF carrier signal remains fixed, with variations in sideband signal power dependent on the modulation signal, the total RF power (carrier plus sidebands) is increased by the presence of modulation. In order to compute this amount of increased RF power in the sideband signals, you must first determine the ratio of the RF carrier and AF modulating signals. This ratio, called *modulation index,* is based on a modulating signal equal to, or less than, the RF carrier signal. Thus, the modulation index should always be between values of 0.0 and 1.0, with varying values of modulating signals; and it is measured by observation of the modulated RF carrier voltage levels.

Modulation Index

Figure 6-5 shows the amplitude-modulated voltage waveform of the RF carrier signal superimposed upon the zero-modulation (*resting* level) signal. The modulation *envelope* shows an equal increase and reduction of the RF carrier voltage above and below the zero-modulation level. The modulation index, here indicated by M, is computed as the ratio of the peak envelope voltage above zero modulation, to the peak zero-modulation RF voltage, which reduces to:

$$M = (V_{\max} - V_{\min})/(V_{\max} + V_{\min})$$

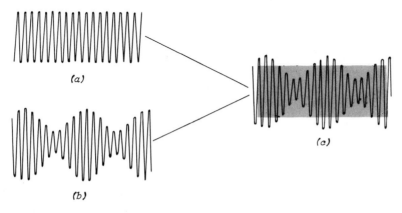

(a)

(b)

(c)

FIGURE 6-5

V_{max} and V_{min} may be measured as either *peak* or *peak-to-peak* values as long as they are consistent throughout the equation. V_{max} is the maximum voltage of the modulated RF carrier waveform; V_{min} is the minimum voltage of the modulated RF carrier waveform. If, for example, an unmodulated RF carrier of 10 V were modulated, yielding a maximum voltage of 15 V and a minimum voltage of 5 V, the modulation index would be:

$$M = (15 - 5)/(15 + 5) = 0.50 \text{ or } 50\%$$

In practice the modulation index for AM systems is often specified as a percentage value.

Once the modulation index is known, the sideband power of an AM signal may be determined as a percentage of the carrier power, with the total power being the sum of the fixed-value carrier power and the variable-value sideband power. The power in the upper sideband and the power in the lower sideband are equal, and they are across the same impedance as the carrier power. The sideband power may be computed as:

$$P_{SB} = M^2 P_C/2 \qquad \text{and} \qquad P_{LSB} = P_{USB} = M^2 P_C/4$$

The total power of the entire AM signal is computed as:

$$P_t = P_C + P_{SB} = P_C(1 + M^2/2)$$

It is recommended that sideband power computations be made using carrier power and modulation index ratios, and that voltage and current computations be made using Ohm's and Watt's laws. You should remember that Watt's law in its basic form requires rms values of voltage and current, whereas the modulation index and V_{max} and V_{min} values are based on either *peak* or *peak-to-peak* signals. In this application, Watt's law may be modified for peak or peak-to-peak voltages as follows:

$$P = (V_{rms})^2/R = (V_{peak})^2/2R = (V_{p-p})^2/8R$$

Sample Problem 1

Given: A 2 MHz RF carrier signal modulated with a 1.5 kHz AF signal, with a measured modulated carrier voltage of 28 V p-p max and 10 V p-p min, across a 100 Ω resistive load impedance.

Determine: The peak, zero-modulation, RF carrier voltage, and RF carrier power; the modulation index; and the sideband signal power, peak voltage, and frequencies.

(*a*) The peak zero-modulation RF voltage is first computed as the average value of the maximum and minimum voltages of the modulated waveform:

$$V_{peak} = (V_{peak\ max} + V_{peak\ min})/2 = (14 + 5)/2 = 9.5 \text{ V (peak)}$$

(b)   The RF power in the carrier is then computed:

$$P = V_{peak}^2/2R_L = 9.5^2/(2 \times 100) = 450 \text{ mW}$$

(c)   The modulation index is then computed as:

$$M = (V_{max} - V_{min})/(V_{max} + V_{min}) = (28 - 10)/(28 + 10) = 0.474$$

(d)   The total sideband signal power is computed as:

$$P_{SB} = P_C(M^2/2) = 450(0.474^2/2) = 50 \text{ mW}$$

so there is 25 mW in *each* sideband.

(e)   The peak voltage for each sideband, LSB or USB, is computed, using Watt's law, as:

$$V_{peak} = \sqrt{2RP} = \sqrt{2 \times 100 \times 0.025} = 2.25 \text{ V}$$

(f)   Finally, the frequencies of the sideband signals are calculated as:

$$f_{LSB} = f_C - f_M = 2 - 0.0015 \text{ MHz} = 1.9985 \text{ MHz}$$
$$f_{USB} = f_C + f_M = 2 + 0.0015 \text{ MHz} = 2.0015 \text{ MHz}$$

MEASUREMENTS

As you may have deduced from the above problem and earlier description of the modulation index, this most important factor is generally obtained by direct measurement.

Time-Domain and Other Techniques

A measurement of the amplitude-modulated RF carrier using a high-frequency time-domain oscilloscope will yield the display shown in Fig. 6-6a. You should be able to quickly observe the modulation index of the waveforms as the ratio of minimum to maximum amplitude. Figure 6-6b shows another popular method of modulation analysis using the scope in an X-Y mode, with the RF signal feeding the vertical input and the AF signal feeding the horizontal (external) input. This second method provides a quick check of the modulation linearity along with the amplitude ratios from which modulation index may be calculated. Of course, the zero-modulation RF carrier power can be computed using the measured voltages and Watt's law (keeping in mind the rms, peak, and peak-to-peak values); and the total RF power (carrier and sidebands) can be computed from the RF carrier power and modulation index values. The time-domain oscilloscope will also provide reasonably accurate measurements of the AF modulation frequency and RF frequency (to the limits of its vertical input and sweep response), from which the sideband frequencies can be computed.

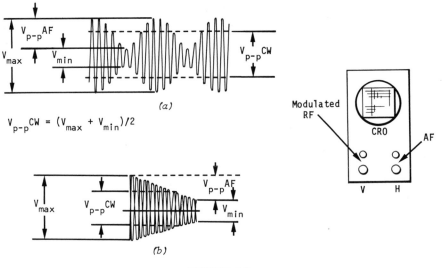

$$V_{p-p}CW = (V_{max} + V_{min})/2$$

FIGURE 6-6

If the frequency of the RF signal is beyond the limits of the oscilloscope, then the RF power and modulation index may be measured indirectly with the use of an RF voltmeter or detector. With this technique the zero-modulation RF carrier power is first measured as an RF voltage across a given load resistance; then the increased total RF power *with* modulation is measured as an increased voltage across the load resistance. If, for example, a 6 V rms signal were measured across a 50 Ω load as zero-modulation carrier power, and a 7 V rms reading were obtained with modulation, the power relationships and modulation index could be computed as follows:

i. First, the zero-modulation RF carrier power is computed as:

$$P_C = V^2/R_L = 6^2/50 = 720 \text{ mW}$$

ii. Then the total power with modulation is computed as:

$$P_t = V^2/R_L = 7^2/50 = 980 \text{ mW}$$

iii. The *total sideband* power is next determined as:

$$P_{SB} = P_t - P_C = 980 - 720 = 260 \text{ mW}$$

iv. Finally, the modulation index is computed by transposing the formula, $P_{SB} = M^2 P_C/2$:

$$M = \sqrt{2P_{SB}/P_C} = \sqrt{2 \times 260/720} = 0.85$$

Unfortunately, this technique is limited because it does not allow for the measurement of sideband frequencies. Special instruments have been developed specifically for the application of the above techniques, however, which will measure carrier and sideband frequencies, as well as carrier and sideband power. The instruments fall into two categories of frequency-domain measurements: frequency selective metering and spectrum analysis.

Frequency-Domain Measurements

The frequency-selective voltmeter (or wave analyzer) is a sort of "poor man's" spectrum analyzer. These instruments were very popular a few years ago when sweep-signal generators and spectrum analyzers were less abundant than they are today. They are nonetheless a basic instrument found in many communications laboratories, and are essentially identical in function to signal-level meters used in communications field tests. The frequency-selective voltmeter is basically a broad-coverage, highly selective radio receiver with very sharp-tuning narrow-bandwidth filtering circuits, with its output connected to an accurately calibrated RF voltmeter. It is used to measure the voltage of RF signals at precise frequencies to the exclusion of all other frequencies. With this instrument the RF carrier and sideband signal voltages are measured, and their frequencies are tuned and identified with the accurately calibrated tuning circuits. To compute modulation index using this instrument, you should obtain a ratio of index values to RF carrier and sideband signal voltages. Since the impedance of the sideband and carrier signals is the same, the voltage ratio is simply the square root of the power ratio of the carrier and *one* (LSB or USB) sideband signal:

$$V_{SB} = MV_C/2, \quad \text{so} \quad M = 2V_{SB}/V_C$$

V_{SB} and V_C can be p-p, peak, or rms values as long as they are consistent within the formulas.

The spectrum analyzer is a sort of automatic frequency-selective voltmeter that graphically displays both frequency and voltage on the face of a CRT. It too is a broad-coverage, highly selective radio receiver of sorts, with sharp *sweep*-tuning narrow-bandwidth filtering circuits. You should recall the detail of the spectrum analyzer as discussed in Chapter 3. The beauty of this instrument compared with the frequency-selective voltmeter is its elimination of the tedium of tabulation and point-to-point plotting of voltage vs frequency curves, as required for analysis of amplitude-modulated signals. Since the spectrum analyzer displays sideband signal voltage levels, the modulation index is calculated using the same formulas that were used with the frequency-selective voltmeter. Let us, for example, consider a spectrum analyzer display showing an RF carrier of 1 V rms across a 75 Ω load at a frequency of 10 MHz, with two 0.2 V (rms) sideband signals (LSB and USB) at frequencies of 9.995 MHz and 10.005 MHz, as in Fig. 6-7b. We can compute the modulation frequency, modulation index, and carrier and sideband power as follows:

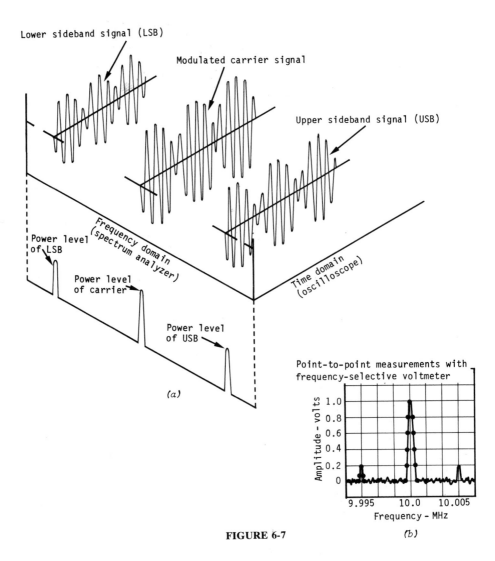

Lower sideband signal (LSB)

Modulated carrier signal

Upper sideband signal (USB)

Power level of LSB

Frequency domain (spectrum analyzer)

Time domain (oscilloscope)

Power level of carrier

Power level of USB

(a)

Point-to-point measurements with frequency-selective voltmeter

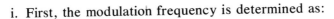

(b)

FIGURE 6-7

i. First, the modulation frequency is determined as:

$$f_M = f_C - f_{LSB} = 10.000 - 9.995 = 0.005 \text{ MHz} = 5 \text{ kHz}$$

ii. Then, the modulation index is computed using voltage ratios:

$$M = 2V_{SB}/V_C = 2 \times 0.2/1 = 0.40$$

iii. The carrier power is then:

$$P_C = V_C^2/R_L = 1^2/75 = 13.3 \text{ mW}$$

iv. The total sideband power is computed as:

$$P_{SB} = M^2 P_C/2 = (0.4^2) \times 13.3/2$$

$$= \begin{cases} 1.07 \text{ mW in both sidebands} \\ 1.07/2 = 533 \ \mu\text{W in each sideband} \end{cases}$$

v. The total power is then computed as:

$$P_t = P_C + P_{SB} = 13.3 + 1.07 = 14.4 \text{ mW}$$

The more sophisticated spectrum analyzers display the RF carrier and sideband signals in dB or dBm through the use of logarithmic amplifiers. This brings two advantages to the measurements: first, low-level sideband signals are increased in proportion to the carrier for better resolution and visual interpretation; and second, power levels can be read directly on the display without resorting to Watt's law and other calculations. Such instruments are complex and expensive, but they quickly pay for themselves by saving time and eliminating errors in measurement. As such, they are a most important tool of the electronic communications field, and you should understand their function to use them effectively.

CIRCUIT TECHNIQUES

Numerous different amplitude modulation circuits represent various ways of achieving the same ends. Many of the older circuits have become obsolete, and thus modified or replaced by circuits using newer improved electronic devices and hardware. Since Part II of this text is concerned with signals, functions, and general processes, rather than the electronic circuit principles covered in Part I, the many varied AM circuits are here reduced to three general modulation technique categories: (1) modulation of the output of an RF power amplifier circuit, (2) modulation of the input to an RF power amplifier circuit, and (3) modulation of low-power circuits preceding the RF power amplifier circuit.

High-Level Modulation

The modulation process can be accomplished only if the circuit where the modulation takes place is *nonlinear*. All the circuits of an electronic system that follow the point of modulation, however, must be *linear* in order to retain the full characteristics of the modulated signal without introducing distortion. Since class C amplifier circuits are most efficient, and incidentally nonlinear, modulation very often takes place at the *final* RF power amplifier circuit of an electronic system. The process is called *high-level modulation*

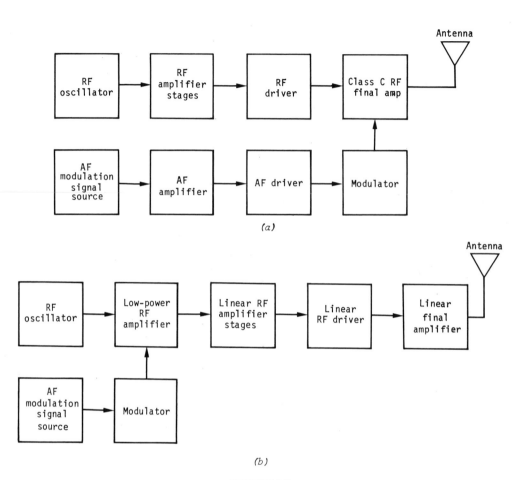

(a)

(b)

FIGURE 6-8

because the power level of the modulating signal must be nearly the same as the RF output signal. An advantage of this approach (Fig. 6-8a) is the use of efficient class C final power amplifier circuits, without the necessity of employing inefficient linear RF power amplifier circuits; a distinct disadvantage is the requirement of high modulating signal power. To place this "tradeoff" in perspective, you might consider that an AM broadcast radio station with 20 kW of RF output power would also require a 10 to 20 kW audio amplifier to modulate the RF carrier signal. So while the high-level modulation approach is simple to achieve, and thus most popular, there are some advantages in favor of lower level modulation (Fig. 6-8b), even though linear RF amplifier circuits are then required.

A reference to power relationships is here required to put the electronic modulation process into perspective. We will use the simple RF amplifier/

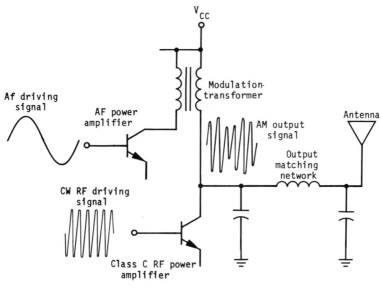

FIGURE 6-9

modulator circuit of Fig. 6-9 as our example of how AM can be accomplished. Given an unmodulated RF carrier signal of 10 V peak across a 10 Ω load, a maximum modulation index of 1.0 will increase the peak carrier voltage up to a maximum of 20 V and down to a minimum of 0 V, and increase the total RF power by a factor of 1.5 times the carrier power. The additional power and control of voltage is a function of the modulator circuit and modulation transformer. The transformer voltage and current must alternately cancel and double the 10 V dc power input to the RF amplifier. Given the dc collector current of the circuit at 500 mA, the dc power input to the RF amplifier is: 10 V × 500 mA = 5 W. If the modulation transformer is to then double and cancel this 10 V, 500 mA level, its *peak* values must also be 10 V and 500 mA. The power in the transformer signal is then computed from its rms values: $0.707^2 \times 10$ V × 500 mA = 2.5 W.

The point to be made here is that while 2.5 W of power is delivered through the modulation transformer, the *average* value of the dc power still remains the same. For example, this supply which is varied from 0 to 20 V, and from 0 to 1 A, still has *average* values of 10 V and 500 mA. In order to account for the power from the modulation transformer we must consider that neither dc nor AF passes through the antenna or other RF circuit which the class C amplifier drives. Only RF signals pass through these circuits, specifically, the carrier and sideband signals. The power transferred—the dc input power and the modulator power—takes the form of these RF signals. Assuming an ideal maximum efficiency (η) of 80% in the class C amplifier device, for example, the RF carrier and sideband power levels would be 20% less than the dc and modulator input powers, respectively.

Low-Level Modulation

Modulating the signal input of an RF power amplifier circuit requires less power than does the high-level modulation technique. This approach is more critical, however, as the input bias of the amplifying device (tube or transistor) must maintain nonlinear device operation for modulation to occur. Figure 6-10 shows the simplified schematic diagram and waveforms associated with input circuit modulation. As you may observe, the amplifying device must be biased at cutoff for proper operation, since only linear mixing (adding) takes place at the input circuit. You should further observe that for 100% modulation the peak-to-peak amplitude of the modulating signal must drive the device from cutoff to full *current*. While the efficient class C amplifier is *always* operating at full current, this type of input modulation *varies* the output current of the amplifier circuit and greatly reduces its efficiency. Input circuit modulation techniques yield efficiency levels of about one-quarter of the class C amplifier efficiency. Considering that our previous example had an optimum 80% efficiency with high-level modulation, it would have about 20% efficiency with input circuit modulation. Even

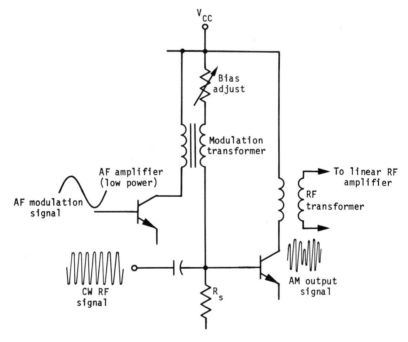

FIGURE 6-10

with such poor output efficiency, input circuit modulation is often used where the power requirements for high-level modulation are excessive.

The modulation power requirements for the input circuit approach depend upon the amplifying device and its output circuit. If we use the class C transistor circuit of our previous example, with 10 V and 500 mA dc power input, and assume an input-to-output current ratio (h_{fe}) of 20, a 5 Ω input impedance (h_{ie}), and a 600 mV cutoff point, we can compute the modulation requirements as follows:

(a)   The peak-to-peak modulation signal current is first computed, based on the peak output current and h_{fe}, as:

$$i_{B(\text{p-p})} = I_C/h_{fe} = 500/20 = 25 \text{ mA}$$

(b)   The p-p modulation voltage is then computed, using Ohm's law, as:

$$v_{\text{p-p}} = i_{\text{p-p}}R_{\text{in}} = 25 \text{ mA} \times 5 = 125 \text{ mV}$$

(c)   The modulation power is computed, using Watt's law for p-p values, as:

$$P_M = v_{\text{p-p}}^2/8R_{\text{in}} = 0.125^2/(8 \times 5) = 391 \ \mu\text{W}$$

(d)   The output power is computed, considering an optimum device efficiency (η_{device}) of 80% as:

$$P_{\text{out}} = \text{rms}(VI)\eta = 0.707^2 \times 10 \times 0.5 \times 0.8 = 2 \text{ W}$$

(e)   The dc input power and efficiency for this modulation scheme are computed as follows:

$$P_{\text{in}} = VI = 10 \times 0.5 = 5 \text{ W}$$
$$\eta_{\text{circuit}} = P_{\text{out}}/P_{\text{in}} = 2/5 = 0.4, \text{ or } 40\%$$

(f)   Finally, the ratio of output power to modulating power is computed:

$$P_{\text{out}}/P_M = 2 \text{ W}/391 \ \mu\text{W} = 5115$$

As you may note from the results of the above calculations, the efficiency of input circuit modulation RF amplifiers is poor and the maintenance of proper bias and circuit operating conditions is critical, but the modulation power requirements are small in comparison to high-level modulation requirements. This modulation technique is thus categorized as *low-level modulation*.

Low-Power Modulation

The third AM technique specifies the modulation of low-power circuits preceding an RF amplifier. This technique is also used in applications where very high power RF amplification is *not* required. As such, this low-level modulation approach is concerned with neither efficiency factors nor power levels, since only low-power and low-signal levels are involved. It is still important, however, to assure that the modulation circuit is nonlinear and that all following circuits *are* linear. The approach is a poor choice for very high power RF systems, because class A and B (linear) RF power amplifier circuits are inefficient. Although it is possible to modulate the input circuits of low-power devices at cutoff, as with the previous class C example, the critical biasing requirements clearly eliminate its use in favor of output-circuit modulation. The simplest approach to output modulation is the control of the dc power supplied to an RF amplifier circuit with the modulating signal. A second approach that is popular with symmetrical-power-supply IC devices is through use of the modulating signal to control the gain of the integrated circuit. Figure 6-11 shows these general approaches to low-power, low-level amplitude modulation. The circuit and modulation power requirements with these approaches can be simply computed with Ohm's and Watt's laws or with IC device gain formulas. It is also possible to use this approach to amplitude modulate the output or gain of RF oscillators, but the modulation signal can shift the frequency and change the feedback, thus reducing the stability of the oscillator circuit. The practice is not recommended except in cases where extreme economies are required in electronic systems.

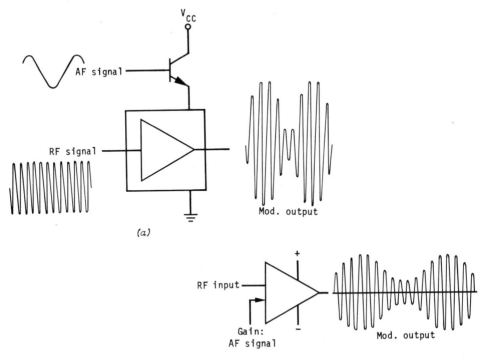

FIGURE 6-11

TYPES OF AMPLITUDE MODULATION

The other AM-related techniques that are used in electronic communications more efficiently utilize the RF power in the sideband signals and/or the amount of radio space occupied by the modulated RF signals. The conventional AM technique is more properly defined as double sideband amplitude modulation (DSB-AM), because the carrier and both sideband signals are transmitted. The amount of radio space occupied by this DSB-AM technique is twice the highest modulating frequency. You should know, for example, that the popular AM broadcast radio stations have a limited bandwidth to as little as 10 kHz, depending on their possible interference with adjacent stations. The highest audio frequency that they may modulate on the RF carrier is only 5 kHz. This is of course why AM radio isn't "hi-fi." If the frequency of modulation were increased to 15 kHz for better sound, each radio station would occupy 30 kHz of radio space, thus reducing the available station allocations by a factor of 3. Figure 6-12*a* and *b* compares these bandwidths for your reference. DSB-AM video modulated TV signals would require over 8 MHz of radio space for just the 4 MHz picture portion of the transmission, as shown in Fig. 6-12*c*. Television systems have thus modified the character of the amplitude-modulated RF signals by filtering out the lower sideband. The U.S. system leaves about 1 MHz LSB signal with a 4.5 MHz USB signal. This *vestige* of the lower sideband signal, shown in dashed lines on Fig. 6-12*d*, is the source of the name applied to the modulation technique: *vestigial sideband amplitude modulation,* abbreviated VSB. Figure 6-12*e* shows the frequency-domain response of full modulation VSB transmission for TV channel 2 according to U.S. standards (National Television Standards Committee, NTSC).

Vestigial-Sideband Modulation

Vestigial sideband modulation is achieved by the conventional DSB-AM techniques and circuits outlined earlier in this chapter, but the higher frequency portion of the modulation on the lower sideband (actually *lower* radio frequency) is filtered using highpass filter circuits. A unique power and modulation condition results with VSB-AM when only a portion of the LSB signal is filtered: the lower modulating frequencies distribute power in both sidebands, while the higher frequencies are present in only *one* (the USB) sideband (see Fig. 6-12*e*). The total RF power with full (100%) modulation is 1.5 times the carrier power for the lower modulating frequencies, and reduces to 1.25 times the carrier power for the higher modulating frequencies. Since the average RF carrier power is a fixed value, however, and all the intelligence signal power is in the sidebands, the higher frequency modulation signals will be only half the intensity of the lower frequency signals.

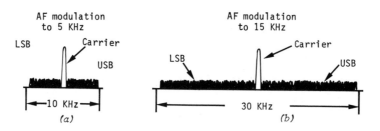

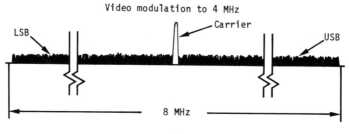

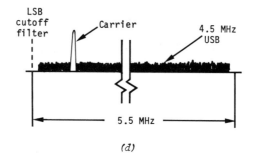

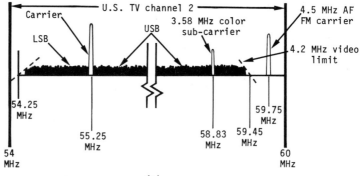

FIGURE 6-12

This nonuniform frequency response must be compensated for in other system components when VSB modulation is employed to reduce bandwidth requirements. Such frequency compensation is achieved in the receiver portion of television systems.

Although the VSB modulation technique effectively reduces the bandwidth of a transmitted RF signal, the approach only slightly reduces the total RF power since there are substantial lower frequency signals in DSB form. Significant power reduction can be achieved, however, by removing the RF carrier and leaving only the sideband signals in the system. This drastic step is not all that severe if you give it some thought. First, the average RF carrier power remains fixed with or without modulation, so it actually "carries" neither the frequency nor the intensity of the modulation signal. Both the frequency and the intensity of the modulation are contained in the sideband signals. Second, at least two-thirds of the total RF power in the conventional DSB-AM system is transmitted in the RF carrier. From these two statements of fact you may predict that the RF carrier is somewhat useless. Well the answer to that prediction is a qualified "yes" and a qualified "no." *Some* RF communications systems need a carrier signal, while others don't; and for those that don't, there is a big power bonus of at least three to one. This "carrier-less" modulation technique is called *double sideband suppressed-carrier amplitude modulation,* abbreviated DSB-SC.

Suppressed-Carrier Modulation

It is a little more difficult to obtain DSB-SC modulation than it is to achieve VSB modulation. Filtering won't work because even the highest Q filters occupy some bandwidth; so the low-frequency modulation in the frequencies adjacent to the RF carrier would be filtered out along with the carrier. The RF carrier can be removed by cancellation, however, through splitting and shifting the phase of one split signal by 180°, and then combining the two signals for cancellation. This neat trick is accomplished with a circuit borrowed from telephone technology: the *balanced modulator.* Figure 6-13 shows two balanced modulator circuits. One approach uses transistors and the other uses diodes in order to obtain the required nonlinearity for mixing the RF and modulation signals. Both circuits are actually bridge networks, with transformer-coupled RF and modulation input signals and with the output transformer tuned to the frequency of the carrier. In the transistor circuit the RF signal is applied to the balance point of the floating emitters. In the absence of a modulation signal the center-tapped output transformer places the collector signals 180° out of phase, and the RF carrier is thus canceled. The modulation signal, however, passes through the center-tapped input transformer, which splits the phase of the inputs to the transistors' bases, thus bringing the collector signals in phase—and mixed with the carrier—to become the USB and LSB components of the RF signal.

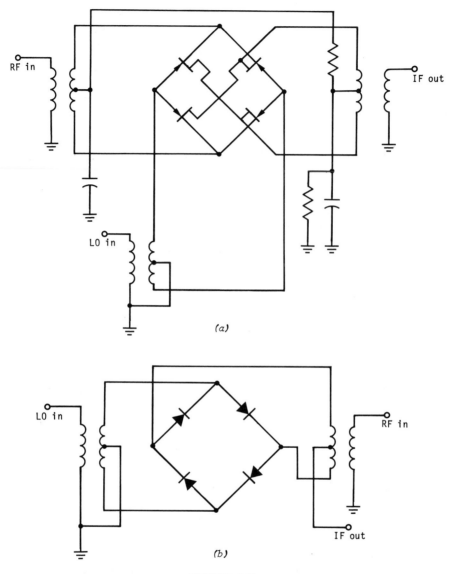

RF in

IF out

LO in

(a)

LO in

RF in

IF out

(b)

FIGURE 6-13

The diode balanced modulator (Fig. 6-13*b*) is often called a *ring modulator* because of the diode "ring." The diodes are basically connected in a bridge-rectifier configuration, but with the diodes reversed to compensate for the center-tapped transformer input of the RF signal. With the absence of modulation and with the RF signal large enough to *switch* the diodes alternately *on* and *off*, the carrier signals will be 180° out of phase across the center-tapped output transformer when either pair (D_1-D_4 or D_2-D_3) of diodes is

conducting; thus the RF carrier is canceled. With the presence of the modulating signal *alone,* the alternately conducting diodes (D_1-D_2 or D_3-D_4) will
bypass the output transformer. Now, when *both* RF *and* modulating signals
are present, however, the carrier switching will alternately turn *off* the shunt
diodes (D_2 and D_3), and the signals will combine in the output transformer in
phase to become the USB and LSB components of the RF signal.

Now that we have reviewed two reduction techniques—one for reducing
bandwidth (VSB), and the other for reducing power (DSB-SC)—we can
proceed to combine the concepts of both approaches for the ultimate of AM
processes, called *single sideband* (SSB). With DSB-SC modulation, the upper
and lower sideband signals are identical; they each contain all the frequencies and intensities of the modulated intelligence signals. Assuming a receiving system that is capable of recovering the intelligence from the transmission of only *one* sideband, the SSB modulation technique can achieve electronic communication equivalent to the full DSB-AM approach, but using
only half the radio space and, incredibly, only one-sixth the power of conventional AM.

Single-Sideband Modulation

The higher power outputs from SSB modulators require linear RF amplifiers
to assure no distortion of the SSB signals. Unlike the DSB-AM system, there
is no output from an SSB circuit without modulation, and the output power
will vary directly with the intensity of modulation. The concept of *dc input
power vs RF output power* does not apply here; instead, SSB power levels are
rated in *peak envelope power* (PEP). In comparison to *real* rms power, peak
envelope power is the product of p-p voltage times p-p current:

$$\text{PEP} = V_{\text{p-p}}I_{\text{p-p}} = 2[1.414^2(V_{\text{rms}}I_{\text{rms}})] = 4P_{\text{rms}}$$

In comparison to a full (100%) DSB-AM power output of 3 W when operating
at 60% efficiency and 5 W dc input power, the SSB equivalent PEP, for
example, would be rated at $4 \times 3 = 12$ W. You should keep in mind that the
12 W is not an equivalent average dissipation value, but only an equivalent
instantaneous "talk-power" value—the *real* (rms) power dissipation is 12/4
= 3 W in this example, or PEP/4.

SSB modulation can be obtained using the combination of a balanced
modulator to suppress the carrier, and a filter to suppress the upper or lower
sideband. Since even the best of bandpass filters have limitations, however,
their use requires other system modifications in order to obtain at least 50 dB
of rejection in as tight a frequency range as 500 Hz. Since the sideband
frequencies are separated by twice the (AF) modulating frequency, the lowest modulating frequencies dictate this separation. If the modulation consists
of a voice channel of 300 to 3000 Hz, for example, the filter will have to pass
that voice channel, but greatly reject the adjacent sideband signal which is

only 600 Hz (twice 300 Hz) above or below the passband. Although filter technology has advanced by the use of specialized crystal and mechanical elements, these must operate at relatively low frequencies in order to provide tight frequency-rejection response. On this basis the *initial* carrier modulation must take place at relatively low radio frequencies, then subsequently be upconverted to the final radio frequency through additional mixer and linear RF amplifier stages. Figure 6-14*b* shows the system block diagram for the balanced-modulator/filter technique of SSB modulation.

A second popular technique for obtaining SSB modulation uses two balanced modulators—one to cancel the carrier, and one to cancel the upper or lower sideband. This approach has the advantage of operation at the final radio frequency, thus alleviating the need for upconversion. Figure 6-14*c* shows this *phase-shift* SSB modulation technique in block diagram form. Both balanced modulators have direct and 90° delayed signals from the RF carrier and modulation signal inputs. (You may refer back, if necessary, to Fig. 6-13 to review the operation of the balanced modulator.) Two DSB-SC signals appear at the outputs of the two balanced modulators, but with one pair (LSB or USB) of the sideband signals 180° out of phase and the other pair in phase. The two signals are then combined in a linear mixer (adder) circuit, where the out-of-phase signals cancel and the in-phase signals double, leav-

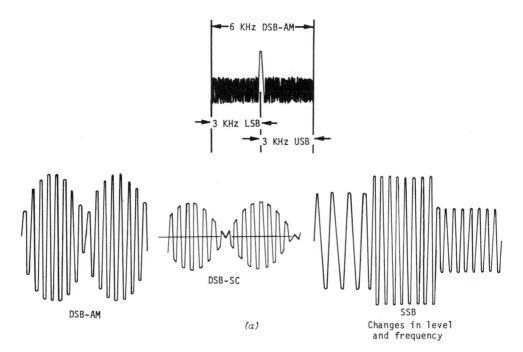

FIGURE 6-14

ing only the single upper or lower sideband signal. With the availability of special IC devices, the phase-shift SSB modulation technique has become more practical to achieve than the filter technique.

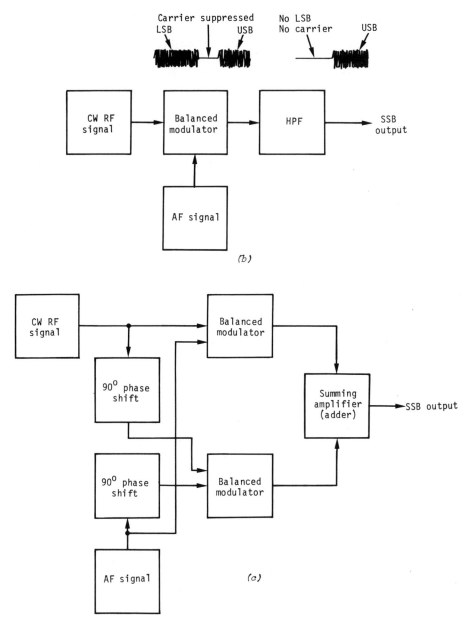

(b)

FIGURE 6-14 (continued)

DEMODULATION

All four modulation techniques that were here reviewed—DSB-AM, VSB, DSB-SC, and SSB—represent only a portion of system requirements for electronic communications. The *correlated* system element for each of the techniques is the detector (or demodulator). Any modulation system is useless without an appropriate detector to extract the intelligence information from the modulated RF signals. While detectors for DSB-AM are relatively simple, those for SSB modulated signals are complex and often critical to tune. There are two general categories of AM detectors—*envelope* detectors and *product* (synchronous) detectors. Envelope detectors are most often used for DSB-AM and VSB modulation, whereas the product detectors are most often used for DSB-SC and SSB modulation. Let us now proceed to the demodulation processes.

Envelope Detection

Envelope detectors, as their name implies, extract the intelligence signal from the modulated envelope of the RF carrier. We earlier advanced a strong argument that the modulated RF signal was necessary for radio transmission through space or wires, but that it was useless for carrying the intelligence signal since the intelligence also appears in the sideband signals. The two reasons for the "loss" of the modulation signal in the RF carrier are here reviewed, since they will serve as the basis for requalifying the virtue of the RF carrier. First, the reason that the RF carrier carries no intelligence signal with DSB-AM and VSB modulation is that the RF signal is "ac" modulated; so for every *increase* in its amplitude, there is an equal, but opposite, *decrease*. The envelope of the modulation signal is thus 180° out of phase above and below the zero line. Second, the reason that the *average* power of the RF carrier is the same, with or without modulation, is that for every *increase* in amplitude there is an equal *reduction* in amplitude of the RF signal as it follows the envelope of the modulation signal. These increases and reductions cancel, leaving the same *average* power in the carrier that it dissipated without modulation. You might better understand these two long justification statements if you review them while referring to Fig. 6-15.

Now the envelope detection process simply modifies the character of the amplitude-modulated RF carrier by eliminating the two reasons for its "uselessness," and it is thereby able to extract an intelligence signal from the carrier's envelope. If the amplitude-modulated RF carrier signal drives a simple half-wave rectifier and filter circuit, the process of envelope detection can be summarized as follows:

1. The AM RF signal passes through a rectifier diode network which eliminates the negative or positive (180° out-of-phase) portion of the modulation envelope.

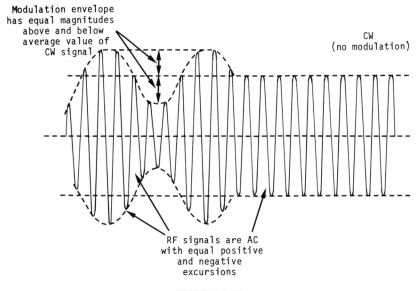

Modulation envelope
has equal magnitudes
above and below
average value of
CW signal

CW
(no modulation)

RF signals are AC
with equal positive
and negative
excursions

FIGURE 6-15

2. The lowpass filter will block the high-frequency rectified signals and average their changing amplitude, leaving the modulation signal that is dc offset by the average RF carrier level.

3. The coupling capacitor will block the dc offset from the carrier, and what remains is the low-frequency signal identical to that which initially modulated the RF carrier.

Figure 6-16 shows a conventional low-cost envelope detector circuit. Although there are limitations to its effectiveness, the circuit is sufficiently popular to demand our attention to practical component selection. The critical components are the detector diode and filter capacitor; and the extent to which this simple circuit may be effective depends upon the maximum value of the modulation index, the loading impedances, and the frequency ratio of the RF and modulation signals. First, conventional diodes have nonlinear regions near their cutoff voltages. Silicon diodes, for example, may cut off at 550 mV and become linear above 650 mV. If an AM signal at 100% modulation is to be detected by such a silicon diode, the average RF carrier level must raise the dc offset so that the minimum modulation envelope amplitude will still be greater than 650 mV. Since the average carrier level can now only equal the peak value of the modulation signal, however, this desired condition cannot occur. The modulation index must either be reduced to accommodate the diode, or an additional dc bias must be applied to increase the offset on the diode to assure linearity. The modulation index would have to be reduced in this case to obtain a difference of 650 mV between the peak

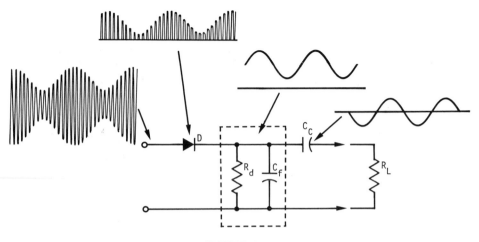

FIGURE 6-16

value of the RF carrier and the peak value of the modulation. If an RF carrier of 2 V peak (which creates the dc offset) is assumed for this example, the modulation signal would be limited to 2 V − 0.65 V = 1.35 V, and the maximum modulation index would be 1.35 V/2 V = 0.65 or 65%. A maximum modulation index of only 0.65, however, is unacceptable in such a detector, as signals of higher modulation index factors will be distorted in the demodulation process.

Without changing the circuit configuration of the detector, it would be possible to bring the modulation index to a most acceptable 0.90 by raising the value of the peak RF carrier voltage to 6.5 V, but this ridiculously high value RF signal is impractical to obtain. A more intelligent choice for the detection of higher modulation indices with low-level RF signals is the use of other diode devices with low-voltage linear response. Germanium diodes are linear at about 250 mV and are a good choice for use in noncritical, low-cost applications such as portable radio receivers. Point contact and hot carrier (Schottky barrier) diodes are the best choice for AM detectors in quality communications systems, however, with linear response as low as 50 mV and 5 mV, respectively. (Zero-bias Schottky barriers are also available, but costly.) Such diodes are also used as detector elements in RF test and measurement setups.

Detector Component Values

The computations for the circuit component values of Fig. 6-16 relate the maximum modulation index to the resistance ratios of the diode load and circuit load. The diode must, of course, have a dc path for operation as a rectifier. This dc load resistance is designated R_d. The ac load on the diode is this resistance (R_d) in parallel with the equivalent-circuit load resistance, designated R_L. If the value of R_L is significantly lower than that of R_d, the operating point on the diode response curve, and consequent dc offset of the modulation signal, will drop upon the application of modulation to the RF carrier. This loading effect will tend to cut off the output voltage at higher modulation levels, as shown in Fig. 6-16. In order to drive practical equivalent load resistance values and maintain proportional dc bias, there is some practical limit to the maximum modulation index that the envelope detector can process without distortion. Since modulation index is a proportional value, it can be related to the resistive load on the diode, as well as the signals across its output, as:

$$M = R_{ac}/R_{dc} = (R_d\|R_L)/R_d = [(R_d R_L)/(R_d + R_L)]/R_d$$

In practice, the value of load resistance (R_L) is usually known, with the value of R_d established based on the practical maximum modulation index of the signal applied to the detector. In this case the above formula is used in its transposed form to solve for R_d:

$$R_d = R_L|(1 - M)/M|$$

If, for example, a maximum modulation index of 0.80 is to be detected and drive a 600 Ω load, the diode load resistor can be computed as:

$$R_d - R_L|(1 - M)/M| = 600 \times (1 - 0.8)/0.8 = 150 \ \Omega$$

Once the values of R_d and R_L are established, the value of the filter (or RF bypass) capacitor, designated C_f, can be computed. Its value must provide a lowpass filter function without excessive loading. This provision can be achieved with a single capacitor if the frequencies of the RF and modulation signals are at least two orders of magnitude apart (100 : 1). If a single capacitor is not practical as a filter, a π-network of two capacitors and a series resistor is used. Since the reactance of the filter capacitor is proportional to the ratio of R_{ac} to R_{dc}, the equivalent magnitude of the modulation index (M), its value can be determined using those terms:

$$C_f = \sqrt{1 - M^2}/\omega R M$$

C_f is the filter capacitance, M is the modulation index, R is the parallel combination of R_d and R_L, and ω is computed using the highest modulation frequency. If we use the same values as in the previous example, and consider the highest modulating frequency to be 5 kHz, the filter capacitance can be computed as follows:

i. First, ω is computed for 5 kHz as:

$$\omega = 2\pi f = 6.28 \times 5 \times 10^3 = 3.14 \times 10^4$$

ii. Then, R is computed as the parallel value of R_d and R_L:

$$R = R_d R_L / (R_d + R_L) = 150 \times 600/(150 + 600) = 120 \ \Omega$$

iii. C_f is computed as:

$$C_f = \sqrt{1 - M^2}/\omega R M$$
$$= \sqrt{1 - 0.8^2}/(3.14 \times 10^4 \times 120 \times 0.8) = 0.2 \ \mu\text{F}$$

Product Detection

If you review the initial mixing processes outlined at the beginning of this chapter along with the techniques for achieving amplitude modulation, you should conclude that since modulation results from a mixing process, the reversal of that process will undoubtedly recover the *original* modulation signal from its mixed RF sideband form. If we remix a full DSB-AM signal with a second fixed-frequency RF signal that is locked-on to the carrier, and obtain the difference frequency through a lowpass filter at the output of the mixer, as in Fig. 6-17, the low-frequency modulation signal will remain. This technique is called *product detection, synchronous detection,* or *heterodyne detection,* because the modulation signal is obtained through the nonlinear mixing process. When compared to the low cost and simplicity of envelope detectors, product detection is impractical for DSB-AM or VSB modulated signals. Product detection is a most practical approach for DSB-SC and SSB, however, where envelope detectors won't work, because neither of these modulation techniques include an RF carrier signal in their transmission.

If a SSB voice-modulated RF signal is received and downconverted to operate between 450.3 kHz and 453 kHz, the voice channel of 300 Hz to 3 kHz can be *product detected* through mixing the sideband signal with a 450 kHz fixed-frequency signal that is locally generated in the receiver. This signal is often generated in a circuit called a *beat-frequency oscillator* (BFO). As with all mixers, four signals are available at the output: the two original frequencies, 450 kHz and the sideband 450.3 to 453 kHz; the sum of the two, 900.3 to 903 kHz; and the difference of the two, 0.3 to 3 kHz. A lowpass filter at the mixer output would reject all signals but the 0.3 to 3 kHz of the original voice modulation. You can incorporate a sense of humor into SSB voice reception with the use of a variable-frequency product detector. A deep male voice of 300 Hz could be changed into a soprano by reducing the locally generated signal frequency by about 800 Hz; and a child's high-pitched voice could be made instantaneously mature by increasing the local signal frequency by about 800 Hz.

It is relatively simple to achieve product detection through the mixing process, especially at the lower signal levels that prevail in demodulation circuits. A circuit as simple as that of Fig. 6-18a, with only two diodes for nonlinearity, will work effectively as a product detector for SSB RF signals.

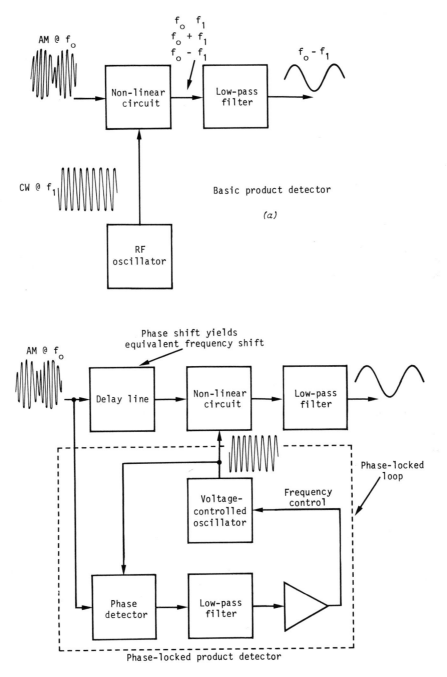

Basic product detector

(a)

Phase shift yields
equivalent frequency shift

Phase-locked
loop

Phase-locked product detector

(b)

FIGURE 6-17

As with all mixers/modulators, the *carrier signal,* which in this case is locally generated, must have at least twice the power and voltage of the sideband signals at the point in the circuit where the mixing takes place. If, for example, the circuit of Fig. 6-18*a* is to be used as a SSB detector for a 2 V p-p input signal, the detection process would yield the waveforms shown in Fig. 6-18*b*, as follows:

1. The rectification function of D_1 would reduce the 2 V p-p input signal to a 1 V peak value at the junction of the local *carrier* signal input between D_1 and D_2.

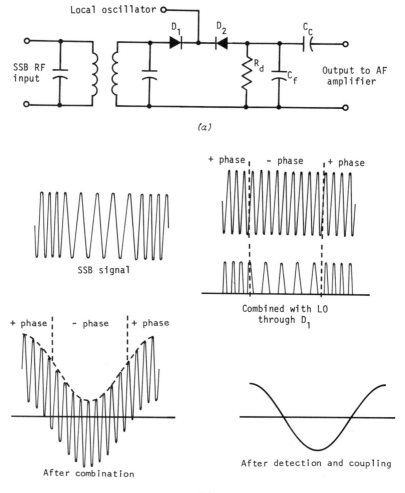

(a)

SSB signal

+ phase   − phase   + phase

Combined with LO
through D_1

+ phase   − phase   + phase

After combination

After detection and coupling

(b)

FIGURE 6-18

2. A local carrier signal of 2 V peak would then combine with the SSB signal, shifting phase at a rate equal to their difference in frequency.

3. When the phase of the two signals combines the zero voltage of the rectified SSB signal with the local *carrier* signal, the full 2 V peak value of the local signal remains.

4. When the phase of the two signals shifts by 180° to combine the positive peak voltage of the rectified SSB signal with the local signal, the dc level shifts and reduces the negative peak value of the local signal to a 1 V level.

5. Since diode D_2 passes only the negative peak values of the combined signals, the output waveform is now similar to that of the *envelope* detector.

6. The lowpass filter function of C_f will average the changing amplitude of the rectified RF signal, leaving the original voice-modulation dc offset by the local *carrier* level.

7. Finally, the coupling capacitor C_c will block the dc offset from the *carrier*, and what remains is the 1 V p-p voice signal.

The procedures of the preceding steps can be applied to the analysis of signal combination in a variety of nonlinear circuits used as product detectors. Figure 6-19 shows some variations of product detector circuits using transistors, JFET's, and a dual-gate MOSFET. A wide variation of circuits can achieve product detection, and choices vary depending upon cost, linearity, and stability requirements. The balanced modulator is fast becoming a most popular product detector circuit. This versatile circuit has such widespread uses that both the transformer and semiconductor elements are mass produced in large quantities, making this the best choice for both modulator and detector applications.

IC APPLICATIONS

The amplitude-modulation and detection *processes,* which are a most important aspect of electronic communications, have been reduced to IC subsystem function components wherever high-volume needs for low-power circuits exist. Among the most popular of these IC elements are diode and transistor arrays arranged to perform as balanced mixer/modulator/demodulator elements in conjunction with RF/AF transformers. The arrays can also be used effectively in simpler mixer-type product detectors that require closely matched semiconductor devices. Figure 6-20a shows a balanced modulator

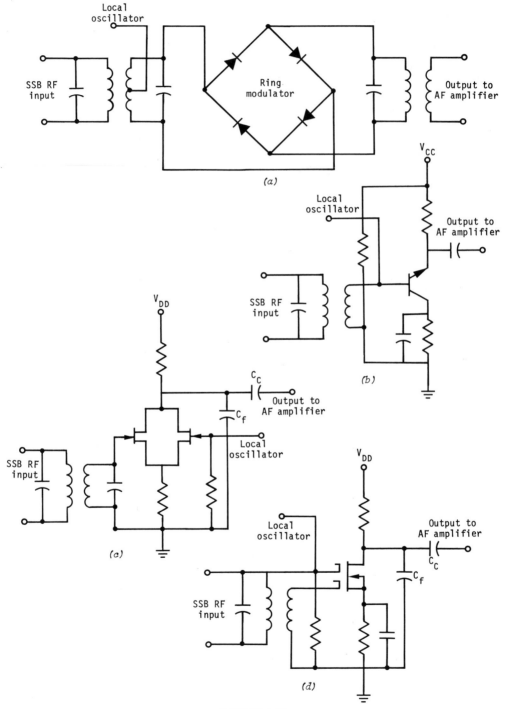

FIGURE 6-19

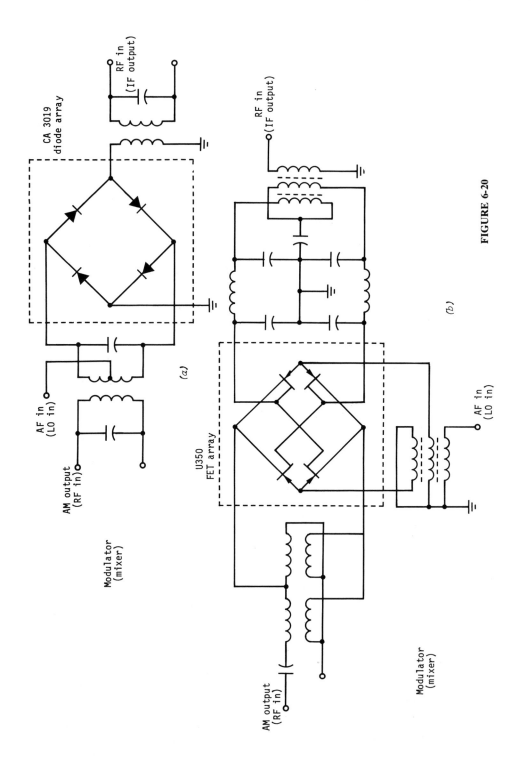

FIGURE 6-20

TABLE 6-1.   COMPARISON OF DIODE AND FET
BALANCED MODULATOR/MIXER SPECIFICATIONS

Specification	Diode Type	FET Type
Frequency range, MHz		
LO	1-750	1-200
RF	1-750	1-200
IF	Dc-750	1-70
Conversion loss, dB	6.5	1.5
Input signal (1 dB	+10	+10
compression level), dBm		
Isolation, dB		
LO-IF	40	25
LO-RF	45	45
Impedance, all ports, Ω	50	50
LO power level, dBm	+17	+17
Two-tone, third-order	−40	−60
intermodulation		
distortion, dB		

circuit that uses a simple IC diode array; Fig. 6-20*b* shows the same basic circuit using an FET array. The performance of the two circuits is compared in Table 6-1.

Balanced Modulator Devices

Recent advances in circuit technology have made available high-volume low-cost subsystem mixer/modulator/demodulator circuits as off-the-shelf components. Although they are not *monolithic,* they are essentially *integrated circuits,* as they are available in standard packages and in a range of performance specifications for both circuit and system use. The choice of using such a complete subsystem unit is often a matter of cost vs production volume vs convenience vs engineering conservatism. It is the author's experience that building your own balanced modulators, even using IC arrays and separate transformers, just isn't worth the *hassle!* Figure 6-21 shows the popular subsystem balanced modulator/mixer components with internal circuit connections.

Effective use of balanced modulator/mixer elements requires knowledge of their important specifications and operating requirements. Among the important performance specifications are conversion loss and isolation, while operating requirements include a specified LO level and frequency range of operation. *Conversion loss* is generally specified in dB and is the ratio of output-to-input signal loss through the device. *Isolation* is also specified in dB and is the ratio of the unwanted carrier output signal-to-carrier input signal through the device. *LO* is the abbreviation for *local oscillator,* which provides the locally generated fixed-frequency *carrier* in modulation and detection

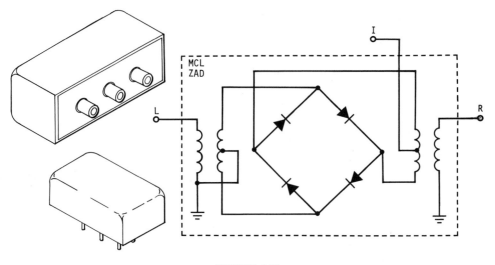

FIGURE 6-21

applications of the balanced mixer. The LO level is generally specified in volts or dBm with a given impedance value. The *frequency range* of operation indicates broadband RF limits at two terminals of the balanced mixer/modulator, with response to dc at the third terminal for the low-frequency modulation signal.

Sample Problem 2

Given: The *phase-shift* SSB modulation subsystem of Fig. 6-14c, using two off-the-shelf component-type double-balanced modulators, single-stage AF and RF amplifier circuits for the 90° phase shift functions, and a differential amplifier circuit (IC) for the linear mixer (adder) function.

Determine: The *optimum signal levels* assuming that all three amplifier circuits have unity gain ($A_v = 1.0$) and that specifications for the balanced modulators include 6 dB conversion loss, 60 dB isolation, and a required +23 dBm/50 Ω LO level.

(a) First, the voltage level for the LO signals is computed in two steps, based on a nominal 50 Ω system impedance, as follows:

$$P = \log^{-1}(\text{dBm}/10) = \log^{-1}(23/10) = 200 \text{ mW}$$
$$V = \sqrt{PZ} = \sqrt{0.2 \times 50} = 3.16 \text{ volts (rms)}$$
$$V_{\text{peak}} = V\,(\text{rms}) \times 1.414 = 3.16 \times 1.414 = 4.5 \text{ volts (peak)}$$

(b) The voltage level for the AF modulating signals is then computed, based on a standard level at 10 dB below that of the LO, as:

$$V_{AF} = V_{LO}/\log^{-1} (dB/20) = 3.16/\log^{-1} (10/20)$$
$$= \begin{cases} 1 \text{ volt (rms)} \\ 1.4 \text{ volts (peak)} \end{cases}$$

(c)   The voltage level for the sideband signals from the balanced modulators is computed, based on a specified level of 6 dB below that of the AF voltage, but combined in phase to twice that value:

$$V_{SB} = V_{AF}/\log^{-1} (dB/20) = 1/\log^{-1} (6/20) = 0.5 \text{ volt (rms)}$$
$$V_{out} = 2V_{SB} = 2 \times 0.5 = 1 \text{ volt (rms)}$$

(d)   Finally, the carrier output and suppression factor are computed, based on a specification that the carrier signal will be 60 dB below that of the LO, and on the fact that the two signals are combined in the linear mixer which adds 6 dB to their value:

$$\text{Carrier isolation} = -60 + 6 = -54 \text{ dB}$$
$$V_{C(out)} = V_{LO}/\log^{-1} (dB/20)$$
$$= 3.16/\log^{-1} (54/20) = 6.3 \text{ mV}$$
$$\text{Suppression} = 20 \log (V_{SB}/V_{C(out)})$$
$$= 20 \log (1/6.3 \times 10^{-3}) = 44 \text{ dB}$$

Linear IC Balanced Modulator

Although the diode/transformer-type double-balanced modulators that we have reviewed are a most popular choice for communication systems use, there is a second, more modern approach that uses matched transistors in a monolithic IC chip. This approach requires no transformers or tuned circuits; and instead of being a lossy (conversion loss) passive circuit element, it has signal gain. An example of this monolithic device is the IC manufactured as a 1496/1596 linear product. (There are many similar dual-differential IC devices with different numbers.) This most versatile IC can be used as a DSB-AM, DSB-SC, and SSB modulator, as well as a DSB-SC and SSB detector. Its internal circuitry and external connections for use as a modulator and demodulator are shown in Fig. 6-22. The balanced-modulator function is achieved with the matched transistors of the differential amplifiers, with the modulating signal applied to control the "diff-amp" emitter current and the carrier signal to *switch* the bases of the diff-amp transistors. The resulting mixing process occurs with the mixing product signals out of phase at the collectors of the diff-amp transistors. Since the 1496/1596 is a dual device, the appropriate combinations of phase are available at the outputs, thus providing versatility of modulator/detector functions. If the IC is to be used as a product detector, as in Fig. 6-22d for example, its operation can be described as follows:

The locally generated carrier signal is applied to the diff-amp bases (pin 7) at a specified 300 mV level for switching operation. The received SSB signal is applied to control the emitters of one diff-amp (pin 1) of the pair at a signal voltage between the specified device limits of 5 mV and 500 mV. The lowpass output filter network of the π-connected RC components removes all signals except the original low-frequency modulation of the SSB signal.

This same circuit connection will allow the 1496/1596 IC to be used as a DSB-AM detector, as may be required in communication systems using both SSB and DSB-AM approaches.

CONCLUSION

This chapter has been concerned with the principles and techniques of amplitude-modulating intelligence signals on RF carrier signals to achieve efficient radio communications. Your working knowledge of AM RF signals, functions, and processes should provide you with a thorough understanding of electronic communication circuits and systems that employ these modulation and detection techniques. These systems concepts will be further expanded in later chapters. The problems that follow will serve to test your knowledge of AM principles and techniques, and measure your achievement of the *learning objectives* specified at the beginning of this chapter. Your further success with these objectives will be enhanced by your review of commercial communication systems literature from manufacturers, and also your experimentation with the circuits and IC devices used in these systems.

Problems

1. Given: A 27 MHz RF carrier signal modulated with a 3.0 kHz AF signal, operating with a range of 30% to 90% modulation index values, and with the RF carrier signal at 2.5 W across a 50 Ω resistive load impedance.

Determine:
- (*a*) Zero-modulation peak RF carrier voltage
- (*b*) Sideband signal frequencies
- (*c*) Sideband signal power at 30% and 90% modulation
- (*d*) Sideband signal voltage at 30% and 90% modulation
- (*e*) Maximum RF carrier voltage at 30% and 90% modulation
- (*f*) Minimum RF carrier voltage at 30% and 90% modulation

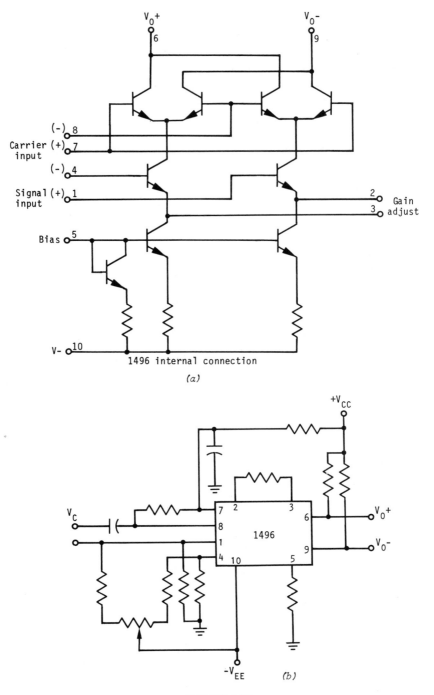

1496 internal connection

(a)

FIGURE 6-22

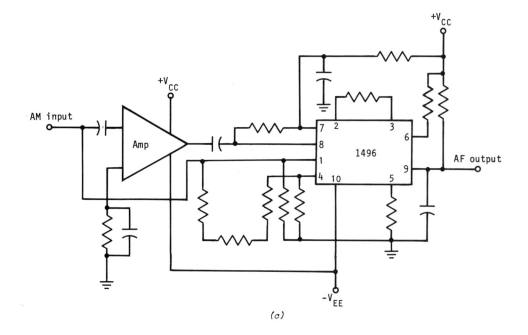

(c)

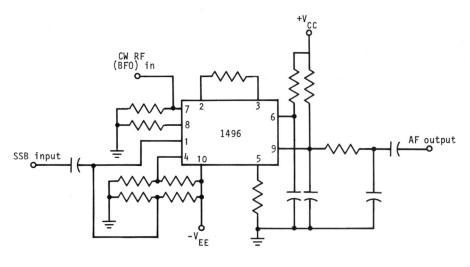

(d)

FIGURE 6-22 (continued)

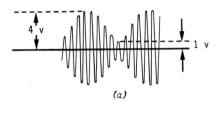

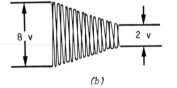

(a)

(b)

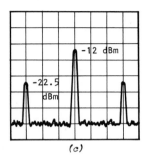

(c)

FIGURE 6-23

2. Given: The measured responses of oscilloscope and spectrum analyzer displays as shown in Fig. 6-23.

Determine:
(a)   RF carrier power
(b)   Maximum RF carrier voltage
(c)   Minimum RF carrier voltage
(d)   Modulation index
(e)   Resistive load impedance
(f)   Sideband signal voltage
(g)   Sideband signal power
(h)   RF carrier frequency
(i)   Modulation frequency

3. Given: A class C RF final amplifier circuit with 10 W dc power input, a maximum efficiency of 75%, and an RF resistive load of 75 Ω; to be high-level amplitude modulated with an AF signal to a maximum modulation index of 1.0.

Determine:
(a)   Full-modulation RF power output
(b)   Zero-modulation RF power output
(c)   Dc input supply voltage
(d)   Peak zero-modulation carrier voltage
(e)   Peak maximum carrier voltage (100% modulation)
(f)   AF modulator output power
(g)   AF modulator rms output voltage

4.   Given: A class C RF transistor final amplifier circuit with 50 W dc power input from a 40 V supply, an h_{fe} of 15, an h_{ie} of 2 Ω, and a maximum efficiency of 80%.

Determine the requirements for low-level AM to a maximum modulation index of 1.0 as follows:

    (*a*)   Peak output current
    (*b*)   Peak-to-peak input modulation current
    (*c*)   Peak-to-peak input modulation voltage
    (*d*)   Input modulation power
    (*e*)   Maximum total RF output power
    (*f*)   RF carrier output power
    (*g*)   Zero-modulation RF output efficiency
    (*h*)   Resistive RF load impedance

5.   Given: The differential amplifier circuit of Fig. 6-24 to be amplitude modulated with the AF transistor Q_1 controlling the base-bias voltage of the diff-amp through the voltage drop across R_1, and at a Q point of 4 V across R_1.

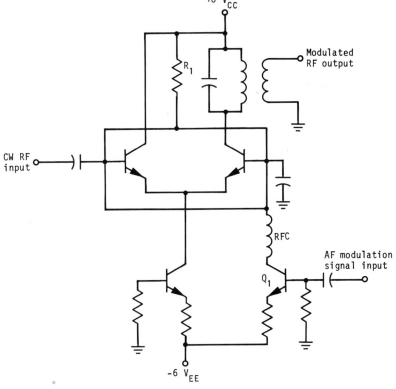

FIGURE 6-24

Determine the circuit conditions for a maximum modulation index of 0.8, as follows:
- (a)   Zero-modulation RF output voltage
- (b)   Maximum RF output voltage
- (c)   Minimum RF output voltage
- (d)   Maximum Q_1 collector voltage
- (e)   Minimum Q_1 collector voltage
- (f)   Modulation signal input to base Q_1

6.   Given: Three identical power transistor circuits to be used as final class B linear RF amplifiers for DSB-SC, SSB, and 20% VSB modulation, with a modulation index of 1.0 and an AF modulating signal at 10 kHz, with dc input power at 20 volts and 1 A, and an RF efficiency of 60% operating at 5 MHz.

Determine the following, and tabulate them on a comparison chart of the three modulation techniques:
- (a)   Operating bandwidth requirements
- (b)   RF carrier power
- (c)   Lower sideband power
- (d)   Upper sideband power
- (e)   Peak envelope power

7.   Given: A +23 dBm double-balanced mixer and 500 kHz mechanical filter to be used in a SSB modulator circuit.

Determine the optimum signal levels at the following points in the circuit if all impedances are 600 Ω:
- (a)   RF *carrier* level in volts
- (b)   AF modulating signal voltage
- (c)   Output level of the RF sideband signal in volts
- (d)   Output level of the RF carrier in volts
- (e)   Carrier suppression in dB
- (f)   Level of 300 Hz (lowest AF) modulating signal on suppressed sideband in volts
- (g)   Minimum sideband suppression (for 300 Hz AF) in dB

8.   Given: Two 1496/1596 IC balanced modulators (see Fig. 6-22) to be used in a *phase-shift* SSB modulation circuit, with unity-gain AF and RF phase-shift circuits and a unity-gain adder.

Determine the optimum signal levels at the following points in the circuit if all impedances are 50 Ω:
- (a)   RF carrier level in volts
- (b)   AF modulating signal voltage
- (c)   Output level of the RF sideband signal in volts

(d)   Output level of the RF carrier in volts
(e)   Carrier suppression in dB
(f)   Output level of suppressed sideband signal in volts
(g)   Sideband suppression in dB

9.   Given: The envelope detector of Fig. 6-16 using a germanium diode (V_{co} = 250 mV) specified to have minimum distortion with modulation levels of 75%, and driving a resistive load impedance of 1000 Ω, and with a highest AF of 10 kHz.

Determine:
(a)   Minimum zero-modulation RF carrier level
(b)   Peak modulation envelope voltage
(c)   Diode resistor value
(d)   Filter capacitor value
(e)   Rms AF output signal voltage

10. Given: The simple two-diode product detector of Fig. 6-18 using silicon diodes (V_{co} = 650 mV), specified to operate with SSB signal levels to 2 volts p-p and AF of 3 kHz, and driving a resistive load impedance of 5000 Ω.

Determine:
(a)   Minimum LO level in volts
(b)   Peak combined (LO + RF) signal voltage
(c)   Minimum combined signal voltage
(d)   Peak output envelope voltage
(e)   Diode resistor value and filter capacitor value
(f)   Rms AF output signal voltage

CHAPTER 7

FREQUENCY AND PHASE MODULATION AND DETECTION

Although amplitude modulation (AM) was historically the most popular means of imposing and extracting intelligence signals upon and from RF carriers, the AM approach has certain disadvantages. First, the power levels of the intelligence (modulating) signals in the sidebands vary with the modulation index, yielding low efficiency, as we reviewed in the previous chapter. Second, AM detectors will respond to *any* amplitude change of the RF signal, including static discharges like lightning, auto ignition, and many other sources of interference and noise. Third, the level of modulation will change with the strength of the RF carrier signal, so the AM communications system is prone to fading.

The technique of *frequency modulation* (FM) was developed to overcome the disadvantages of AM approaches. The advantages of FM communications systems include power efficiency without fading and susceptance to noise and static interference. There are, of course, some tradeoff disadvantages with FM techniques, including increased radio-space requirements for sideband signals and more complex modulation and detection circuits. With our advanced circuit technology and increased availability of radio space, however, frequency modulation has become a most popular means for efficient RF communications. You can make the obvious comparison of FM to AM by switching bands on a high-fidelity receiver (or tuner) in an audio system. The comparison can also be made by tuning your TV receiver to a distant station (or disabling the receiving antenna while tuning a local station) and observing the fading of the AM picture signal while the FM sound signal remains constant.

The frequency-modulation approach uses the modulating signal to control the *frequency* of the RF carrier signal, while the amplitude of the RF signal remains constant. If the carbon microphone used in the early AM radio transmitters, for example (see Fig. 6-1), were replaced with a condenser (capacitor) microphone and connected as a part of the resonant antenna circuit, frequency modulation of the RF signal would result. Figure 7-1a and *b* compares the crude AM and FM modulators, which incidentally develop

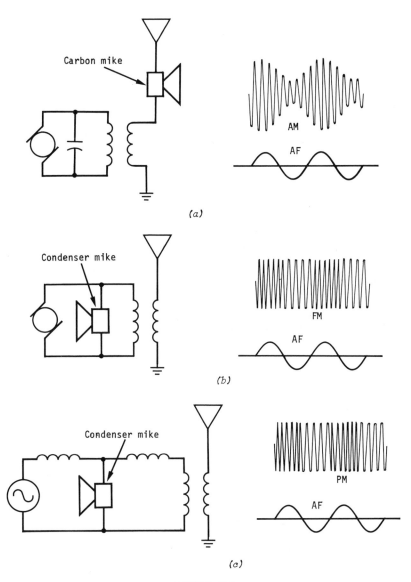

Carbon mike

AM

AF

(a)

Condenser mike

FM

AF

(b)

Condenser mike

PM

AF

(c)

FIGURE 7-1

the same RF waveform as that transmitted from modern radio communications systems. The condenser microphone becomes a voice-controlled capacitor in this example, with its internal capacitance varying with the application of sound pressure to its diaphragm. With no sound input, the moderate capacitance value of the microphone will yield a continuous RF antenna signal at a *center*, or *resting*, frequency. With a sound signal applied to the microphone, the diaphragm will vibrate at the AF rate and vary the frequency of the RF signal at that AF rate. A loud AF signal will increase the *change* in capacitance, yielding a greater proportional *change* in frequency than would result from a quiet AF signal. Thus, the *magnitude* of the modulating signal controls the change, or *frequency deviation,* of a frequency-modulated RF signal, whereas the *frequency* (AF) of the modulating signal controls the *rate* of the frequency deviation.

Phase modulation (PM) is a technique that is closely related to FM. As its name implies, PM uses the modulating signal to control the *phase* of the RF carrier signal, while the RF amplitude remains constant. If we continue with our simplistic comparisons of the early radio modulation techniques, we can replace the microphone in the antenna circuit of Fig. 7-1*b* with a stable *fixed*-frequency signal source. The signal source is then connected to a resonant circuit made up of coils and condenser microphone, as in the T-network of Fig. 7-1*c*, to achieve phase modulation. With no sound input, the moderate capacitance of the microphone will apply 0° phase shift to the RF signal passing through the network. With a sound signal applied to the microphone, the diaphragm will vibrate at the AF rate, increasing and decreasing its capacitance, and thus varying the phase shift through the T-network—leading and lagging in step with the changing capacitance. A loud AF signal will increase the change in capacitance and yield a greater proportional change in phase than would result from a quiet AF signal. The *magnitude* of the modulating signal thus controls the *phase deviation* of a PM RF signal, whereas the frequency (AF) of the modulating signal controls the *rate* of the phase deviation.

As you may conclude, the fundamental difference between FM and PM is that the frequency of the signal source is varied directly with FM, whereas the frequency of the signal source remains fixed with PM. The effect of a changing *LC* network on a fixed-frequency signal source will result in a changing phase shift of the RF signal as it passes through the network. Phase modulation thus affords the advantage of maintaining a stable frequency in a radio transmitter to assure that it will not drift to interfere with adjacent channels. The transmitted FM and PM signal waveforms are virtually identical, so they are received and detected with identical circuits. The FM (or PM) detection process and circuits are more complicated, however, compared to the simple AM diode detection process. For this reason the adoption of FM techniques for electronic communications had to wait for the development of efficient detection circuits to challenge the popularity of AM radio. The FM detection process requires conversion of the frequency (or phase) *deviation* of the RF carrier signal to a varying *voltage level.* The process was first accomplished with diodes and transformers in tuned networks, but recent advances with IC technology employ electronic circuit functions to achieve FM detection. A simplified FM *slope detector* is shown in Fig. 7-2 to demonstrate the concept of converting frequency deviation to voltage.

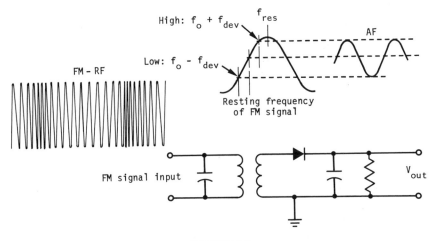

FIGURE 7-2

The content of this chapter will emphasize the functional principles and signal characteristics of FM and PM systems, and support this emphasis with examples of signal-processing circuits.

LEARNING OBJECTIVES

Upon completing this chapter, including the problem set, you should be able to:

1. Understand the concepts of frequency and phase deviation of an RF signal.
2. Understand the principles of frequency modulation and compute modulation index.
3. Understand and compute sideband signal frequencies and power distribution in frequency-modulated RF signals.
4. Understand the principles of noise reduction through the processing of modulation signals with the technique called *preemphasis,* and calculate component values necessary to achieve specified preemphasis values.
5. Understand the concepts of electronic measurements as applied to FM and PM.
6. Use specific modulation circuit techniques to calculate critical signal and circuit values for FM and PM modulators.
7. Understand the principles of phase and slope detection of frequency-modulated RF signals and the elimination of amplitude variation through limiting.

8. Compute the component values for specific detector circuits, including balanced slope detectors, discriminators, quadrature and ratio detectors.
9. Use the phase-locked loop (PLL) techniques of frequency and phase detection to compute circuit values using PLL elements.
10. Adapt to specific requirements the integrated circuit devices used in frequency and phase modulation and detection functions.

FREQUENCY AND PHASE DEVIATION

Frequency and phase modulation of RF carrier signals requires deviation of the RF by the modulating signal. The practical limits of deviation can be considered to be the -3 dB bandwidth of the RF circuits that are a part of the FM or PM system, but international radio (and Federal Communications Commission) standards for this service limit the maximum deviation to specified values. The familiar FM broadcast service specification, for example, includes a maximum deviation of 75 kHz with RF carrier signals in the 88 to 108 MHz frequency band. When deviation is so specified, it means that the RF signal can shift 75 kHz above *and* below the resting (center) frequency at zero modulation, as shown in Fig. 7-3*a*. This deviation value can be translated into percentage of frequency shift (picking a zero-modulation RF signal of 100 MHz, for example) as follows:

$$\text{Percent deviation} = f_{\text{dev}}/f_{\text{RF}} = 75 \text{ kHz}/100 \text{ MHz}$$
$$= 7.5 \times 10^{-4} = 0.075\%$$

This percentage of frequency deviation can now be related to the *change* in L or C values of a resonant network connected to *produce* the deviation if we apply the formulas, $C = 1/\omega^2 L$ and $L = 1/\omega^2 C$. Thus the change in L or C values will vary as the square of the change in frequency (deviation) that they affect. If we apply this relationship to our previous example, we can compute the percentage of change in L or C values as follows:

$$\text{Percent } L \text{ or } C = ((f_{\text{RF}} + f_{\text{dev}})/f_{\text{RF}})^2 - 1 = (1 + (\%)_{\text{dev}})^2 - 1$$
$$= (1 + 7.5 \times 10^{-4})^2 - 1 = 1.5 \times 10^{-3} = 0.15\%$$

If a capacitor even as large as 100 pF were used for the 100 MHz resonant network, for example, the frequency deviation of 75 kHz would require a change in capacitance of only 0.15 pF based on the above calculations. Such a small change in small L or C values makes this *direct* approach to frequency deviation (at 100 MHz) extremely impractical.

To achieve more practical frequency deviation, the RF carrier frequency may be reduced to increase the L and C values, while maintaining the same *percentage* of deviation. This approach is somewhat like the low-level modulation of AM systems, as the modulation takes place *ahead* of the final RF

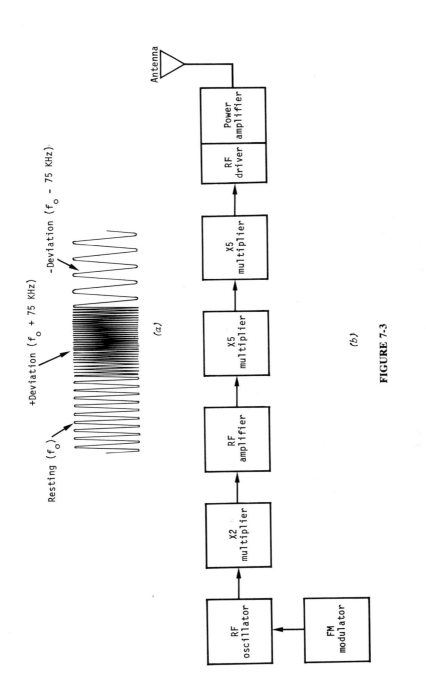

FIGURE 7-3

power amplifier circuits. In this case, however, low-level frequency modulation is achieved with *both* lower power and lower frequency circuits. If we thus reduce the carrier signal frequency of our previous example by a factor of 50 (from 100 MHz to 2 MHz), we can use resonant-network capacitor values as high as 20 nF (0.02 μF), yielding a 30 pF change in capacitance for a 0.075% deviation. (From the calculations of the previous paragraph, 0.075% frequency deviation requires 0.15% change in C.) After modulation is achieved, the 2 MHz frequency-modulated carrier signal is passed through $\times 50$ frequency-multiplier circuitry (as in Fig. 7-3b) to bring it to the required 100 MHz RF with 75 kHz deviation. If greater values of C (or L) or smaller percentages of deviation are required, lower RF modulation frequencies can be employed with higher frequency multiplication factors to achieve stable *direct* FM.

Frequency Modulation Derived from Phase Modulation

Phase modulation is often referred to as *indirect* FM because the technique develops a frequency-modulated signal from a fixed-frequency signal source. The advantages of the PM approach include the use of crystal-stabilized RF oscillators in order to maintain zero-modulation carrier signals with minimum frequency drift. Phase shifting (deviation) of an RF carrier requires a greater change in L or C values, however, to achieve the equivalent deviation of direct FM. Phase deviation may be computed based on the equivalent deviation frequency *and* modulating frequency. You may better understand the concepts of phase deviation by first reviewing a simple frequency-deviation example, and then translating the relationships to phase shift. If, for example, an RF carrier signal is to be deviated to 1 kHz by a modulating signal of 100 Hz, the total frequency deviation will be achieved in 10 cycles of the modulating signal, with a deviation of 100 Hz per cycle (10×100 Hz = 1 kHz). If the same RF signal were to be phase modulated, the total phase deviation would *also* be achieved in 10 cycles of the modulating signal. So the phase shift per cycle can be computed as follows:

$$\phi_{\text{dev}} = f_{\text{dev}}/f_{\text{mod}} \quad \text{and} \quad f_{\text{dev}} = \phi_{\text{dev}} f_{\text{mod}}$$

ϕ_{dev} is the phase deviation in radians, f_{dev} is the equivalent frequency deviation of the RF carrier, and f_{mod} is the modulating frequency.

If you examine the transposed form of the previous phase deviation formula, you should observe that the magnitude of deviation (ϕ) varies with the modulating signal frequency, and you should recall that the deviation *also* varies with the modulating signal amplitude. So here is the complicating factor of phase modulation: to deviate the RF carrier with a range of modulation signal frequencies and obtain the equivalent of FM, the modulation signals must be attenuated through their frequency range at a slope of 6 dB/octave (20 dB/decade). This attenuation slope of the modulating signals will eliminate phase deviation due to a changing modulation frequency, and leave only phase deviation proportional to the changing modulation amplitude, thus yielding identical PM and FM signals. Figure 7-4*a* shows a graphical comparison of *direct* FM, phase modulation (PM), and *indirect* FM using PM with frequency/slope attenuation of the modulating signals.

With the elimination of phase deviation due to varying modulation frequencies, we can now proceed to relate phase deviation to equivalent frequency deviation for FM communications systems. First, the maximum practical phase deviation that can be achieved with *L-C* resonant networks is limited to about 45°, based on the −3 dB network response where $X = R$ and the phase angle of the impedance is 45°. Phase deviation (ϕ) is properly specified and computed in radians (1 rad = 360°/2π = 57.3°), however, so the equivalent deviation for a 45° phase shift would be:

$$\phi = (°)_{\text{dev}}/\text{rad} = 45/57.3 = 0.79 \text{ radian}$$

Second, the maximum modulation amplitude will occur at the lowest modulation frequency, since slope attenuation is applied to higher frequencies; so the maximum equivalent deviation frequency can now be computed as (picking 200 Hz as the lowest modulating frequency):

$$f_{\text{dev}} = \phi_{\text{dev}}f_{\text{mod}} = 0.79 \times 200 = 157 \text{ Hz}$$

Although the computation is made at the lowest modulating frequency, this small value of equivalent frequency deviation is the maximum for all modulating signal frequencies. If this technique were applied to broadcast FM service, it would require considerable frequency multiplication in order to achieve the 75 kHz deviation as specified. The frequency multiplication factor can be computed as:

$$f_{\times \text{mult}} = f_{\text{dev(specified)}}/f_{\text{dev}} = 75 \text{ kHz}/157 \text{ Hz} = 477\times$$

It is impractical to achieve such a high degree of frequency multiplication, so phase modulation is best used for narrowband FM communications services that specify a maximum frequency deviation of 15 kHz and a lowest modulating frequency of 300 Hz (commercial telephony, designation 36F3). Typical phase-modulated commercial FM communications systems employ frequency multiplication factors of 50 to 100 (Fig. 7-4*b*) to achieve the specified 15 kHz deviation.

The phase deviation can now be related to the change in *L* or *C* values of a resonant network connected to produce deviation on a fixed-frequency RF signal. You may recall from Chapter 2 that the reactance at the −3 dB point frequencies varied by a factor of $\pm X_o/2Q$. If we reverse the approach by shifting the resonant frequency for phase modulation, the change in *L* or *C* values will also equal $1/2Q$ at the 45° (−3 dB) points. If, for example, the *Q* of the phase-modulation network were 25, the change in *L* or *C* values required to achieve the PM would be computed as:

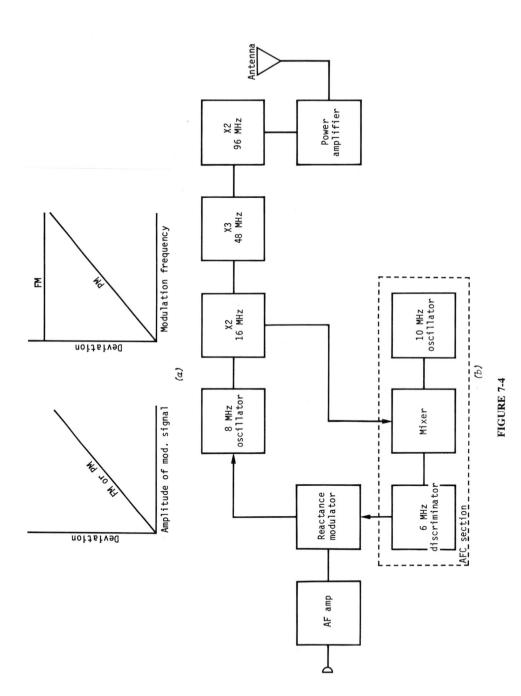

(a)

(b)

FIGURE 7-4

Percent L or $C = 1/2Q = 1/(2 \times 25) = 0.02 = 2\%$

As you may here observe, the phase-modulation approach requires a greater change in L or C values to achieve FM than the *direct* FM approach. This factor may, or may not, be an advantage depending on the circuit used for modulation.

Phase and Frequency Modulation Comparison

The advantages and tradeoff disadvantages of FM and PM are here summarized for your convenience:

1. Direct FM techniques have the advantages of (*a*) possessing simpler circuitry, (*b*) requiring lower frequency multiplication factors, and (*c*) achieving higher frequency deviation with no requirements for slope attenuation of the modulating signal frequencies. On the other hand, they are susceptible to frequency drift and instability, and require very small, but linear, changes in network L or C values that may be difficult to achieve.

2. PM (indirect FM) techniques have the advantages of using independently stabilized RF oscillators and requiring larger, more practical, changes in network L or C values. They also require slope attenuation of the modulating signal frequencies, higher frequency multiplication factors, and have limited practical equivalent values of frequency deviation.

Sample Problem 1

Given: A 150 MHz commercial FM communications system with a specified 15 kHz deviation of the RF carrier, and a lowest voice-modulating frequency of 300 Hz.

Determine: The percent frequency deviation and percent change in L or C values for direct FM, and determine the frequency multiplication factor and FM oscillator frequency if the maximum *change* in C of the simple LC resonant network is ± 10 pF, with a Z_p of 100 Ω and a Q of 25.

(*a*) The percent of frequency deviation is first computed as:

$(f_{\text{dev}}/f_{\text{RF}}) \times 100 = (15 \text{ kHz}/150 \text{ MHz}) \times 100 = 0.01\%$

(*b*) The percent of change in capacitance (C) is then computed as:

$[(1 + \text{dev})^2 - 1] \times 100 = [(1 + 10^{-4})^2 - 1] \times 100 = 0.02\%$

(*c*) Then, the total required capacitance (C_T) is computed, based on a practical value of 10 pF deviation in capacitance being 0.02% of C_T:

$$C_{dev}/(\%)_{dev} = 10 \text{ pF}/(2 \times 10^{-4}) = 5 \times 10^{-8} \doteq 50 \text{ nF}$$

(d) The required network reactance is now computed, based on the Z_p and Q values, as:

$$X = Z_p/Q = 100/25 = 4 \ \Omega$$

(e) The frequency at which the 50 nF computed capacitance value will have a 4 Ω reactance, and resonate with an inductor with a 4 Ω reactance, is computed as:

$$f_1 = 1/2\pi CX = 1/(6.28 \times 5 \times 10^{-8} \times 4) = 796 \text{ kHz}$$

(f) Finally, the frequency multiplication factor is computed for direct FM, as:

$$f_{\times mult} = f_{RF}/f_1 = 150 \text{ MHz}/796 \text{ kHz} = 188$$

Sample Problem 2

Given: The same 150 MHz commercial FM communications system specified in the above problem, but to be indirectly frequency modulated using phase-modulation techniques.

Determine: The equivalent frequency deviation for a maximum phase deviation of 0.6 radian, the frequency multiplication factor, the low RF modulation frequency, the deviation network phase shift, and the change in capacitance for a resonant network with a Z_p of 100 and a Q of 25.

(a) The equivalent deviation frequency is first computed as:

$$f_{dev} = \phi f_{mod} = 0.6 \times 300 = 180 \text{ Hz}$$

(b) Then, the phase-modulation frequency-multiplication factor is calculated:

$$f_{\times mult} = f_{dev(specified)}/f_{dev} = 15 \text{ kHz}/180 \text{ Hz} = 83.3$$

(c) The low radio frequency is then computed, based on the above frequency multiplication factor, as:

$$f_1 = f_{RF}/f_{\times mult} = 150 \text{ MHz}/83.3 = 1.80 \text{ MHz}$$

(d) The deviation network phase shift, equal to 0.6 radian, is:

$$\text{Degrees deviation} = \phi \text{ rad} = 0.6 \times 57.3° = 34.4°$$

(e) The change in capacitance, based on the 34.4° phase shift, is computed using the $\times/2Q$ technique of Chapters 2 and 3 (where $\times$ is the multiplying factor):

$$\begin{aligned} \text{Percent } C &= \tan (°)_{dev}/2Q \ (\times 100) \\ &= |\tan 34.4°/(2 \times 25)| \times 100 = 1.37\% \end{aligned}$$

(*f*) The total capacitance is then determined based on the Z_p/Q reactance value of 4 Ω, as:

$$C_T = 1/\omega X = 1/(6.28 \times 1.81 \times 10^6 \times 4) = 22 \text{ nF}$$

(*g*) Finally, the change in capacitance is computed as:

$$C_{\text{dev}} = C_T(\%)C = 22 \text{ nF} \times 1.37 \times 10^{-2} = 303 \text{ pF}$$

From the results of the above problems you should observe the relative practicality of the two approaches to FM. The most practical approach for the above conditions is clearly indirect FM from PM, because of the lower frequency multiplication factor. Incidentally, since the frequency multiplication cannot yield fractional factors, the actual multiplication must be the next-lower even-numbered factor. In the case of Sample Problem 2, the actual frequency multiplication factor should have been reduced from 83.3 to 82, thus increasing f_1 to 1.83 MHz and proportionally reducing the capacitance values.

FM AND SIDEBANDS

Unlike the relatively simple RF carrier and sideband structure of AM signals, the spectral response of FM RF signals is quite complex. Rather than just *one* pair of sideband signals for a single modulating frequency, FM RF carriers will develop *many* pairs of sideband signals that vary in both amplitude and phase with respect to an amplitude- and phase-varying carrier signal. If this description portrays a rather complicated spectrum for FM signals, don't become discouraged. If you can understand phase modulation, you can surely understand the complications of FM sidebands. If you have difficulty with both topics, rest assured that a thorough understanding of frequency modulation *can* be obtained by your persistence with this chapter.

Modulation Index

The FM *modulation index* can theoretically be any number from zero to infinity, and the idea of 100% modulation does not exist with FM. *Full modulation*, on the other hand, means that the RF carrier is deviated to the limits specified by standards for the broadcast or communications service— not 100% modulation. So full modulation for commercial FM broadcast (designated 180F3), for example, is a deviation of 75 kHz, whereas full modulation for commercial telephony (36F3) is 15 kHz in accordance with the specifications of international technical standards. The modulation index for FM is determined using the same factors that deviate the carrier: deviation frequency and modulation frequency. The modulation value, or index number, is thus computed using the following formula, which is incidentally the same formula used to determine phase deviation:

$$M_{\text{FM}} = f_{\text{dev}}/f_{\text{mod}} = \phi$$

If the modulation index for commercial FM broadcast service is computed, based on the maximum deviation of 75 kHz, with a maximum modulation frequency of 15 kHz, full modulation is achieved with a modulation index of 5. As the *deviation* is reduced, the modulation index is also reduced until zero modulation is reached ($M_{\text{FM}} = 0$), yielding zero carrier deviation. As the *modulation frequency* is reduced, while maintaining maximum deviation however, the modulation index is increased to a high value. Using commercial broadcast FM as an example again, the lowest modulation frequency of 50 Hz will yield a modulation index of:

$$M_{\text{FM}} = f_{\text{dev}}/f_{\text{mod}} = 75 \text{ kHz}/50 \text{ Hz} = 1500$$

The *multiple* sideband frequency pairs occur at multiples of the modulation frequency; e.g., the first pair is at $f_o \pm f_{\text{mod}}$, the second pair is at $f_o \pm 2f_{\text{mod}}$, the third pair is at $f_o \pm 3f_{\text{mod}}$, and so on. At some higher modulation frequency multiple, depending on the modulation index value, the power level of the sideband pair signals becomes insignificantly small. At this point the *usable* bandwidth for FM signal transmission is established. In our earlier example, with a modulation frequency of 15 kHz and a modulation index of 5, the number of significant sideband signal pairs is 8, so the required FM bandwidth under these conditions is computed as:

$$BW_{\text{FM}} = \text{number of sideband pairs} \times f_{\text{mod}} \times 2$$
$$= 8 \times 15 \text{ kHz} \times 2 = 240 \text{ kHz}$$

Even though the frequency deviation for commercial broadcast FM is $f_o \pm 75$ kHz $= 150$ kHz, the radio space occupied by the sideband signals is $f_o \pm 120$ kHz.

People often confuse deviation with bandwidth in FM systems. These two parameters are related in the sense that deviation determines modulation index, which in turn determines the number of sideband pairs at significant power levels. The bandwidth is computed from sideband signal frequencies, however—not from deviation frequency.

Sideband Structure

The *amplitude* and *phase* of the sideband signals are mathematically determined from the multiple phase combinations that result from mixing the deviating RF carrier and modulation frequencies. As you can imagine, the mathematics relating this combination of signals is tedious. Fortunately, the astronomer Fred Bessel and his followers have simplified the mathematical headaches, so we mortals can proceed without undue difficulty. Bessel's mathematical functions relate the amplitude and phase of FM sideband signal pairs to modulation index values. Table 7-1 relates the RF carrier and 16

TABLE 7-1. BESSEL FUNCTIONS SPECIFYING CARRIER LEVELS (J_0) AND SIDEBAND LEVELS (J_1 THROUGH J_{16})

x (m_f)	J_0	J_1	J_2	J_3	J_4	J_5	J_6	J_7	J_8	J_9	J_{10}	J_{11}	J_{12}	J_{13}	J_{14}	J_{15}	J_{16}
0.00	1.00	—	—	—	—	—	—	—	—	—	—	—	—	—	—	—	—
0.25	0.98	0.12	—	—	—	—	—	—	—	—	—	—	—	—	—	—	—
0.5	0.94	0.24	0.03	—	—	—	—	—	—	—	—	—	—	—	—	—	—
1.0	0.77	0.44	0.11	0.02	—	—	—	—	—	—	—	—	—	—	—	—	—
1.5	0.51	0.56	0.23	0.06	0.01	—	—	—	—	—	—	—	—	—	—	—	—
2.0	0.22	0.58	0.35	0.13	0.03	—	—	—	—	—	—	—	—	—	—	—	—
2.4	0	0.52	0.43	0.20	0.06	—	—	—	—	—	—	—	—	—	—	—	—
2.5	-0.05	0.50	0.45	0.22	0.07	0.02	—	—	—	—	—	—	—	—	—	—	—
3.0	-0.26	0.34	0.49	0.31	0.13	0.04	0.01	—	—	—	—	—	—	—	—	—	—
4.0	-0.40	-0.07	0.36	0.43	0.28	0.13	0.05	0.02	—	—	—	—	—	—	—	—	—
5.0	-0.18	-0.33	0.05	0.36	0.39	0.26	0.13	0.05	0.02	—	—	—	—	—	—	—	—
5.5	0	-0.34	-0.12	0.26	0.40	0.32	0.19	0.09	0.03	0.01	—	—	—	—	—	—	—
6.0	0.15	-0.28	-0.24	0.11	0.36	0.36	0.25	0.13	0.06	0.02	—	—	—	—	—	—	—
7.0	0.30	0.00	-0.30	-0.17	0.16	0.35	0.34	0.23	0.13	0.06	0.02	—	—	—	—	—	—
8.0	0.17	0.23	-0.11	-0.29	-0.10	0.19	0.34	0.32	0.22	0.13	0.06	0.03	—	—	—	—	—
8.65	0	0.27	0.06	-0.24	-0.23	0.03	0.26	0.34	0.28	0.18	0.10	0.05	0.02	—	—	—	—
9.0	-0.09	0.24	0.14	-0.18	-0.27	-0.06	0.20	0.33	0.30	0.21	0.12	0.06	0.03	0.01	—	—	—
10.0	-0.25	0.04	0.25	0.06	-0.22	-0.23	-0.01	0.22	0.31	0.29	0.20	0.12	0.06	0.03	0.01	—	—
12.0	0.05	-0.22	-0.08	0.20	0.18	-0.07	-0.24	-0.17	0.05	0.23	0.30	0.27	0.20	0.12	0.07	0.03	0.01
15.0	-0.01	0.21	0.04	-0.19	-0.12	0.13	0.21	0.03	-0.17	-0.22	-0.09	0.10	0.24	0.28	0.25	0.18	0.12

sideband-pair signal amplitude and phase to selected typical FM modulation index values from zero to 15. The J values across the table represent the numbered sideband pairs, the modulation index values are tabulated down the first column of the table, and the values in the body of the table represent the relative amplitude and phase of the sideband-pair signal voltages. If the zero-modulation RF carrier signal amplitude were 1.0 V, for example, FM'ing the RF to a modulation index of 0.5 would reduce the carrier amplitude to 0.94 V and develop two significant pairs of sideband signals of 0.24 and 0.03 V, respectively. These two sideband signals would, of course, be separated from the carrier and each other by the modulation frequency. Figure 7-5 shows the graphical distribution of sideband signals for modulation index values of 1.0, 2.0, and 5.0, respectively.

The table is extended to include sideband signal levels above those pairs that reach 0.01 or 1% of the zero-modulation carrier voltage. The 1% value represents a voltage ratio of 40 dB; but you should observe that the carrier voltage varies also, yielding a range of less than 40 dB (100:1) with given modulation indices. Using the previous example, the carrier level is 0.94, while the second sideband-pair level is 0.03, so the dB ratio of the signals is:

$$20 \log V_1/V_2 = 20 \log (0.94/0.03) = 29.9 \text{ dB}$$

With a modulation index value of 5, the highest signal level is not in the carrier, but rather in the fourth (J_4) sideband-pair signals, at 0.39, with the

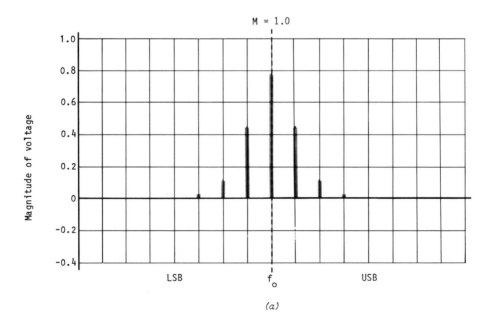

(a)

FIGURE 7-5

eighth (J_8) sideband-pair signal level at 0.02. The dB ratio of these signals is then:

20 log (0.39/0.02) = 25.8 dB

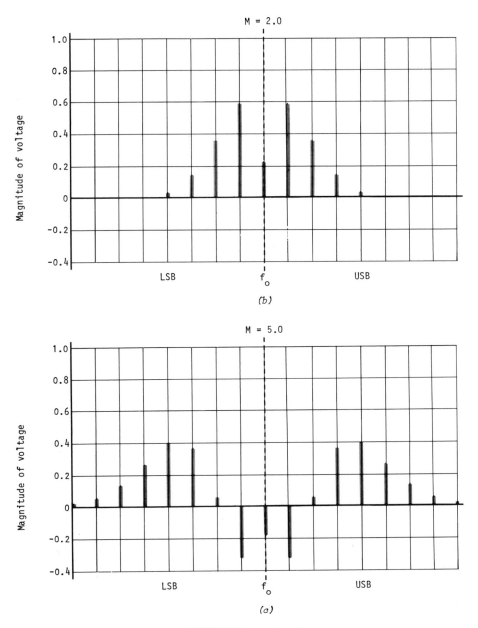

FIGURE 7-5 (*continued*)

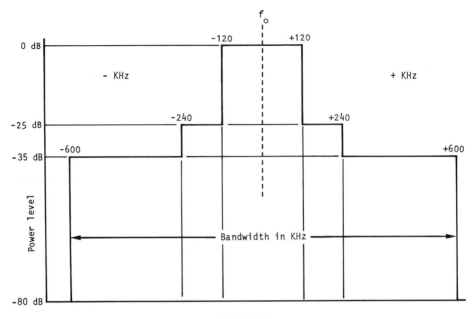

FIGURE 7-6

The point of comparing these signal ratios in dB is to establish a reference to the real world of FM communications. A signal difference of 25 dB is *not* enough to prevent interference from FM sideband signals that extend beyond the "significant" limits of Table 7-1. If we again use commercial broadcast FM as an example, the specifications include accounting for sideband signal pairs at 35 dB signal ratios, extending to 600 kHz above and below the frequency of the zero-modulation carrier, with sideband signal pairs beyond these frequency limits at 80 dB signal ratios. At 80 dB the level of the sideband signal pairs is factually insignificant. If you are setting up or making tests and measurements of the FM spectrum, you will require a far more extensive table than Table 7-1 or you will need to enlist the aid of a computer. Figure 7-6 shows the spectral relationships for commercial broadcast FM with specified bandwidths beyond 1.2 MHz, while the allocated "channel" is specified as 200 kHz (180 kHz with 10 kHz "guard bands").

Unlike AM the *total* RF power in an FM signal remains constant regardless of the modulation index. The carrier power varies, however, depending on the modulation index; and it is actually zero at modulation index values of 2.4, 5.8, 8.6, and points beyond, where the Bessel function of that value crosses zero. If you *square* the values in Table 7-1 to indicate proportional power ratios, you can determine the carrier and sideband power distribution for a given modulation index. With a modulation index of 2.0, for example, the carrier power is $(0.22)^2$, or only 4.84% of the total power. The power in

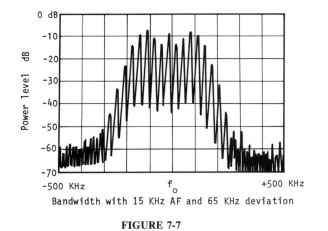

Power level dB

0 dB

−10

−20

−30

−40

−50

−60

−70

−500 KHz f_o +500 KHz

Bandwidth with 15 KHz AF and 65 KHz deviation

FIGURE 7-7

the first *pair* of sideband signals is *twice* $(0.58)^2$, or $2 \times 33.64\% = 67.3\%$, since there are *two* sideband signals. The power in the remaining sideband pairs is $2 \times (0.35)^2 = 24.5\%$, $2 \times (0.13)^2 = 3.38\%$, and $2 \times (0.03)^2 = 0.18\%$, respectively, for the second, third, and fourth sideband pairs. The percentages should total nearly 100%, with 97+% of the total power within the frequency range of the first two sideband pairs. If you now proceed with this same exercise for commercial broadcast FM, but reverse the procedure by computing only the seventh and eighth sideband pairs, you will observe that less than 1% of the total power in the FM signal extends beyond the sixth sideband pair. With a maximum modulation frequency of 15 kHz, virtually all of the signal power is distributed in the bandwidth of $f_o \pm 90$ kHz, yielding the 180 kHz bandwidth of the allocated channel specification of Fig. 7-6. In actual practice, of course, the complications of "real" equipment and circuits result in slight variations of the sideband signal values, but the general pattern of the Bessel functions is maintained. Figure 7-7 shows the actual measured response of a broadcast FM signal with a 15 kHz signal deviating the RF carrier ±65 kHz.

Sample Problem 3

Given: An FM RF signal with a 3 kHz modulation frequency and a deviation of 7.5 kHz.

Determine: The modulation index, voltage, and power of the carrier and significant sideband signals; the dB ratio of the highest to lowest significant-level signals; the bandwidth of the significant-level signals; and the bandwidth with greater than 95% of the signal power.

(a)   First, the modulation index is computed as:

$$M_{FM} = f_{dev}/f_{mod} = 7.5/3 = 2.5$$

(b)   The relative voltage levels are obtained for the carrier and five significant sideband-pair signals from the Bessel-function Table 7-1:

Carrier = 0.05
First sideband pair = 0.5
Second sideband pair = 0.45
Third sideband pair = 0.22
Fourth sideband pair = 0.07
Fifth sideband pair = 0.02

(c)   The relative power levels are computed by squaring the carrier value
and adding it to twice each squared sideband value obtained from the
list in part b above:

$$P_T (\%) = 0.05^2 + (2 \times 0.5^2) + (2 \times 0.45^2)$$
$$+ (2 \times 0.07^2) + (2 \times 0.02^2) = 101.5\%$$

(The sum exceeds 100% because of rounding off.)

(d)   The dB ratio of the fourth and fifth sideband pairs to the zero-
modulation carrier value (1.0) is computed as follows:

$$20 \log V_1/V_2 = \begin{cases} 20 \log (1/0.07) = 23.1 \text{ dB} \\ 20 \log (1/0.02) = 34 \text{ dB} \end{cases}$$

(e)   The bandwidth of the five significant sideband-pair signals is computed
as:

$$BW = SB's \times f_{mod} \times 2 = 5 \times 3 \text{ kHz} \times 2 = 30 \text{ kHz}$$

(f)   Finally, the 20 dB bandwidth, where the sideband pairs greater than 20
dB below the zero-modulation carrier level are considered insignificant,
is computed for three sideband-pair signals:

$$BW = SB's \times f_{mod} \times 2 = 3 \times 3 \text{ kHz} \times 2 = 18 \text{ kHz}$$

The calculations and results of this sample problem should emphasize two
important FM relationships. The first, and often confusing, relationship is
that of bandwidth to deviation. In this example the power bandwidth of 18
kHz is not apparently related to the deviation of 7.5 kHz. This observation is,
in fact, true, as both concepts appear to be separate. They are both indirectly
related, however, through their effect on the modulation index: the deviation
affects a specific modulation index, which in turn affects a given bandwidth,
with both factors dependent upon the frequency of the modulating signal.
The second relationship is that of the power distribution within the sideband
structure, with supposedly insignificant sideband signals 20, 30, and more dB

below the level of the most intense carrier signal. These small signal values may not, in fact, be insignificant, as weak sideband signals can interfere with adjacent channel signals from distant stations.

NOISE AND PREEMPHASIS

One of the major advantages of FM RF communications systems is an immunity to amplitude-varying noise and interference signals. Since the initial deviation of the FM signal is developed from an amplitude-varying (modulation) signal, however, noise or interference can be introduced into an FM system at *this* point, or in any other amplitude-controlled phase or frequency-sensitive circuit throughout the system. You should recall that a changing amplitude in a reactive (*L* or *C*) circuit will also yield a change in phase. Although such phase- or frequency-varying signals that produce noise and interference in FM systems are considerably smaller than those encountered in AM systems, they can become sufficiently large to produce an annoying "hiss" in sound reception, and actually mask higher-frequency video and pulse modulation signals.

Since a high proportion of the noise affects *phase* modulation of the RF carrier, without the 6 dB/octave frequency compensation discussed earlier in this chapter, the phase deviation and the relative FM modulation index increase as the frequency of the noise signal increases. The result of this noise modulation is an increase in noise level with frequency, but the modulation signal level remains fixed as its frequency increases. The consequent signal-to-noise ratio becomes smaller (greater noise) at the higher frequencies. Figure 7-8 shows the graphical response of noise level with increasing frequency, called the *noise triangle*. Note that the proportion of noise is less with the increased modulation index.

In addition to the problem of noise level increasing with frequency, there is a natural decrease in signal with frequency, especially if the modulating signal is from voice or music sources. The higher modulating-frequency

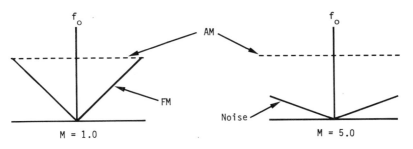

FIGURE 7-8

signal-to-noise *ratio* is thus lower (higher noise, lower signal)—due *also* to the decrease in modulating signal levels—making the noise level *apparently* higher. Acceptable signal-to-noise ratios vary depending on particular radio-service requirements, but *optimum* signal-to-noise ratios are best achieved where signal levels are maintained at their maximum values. Maximum modulation signal levels in FM systems result from maximum deviation of the RF carrier. Higher frequency modulation signals, however, are generally *not* intense. You should obtain some perspective of the lower level, higher frequency modulating signals by saying "lisp," and comparing the intensity of the "li" sound to the "sp" sound. The high frequencies of most sounds are naturally less intense than the low frequencies, and most other modulation signal waveforms have their most intense signal levels in the lower frequency portion of their spectrum. So here is the problem: higher frequency modulation signals should deviate the RF carrier to its maximum value, but they are less intense than the lower frequency signals which may already be deviating the RF carrier to full modulation.

The solution to the FM noise problem, stated as being due to reduced higher modulation-frequency signal levels, is rather simple: it requires frequency compensation of the modulating signal in order to intensify the higher modulating frequencies. This form of compensation is called *preemphasis,* and it defines the amplification response of the higher frequency modulation signals, which must in turn be attenuated after the signals are detected (demodulated) in the receiver portion of the FM system. This opposite compensation (attenuation) in the receiver is aptly called *deemphasis*; and it corresponds inversely to the response curve applied in the modulation process. So while the deemphasis circuit reduces the excess signal level at the higher frequencies, it also reduces the uncompensated noise level. The most popular response-curve standards are identified by the circuit component values in the frequency compensation circuits. These are either *RC* or *RL* networks, with the ratio of the two components determining the slope (6 dB/octave, 20 dB/decade) of the response curve. Figure 7-9 shows typical preemphasis and deemphasis networks, with their corresponding response curves.

Time Constants

For some strange reason, the response curves for preemphasis and deemphasis are identified in *time* units, based on the *RC* or *L/R* time constants of the ratio of the network components, e.g., 25 μs, 50 μs, and 75 μs. While frequency response is scientifically referenced to sine waves, with ω used as the basis for the reactance and impedance of networks that produce frequency-response curves, the curves are here instead defined in time constants—the units that apply to step voltages and pulse signals. The standard U.S. preemphasis/deemphasis response for FM broadcast is 75 μs, whereas most of the other countries throughout the world use the European

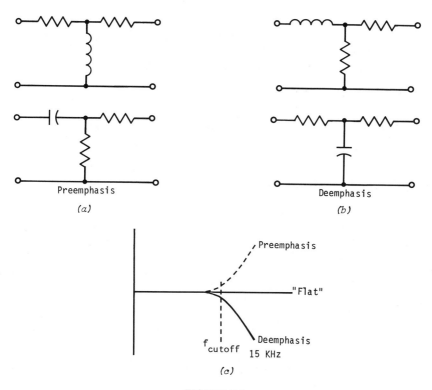

FIGURE 7-9

50 μs standard. Some countries have recently adopted a 25 μs preemphasis/deemphasis standard in conjunction with the "Dolby" system of noise reduction.

The 75, 50, and 25 μs terminology specifies the effective values of RC or LC components used to provide the frequency-response curves. If, for example, an AF network with 10 kΩ resistance were to provide a 75 μs response curve, the value of required circuit capacitance or inductance could be computed as follows:

$T = RC$, thus $C = T/R = 75\ \mu\text{s}/10\ \text{k}\Omega = 7.5\ \text{nF}$

or

$T = L/R$, thus $L = TR = 75\ \mu\text{s} \times 10\ \text{k}\Omega = 0.75\ \text{H}$

In general practice, RC networks are more prevalent than RL networks, with electronic devices simulating "effective" values of R and C in active circuits. While this approach and definition may be somewhat satisfactory for computing the effective values of components, the approach is both inadequate and misleading. A better understanding can be obtained by applying the proper frequency analysis to the preemphasis/deemphasis specifications.

Let us begin with the basic relationship from Chapter 2 that relates to the 3 dB response of simple resistive-reactive networks: When the reactance and resistance are equal, the circuit response is 0.707 or 1.414 times the voltage or current, and there is a 45° phase shift between the two quantities. If you apply this equality to the transposed formula, $f = 1/2\pi CX$, and then substitute $R = X$, the formula translates to $f = 1/2\pi CR$. If the RC time-constant concepts are applied, the formula translates to $f = 1/2\pi T$. The 3 dB frequency, with 75 μs preemphasis/deemphasis for example, can be computed as:

$$f = 1/2\pi T = 1/(6.28 \times 75 \times 10^{-6}) = 2.12 \text{ kHz}$$

Repeating the computation for 50 and 25 μs preemphasis/deemphasis will yield 3 dB frequencies of 3.18 and 6.37 kHz, respectively. The frequency response of the RC circuit beyond the 3 dB point is approximately 6 dB/octave and 20 dB/decade. The amplitude of signals at these frequencies can be computed as:

$$V_2 = V_1[20 \log (1.414 \ f_2/f_1)]$$

where V_1 is the maximum low-frequency signal voltage without preemphasis/deemphasis, V_2 is the output level with preemphasis/deemphasis, f_1 is the 3 dB frequency for a given time constant, and f_2 is the higher operating frequency with preemphasis/deemphasis.

Although effectively reducing noise, the preemphasis/deemphasis technique creates increased-amplitude sideband signals at frequencies further removed from the zero-modulation RF carrier frequency. This situation results in an increase in the required bandwidth of the FM transmission. Using the 75 μs preemphasis response, as an example, we can compute the increase in amplitude at 15 kHz as follows:

$$V_2/V_1 = 20 \log (1.414 \ f_2/f_1) = 20 \log (1.414 \times 15 \text{ kHz}/2.12 \text{ kHz})$$
$$= 20 \text{ dB}$$

Given a full-deviation FM signal modulated with a 15 kHz, 75 μs preemphasized AF signal, the level of the seventh and eighth sideband pairs will be raised from the insignificant 0.05 and 0.02 to 0.5 and 0.2, respectively. (See Table 7-1 and Fig. 7-6.) This increase, of course, accounts for the 240 kHz bandwidth of U.S. broadcast FM, which prohibits adjacent-channel utilization because the specified channel bandwidth is only 200 kHz. With the alternative Dolby noise-reduction technique, the 25 μs preemphasis will reduce bandwidth utilization and potential interference.

FM Signal-to-Noise

The signal-to-noise improvement due to preemphasis and deemphasis depends on a number of factors, including the characteristics of the modulating signals and the receiver bandwidth and demodulation circuits. For a single modulating frequency the approximate improvement in S/N ratio can be calculated based on modulation frequency and preemphasis/deemphasis time as follows:

$$dB_{S/N \text{ improved}} = 20 \log (\omega T)^2/3$$

Applying this formula to audio modulating frequencies yields a low S/N improvement of about 1.5 dB at 4 kHz and a high S/N improvement of about 24 dB at 15 kHz. Single frequencies are generally *not* transmitted in the form of modulation signals on an RF carrier, however, so the *real* dB improvement will vary depending on the intensity and distribution of the audio modulation frequencies within the range of 4 kHz to 15 kHz.

MEASUREMENTS

Measurements of frequency-modulated RF signals are more complex than their AM counterparts. The important measurements of FM signals include zero-modulation signal *frequency and power,* frequency *deviation and modulation index,* and *frequency and power distribution* of the frequency-modulated signals. Obviously, the zero-modulation radio frequency and power are measured directly regardless of which modulation scheme is used; but with FM systems the total RF power does not increase with modulation, as it does with AM systems. The total power with modulation is distributed in the carrier and sideband frequencies in proportion to the relative magnitudes of these signals for a given mod index; and when the power levels of these signals are totaled, their sum is equal in magnitude to the easily measured zero-modulation power level of the RF carrier.

Since the modulation index of FM signals depends on both the modulating frequency *and* the deviation of the RF carrier, a simple and direct measurement of modulation index is not practical. In practice, a fixed-frequency modulating signal is applied to the FM system, with the resulting RF deviation measured with an FM deviation meter (or monitor). Such a measurement of RF deviation provides the data for calculating modulation index along with an indication of the limits of transmitter frequency utilization, while simultaneously monitoring how far off frequency the transmitter RF carrier may be when the modulating signal is turned off. A popular instrument used for this purpose is a *heterodyne* frequency-*deviation meter* which measures the maximum audio difference frequency of two mixed RF signals—one from the FM transmitter under test, and the other from a fixed-

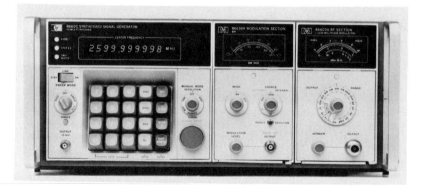

FIGURE 7-10

frequency standard (usually a crystal stabilized or phase-locked oscillator) in the measuring instrument. Often the deviation meter is one function of a multifunction RF signal generator or test set used to generate or measure FM signals. Figure 7-10 shows an example of both the FM deviation meter and the multifunction signal generator used for FM tests and measurements. You should note the control functions and meter-scale nomenclature (marked designations) and relate these to the FM and PM principles described earlier in the chapter.

Bessel-Zero Method

As with most electrical and electronic techniques, there is also a "poor man's" approach to FM measurements using a frequency-selective voltmeter, or even a high-selectivity radio receiver. The approach is, of course, inconvenient and requires some calculation, but it will provide accurate measurements without the expense of specialized FM instrumentation. This FM

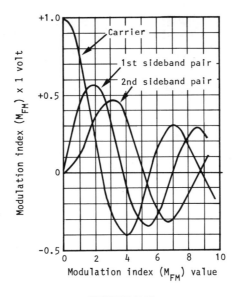

FIGURE 7-11

measurement technique is called the *Bessel-zero method*. As the name suggests, the measurement involves the Bessel-function analysis of the FM
carrier and sideband signals as described earlier in this chapter. If you went
to the trouble to graph the response of the RF carrier and first two sideband
pairs of an FM signal using Bessel's calculations (refer to Table 7-1), you
would obtain a continuous graph of signal amplitude and phase vs modulation index values, as shown in Fig. 7-11. Referring to the curves in Fig. 7-11,
you should note that at some modulation index values the signals actually
cross the zero-amplitude line and thus have no magnitude—they disappear.
For the RF carrier this zero magnitude occurs at modulation index values of
2.405, 5.52, and 8.654, etc. If you now transpose the modulation index for-
mula, $M_{FM} = f_{dev}/f_{mod}$, to $f_{dev} = M_{FM}f_{mod}$, and apply the 2.405 zero-magnitude
modulation index value, you should see a direct relationship between the
easily measured modulation frequency and the deviation frequency. The FM
deviation can thus be measured as follows:

1. The selective voltmeter or receiver is tuned to the zero-modulation RF
   carrier for maximum output.

2. The required audio modulating-signal frequency is computed using the
   transposed formula, $f_{mod} = f_{dev}/M_{FM}$, where f_{dev} is the maximum
   specified deviation for the particular communications service (75 kHz
   for FM broadcast, 25 kHz for TV audio, 15 kHz for commercial telephony, etc.); and M_{FM} is the carrier zero-magnitude modulation index
   value (2.405, 5.52, 8.654, etc., from Bessel curves).

3. The FM carrier is modulated with the computed modulation frequency from step 2 above and the amplitude of the modulating signal is increased until the RF carrier is reduced to zero, as read on the selective voltmeter or receiver.

4. The *voltage* of the modulating signal is measured as that value required to deviate the RF carrier to the maximum specified deviation (see step 2 above). You may recall that deviation is proportional to the *magnitude* of the modulating signal.

Based on the above measurements, and of course assuming linearity of deviation, there is a direct ratio of audio modulating signal voltage to RF carrier deviation. A simple comparative relationship may be obtained by identifying the voltage of step 4 as 100% deviation, with reduced voltages yielding proportionally reduced deviation values.

If you were to test a 154 MHz commercial FM telephony transmitter, for example, using the Bessel-zero method, you could proceed as follows:

1. The selective voltmeter (or receiver with BFO or selective lowpass filter) would be tuned to the 154 MHz zero-modulation RF carrier signal for peak response.

2. The audio modulating signal frequency would be calculated as $f_{mod} = f_{dev}/M_{FM} = 15\ kHz/2.405 = 6237\ Hz$. Since this frequency is above the audio response of the modulator (the telephony AF range is typically 300 Hz to 3 kHz), the next carrier-zero modulation index value would be used in computing: $f_{mod} = 15\ kHz/5.52 = 2717\ Hz$.

3. The 2717 Hz audio modulating signal would be applied to the transmitter at low level and increased until the carrier signal voltage was canceled at the selective voltmeter (or receiver).

4. The AF modulating signal voltage would be measured and specified as a 100% deviation level, and a calibration chart would be attached to the transmitter indicating percent deviation, deviation frequency, and AF modulating signal voltage, as in the example of Table 7-2. (If this technique were applied to FM broadcast measurements, the frequency response of the AF voltmeter would be modified to cancel the 50 μs or 75 μs preemphasis, with the same deemphasis response as used in the broadcast FM receiver.)

Spectrum Analyzer

Without a doubt the spectrum analyzer is *the* most useful instrument for communications measurements. When it is used for the analysis of FM signals, the spectrum analyzer can provide *direct* measurements of zero-

TABLE 7-2. SAMPLE CALIBRATION CHART
FOR BESSEL-ZERO METHOD

Audio Frequency, Hz	Deviation Produced, kHz		
	1st Null	2nd Null	3rd Null
905.8	±2.18	±5.00	±7.84
1000.0	±2.40	±5.52	±8.65
1500.0	±3.61	±8.28	±12.98
1811.0	±4.35	±10.00	±15.67
2000.0	±4.81	±11.04	±17.31
2079.2	±5.00	±11.48	±17.99
2805.0	±6.75	±15.48	±24.27

modulation carrier frequency and power, along with the bandwidth, frequencies, and power distribution of FM RF signals through the full range of modulation frequencies and deviation levels. The instrument can also provide convenient *indirect* measurements of RF deviation and modulation index values. You may review the basic principles of the spectrum analyzer in Chapter 3, and its application to AM measurements in Chapter 6. When the spectrum analyzer is used to measure FM signals, it will display a power (in dBm) or voltage vs frequency trace similar to those of Fig. 7-12. The figures display a 100 MHz carrier signal at the center of the graticule, with the horizontal calibration of the spectrum analyzer at 20 kHz/division, and the vertical calibration at 10 dB/division below a 0 dBm reference. You should be able to observe the following from Fig. 7-12a:

1. The approximate modulating frequency is the calibrated horizontal distance between the sideband signals: 10 kHz.

2. The number and magnitudes of the carrier and significant sideband-signal pairs identify the power and frequency distribution and bandwidth of the FM signal (see Table 7-1): four sideband-signal pairs above the −40 dB (1% voltage) level distributed over an 80 kHz bandwidth; and the power levels of −13.15 dBm (carrier), −4.73 dBm (first sideband pair), −9.12 dBm (second sideband pair), −17.72 dBm (third sideband pair), and −30.46 dBm (fourth sideband pair).

3. The total power (which is equal to the zero-modulation RF carrier power) is the sum of the carrier and sideband power levels (as converted from step 2 above): 0.05 mW + (2 × 0.34 mW) + (2 × 0.12 mW) + (2 × 0.02 mW) = 1.0 mW = 0 dBm.

With control of the frequency and amplitude of the modulating signal, the total RF power can be more quickly obtained by removing the modulation, resulting in the carrier level of Fig. 7-12b.

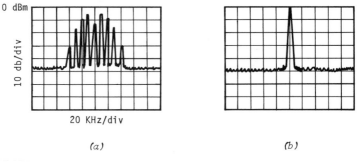

(a)   *(b)*

FIGURE 7-12

The Bessel-zero method can also be employed with the spectrum analyzer by adjusting the voltage and frequency of the modulating signal until the RF carrier level is reduced to zero. You can identify which carrier-zero modulation index is displayed by observing the relative levels and number of sideband-signal pairs (above the −40 dB level as noted in Table 7-1). For the 2.405 carrier zero-modulation index there are 5 sideband-signal pairs above the −40 dB level; for the 5.52 modulation index there are 9 pairs; and for the 8.654 modulation index there are 12 pairs. With the carrier-zero modulation index identified, and the modulation frequency identified as the distance between the sideband signals, the deviation may be easily computed using the formula: $f_{dev} = M_{FM}f_{mod}$. A quick measurement of the modulating signal voltage (refer to step 4 of the Bessel-zero procedure) will then provide the proportional modulation voltage vs carrier deviation. If, for example, the display on the spectrum analyzer were as shown in Fig. 7-13, you could identify the modulation index as 5.52, because the carrier level is zero and there are nine sideband-signal pairs above the −40 dB level. You could also identify the modulating frequency as approximately 2 kHz from the horizontal spacing of the sideband signals along the 5 kHz/division calibrated X-axis of the analyzer. It is thus possible to obtain the information necessary to directly calculate the frequency deviation of the RF carrier as follows:

$$f_{dev} = M_{FM}f_{mod} = 5.52 \times 2 \text{ kHz} = 11.04 \text{ kHz}$$

CIRCUIT TECHNIQUES

Frequency-modulation circuits generally employ either *direct FM* or *PM* (*indirect FM*) techniques at reduced radio frequencies, which are then multiplied up to the final operating frequency. Since linear amplifiers are not required in either the multiplier or final stages of FM transmitters, the power efficiency of class C RF amplifier circuits can be used to advantage in FM

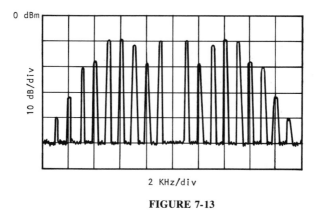

0 dBm

10 dB/div

2 KHz/div

FIGURE 7-13

systems. A further advantage of FM systems is that modulation, including preemphasis or other signal processing, occurs at lower power levels where electronic circuit functions can be easily achieved.

Reactance Modulators

Direct FM is achieved by varying the frequency of an oscillator circuit. A most popular circuit for direct FM is the *reactance modulator,* which provides a varying capacitance or inductance in a resonant oscillator circuit in order to achieve frequency deviation. The "varying reactance" of the circuit was historically achieved with a vacuum tube; but modern circuit approaches use transistors, IC's, and varicap diodes to vary the resonant frequency of the circuit. The electronic device (transistor, etc.) can be connected as either a variable inductance *or* a variable capacitance in parallel with the resonant *LC* circuit of an RF oscillator. The critical requirement of the circuit is that the reactance modulator must be completely reactive, i.e., it must have either a $+90°$ or a $-90°$ phase angle. Any resistive ($0°$) component in the reactance modulator will introduce amplitude variations (AM) at its output in proportion to the resistance-reactance ratio. This undesirable AM usually occurs with the use of bipolar transistors and requires further signal processing with limiter amplifier circuits in order to eliminate the AM from the FM signal. MOS field-effect transistors, however, like vacuum tubes, yield a minimum resistive component and are thus better suited for use in reactance modulator circuits.

Figure 7-14 shows a typical reactance modulator circuit using a MOSFET as the electronic (amplifying) device, with its output connected across (in parallel with) the resonant *LC* circuit (L_2, C_3) of an RF oscillator. The required $90°$ phase shift of the MOSFET signal is achieved by its input network, R_1, C_1. If you consider that the reactance (X_C) of C_1 is much greater

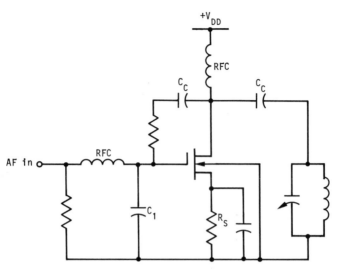

FIGURE 7-14

than the resistance of R_1, nearly $-90°$ phase shift of the RF signal will be coupled through C_2 to the input network. Thus the input signal at the gate of the MOSFET is almost at $-90°$ with respect to the resonant circuit RF signal. In the common-source connection the MOSFET will amplify the input signal and maintain oscillation in the LC resonant circuit. It will also shift its input signal by $+180°$ to its output (drain) terminal. The amplified feedback signal appears at the drain of the MOSFET and across the LC resonant circuit with a phase shift of $+90°$ $[(-90°) + (+180°) = +90°]$, making the MOSFET appear to be an inductor (coil). With zero modulation the resonant frequency of the oscillator is determined by C_3 and by the parallel combination of L_2 and the effective *inductance* of the MOSFET. When an AF modulating signal is applied to the gate of the MOSFET through the RF choke (L_1), it will vary the FET transconductance. The resulting AF output current at the drain will vary the magnitude of effective MOSFET inductance, and thus frequency modulate the oscillator circuit.

The example of Fig. 7-14 describes one of four possible reactance-modulator circuit connections that employ a feedback network to achieve 90° phase shift at the input of an electronic device. Interchanging the RC components of the feedback network will yield an output capacitance instead of an inductance. Two connections of RL feedback networks will also yield device output inductance or capacitance across a resonant circuit. The four circuit connections and their resulting output L and C values are shown in Fig. 7-15.

If we consider the circuit of Fig. 7-14, and assume a MOSFET G_{fs} of 1000 μS, the effective parallel inductance can be computed as:

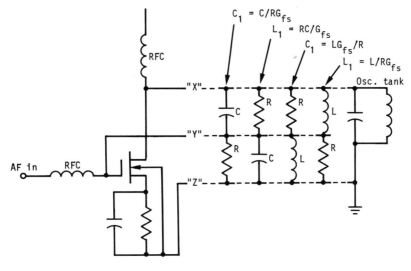

FIGURE 7-15

$$L = RC/G_{fs} = 27 \times 10^3 \times 6 \times 10^{-12}/10^{-3}$$
$$= 1.62 \times 10^{-4} = 162 \ \mu\text{H}$$

The resonant frequency at which the 6 pF has a reactance of $R/10$ can be computed in order to assure a phase shift approaching 90° for this circuit connection where R should be at least $10X$:

$$f = 1/2\pi CX = 1/(6.28 \times 6 \times 10^{-12} \times 2.7 \times 10^3)$$
$$= 9.83 \times 10^6 = 9.83 \ \text{MHz}$$

The frequency of deviation for this circuit can be determined based on the multiplication factor, as reviewed earlier in this chapter. The consequent required percent deviation in the value of equivalent inductance affected by the MOSFET reactance modulator can then be applied to a percent change in G_{fs}, which in turn depends on the amplitude of the applied modulation signal.

The choice of reactance modulator circuit connection will depend on a number of factors, including the device's (transistor, FET, etc.) internal impedances and the practicality of component values. Practical equivalent output inductance or capacitance values, however, should represent only a *small* part of the total resonant circuit L or C in order to assure stable operation of the adjacent RF oscillator at approximately the same frequency, with or without an active reactance modulator. As a general rule the ohmic value of the feedback resistor (R_1 in Figs. 7-14 and 7-15) should be established between two limits:

1. It should be less than 1/10 of the device terminal impedances (Z drain-gate or Z gate-source in the MOSFET, for example).

2. It should be 10 times greater than the reactance of the feedback coil or capacitor (X_{L1} or X_{C1}) when it connects to the high side of the resonant circuit, and 1/10 the reactance of the feedback coil or capacitor (X_{L2} or X_{C2}) when it connects to the low side of the resonant circuit.

Based on the above limits you should deduce that often the final decisions as to which circuit connection of Fig. 7-14 is to be used, are not made until the connections of *all* four connections are evaluated.

Varicap Modulators

A simpler type of reactance modulator is achieved with a *varactor* diode (often called *varicap,* or specified as *voltage-variable capacitance* diode). Such a reverse-biased diode will change its junction capacitance in proportion to the applied voltage. The varactor diode can be applied directly to a resonant circuit with few additional components, making the varactor-diode reactance modulator the simplest means of achieving FM. Figure 7-16a shows a Hartley RF oscillator circuit that uses the varactor diode as the total capacitance of the resonant network. Figure 7-16b shows a simpler Colpitts RF oscillator circuit using a varactor diode. The resonant frequency of these RF oscillator circuits depends on the dc bias applied to the diode, while the frequency modulation results from the audio-frequency modulating signal applied through the RF choke (L_1). Varactor diodes are typically specified by their maximum capacitance and capacitance ratio at the voltage limits of their operation. While they are available with various response curves, the tuning diodes change capacitance as the inverse square of the applied voltage: $V = 1/C^2$. When the varactor diodes are operated at higher radio frequencies (above about 50 MHz), the distributed resistance and inductance become significantly large to complicate the diodes' response. For use in FM reactance modulator circuits, varactor diodes are primarily capacitive devices. Thermal drift and instability are among the disadvantages of varactor-diode reactance modulators, but they are used in conjunction with feedback AFC (automatic frequency control) circuits in order to achieve stability in many FM applications.

The linearity of reactance modulator circuits is specified as the proportional change in resonant frequency compared to a change in the modulation signal *level*. Factors that contribute to nonlinearity include stray capacitances and electronic device (the transistor, FET, diode, vacuum tube, etc.) response. A strictly *linear* response can only be achieved in a resonant circuit if the capacitance or inductance changes as the *square* of the applied voltage, as you may recall from earlier chapters:

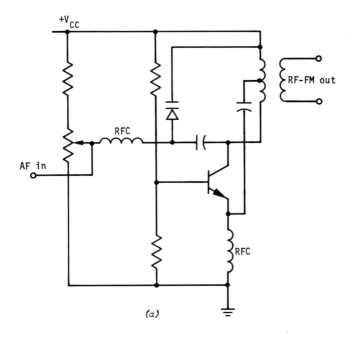

(a)

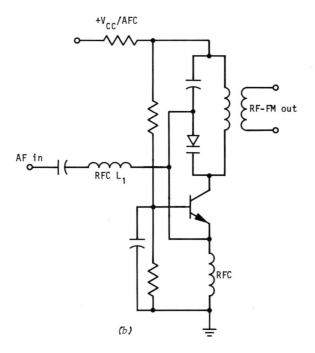

(b)

FIGURE 7-16

$$f = 1/\sqrt{LC}$$

so if L or C is squared, f will change in direct (or inverse) proportion.

One useful characteristic of the MOSFET reactance modulator is the device square-law response, where its output current changes as the *square* of its input voltage. Since the effective inductance or capacitance of the reactance modulator varies with its output current, the output inductance or capacitance of the MOSFET changes as the square of its input modulation voltage, yielding a direct linear FM response. The varactor diode is also available with square-law response. When it is used in a reactance modulator circuit, its capacitance will change as the inverse square of its input voltage, also yielding a direct linear FM response.

As noted earlier in this chapter, phase modulation basically differs from frequency modulation by function: FM circuits vary the *frequency* of RF oscillators, and PM circuits vary the *phase* of independently generated fixed-frequency RF signals. The same reactance modulator circuits here described may also be used for phase modulation of RF signals without the necessity of providing feedback to maintain oscillation. Figure 7-16*b* shows a representative modification of the FM modulator of Fig. 7-16*a* for use as a phase modulator.

Sample Problem 4

Given: The bipolar transistor Hartley oscillator/varactor reactance modulator circuit of Fig. 7-16*a*, operating at 10.7 MHz with a deviation of 100 kHz and using a varicap diode specified as having a maximum capacitance of 100 pF and an inverse square capacitance ratio from 4 to 40 V.

Determine: The required resonant circuit capacitance (C_o) and the AF modulation voltage for 100 kHz deviation.

(*a*) First, the resonant circuit capacitance is computed as:

$$C_o = 1/\omega^2 L = 1/(6.28 \times 10.7 \times 10^6)^2 \times 4 \times 10^{-6} = 55.4 \text{ pF}$$

(*b*) The varactor bias voltage is computed for operation at 55.4 pF:

$$V_o = [(C_o/C_{max})^2(V_{max} - V_{min})] + V_{min}$$
$$= [(55.4/100)^2 \times (40 - 4)] + 4 = 15.05 \text{ V}$$

(*c*) The capacitance is computed for the 100 kHz deviation by ratio and proportion:

$$C_{hi} = C_o f_o/(f_o + f_{dev}) = 55.4 \times 10.7/(10.7 + 0.1) = 54.9 \text{ pF}$$
$$C_{lo} = C_o f_o/(f_o - f_{dev}) = 55.4 \times 10.7/(10.7 - 0.1) = 55.9 \text{ pF}$$

(*d*) The varactor bias voltages are then computed for the above deviation limits:

$$V_{hi} = [(C_{hi}/C_{max})^2(V_{max} - V_{min})] + V_{min}$$
$$= [(54.9/100)^2 \times (40 - 4)] + 4 = 14.85 \text{ V}$$
$$V_{lo} = [(C_{lo}/C_{max})^2(V_{max} - V_{min})] + V_{min}$$
$$= [(55.9/100)^2 \times (40 - 4)] + 4 = 15.25 \text{ V}$$

(e) Finally, the p-p AF modulation voltage for 100 kHz deviation about an operating-point voltage (V_o) is computed as:

$$V_{lo} - V_{hi} = 15.25 - 14.85 = 400 \text{ mV p-p}$$
$$V_o = (V_{lo} + V_{hi})/2 = (15.25 + 14.85)/2 = 15.05 \text{ V}$$

Sample Problem 5

Given: The reactance modulator circuit of Fig. 7-14 operating at a frequency of 10.7 MHz with a deviation of 100 kHz, and using a MOSFET with 6 pF input capacitance (C_{gs}) and a 1.0 V p-p input modulating signal that varies the drain current from 2 to 6 mA and varies the transconductance (G_{fs}) from 6000 to 9000 μS.

Determine: The zero-modulation drain current and transconductance, the zero-modulation equivalent output inductance, the change in output inductance, and the required resonant circuit L and C values.

(a) First, the zero-modulation drain current is computed based on the square-law response of the MOSFET:

$$I_{Do} = I_{D(min)} + \sqrt{(I_{D(max)} - I_{D(min)})^2/2}$$
$$= 2 + \sqrt{(6 - 2)^2/2} = 4.83 \text{ mA}$$

(b) The zero-modulation square-law response transconductance is computed as:

$$G_{fso} = G_{fs(min)} + \sqrt{(G_{fs(max)} - G_{fs(min)})^2/2}$$
$$= 6 \text{ k} + \sqrt{(9 \text{ k} - 6 \text{ k})^2/2} = 8120 \text{ S}$$

(c) The zero-modulation output inductance is computed:

$$L_{eq} = RC/G_{fs} = 27 \times 10^3 \times 6 \times 10^{-12}/(8.12 \times 10^{-3}) = 20 \text{ } \mu\text{H}$$

(d) The maximum and minimum effective inductances at the MOSFET output are computed by ratio and proportion:

$$L_{eq(max)} = L_{eq}G_{fs(max)}/G_{fso} = 20 \times 9/8.12 = 22.17 \text{ } \mu\text{H}.$$
$$L_{eq(min)} = L_{eq}G_{fs(min)}/G_{fso} = 20 \times 6/8.12 = 14.78 \mu\text{H}$$

and thus the change in equivalent output inductance is:

$$22.17 - 14.78 = 7.39 \text{ } \mu\text{H}$$

(e)   The ratio of inductance changes with the square root of frequency, thus RF carrier deviation is now computed as:

$$\Delta f = 1/[(f_{hi}/f_{lo})^2 - 1] = 1/[(10.8 \text{ MHz}/10.6 \text{ MHz})^2 - 1]$$
$$= 26.25 \text{ (or } 26.25 : 1)$$

(f)   The parallel-resonant circuit inductance (L_2 of Fig. 7-14), as connected in *parallel* with the changing L_{eq}, that is required to shift resonance from 10.6 to 10.8 MHz is now computed as follows:

$$L = L_{eq(max)} L_{eq(min)}/[(L_{eq(max)}\Delta f) - (L_{eq(min)} (\Delta f + 1))]$$
$$= 22.17 \times 14.78/[(22.17 \times 26.25) - (14.78 \times 27.25)]$$
$$= 1.83 \ \mu\text{H}$$

(g)   Then the total zero-modulation inductance is computed as the parallel combination of L and L_{eq}:

$$L_T = L L_{eq}/(L + L_{eq}) = 1.83 \times 20/(1.83 + 20) = 1.68 \ \mu\text{H}$$

(h)   Finally, the resonant circuit capacitance is computed as:

$$C = 1/\omega^2 L = 1/(6.28 \times 10.7 \times 10^6)^2 \times 1.68 \times 10^{-6}$$
$$= 1.32 \times 10^{-10} = 132 \text{ pF}$$

A modern approach to frequency modulation uses versatile circuits called *phase-locked loops* (PLL's). These circuits are becoming increasingly popular with their availability in integrated (IC) form. PLL's are used in frequency-modulation and phase-modulation circuits, as well as FM detector and processing circuits. Their functional operation, however, requires an understanding of frequency and phase detectors which are yet to be introduced in this chapter; so the PLL circuits are placed near the end of the chapter for your better understanding.

FM DETECTION

Slope Detectors

As you may recall from the introductory part of this chapter, *FM detection* is the process that converts the *deviation* of an RF carrier signal to a varying *voltage level*. While the concept of slope detection may be adequate for your understanding of the FM detection process, slope detector circuits are not commonly used in quality FM systems because they are nonlinear and do not reject AM. Because of requirements for circuit simplicity at higher RF and microwave frequencies, however, the *balanced slope detector* is used where the frequency response of more complex circuits is limited. This popular FM detector circuit (Fig. 7-17) is essentially two simple slope detectors, with frequency and phase response set to oppose each other. Although the balanced

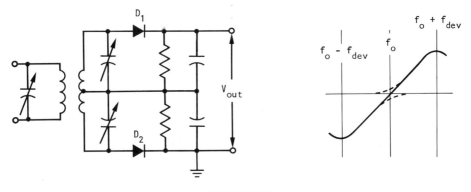

FIGURE 7-17

slope detector is not popular, it is the first detector circuit to be here ana-lyzed, as its simple operation may provide you with a perspective on the operation of the more complex *popular* detector circuits.

If you consider the circuit of Fig. 7-17 as two mirror-image envelope detectors driven by the center-tapped RF transformer, you should conclude that the currents through R_1 and R_2 would oppose each other; thus there would be zero dc output voltage across the terminals, *A-C*. This zero-output voltage condition would exist if both halves of the tuned-secondary RF trans-former were set at the *same* frequency, or if the RF signal were operating at a frequency exactly between two *differently tuned* halves of the RF trans-former. It is this second, differently tuned, condition by which the circuit of Fig. 7-17 is able to convert an incoming RF signal deviation to a varying output voltage, and thus demodulate an FM signal. In operation the two halves of the RF transformer secondary are tuned to frequencies equally above *and* below the RF signal frequency to which the transformer primary is tuned. The frequency-response curves are selected, based on the Q of the tuned circuits, to accommodate the amount of signal deviation along the linear portion of the curves. For the FM broadcast service this deviation is 75 kHz; so one-half of the transformer secondary winding is tuned (with C_2) to peak at about 100 kHz *above* the zero-modulation RF signal, and the other half of the transformer secondary winding is tuned (with C_3) to peak at about 100 kHz *below* the zero-modulation RF signal. When the incoming RF signal deviates *above* the zero-modulation frequency, there is a greater RF level across C_2-L_2, causing an increased dc through D_1-R_1, with the resulting positive-polarity output voltage level that varies with the *amount* of devia-tion approaching the peak tuning of C_2-L_2. When the incoming RF signal deviates *below* the zero-modulation frequency, there is a greater RF level across C_3-L_3, causing an increased dc through D_2-R_2, with the resulting negative-polarity output voltage level that varies with the *amount* of devia-tion approaching the peak tuning of C_3-L_3. So a deviating RF signal into the balanced slope detector circuit will result in an alternating voltage output that is in step with the RF deviation.

One of the limitations of the balanced slope detector is its inability to exclude any amplitude variations from the incoming FM RF signal. Such amplitude variations would appear as a varying voltage signal at the output of the detector along with the varying output from the RF deviation. In order to use balanced slope detectors effectively, all amplitude variations must be eliminated from the incoming FM signal prior to detection. The circuit function that achieves such removal of AM from FM signals is called *limiting*. A limiter circuit is simply an RF amplifier that is driven with a large input signal, causing its output signal to clip at a given voltage level. The limiting function can be achieved with separate RF circuits, or as a part of the intermediate-frequency (IF) section of the FM receiver system.

Modern IC devices used for FM IF amplifier applications include limiter functions within the integrated circuit. Since the maximum practical input signal levels to the detector stage of FM receiver systems fall below 2 V p-p, limiting is most simply accomplished with diodes. Figure 7-18 shows the limiting function achieved with two series, forward-biased diodes, clipping the RF signal at 1.4 V p-p. When the clipped dc-offset waveform is applied to a resonant *LC* network, it, of course, develops an ac sine-wave signal that is RF transformer coupled to the detector circuit, thus retaining the frequency variations (FM), while removing any amplitude variations (AM) from the RF signal.

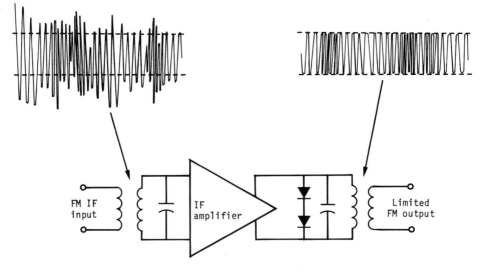

FIGURE 7-18

From your understanding of the balanced slope detector you should note that its linearity and accuracy depend on *both* the critical tuning of three separate resonant circuits and the Q of the three circuits. If FM systems had to depend on such critical circuitry, low-cost FM broadcast receivers would be virtually impossible to build in high production volume. Recognizing this problem, circuit designers brought about solutions for less critical and easier-to-tune circuits for FM detection. Such circuit types include discriminators, ratio detectors, quadrature detectors, and phase-locked-loop detectors.

Discriminators

The discriminator circuit samples a portion of the input RF signal which serves as a reference for comparing the phase and magnitude of frequency deviation. Figure 7-19 shows two discriminator circuits: one where sampling is achieved by a coupling capacitor from the high side of the transformer primary to the center tap of its secondary; and the other where sampling is

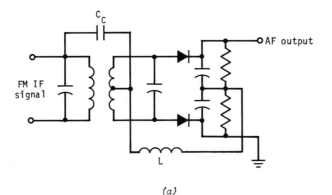

(a)

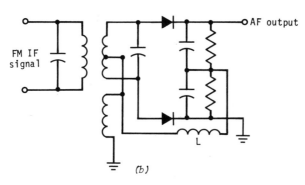

(b)

FIGURE 7-19

achieved by coupling through an additional winding on the RF transformer. An additional inductance (L_3) is required of the capacitor-coupled circuit in order to establish the appropriate phase relationship at the center of the secondary network (point A). In the discriminator both primary and secondary transformer windings are tuned for resonance at the zero-modulation frequency of the FM RF signal, so the signal voltages applied to the diodes (D_1 and D_2) are equal in magnitude but out of phase. The resulting currents through R_1 and R_2, filtered by the capacitors C_1 and C_2, are equal in magnitude but 180° out of phase, yielding zero dc output voltage across the terminals, B-C.

The current flowing through the inductor (L_3 or the auxiliary transformer winding) is the return path for the currents through R_1 and R_2; and the voltage developed across the inductor is 90° out of phase with the signal applied to the diodes (of course, you remember *ELI the ICEman* from the early chapters of this text). Since the signal voltages applied to the diodes are at the transformer terminals, 180° out of phase with each other, they are only alternately +90° and −90° out of phase with the return signal voltage at the transformer center tap. Now the center tap of the transformer is in turn coupled to the high side of the transformer primary, and so the return signal voltage is *also* in phase with the primary input signal, making the transformer *terminal voltages* alternately +90° and −90° with respect to the input (primary) signal voltage. (If you have difficulty in following this explanation, it is recommended that you reread it while tracing the circuit until you are completely confident about the circuit's operation with a zero-modulation applied signal. Do not attempt the next paragraph until you fully understand the circuit operation as just described.

To understand the concepts of phase detection—the basis on which the discriminator converts FM deviation to varying dc voltage levels (or AF modulation signals)—you must be able to keep track of the voltages across each of the discriminator components. These voltages are the transformer primary (V_{in}), each *half* of the transformer secondary (V_1 and V_2), each of the diodes (V_{D1} and V_{D2}), and the inductor or winding (V_L). Vector diagrams are a most convenient way to show phase relationships, especially in this case where there are six quantities to observe. Figure 7-20 shows three vector diagrams that describe the operation of the discriminator: (1) the zero-modulation center-frequency operation, (2) the results of deviation *above* the center frequency with FM, and (3) the results of deviation *below* the center frequency with FM. You should alternately refer to the schematic diagram of Fig. 7-19 and the vector diagram of Fig. 7-20 as you proceed through the text of the following paragraph in order to relate the description to the vectors, and then to the actual circuit.)

Figure 7-20a shows the following steps for a zero-modulation center frequency:

1. The induced transformer primary voltage (V_{in}) of the parallel-resonant circuit is in phase with the circulating current but 180° out of phase with the *applied* voltage.

2. The applied voltage is across the inductor or winding (V_L), from the transformer center tap to ac ground through the low-reactance capacitors. It has the same amplitude, but opposite phase, as the induced primary voltage.

3. One-half of the transformer secondary voltage (V_1) is at +90° with respect to the center tap. The other half of the transformer secondary voltage (V_2) is at −90° with respect to the center tap. (V_1 and V_2, of course, alternate with the applied RF signal, but the phase relationship with the center tap remains constant, as it too is alternating with the applied RF signal.)

4. The diode voltage (V_{D1}) is the resultant of the inductor (V_L) and transformer voltage (V_1), as these voltages are across D_1. The diode voltage (V_{D2}) is the resultant of the inductor voltage at 0° and the respective half-transformer voltage at 90°.

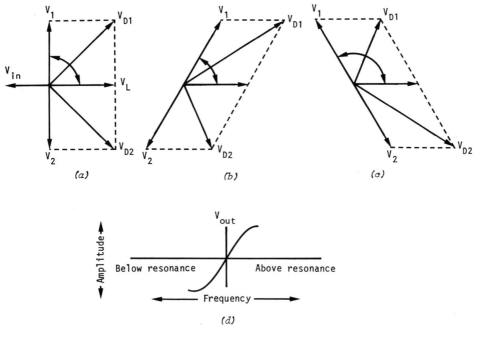

(a)   (b)   (c)

(d)

FIGURE 7-20

5. The varying dc output current, that is, the AF (modulation) signal which develops the output signal voltage across resistors R_1 and R_2, is a function of both diode voltages V_{D1} and V_{D2}, with the resultant output signal at the vector sum of these two signals.

$$V_{out} = [(V_{D1} - V_L) - (V_{D2} - V_L)]$$

(The above formula assumes zero voltage drop across diodes D_1 and D_2; so with *real* circuits, both the forward voltage drop and the internal diode resistance must be considered in determining the *actual* output voltage.)

The vector diagram of Fig. 7-20*b* shows the results of the RF input signal deviation *above* the center (zero-modulation) frequency. Both steps 1 and 2 above remain the same, but the resultant *magnitudes* of steps 3, 4, and 5 change for the following reason:

At frequencies above resonance the tuned *LC* network (the transformer inductance and tuning capacitor, C_s) becomes inductive and shifts the phase of the applied transformer voltages, V_1 and V_2 (ELI the ICEman). The resulting positive (+) phase shift yields a greater voltage across D_1 and a smaller voltage across D_2, because the phase of the initial +90° transformer half shifts *toward* the phase of the inductor voltage. If you now apply a greater V_{D1} and smaller V_{D2} to the formula for obtaining the detector output voltage in step 5 above, you will observe that a positive V_{out} will result in direct proportion to a positive phase and frequency shift.

The vector diagram of Fig. 7-20*c* shows the results of the RF input signal deviation *below* the center (zero-modulation) frequency, with changes in the *magnitudes* of steps 3, 4, and 5 for the following reason:

The tuned *LC* network becomes capacitive at frequencies below resonance, and shifts the phase of the applied transformer voltages, V_1 and V_2. The resulting negative (−) phase shift yields a smaller voltage across D_1 and a greater voltage across D_2, because the phase of the initial +90° transformer half shifts *away from* the phase of the inductor voltage. Applying a smaller V_{D1} and greater V_{D2} to the detector output formula will now result in a negative V_{out} in direct proportion to a negative phase and frequency shift.

Figure 7-20*d* shows the frequency response of the discriminator circuit, where the output voltage level is proportional to the change in frequency about a zero-modulation center frequency. This response curve is often referred to as an *S-curve* because of its shape. The S-curve will result from swept-frequency measurements yielding a frequency vs voltage (frequency domain) display of FM detector response on an oscilloscope. As with all tuned circuits, the efficiency of the discriminator is directly proportional to Q and inversely proportional to bandwidth. For maximum efficiency the tuned circuit bandwidth should be approximately twice the FM deviation. The output of the discriminator will remain linear to approximately the -3 dB frequency response of the primary and secondary tuned circuits of the RF transformer that are peak tuned to the zero-modulation center frequency of the FM RF signal. Although the discriminator is less critical to tune and remains more linear than the balanced slope detector, it also requires limiting of the FM RF input signal, because it will not reject AM.

The explanations for the discriminator may have been painfully tedious, and possibly confusing, but the concepts advanced in the preceding pages are the basis for a wide variety of circuit functions. Your thorough understanding of the *phase detector process* on which the discriminator operates will pay dividends in allowing you to more easily understand the FM detector and phase-locked-loop circuits that follow. Because the complicated circuit action of the phase detector will not be reviewed elsewhere in the text, you should refer to the preceding section if you are in any way confused about phase detection.

Ratio Detectors

The ratio detector circuit was developed in response to the need for an FM detector *with* AM rejection. The use of this circuit reduces the need for *hard* limiting circuits ahead of the detector in many noncritical FM systems. The ratio detector operates on the same phase detector basis as the discriminator, but its output is taken as the *ratio* of diode voltages (or resistor currents) instead of *direct* magnitudes, as with the discriminator. An amplitude-varying FM input signal will maintain the same ratio of diode voltages, even though the magnitudes may vary. The resulting detected output signal would *not* vary with input-signal *intensity* although still providing a varying output level from input-signal frequency deviation. The tradeoff limitation of the ratio detector circuit is its efficiency, because the *ratio* of diode voltages will always be substantially less than the *total* of the voltages.

Figure 7-21 shows two ratio detector circuits: one with sampling of the input RF signal through a coupling capacitor, and the other through an auxiliary winding on the RF transformer. As with the discriminator, an additional inductance is required of the capacitor-coupled circuit in order to establish the appropriate phase relationship at the center tap of the transformer secondary. Both the primary and secondary windings of the transformer are also tuned for resonance at the zero-modulation frequency of the FM RF signal. The ratio detector circuitry differs from the discriminator as follows: The output signal is taken from the junction of the diode load resistors and the inductor (L_3); a large capacitor across the diodes stores and maintains a dc voltage level; and the diodes (D_1 and D_2) are opposing, wired

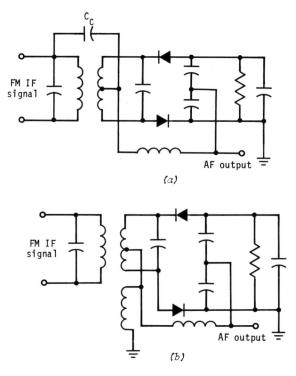

(a)

(b)

FIGURE 7-21

for series operation of the circuit. This diode connection makes the voltage drops across C_1 and C_2, as well as the voltage drops across R_1 and R_2, *series additive,* instead of series opposing as with the discriminator. With a zero-modulation FM RF signal applied to the input of the ratio detector circuit, it operates like the discriminator except for the varying dc output. If you review Fig. 7-20 and the first four steps of descriptive text that identify the voltage vectors, you should be able to apply that information to the output networks of the ratio detector.

In reference to Fig. 7-21 you should note that even the AF modulation signals out of both diodes are shorted by the large filter capacitor, C_3, so there is only smooth dc at the output of the diodes and flowing through the resistors, R_3 and R_4. The output resistors thus divide the dc voltage across the filter capacitor and establish a dc voltage reference across which the output signal is obtained. Since the diodes are connected for series operation, the applied voltages, as shown by vectors V_{D1} and V_{D2} in Fig. 7-20, will *add* in phase rather than cancel as with the discriminator. This addition will serve to maintain a fixed voltage across the filter capacitor, C_3, even with a deviating FM signal:

$$V_{C3} = V_{D1} + V_{D2} \quad \text{and} \quad V_{R3} = V_{R4} = V_{C3}/2$$

The deviating FM RF signal yields the same phase shift in transformer voltage and current with the ratio detector as occurs with the discriminator. With frequencies *above* the zero-modulation RF center frequency, the transformer-tuned circuit becomes inductive, and the induced current lags the induced voltage. At frequencies *below* the center frequency, the tuned circuit becomes capacitive, and the induced current leads the induced voltage. The vector sum of voltages that result from this varying phase shift appears at the inductor terminal opposite the transformer's center tap. Since this point is *not* grounded as it is in the discriminator circuit, the varying dc output signal (AF modulation signal) is obtained from this point of the circuit. The frequency response of the ratio detector circuit results in the same S-curve as that obtained from the discriminator. The limits of the response depend upon the Q/bandwidth of the LC tuned circuits at the primary and secondary of the RF transformer. The efficiency of the ratio detector, however, is less than that of the discriminator, and the output can be computed as:

$$V_{out} = \{[(V_{D1} - V_L) + (V_{D2} - V_L)]/2\} - (V_{D2} - V_L)$$

The *extent* to which the ratio detector *eliminates* the need for limiting depends upon the FM service in which the detector is used. In no way will the performance of the ratio detector alone compete with a limiter-discriminator (or limiter-ratio detector) circuit combination. First, the time constant of the diode load resistors (R_1 and R_2) and the filter capacitor (C_3) is too slow to allow limiting of short-duration amplitude changes like static, etc.; and it is not slow enough to respond to long-duration changes like fading, etc. Second, the amplitude linearity of the circuit will introduce distortion into the signal, as it will not detect weak and strong signals equally. For many years ratio detector circuits were a popular choice for TV and less-expensive FM radio receivers, where the addition of discrete limiter circuits was costly. With modern IC technology, however, most all FM detector circuits include limiting functions.

Quadrature Detectors

Quadrature is a fancy word that refers to two signals that are 90° out of phase. The quadrature phase detector is a popular FM detector circuit that is available in IC form and requires only one external LC tuned circuit for operation. The circuit basically produces a varying dc output level (or AF modulation signal) that is dependent on the phase deviation of two RF input signals which are 90° out of phase with each other. The two signals are obtained from the single deviating FM RF signal that has been amplified and limited. Figure 7-22 shows a typical circuit for the quadrature detector, where the deviating FM RF signal switches (remember the limited FM signal is a square wave) the transistor current source (Q_1) of the differential

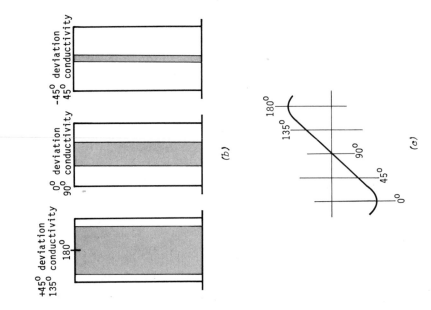

(b)

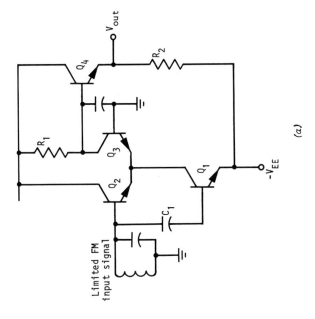

(a)

FIGURE 7-22

amplifier, Q_2 and Q_3. The same limited FM RF signal is coupled to the resonant circuit, L_1-C_2. Since the signal is a square wave, the 90° phase shift occurs across the L_1-C_2 network. Even though the signal *into* the resonant circuit is deviating, the signal *out of* the circuit (L_1-C_2) is at a fixed frequency due to the high resonant Q. The fixed-frequency/fixed-phase RF signal is fed to one input of the differential amplifier at the base of Q_2, where it is combined with the switching signal from the current source (Q_1). The conduction through Q_3 thus depends on the coincident phase relationships of the two input signals.

Figure 7-22*b* shows the effects of combining the quadrature signals and the demodulation of the deviating FM signal. If you assume that the direct FM signal is deviating in frequency to yield an equivalent of ±45° of phase shift, for example, the conduction period where *both* signals are positive would vary depending on the coincident phase relationships of the two signals. One drawing of the three signals in Fig. 7-22*b* shows the direct FM signal shifted by +45°, yielding 135° of conduction (where full conduction is 360°), so the output current integrates to 135/360 of $I_{C(max)}$. The second drawing shows the direct FM signal at the zero-modulation center frequency with 0° phase shift, yielding 90° of conduction for 90/360 of $I_{C(max)}$. The third drawing shows the direct FM signal shifted by −45°, yielding 45° of conduction for 45/360 of $I_{C(max)}$.

The capacitor C_3 acts as an RF filter and integrating network with R_1, but with a short C_3-R_1 time constant with respect to the modulation (AF) signal. The RF pulses thus establish the varying dc level of the AF modulation, and so bias the output transistor, Q_4. The value of R_2 is adjusted to bring the output terminal to exactly zero volts with 90° of conduction, when the FM RF signal is at the zero-modulation center frequency. A positive phase deviation of the FM signal will thus result in a positive dc level at the output; a negative phase deviation will result in a negative dc level at the output. This frequency/phase response yields the same S-curve as is obtained from the discriminator and ratio detector, with the limits of response dependent on the Q of the tuned circuit (L_1-C_2) and with a maximum phase shift of 90°. Since the average (zero-modulation) output of the quadrature detector results from only 90/360 of the available signal-current cycle, it operates at less than 25% efficiency. Amplification of this signal is easily achieved, however, since active amplifying devices (transistors, etc.) are employed for the circuit operation.

Sample Problem 6

Given: The phase discriminator circuit of Fig. 7-19*a*, operating at a frequency of 10.7 MHz, driven with a 2 V p-p RF signal that is deviating ±45° at the primary of the 1 : 1 (center-tapped secondary) RF transformer, and driving an output load resistance of 10 kΩ.

Determine: The component values R_1-R_2, C_1-C_2, and the peak-to-peak AF output signal voltage.

(a)   First, the matched resistances R_1 and R_2 and the matching reactances X_{C1} and X_{C2} are computed based on their total equaling R_L:

$$R_1 = R_2 = X_{C1} = X_{C2} = R_L/2 = 10 \text{ k}\Omega/2 = 5 \text{ k}\Omega$$

(b)   The capacitance values for C_1 and C_2 are then computed at the highest AF modulation frequency of 15 kHz:

$$C = 1/\omega X = 1/(6.28 \times 15 \times 10^3 \times 5 \times 10^3)$$
$$= 2.12 \times 10^{-9} = 2.12 \text{ nF}$$

(c)   A check of the calculated values of R and C (from steps a and b above) is now made to assure that the RC time constant is greater than the period of one cycle at the highest modulating frequency (15 kHz):

$$t = 1/f = 1/15 \text{ kHz} = 1/(15 \times 10^3) = 66.7 \ \mu s$$

and

$$t = RC = 5 \times 10^3 \times 2.12 \times 10^{-9} = 10.6 \ \mu s$$

so within one cycle at 15 kHz there are $66.7/10.6 = 6.29$ time constants, which exceed requirements for circuit operation.

(d)   Next, the reactance values for C_1 and C_2 are computed at the applied RF (10.7 MHz) signal frequency to assure at least 40 dB (100 : 1) RF rejection:

$$X = 1/\omega C = 1/(6.28 \times 10.7 \times 10^6 \times 2.12 \times 10^{-9}) = 7.02 \ \Omega$$

and the rejection ratio is:

$$X_{\text{mod}}/X_{\text{RF}} = 5 \text{ k}\Omega/7.02 = 712 : 1 \quad \text{and} \quad 20 \log 712 = 57.1 \text{ dB}$$

(e)   The output AF signal is computed based on the relationship of the phase and magnitude of the applied signals V_{D1}, V_{D2}, and V_L:

$$V_{\text{out}} = (V_{D1} - V_L) - (V_{D2} - V_L)$$

With 45° deviation the phase shift in the transformer voltages V_1 and V_2 will alternately increase and reduce the magnitudes of V_{D1} and V_{D2}:

$$V_{D1} \text{ or } V_{D2} = V_1 \sin \theta/\sin \tan^{-1}[V_1 \sin \theta/(V_L + V_1 \cos \theta)]$$
$$= 1 \sin 45°/\sin \tan^{-1}[1 \sin 45°/(2 + 1 \cos 45°)]$$
$$= 2.8 \text{ V}$$

and

$$V_{D1} \text{ or } V_{D2} = V_1 \sin \theta/\sin \tan^{-1}[V_1 \sin \theta/(V_L - V_1 \cos \theta)]$$
$$= 1 \sin 45°/\sin \tan^{-1}[1 \sin 45°/(2 - 1 \cos 45°)]$$
$$= 1.47 \text{ V}$$

(f)   The above-computed magnitudes of V_{D1} and V_{D2} are now applied to determine the peak-to-peak AF output signal voltage, as follows:

$$V_{out(peak)} = (V_{D1} - V_L) - (V_{D2} - V_L) = (2.8 - 2) - (1.47 - 2)$$
$$= 1.33 \text{ V (positive peak)}$$

and $$V_{out(peak)} = (1.47 - 2) - (2.8 - 2) = -1.33 \text{ V (negative peak)}$$

so $$V_{out(p-p)} = (V_{out+}) - (V_{out-}) = 1.33 - (-1.33)$$
$$= 2.66 \text{ V p-p}$$

Phase-Locked Loops

Phase-locked loops (PLL) are becoming more widely used for both frequency modulation and detection functions. Although PLL circuit concepts were developed in the 1930's, their initial complexity prevented their widespread use in communications systems. Recent advances in modern IC technology, however, have made PLL's both a practical and a popular choice for critical FM systems applications. PLL's are multifunction devices consisting of *phase detector*, lowpass *filter*, dc *amplifier*, and voltage-controlled *oscillator* (VCO) elements.

Figure 7-23 shows the block diagram and functional phase-detector operation of the PLL. You can better understand PLL operation if you refer to Fig. 7-23b and recall the operation of the traditional phase detector (discriminator and ratio detector) circuits, where the incoming RF signal was connected to the center tap of the RF transformer secondary in order to establish the $\pm 90°$ (zero output) relationship at the transformer terminals. Any variation in phase shift, you recall, resulted in a proportionally varying dc output from the phase detector. The PLL phase detector operates in much the same way as the phase discriminator, but it employs a dual differential amplifier circuit as in Fig. 7-23c and does not require the tuned *LC* circuits of the traditional phase detectors. Tuning is established through the important feedback loop/VCO elements of the PLL, where the VCO signal, not the incoming RF signal, establishes the zero-output phase relationship.

When the VCO signal is at the same frequency (and phase) as the incoming RF signal, the two frequencies cancel and only a dc output is available from the PLL phase detector. As the incoming frequency shifts, there is a resulting phase shift in the phase detector, with an output signal voltage proportional to that phase shift. The differential amplifier circuitry of the PLL phase detector operates like that of the quadrature phase detector, where the dc output is proportional to the phase shift of two incoming RF signals. Since two RF signals are being combined in the diff-amp circuit of the PLL, and since no tuned circuits are employed, the output signals include the two original input frequencies *and* the mixing product frequencies of the two input signals. The lowpass filter element of the PLL may be a simple

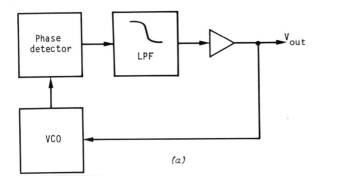

(a)

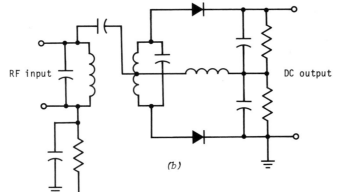

(b)

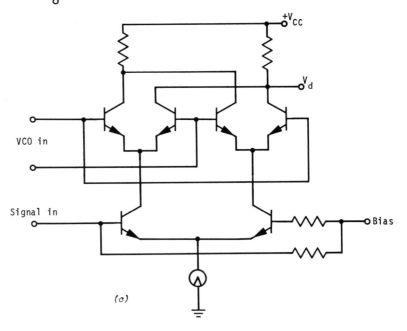

(c)

FIGURE 7-23

(single-pole) *RC* network, or a more complex circuit, which effectively removes all signals but the difference frequency—a varying dc output—from the diff-amp circuit.

The varying dc output from the lowpass filter of the PLL drives the frequency-determining circuitry of the VCO element of the PLL through a dc amplifier. The complete feedback loop effectively yields an RF output signal from the VCO that varies directly with the incoming RF signal. The VCO thus "locks on" to the input signal and follows any variation in its phase and frequency. This function of the phase-locked loop makes it suitable for a variety of applications, including FM detection, frequency modulation, frequency synthesis, synchronization, signal conditioning, and even AM detection. Our major concern in this chapter is FM, so only those applications will be here reviewed.

The PLL is ideally suited for use as a detector of previously limited FM signals. Its major advantages in this application include wide tracking bandwidth, greater linearity than other detectors, and no *LC* circuit tuning requirements. The most popular PLL IC's require a selected number and value of external components in order to optimize their operation for the particular characteristics of an applied FM RF signal. Figure 7-24 shows a typical FM signal limiter and PLL detector circuit. The "outboard" components, C_1, C_2, C_3, and R_1, are chosen to achieve the required VCO frequency, bandwidth, deemphasis response, and lock range and threshold sensitivity. The specific formulas used for the computation of these component values will vary based on the particular PLL type IC used; such data must be obtained from the IC manufacturer in each specific case. For the Signetics 560, a popular PLL IC, C_1 would be computed for a VCO center frequency using the formula, $C = (3 \times 10^{-4})/f_0$, where f_0 is the center frequency of the incoming FM RF signal. The value of C_1 for the 560 can be computed for a typical 10.7 MHz FM RF signal as follows:

$$C_1 = 3 \times 10^{-4}/(10.7 \times 10^6) = 2.8 \times 10^{-11} = 28 \text{ pF}$$

C_2 operates as a lowpass filter component that controls the selectivity or bandwidth of the 560 PLL circuit. The value of C_2 can be computed for a typical *audio* bandwidth of 15 kHz using the manufacturer's data:

$$C_2 = (13.3 \times 10^{-6})/f = 13.3 \times 10^{-6}/(15 \times 10^3)$$
$$= 8.87 \times 10^{-10} = 887 \text{ pF}$$

C_3 performs the deemphasis function of the detector. For the standard FM broadcast time constant of 75 μs, the value of C_3 can be computed for the 560 as follows (the factor 8×10^3 is the equivalent 8 kΩ of internal resistance at terminal 10 of the 560):

$$C_3 = T_c/(8 \times 10^3) = 75 \times 10^{-6}/(8 \times 10^3)$$
$$= 9.38 \times 10^{-9} = 9.38 \text{ nF}$$

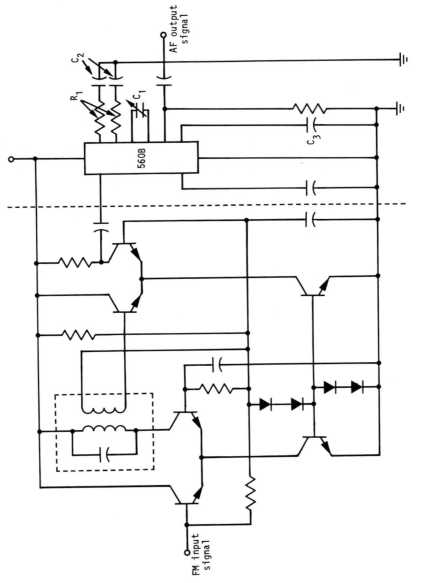

FIGURE 7-24

The value of R_1 is typically specified for a particular PLL device to control the lock range and threshold sensitivity of the VCO element. These characteristics specify the extent to which the VCO can track the frequency of an incoming RF signal. The control voltage for the VCO is developed at the output of the PLL's phase detector element; and its value is, in turn, dependent on the phase difference *and* amplitude of the incoming RF signal as compared to the VCO signal. This loop function *does* require some minimum signal level for operation, however, in order to achieve VCO tracking (phase lock). This minimum limit of control voltage for the VCO is called its *threshold sensitivity*. Since the range of output voltage from the phase detector element is proportional to the deviation of the incoming FM RF signal, the PLL requires higher RF input levels with wide deviation signals. For a given input RF signal level, the range of deviation that provides control voltage levels above the threshold sensitivity of the PLL is called its *lock range* (or tracking range). The lock range is thus inversely proportional to the input RF signal level of the PLL. Lower values of R_1 provide an optimum lock range for a particular PLL device, with higher values of R_1 reducing, or further limiting, the lock range to perform only to the limits of a specified value of RF deviation.

When the 560 PLL is used as an FM detector, the incoming FM RF signal is limited to a relatively high fixed amplitude, far in excess of the device's threshold sensitivity. The *deviation* of the incoming signal is the only factor that will vary the output of the PLL phase detector element. The 560 is specified to have an optimum lock range of $\pm 15\%$ of f_o. The standard broadcast FM deviation is 75 kHz, with an RF (IF) frequency of 10.7 MHz, requiring a PLL lock range of only slightly less than 1%—not 15%. For use as a broadcast FM detector, the lock range is proportionally reduced by increasing the value of R_1, thus reducing the threshold sensitivity. The value of R_1 can be computed based on the threshold sensitivity—lock range reduction factor (which we call Y) and the equivalent internal resistance at the *control* terminals of the PLL. In our example, using the 560, the reduction factor is 15:1, and the internal resistance at pins 14 and 15 is 6 kΩ. R_1 is computed as follows:

$$R_1 = 2R_{int}/(Y - 1) = 2 \times 6 \times 10^3/(15 - 1) = 857\ \Omega$$

PLL FM Signal Generation

Phase-locked loops are ideally suited for the generation of FM signals. They do not require the critical L and C components of the reactance modulators, and they can be frequency stabilized for near drift-free operation. In its basic FM function the VCO element of the PLL is driven by the modulating signal, while it is set for operation at some appropriate zero-modulation radio frequency. If we use the same Signetics 560 PLL, the basic frequency modula-

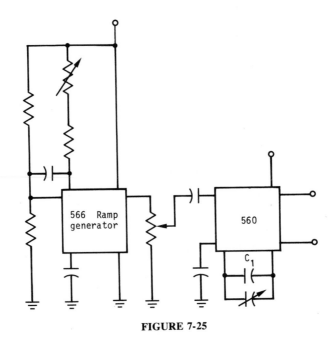

FIGURE 7-25

tion circuit would be connected as shown in Fig. 7-25. Here the modulating signal is injected at the lowpass filter input, where it is amplified to drive (deviate) the VCO. The FM RF output signal is thus obtained from the VCO output of the 560 PLL. This direct voltage-to-frequency conversion is the simplest way to achieve direct FM at frequencies below about 20 MHz (the frequency limit of popular PLL's). In this basic FM application the only critical component in the circuit is C_1, the VCO tuning capacitor, which determines the zero-modulation (center) RF of the PLL. As with the PLL used as a detector element, C_1 is computed from manufacturers' data. Using the 560 as our example for use as a 4.5 MHz TV FM signal generator, C_1 may be computed as follows:

$$C_1 = 3 \times 10^{-4}/f_o = 3 \times 10^{-4}/(4.5 \times 10^6) = 66.7 \text{ pF}$$

Although the generation of FM signals using the basic PLL circuit of Fig. 7-25 may be adequate for many applications, there is no feedback circuitry to eliminate drift of the zero-modulation RF signal. A more complex circuit connection for the PLL is recommended, however, which does provide excellent RF stability through the use of feedback signals. This circuit, shown in block diagram and schematic form in Fig. 7-26, is called a *translation loop*. The connection requires the use of two PLL's, but with only one VCO

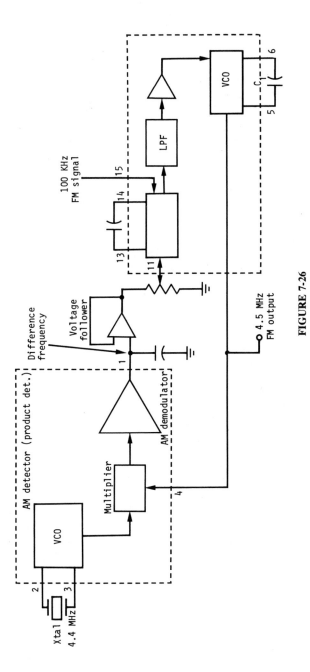

FIGURE 7-26

function, and an AM product-detector function. To achieve the functions for the translation loop requires two more versatile PLL's: the 561 with AM output, and the 562 with internal phase-comparator bias. The overall circuit function for use as a 4.5 MHz TV FM signal generator, for example, uses a 4.4 MHz (4400 kHz) crystal, in conjunction with the VCO element of the 561 PLL, to develop a stable, fixed-frequency RF source. A second RF signal of 100 kHz (0.1 MHz), which is frequency modulated, is combined in the phase-detector element of the 562 PLL to obtain the sum frequency of 4.5 MHz. The difference frequency of these combined signals accurately drives the VCO of the 562 PLL for stable operation, with FM, at 4.5 MHz. The same 4.5 MHz FM RF output signal is also fed back to mix with the original 4.4 MHz signal of the 561 VCO, yielding the 100 kHz difference frequency which is combined, again, with the 100 kHz FM RF signal at the phase-detector element of the 562 PLL. The difference frequency (or phase) of these two 100 kHz signals drives and deviates the VCO of the 562. The virtue of this translation loop is that an unwanted frequency drift of 5% from the 100 kHz FM RF source (yielding 5 kHz drift) will *translate* to only 0.1% drift at 4.5 MHz [(5 kHz/4.5 MHz) × 100].

CONCLUSION

This chapter has been concerned with the principles and practices of frequency modulation and detection—the means by which intelligence signals are applied to, and removed from, RF carrier signals. Although FM techniques are more complicated than their AM counterparts, as reviewed in the previous chapter, FM extends the performance of radio communications beyond what AM can provide. Your direct comparison of high-quality FM broadcast signals to AM broadcast radio should quickly identify the static- and fade-free, low distortion, and low noise advantages of FM. FM is, of course, used in a wide range of communications services beyond broadcast radio and TV sound. Your working knowledge of FM signals, functions, and processes, as reviewed in this chapter, should provide you with the capability of applying these concepts to most all FM communications services. You should also be equipped to apply your knowledge of FM to both electronic circuits and systems reviewed in other sections of this text. The problems that follow will serve to test your knowledge of FM principles and techniques, and measure your achievement of the *learning objectives* as specified at the beginning of this chapter. As with previous chapters, it is recommended that you review communications systems and IC applications literature from manufacturers, along with direct experimentation using the circuits and IC devices typically found in FM systems.

Problems

1. Given: A 92.5 MHz broadcast FM radio transmitter with a full deviation of 75 kHz and a voice-modulation frequency range of 50 Hz to 15 kHz, using a basic RF oscillator/modulator circuit characterized by an *LC* resonant-network Z_p of 400 Ω and a *Q* of 50, and a maximum change in *C* of 20 pF.

Using the direct FM approach to transmitter design, determine:
(*a*)   Percent frequency deviation
(*b*)   Percent change in *LC* network capacitance
(*c*)   *LC* network capacitance
(*d*)   *LC* network inductance
(*e*)   Basic RF oscillator/modulator frequency
(*f*)   Required frequency multiplication factor

2. Given: A 162.5 MHz FM "weather band" radio transmitter with a full deviation of 5 kHz and a voice-modulation frequency range of 300 Hz to 3 kHz, using a basic RF oscillator/modulator circuit characterized by an *LC* resonant-network Z_p of 400 Ω and a *Q* of 50, and a maximum phase deviation of 0.75 rad.

Using the indirect FM (phase modulation) approach to transmitter design, determine:
(*a*)   The equivalent deviation frequency for 0.75 rad at 300 Hz
(*b*)   Required frequency multiplication factor
(*c*)   Basic RF oscillator/modulator frequency
(*d*)   Percent change in *LC* network capacitance
(*e*)   *LC* network capacitance
(*f*)   *LC* network inductance

3. Given: The FM transmitter designs of Problems 1 and 2 above.

Comment on the results, effects, and practicality of:
(*a*)   Reversing the transmitter design approach of Problems 1 and 2 to indirect FM and direct FM, respectively
(*b*)   Increasing the change in *C* with Problem 1 (direct FM) from 20 pF to 100 pF
(*c*)   Reducing the *LC* network *Q* (in both problems) from 50 to 20
(*d*)   Reducing the maximum phase deviation in Problem 2 (indirect FM) from 0.75 to 0.3 rad

4. Given: A 4.5 MHz TV-sound FM RF subcarrier signal with a 5 kHz AF modulation signal at a full deviation of 25 kHz.

Determine:
(*a*)   Modulation index

(*b*)   Percent of the voltage and power of the RF carrier and each of the *significant* sideband pairs

(*c*)   dB ratio of the highest to lowest significant-level signals

(*d*)   Bandwidth of the significant-level signals

(*e*)   Bandwidth using greater than 95% of the signal power

5.   Given: An FM transmitter with a 50 μs *RL* AF preemphasis network for noise reduction.

Determine:

(*a*)   The value of *L* if the network *R* is 5 kΩ

(*b*)   The 3 dB preemphasis frequency

(*c*)   The amount of preemphasis, in dB, at 8, 12, and 15 kHz

(*d*)   The single-frequency noise reduction at 8, 12, and 15 kHz

6.   Given: The measured spectrum analyzer display as shown in Fig. 7-27, and using the Bessel-zero technique.

Determine:

(*a*)   Zero-modulation RF carrier frequency and power

(*b*)   FM AF modulation frequency

(*c*)   Significant sideband signal pairs and FM signal bandwidth

(*d*)   Modulation index and frequency of deviation

7.   Given: A bipolar transistor Hartley oscillator/varactor reactance modulator, operating at 18 MHz for $\times 9$ frequency multiplication to 162 MHz, and with a deviation of 2 kHz, in an *LC* network with a fixed capacitance in parallel with the varicap diode, and with a Z_p of 800 Ω and a *Q* of 20.

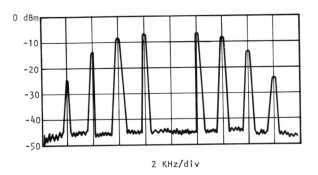

2 KHz/div

FIGURE 7-27

If the varicap diode has a maximum capacitance of 20 pF and an inverse-square capacitance ratio from 4 to 20, determine:

(a)   Resonant network L and C values
(b)   The change in C for 2 kHz deviation
(c)   Varactor bias voltage for a 10 pF zero-modulation capacitance
(d)   Varactor bias voltage for the deviation limits (the 10 pF $\pm$ the change in C from answer b above)
(e)   Peak-to-peak modulation voltage for 2 kHz deviation
(f)   The deviation frequency and p-p modulation voltage for a 5 kHz deviation at 162 MHz (after $\times 9$ frequency multiplication)

8.   Given: A reactance modulator circuit operating at 18 MHz for $\times 9$ frequency multiplication to 162 MHz, and using a MOSFET to obtain a varying inductance utilizing its 10 pF input capacitance (C_{gs}) and a 10 kΩ resistor from drain to gate.

If a 1 V p-p input signal to the MOSFET yields a variation in G_{fs} from 2000 to 2600 μS (LC circuit $Z_p = 800$ Ω and $Q = 20$), determine:

(a)   Zero-modulation square-law response transconductance
(b)   The effective zero-modulation MOSFET output inductance
(c)   The maximum and minimum inductance values with the 1 V p-p modulation signal
(d)   The required total inductance of the LC resonant circuit
(e)   The maximum and minimum *total* inductance values with the 1 V p-p modulation signal
(f)   The resulting frequency deviation with the 1 V p-p modulation signal
(g)   The deviation frequency and p-p modulation voltage for a 5 kHz deviation at 162 MHz (after $\times 9$ frequency multiplication)

9.   Given: A Foster-Seely phase discriminator circuit operating at a zero-modulation frequency of 4.5 MHz, driven with a 1 V p-p RF signal that is deviating at a full modulation of 25 kHz with an AF of 5 kHz, at the tuned primary of a 1 : 1 center-tapped secondary RF transformer, and driving an output load resistance of 20 kΩ.

For a resonant-circuit Z_p of 6000 Ω and a Q of 60 (assume a coupling coefficient of 1.0), determine:

(a)   Transformer primary and center-tapped secondary inductance values
(b)   Tuned circuit capacitance values
(c)   Divider load resistance values (R_1 and R_2 in Fig. 7-19)
(d)   Divider bypass capacitance values (C_1 and C_2 in Fig. 7-19)
(e)   The time constant with respect to one period of the 5 kHz AF signal

(*f*)   The RF rejection in dB
(*g*)   The value of load inductance (*L* in Fig. 7-19) if $X_L = 100\, X_{C1}$ at 4.5 MHz
(*h*)   The equivalent phase shift of the carrier deviation
(*i*)   The p-p AF output signal voltage

10. Given: A 560 PLL IC for use as an FM detector with the same input-signal characteristics as in Problem 9 above, 1 V p-p 4.5 MHz RF signal deviating 25 kHz with an AF of 5 kHz and preemphasis of 75 μs.

Referring to Fig. 7-24, determine:
(*a*)   The VCO tuning capacitor (C_1)
(*b*)   The lowpass filter capacitor (C_2)
(*c*)   The deemphasis capacitor (C_3)
(*d*)   The threshold sensitivity—lock range reduction factor (*Y*)
(*e*)   The lock-range resistance (R_1)

CHAPTER 8

RADIO SIGNALS AND RECEIVER FUNCTIONS

In studying the two preceding chapters you learned about systems of modulation and demodulation—the ways that *intelligence* (message) *signals* are superimposed upon, and extracted from, RF carrier signals. This chapter concerns the combination of circuit functions that receive and process modulated RF carrier signals. This chapter also introduces the concept of a complete *electronic communications system,* where a number of specialized circuit functions are combined to process RF communications signals and convert them to some aural, visual, or mechanical output.

   In its simplest form a *receiver* of modulated RF signals employs a resonant circuit to tune to the frequency of the RF signal and to transfer the resulting current to a demodulator circuit. Further processing of the demodulator output signal will depend upon the requirements of the transducer that finally converts that electrical signal to some aural, visual, or mechanical form. The earliest broadcast radio receivers simply used a resonant circuit and a diode AM detector to achieve a complete receiver function, where the output from the demodulator (detector) was of sufficient amplitude to drive a pair of earphones. Operation of this early receiver thus depended upon a strong RF signal from a long-wire antenna and a powerful radio transmitting station. Such radio receivers are available today as toys; they will also provide hi-fi AM reception when connected to the low-level input of a good audio amplifier/speaker system. Figure 8-1a shows a *systems-level* block diagram of this simplest receiver. Although the diagram is functionally basic, it covers a wide range of RF receiving systems, including the sophisticated products of our most recent technology.

   The historical progress in *electronic* amplification was first applied to the demodulated AF output signals of the simple radio in order to drive loudspeakers. Later developments in RF amplification of very low antenna currents extended the *sensitivity* of the radio tuner section. The improvements in radio communications through progress in RF amplification, however, created other problems that were to complicate the simple form of the receiver. Of course, some control of RF gain was required to prevent the

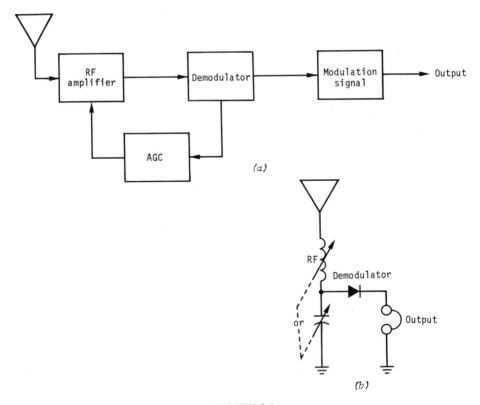

FIGURE 8-1

strong signals of powerful close-proximity radio stations from saturating or overdriving the circuitry of the receiver, while at the same time amplifying the weak signals from distant radio stations to a sufficient level to drive the receiver's demodulator. This RF gain control also had to be automatic (self-adjusting), because the weak signals would fade, and tuning the radio past a strong signal with the gain *up* would shock the listeners into momentary deafness!

A coincident problem created by the *automatic gain control* (AGC) is that it will desensitize a receiver that is set to tune a weak RF signal in the presence of a strong adjacent frequency signal. With lower Q and wider bandpass the early radio tuner sections often simultaneously received two stations. When the center frequency of such a radio is tuned to a weak station, with an adjacent strong station off resonance (down on the response curve of the tuner), the stronger signal will often energize the AGC circuit and thus prevent the radio from receiving the desired weak signal. You can experience this "masking" phenomenon with inexpensive modern radio receivers by noting a greater sensitivity in urban areas at 3 a.m., when most

local stations are off the air. An early technique that was used to successfully separate the weaker signals of distant radio stations from strong
adjacent-channel signals will also serve to prevent masking by the AGC circuit. The approach employs multistage tunable *LC* networks to increase the
Q and reduce the bandpass of the radio's tuner section. Careful tuning of
these networks will reject most adjacent frequency signals, thus making the
receiver highly *selective*. Unfortunately, this solution created yet another
problem: to carefully tune five *LC* networks every time you want to change
radio stations is a real nuisance!

If you were to design and manufacture a specialized radio for a single
fixed RF signal (such as one of the WWV stations of the U.S. National
Bureau of Standards) at 10.0 MHz, for example, tuning would not be a
problem; so the simple receiver format noted in Fig. 8-1 could be expanded
to yield an excellent product. Both automatic and adjustable RF gain controls could be employed, and high selectivity and coincident sensitivity
could be obtained with four RF amplifier transistors or IC's between the
five *LC* networks that were factory aligned for peak response at exactly
10.0 MHz. Such receivers are, in fact, manufactured for specialty purposes
under their original circuit classification as tuned radio-frequency (TRF) receivers. In fixed-frequency operation the TRF circuit will perform as an excellent receiver of even the weakest RF signals. Figure 8-1*b* relates the
functions of a modern TRF receiver to the primitive form of Fig. 8-1*a*,
showing that the only functional addition in the modern TRF is the AGC
circuit.

A giant step in the technical progress of radio came with the development
of the *superheterodyne* receiver. Figure 8-2 is a block diagram of this receiver. In a sense the "superhet" uses the basic form of the TRF receiver,
but includes, in addition, a frequency-converter circuit ahead of the TRF
section for ease of tuning. With the superhet circuit, the five *LC* networks
can still be factory aligned for peak response, while the radio can be tuned

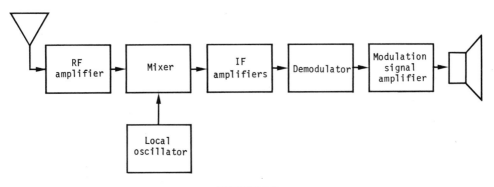

FIGURE 8-2

from station to station with one control. The converter section of the receiver uses nonlinear mixing (*heterodyning*) of two signals: the incoming RF signal from the radio station and a locally generated signal at some frequency above (*super*) that of the incoming RF signal. The difference-frequency mixing product of the above two signals is then fed to the TRF section of the receiver. If the TRF section is tuned to 10 MHz, for example, conversion of an incoming 90 MHz RF signal can be achieved with a *local oscillator* (LO) signal at 100 MHz. Also, conversion of a 92 MHz RF signal can be achieved with a 102 MHz LO signal, and so on. Basically, changing the LO frequency is what *tunes* the superheterodyne radio to different stations, whereas the TRF section provides for the radio's basic sensitivity and selectivity. In the superhet design the TRF section is identified as the *intermediate frequency* (IF) amplifier section, as its frequency of operation is intermediate (between) to the incoming RF signal and final demodulated signal frequency.

The idea of tuning by changing the frequency of a locally generated RF signal in a radio receiver led to yet another advantage of the superhet circuit. A feedback signal can be applied to the LO and change its frequency to correct for any slight drift or mistuning of the receiver, thus "locking it on" to the center frequency of the incoming RF signal. So *automatic frequency control* (AFC) is another circuit function available with the superhet design. Wide-range voltage tuning of the LO adds yet another function to the superhet and provides the RF receiver with remote tuning capability. Combining the wide-range voltage tuning with the AFC function of the receiver will add the capability of locking onto and following a changing incoming radio frequency—a common technique now used to assure security in radio communications. The sweeping LO in the superhet circuit is also the basis of operation for modern spectrum analyzers.

As with each earlier improvement in radio receivers, the superhet design also creates some tradeoff problems. First of all, the nonlinear mixer produces many other intermodulation and harmonic distortion product frequencies, in addition to the desired difference frequency of the LO and incoming RF (station) signals. All undesired frequencies must be thus rejected by filtering or other techniques requiring additional circuit and network components. Second, the nonlinear mixer is not selective in delivering incoming RF signals to the IF section of the receiver. If you carefully review the earlier example, where the 10 MHz IF signal was converted from the mixing of a 102 MHz LO signal with the 92 MHz incoming RF signal, you should deduce that the 102 MHz LO signal could *also* mix with a 112 MHz incoming RF signal and deliver a difference frequency of 10 MHz to the IF section. In the presence of two RF signals that are at equal IF frequencies above and below the LO frequency, the mixer will convert both signals for amplification in the IF section—one from the desired radio station and one from some undesired radio station—and the radio will pick up two stations at once. The undesired incoming RF signal is called the *image frequency*. The image frequency is suppressed in the superheterodyne receiver by adding tuned circuits and tunable RF amplifier stages ahead of the mixer section. The possibility of receiving image-frequency signals, however, has been further reduced by the standards set by international radio committees, who carefully selected IF frequency standards and radio service/station (*channel*) frequency allocations. All these factors have contributed to the superior performance and resulting popularity of the superheterodyne receiver. Almost all modern radio, television, and specialized electronic communications receivers use some form of the superhet circuit.

Thus far we have followed the development of what appears to be the ultimate in functional radio receiver design. We have also identified some major factors in the operation of modern electronic communications receiver systems: sensitivity, selectivity, RF bandpass, AGC action, RF tuning, AFC and voltage-controlled oscillator (VCO) tuning, distortion products, and image frequencies. The final portion of this introductory section will deal with a most important factor of radio communications—*noise*. Electrical noise is produced by the random motions of electronic particles in our gaseous atmosphere and within electronic components and circuitry. The magnitudes of noise signals are very small, but their frequencies are distributed across the full radio spectrum.

Noise signals are received and amplified by radio receivers that are sensitive enough to detect their minute currents. Noise signal levels may often exceed extremely weak modulated RF signals that are transmitted over great distances, thus masking their presence and preventing their demodulation. The limit of RF receiver sensitivity is based on the level of *noise* above which a signal must be present to be demodulated. The total noise power that is amplified and delivered to the demodulator of a receiver depends also on the bandpass of the IF section, with wideband circuits amplifying a greater amount of the random noise frequencies that are distributed across their operating range. Temperature is another factor that increases the noise level from both gaseous (atmospheric) and electronic circuit elements, so receiver sensitivity can be improved by operating in cold environments. For optimum performance the ratio of desired signal to unwanted noise (S/N) must achieve certain standards for acceptance. For example, background noise (hiss) is less acceptable for broadcast music than for broadcast speech. Perceptible snow in a TV image is annoying to a viewer. Moreover, high noise levels in digital communications systems will produce unacceptable errors in the decoding of the demodulated signals.

The purpose of this extended introduction to RF receiving systems is to develop your perspective of the important factors of receiver circuit functions and the signals that they process. Your further study of the development of radio communications will expand your perspectives on the development of modern radio technology.

LEARNING OBJECTIVES

Upon completing this chapter, including the problem set, you should be able to:

1. Understand the sources of electrical noise, and compute thermal agitation noise, electronic circuit noise figure, and signal-to-noise ratio in receiving systems.

2. Understand the functional operation of the superheterodyne receiver system, including single- and multi-conversion processes, and compute the range of image-frequency signals.

3. Understand the functional operation of other specialized receiver systems.

4. Understand the concepts of signal mixing products as applied to the conversion of RF signals in heterodyne receivers, and apply standard mixer circuits and modules to receiver operation.

5. Understand, and apply, bandpass filters and LC networks to tunable RF amplifiers and fixed-frequency IF amplifiers used in modern communications receivers.

6. Understand the operation and characteristics of AFC, VCO, and digitally synthesized oscillator circuits in receiver applications.

7. Understand the operation and characteristics of AGC circuits used to vary the RF or IF gain in receiver applications.

8. Understand the operation and characteristics of automatic noise limiting (ANL) and squelch circuits used in receiver applications.

9. Understand and specify measurements of the important receiver characteristics, including sensitivity, selectivity, frequency response, bandpass, tracking, image rejection, distortion factors, AFC lock range, AGC response, noise figure, and signal-to-noise ratio.

NOISE

External Noise Sources

Radio receivers process signals, including *noise signals*. Effective radio communication requires that the predictable effects of noise be factored into the signal processing functions of receivers in order to determine the practical range (distance) over which such communication is reliable. The writer will never forget the $300 radio-controlled model airplane that flew off into the sunset because it was operated just *beyond* the "practical range of reliable RF communications." Although there are some ultrasophisticated techniques of extracting signals buried in noise, we will here use the *level* of noise signals as the threshhold of communications. Noise signals, as mentioned earlier, result from the random movement of electrons in atoms. The signals can originate from any number of sources, which some physicists have spent lifetimes pursuing. Such noise is categorized as *man-made* (from static sources on earth), *atmospheric* (from the electron motions of our atmospheric gases), and *galactic* (from the noise sources beyond our atmosphere, such as the sun and other sources in our galaxy). Figure 8-3 graphs the typical noise power from various sources, and shows terrestrial and radio-sky noise maps that are regularly published, like weather maps, to guide and inform communications specialists. As you can imagine, the study of noise is a diverse and complex specialty.

In the engineering practice applied to radio receivers, noise from the above three sources is received at the antenna of a communications system and is amplified through the receiver. The highest level noise from these sources, as you can see from the graph of Fig. 8-3, is man-made. You can hear or see the random and variable effects of man-made noise by tuning your AM radio and TV sets off channel (listening to hiss and looking at snow), with the audio and video (contrast) gain controls turned *up*. Since this man-made noise pollution will vary considerably with time and location, there is really no way of quantifying its effects other than averaging a series of measurements of the phenomena. The graph also shows most of the atmospheric noise below the VHF region, most of the galactic noise below the UHF region, and man-made noise as spotty and weak in the UHF region and above.

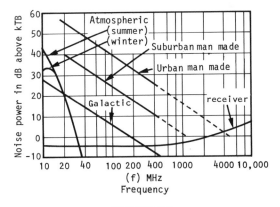

FIGURE 8-3

Internal Noise Sources

In addition to the three sources of noise that are incident on the antenna of the radio receiver, there is another source that is created and amplified within the receiver circuitry. This receiver (or RF circuit) noise is identified as *thermal-agitation* noise; it is a most important factor in engineering practice, because its value is *added* to whatever external noise is transferred to the receiver from the antenna. Figure 8-3 shows that thermal-agitation (receiver) noise becomes increasingly significant as the operating signal frequency increases through the VHF, UHF, and SHF regions of the spectrum. There is thus great concern for the application of low-noise electronic circuitry in higher frequency radio receivers. Fortunately, there is more control and predictability of receiver thermal-agitation noise than of noise from external sources, so the application of appropriate noise-reduction techniques to higher frequency receivers will lower the noise threshold and thus extend the range of reliable RF communications.

Noise Power

Thermal-agitation noise is computed based on a theoretical receiver input resistance with an ohmic *value* equivalent to the noise power that would be developed if a *real* resistor were used as a noise source. If a resistor is placed across the terminals of an extremely sensitive ac voltmeter, there will be a measurable output voltage from the random motion of the electrons within the atoms of the resistor. If the resistor is heated, the electron motion and inertia will increase; and the output voltage will increase in proportion to the change in temperature. If the resistor's ohmic value is increased, the output voltage will also increase in proportion to the change in resistance. If the frequency response (bandwidth) of the voltmeter is increased, the measured output voltage from the resistor will also increase in proportion to the change in bandwidth. Physicists have reduced these variables to a general formula that specifies the thermal-agitation noise power developed in a resistor:

$$P_n = kTB$$

P_n is the noise power in watts, T is the temperature in kelvin ($0°C = 273°K$ and $0°K = -273°C$), B is the bandwidth of the circuit processing the noise signal, and k is the proportionality constant (1.38×10^{-23} J/$°K$) as ascribed to the physicist Boltzmann.

In practical applications the joules are converted to watts by factoring in the time in hertz (formerly cycles per second) (1 joule = 1 watt-second), and the temperature is taken to be 290°K (20°C), the typical room temperature for operation of most electronic equipment. The product of $k \times T$ thus yields a simpler proportionality constant for approximate room temperature applications as:

$$k' \text{ (at 20°C)} = kT = 1.38 \times 10^{-23} \times 293 = 4 \times 10^{-21} \text{ W/Hz}$$
$$= 4 \times 10^{-15} \text{ W/MHz} = 4 \text{ fW/MHz}$$

The value of 4 fW (femtowatts) per MHz provides a convenient reference from which thermal-agitation noise levels can be quickly approximated. If, for example, a given resistor were connected to the matched-impedance input of a 200 kHz-bandwidth RF amplifier, the noise power available for amplification at the output of the resistor could be computed as:

$$P_n = k'B = 4 \times 10^{-21} \times 2 \times 10^5 = 8 \times 10^{-16} = 0.8 \text{ fW}$$

The noise power can also be approximated, in fW, as the ratio of bandwidth to 1 MHz as follows:

$$P'_n \text{ (in fW)} = (B/\text{MHz})k' = (0.2/1) \times 4 \text{ fW} = 0.8 \text{ fW}$$

Noise Resistance

The concept of thermal-agitation noise is applied to the input of an RF amplifier as though its theoretical noise-resistive input impedance is equal to a noise-generating source resistor. As such, the receiver input is treated as a load, and the power generated in the source resistor will transfer, and be equal to, the receiver (load) input-noise power. Figure 8-4 shows the simple circuit relationship on which this concept is based. However, 100% power transfer with matched source and load resistances results in half the open-circuit source voltage appearing across the load. On this basis the open-circuit noise voltage is twice the value that appears across a theoretical matched load, and its value is computed as:

$$v_n = \sqrt{4kTBR} = \sqrt{4k'BR} = \sqrt{4P_nR} = 2\sqrt{P_nR}$$

v_n is the open-circuit noise voltage and R is the equivalent noise resistance of an RF amplifier input. We can now proceed to compute the open-circuit

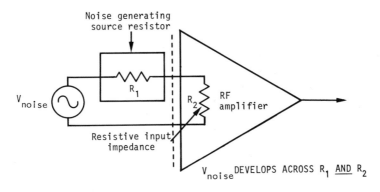

FIGURE 8-4

noise voltage for a 200 Ω *equivalent noise resistance* for the 200 kHz RF amplifier of our previous example as follows:

$$v_n = 2\sqrt{P_n R} = 2\sqrt{8 \times 10^{-16} \times 200} = 8 \times 10^{-7} = 0.8\ \mu\text{V}$$

Two general sources of thermal-agitation noise make up the equivalent input-noise resistance of an RF circuit: the *real* resistive input impedance of the circuit, plus some theoretical resistive value that accounts for the noise that is generated *within* the circuit. Transistors, diodes, vacuum tubes, and other active and passive circuit elements all produce some noise. Although it is impractical to account for the individual noise sources within each circuit element, it is useful to combine the noise voltages that circuit elements produce and reflect these values back to the circuit input. The combined noise can be reflected back to the circuit input as a theoretical resistance value, which may then be added to the real input resistance for computation of input noise voltage. In the example of the previous paragraph, the 200 Ω equivalent noise resistance might have comprised a 50 Ω amplifier input resistance, plus another 150 Ω of equivalent circuit noise reflected back to the input as a noise resistance.

Input, Reflected, and Output Noise

If we expand the above 200 Ω example to account for the *proportion* of added noise that the 150 Ω of reflected noise represents, we can compute a *noise factor* based on the relationship of *real* and *reflected* noise resistance values. When noise signals are combined, their rms value must be computed based on the *square root* of the *sum* of the *squares* of the individual noise levels. For our previous example with a v_{nT} of 0.8 μV, the proportions can be computed as:

$$
\begin{aligned}
v_{nT} &= \sqrt{4P_n R_T} = \sqrt{4P_n(R' + R'')} = \sqrt{(4P_n R') + (4P_n R'')} \\
&= \sqrt{(3.2 \times 10^{-15} \times 50) + (3.2 \times 10^{-15} \times 150)} \\
&= \sqrt{(1.6 \times 10^{-13}) + (4.8 \times 10^{-13})} = \sqrt{v_n'^2 + v_n''^2} = 0.8\ \mu\text{V}
\end{aligned}
$$

v_{nT} is the total noise voltage, v_n' is the equivalent noise voltage produced by the *real* resistive input impedance, and v_n'' is the equivalent noise voltage as reflected back to the circuit input; while R_T is the total equivalent input noise resistance, R' is the *real* resistive input impedance, and R'' is the equivalent noise resistance as reflected back to the circuit input. If you observe the relationship, $v_{nT} = \sqrt{v_n'^2 + v_n''^2}$, you should note its Pythagorean origin, which, in our time of abundant handheld calculators, is more conveniently computed with trigonometric relationships as:

$$
\begin{aligned}
\tan \angle\theta &= v_n'/v_n'' = \sqrt{R'/R''} \\
\sin \angle\theta &= v_n'/v_{nT} = \sqrt{R'/R_T} \\
\cos \angle\theta &= v_n''/v_{nT} = \sqrt{R''/R_T}
\end{aligned}
$$

For our previous example with known noise resistance values the respective noise voltages can be quickly computed based on the noise voltage developed in the total equivalent input resistance as follows:

$$\angle \theta = \tan^{-1} \sqrt{R'/R''} = \tan^{-1} \sqrt{50/150} = 30°$$
$$v_n' = v_{nT} \sin \angle \theta = 0.8 \sin 30° = 0.4 \ \mu V$$
$$v_n'' = v_{nT} \cos \angle \theta = 0.8 \cos 30° = 0.693 \ \mu V$$

As you may observe, the *key* to noise relationships is that the six factors of equivalent noise input are proportionally related.

Noise Figure

The proportion of noise that is added to the input of a circuit by the equivalent noise resistance (R'') is the *noise factor*. In logarithmic form the noise factor is called *noise figure* (NF). Noise figure, as specified and measured in decibels, is one of the most important specifications for RF components, subsystem elements, and receiver systems. Typical measured noise figure specifications for RF products are available as a part of the applications information supplied by manufacturers. Quality RF transistors and diodes have typical noise figures that fall between 1.5 and 6 dB for VHF and UHF applications. Comparable high-quality multistage and IC RF subsystem elements have typical noise figures that fall between 2 and 12 dB, depending on the operating frequency and cost of the elements. The noise factor and noise figure can be computed, based on the trigonometric relationships of the previous paragraph, as:

$$N_f = v_{nT}/v_n' = \sqrt{R_T/R'} = 1/\sin \angle \theta$$
$$NF = 20 \log N_f$$

N_f is the noise factor and NF is the most-often-specified noise figure value. Now the noise figure value of our previous example can be quickly computed as follows:

$$NF = 20 \log \sqrt{R_T/R'} = 20 \log \sqrt{200/50} = 6 \ dB$$

If the above values were reversed to simulate a typical noise analysis procedure, the resistive input impedance (R') and noise figure of this circuit would be specified as $R_{in} = 50 \ \Omega$ and NF = 6 dB. In application you would compute the total equivalent input noise voltage for the circuit as follows:

$$v_n' = \sqrt{4k'BR'} = \sqrt{4 \times 4 \times 10^{-21} \times 2 \times 10^5 \times 50} = 0.4 \ \mu V$$
$$NF = 20 \log v_{nT}/v_n'$$

Thus $v_{nT}/v_n' = \log^{-1} NF/20 = \log^{-1} 6/20 = 2 = N_f$

Therefore $v_{nT} = N_f v_n' = 2 \times 0.4 \ \mu V = 0.8 \ \mu V$

Since noise figure values and resistive input impedance values are available as either published or measured specifications, you can usually neglect the calculation of equivalent noise resistance and simplify the procedures to those of the above calculations.

Low-Noise Circuits

Optimum performance of radio receivers can be achieved with the careful application of low-noise circuits where they can most effectively increase sensitivity. Since the receiver input noise levels are amplified by a greater factor than any noise generated in the intermediate stages of signal amplification, and since the signal levels are lowest at the receiver input, the receiver input circuits are the most critical for noise reduction. This effect is most pronounced in cascaded (multistage) amplifier stages, where quite noisy output circuits have virtually no effect on the signal-to-noise ratio. Because the basis of noise calculations assumes matched resistive values, the noise resistance at the output of an amplifier stage will reflect back to the input as the *inverse square* of the voltage gain. Even a high noise level in a component at the output of an amplifier stage will be significantly reduced when reflected back as an equivalent input noise resistance (R''). This relationship is summarized in the following formula:

$$R_2'' = R_{T2}/(A_{v1})^2$$

R_2'' is the equivalent *input* noise resistance reflected back from a noisy component at the *output* of an amplifier stage, R_{T2} is the total equivalent noise resistance at the *output* of an amplifier stage, and A_{v1} is the voltage gain of the amplifier stage. If, for example, an RF amplifier with a voltage gain of 4 drove a mixer stage with a high total equivalent noise resistance of 800 Ω, the equivalent noise resistance at the input of the amplifier would be computed as:

$$R_2'' = R_{T2}/(A_{v1})^2 = 800/4^2 = 50 \ \Omega$$

This value (R_2'') must then be added to the equivalent input noise resistance in order to determine the total equivalent input noise of the amplifier circuit. If we proceed to our original circuit with $R' = 50 \ \Omega$ and $R'' = 150 \ \Omega$, and add the above $R_2'' = 50 \ \Omega$, we can compute the equivalent input noise of both parts of the amplifier stage as follows:

$$N_f = \sqrt{R_T/R'} = \sqrt{(R' + R'')/R'}$$
$$N_{f2} = \sqrt{(R' + R'' + R_2'')/R'} = \sqrt{(50 + 150 + 50)/50} = 2.24$$
$$\text{NF} = 20 \log N_f = 20 \log 2.24 = 7 \text{ dB}$$

If the real input resistance (R') of the mixer stage were 50 Ω, while the total equivalent resistance (R_T) were 800 Ω, the noise figure of the mixer stage alone would be computed as:

$$\text{NF} = 20 \log \sqrt{R_T/R'} = 20 \log \sqrt{800/50} = 12 \text{ dB}$$

The results of these calculations identify the benefits of employing low-noise first RF amplifier stages in electronic systems. Whereas the RF amplifier of our example had a 6 dB noise figure, the second-stage mixer had a 12 dB noise figure; but by cascading the RF amplifier ahead of the mixer, the total noise figure of the two-stage subsystem increased to only 7 dB. If you review the results of the preceding calculations, you should quickly conclude that there is little change in the noise figure after two stages of RF amplification. With the use of mixer or other lossy stages, however, the voltage gain is less than 1, so the actual value of R''_2 will be greater for the particular circuit. In the superheterodyne receiver circuit, noise calculations are typically made for the cascaded RF amplifier, mixer, and first IF amplifier stages and reflected back to the input as an equivalent input noise voltage, or a noise figure value.

Sample Problem 1

Given: A superhet receiver circuit using modular RF and first IF amplifier stages that are each specified for 12 dB gain and 3 dB noise figure, and a mixer module with 6 dB conversion loss and 5 dB noise figure, and with all module input and output (resistive) impedances of 75 Ω and a system bandpass of 6 MHz.

Determine: The equivalent input noise voltage and overall noise figure of the three-stage subsystem.

(*a*)   The total equivalent input noise resistance for the IF amplifier stage is computed as follows:

$$\begin{aligned}
A_v &= \log^{-1} \text{dB}/20 = \log^{-1} 12/20 = 4 \\
R''_2 &= R_{T2}/(A_v)^2 = 75/4^2 = 4.69 \ \Omega \\
N_f &= \log^{-1} \text{NF}/20 = \log^{-1} 3/20 = 1.41 \\
R_T &= N_f^2 R' = (1.41)^2 \times 75 = 150 \ \Omega \\
R'_{T2} &= R_T + R''_2 = 150 + 4.69 = 154.7 \ \Omega
\end{aligned}$$

(*b*)   The total equivalent input noise resistance for the mixer stage input is computed as follows:

$$\begin{aligned}
A_v &= \log^{-1} \text{dB}/20 = \log^{-1} - 6/20 = 0.5 \\
R''_2 &= R_{T2}/(A_v)^2 = 154.7/(0.5)^2 = 618.8 \ \Omega \\
N_f &= \log^{-1} \text{NF}/20 = \log^{-1} 5/20 = 1.78 \\
R_T &= N_f^2 R' = (1.78)^2 \times 75 = 237.6 \ \Omega \\
R'_{T2} &= R_T + R''_2 = 237.6 + 618.8 = 856.4 \ \Omega
\end{aligned}$$

(c)   The total equivalent input noise resistance for the RF amplifier stage input is computed as follows:

$$R'_2 = R_{T2}/(A_v)^2 = 856.4/4^2 = 53.5 \ \Omega$$
$$\text{So} \quad R_T = N_f^2 R' = (1.41)^2 \times 75 = 150 \ \Omega$$
$$R_{T2} = R_{in}(eq) = R_T + R''_2 = 150 + 53.5 = 203.5 \ \Omega$$

(d)   The total equivalent noise voltage and the overall three-stage noise figure are computed as follows:

$$v_{nT} = \sqrt{4k'BR_T} = \sqrt{16 \times 10^{-21} \times 6 \times 10^6 \times 203.5} = 4.42 \ \mu V$$
$$NF = 20 \log \sqrt{R_T/R'} = 20 \log \sqrt{203.5/75} = 4.34 \ dB$$

The amplitude characteristics of thermal-agitation noise have been described as an important factor of radio receiving systems. This type of noise is primarily amplitude varying and produces noise signals that can be demodulated in an AM detector and reproduced as a hiss in audio systems or random pulses in digital systems. True FM demodulators, or those with previous limiting, are thus quite insensitive to these noise signals. The degree, or deviation, of FM noise produced by thermal agitation is extremely small as compared to the deviation of FM carrier signals received at the input of an FM receiver system. Noise reduction in FM systems is achieved by *limiting* both the amplitude of intelligence signals and the amplitude of thermal-agitation noise signals. This technique of noise reduction in FM systems is called *quieting*.

SENSITIVITY

The sensitivity of a radio receiving system depends upon three factors: RF gain, bandpass, and noise. As you now know, the bandpass of a receiver adversely affects the equivalent noise voltage at its input, and the total RF gain (including IF gain) of the receiver directly affects the noise voltage at its output. The concept of receiver *sensitivity* is a specification of its ability to deliver an intelligible (decodable) information signal to its detector (demodulator) stage. If the signal-to-noise ratio (S/N) at the demodulator input is low, i.e., there is a high level of noise in the intelligence signal voltage, the intelligence cannot be accurately demodulated. For the demodulation of speech signals a 10 dB S/N is satisfactory, even though an annoying hiss is heard in the background of the reproduced speech. For the demodulation of music, however, a 30 dB S/N has been established as the minimum tolerable hiss in the background of reproduced music. Most critical listeners will object to even a 50 dB S/N in some music, however. The S/N required for the accurate demodulation of digital communications signals (see Chapter 9) will vary, depending on the ratio of radio frequency to baud rate, from 10 dB to 50 dB, based on appropriate bandpass filtering. A low S/N will result in decoding errors in the demodulation of digital signals, so higher S/N's are desirable.

The process of computing receiver sensitivity will first require specifications for the minimum modulated signal voltage and bandwidth at the input of the demodulator stage of the system. Typical AM envelope and product detectors that demodulate noise *and* intelligence signals require RF signal voltages in the 1 to 3 V rms range. The bandwidth of the signal delivered to the demodulator stage, of course, depends on the character and form of the modulation signal used for communications, e.g., 8 kHz for 8A3 DSB-AM, 3 kHz for 3A3J SSB, and 5.75 MHz for 5750A5C VSB-AM (TV). The input sensitivity and total RF gain required for an AM superhet aircraft receiver, for example, can be computed, based on the known specifications of 10 kHz bandpass, a 1 V rms signal required for detector demodulation, a 10 dB output S/N, an 8 dB NF, and a 50 Ω resistive input impedance, as follows:

1. The resistive *input* noise voltage and input S/N are computed as:
$$v'_n = \sqrt{4k'BR'} = \sqrt{16 \times 10^{-21} \times 10^4 \times 50} = 0.09 \ \mu V$$
$$S/N_{in} \text{ (in dB)} = S/N_{out} + NF = 10 + 8 = 18 \text{ dB}$$
$$S/N_{in} \text{ (factor)} = \log^{-1} S/N \text{ (dB)}/20 = \log 18/20 = 7.94$$

2. The rms input signal voltage and total RF receiver gain are computed as:
$$v_{in} = v'_n (S/N_{in}) = 0.09 \times 7.94 = 0.715 \ \mu V$$
$$A_v = v_{out}/v_{in} = 1/(7.15 \times 10^{-7}) = 1.4 \times 10^6$$
$$A_r \text{ (in dB)} = 20 \log A_r = 20 \log 1.4 \times 10^6 = 123 \text{ dB}$$

The sensitivity of the above receiver would thus be specified as 0.715 μV for 10 dB S/N. If the above receiver system had a typical 7 dB conversion loss in the mixer stage, the requirements for RF and IF gain would be 130 dB. To get 130 dB of gain with only 8 dB NF requires quality circuitry. If a low-noise RF amplifier were added ahead of the existing circuitry, however, it would not be too difficult to reduce the overall NF to 5 dB. This improvement would, in fact, increase the receiver sensitivity by 3 dB and assure communications equivalent to what would result from *doubling* (+3 dB) the power of an RF transmitter.

In addition to the above S/N sensitivity, receivers are often specified in $(S + N)/N$, SINAD (signal plus noise and distortion), and quieting sensitivity values. These are measured specifications that more accurately define the sensitivity of the *total* receiver circuitry, including the noise developed in the demodulator and audio or baseband amplifiers. If we apply the $(S + N)/N$ measurement to our example, the sensitivity figure would slightly increase from 0.715 μV for 10 dB S/N, to 0.55 μV for 10 dB $(S + N)/N$. The techniques used for performing these and other important receiver specification measurements will be reviewed in a later portion of the chapter.

SUPERHET RECEIVER FUNCTIONS

As reviewed in the introductory section of this chapter, the superheterodyne radio circuit is the most popular and functional of RF receiver designs. The basis of superhet operation is, of course, the local oscillator-mixer function which downconverts the frequency of an RF carrier signal to fixed-frequency, high-gain IF amplifier stages. The IF amplifier section is often called the *IF strip,* as it was historically composed of a strip of cascaded vacuum-tube amplifier stages. With the application of modern IC devices the strip may, in fact, be only one or two stages of high-Q, high-gain circuits. The most popular IF frequencies used in modern receivers are 455 kHz for AM broadcast radio, 10.7 MHz for FM broadcast radio, and 44 MHz for broadcast television. When used in FM applications the IF strip usually includes a limiting function in the final stage(s) of its amplification. The bandpass and gain of IF amplifier circuits will vary considerably depending on the application for which they are used. Figure 8-5 shows popular IF strips, and Fig. 8-6 shows a typical schematic diagram of a modern IF amplifier section.

FIGURE 8-5

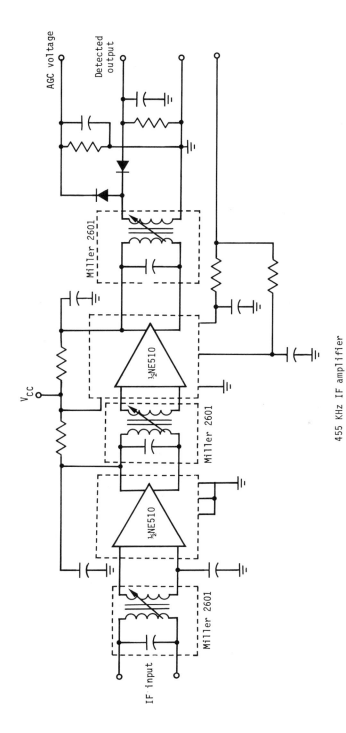

455 KHz IF amplifier

(a)

FIGURE 8-6

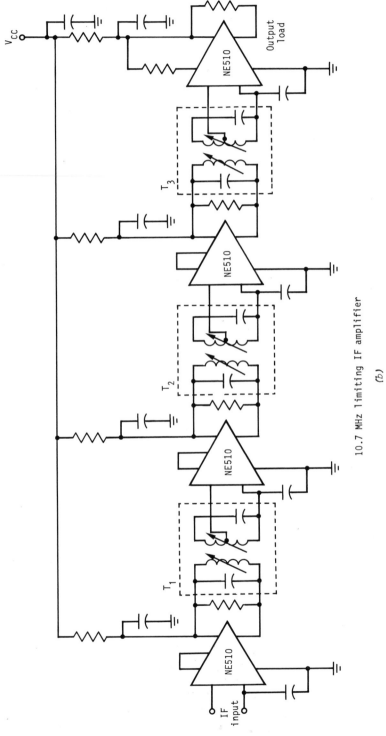

10.7 MHz limiting IF amplifier

(b)

FIGURE 8-6 (*continued*)

Local Oscillator

The local oscillator (LO) in the superhet receiver must provide a stable, low-harmonic, variable-frequency RF signal to the mixer circuit. The LO must tune a frequency range necessary to convert the incoming RF signals to the IF frequency. The LO tuning range is thus the same as the RF amplifier or mixer inputs, but it is offset above the RF signal frequency by the IF frequency. The relationship of the LO, RF/mixer, and IF frequencies can be computed as follows:

$$f_{LO} = f_{RF} + f_{IF}$$

For the standard AM broadcast radio that tunes the range of 540 to 1600 kHz, for example, the LO tuning range is

$$540 + 455 = 995 \text{ kHZ} \quad \text{to} \quad 1600 + 455 = 2055 \text{ kHz}$$

Based on the frequency ratio required to tune the range from 995 to 2055 kHz, a variable capacitor or inductor is normally employed to tune the LO. The L or C element of the resonant circuit must vary as the square of the frequency. For the above example the required change in L or C would be computed as:

$$(f_{hi}/f_{lo})^2 = (C_{lo}/C_{hi}) = (L_{lo}/L_{hi}) = (2055/995)^2 = 4.27 : 1$$

In most AM broadcast radio receivers the LO is tuned with a ganged capacitor that simultaneously tunes the RF amplifier and mixer inputs. The inductor of the resonant circuit is thus fixed-tuned. Since most of the standard variable capacitors can yield up to a 10 : 1 variation in capacitance from their meshed-to-open rotation, additional capacitors are usually connected in series and parallel with the tuning capacitor in order to "pad" (reduce) its range. The series capacitor, identified as a *padder*, is often a fixed-tuned variable or trimmer component; while the parallel capacitor is usually a smaller-value trimmer. Both capacitors are adjusted to obtain LO "tracking," which allows compression or expansion of the frequency coverage vs rotation of the receiver's LO tuning unit. The padder capacitor is often omitted in inexpensive receivers where tracking is not critical. Figure 8-7 shows a typical receiver LO section with ganged tuning capacitors, and padder and trimmer units, along with its associated schematic diagram. Figure 8-8 shows some common variations of LO circuits in schematic form.

Mixers

The LO output signal level is specified by the requirements of the mixer stage in the superhet receiver. The mixer requirements can probably be better understood if you first consider its operation as an AM modulator. (You may review the section on *mixing and modulation* in Chapter 6, if

FIGURE 8-7

necessary.) The LO signal is a fixed-frequency, fixed-level RF signal much like the RF carrier applied to a modulator circuit. The incoming receiver RF signal can here be considered to be analogous to the *modulation* signal. The resulting IF signal is thus like the lower sideband signal from the output of an AM modulator. The idea of 100% modulation can be applied to the LO level as being at least equal to the *highest* anticipated RF input level. Since the mixer is, by function, a nonlinear circuit, the extent to which it is overdriven by the LO or incoming RF signal will yield increasing levels of unwanted harmonic and intermodulation distortion products. For this reason, wide-range AGC is applied in higher quality receivers to maintain a relatively constant RF signal level at the input to the mixer stage. Figure 8-9 relates the modulator and mixer functions as applied to the superhet receiver.

In addition to their design and fabrication as an integral part of manufactured radio receivers, mixer modules are available for application as subsystem elements for use in a wider range of receiver and related systems. Most subsystem mixer modules use diode arrays in conjunction with RF transformers to achieve the mixing function. As such, they do not amplify the signals; they only mix them. If two or four transistors are used for mixing, their

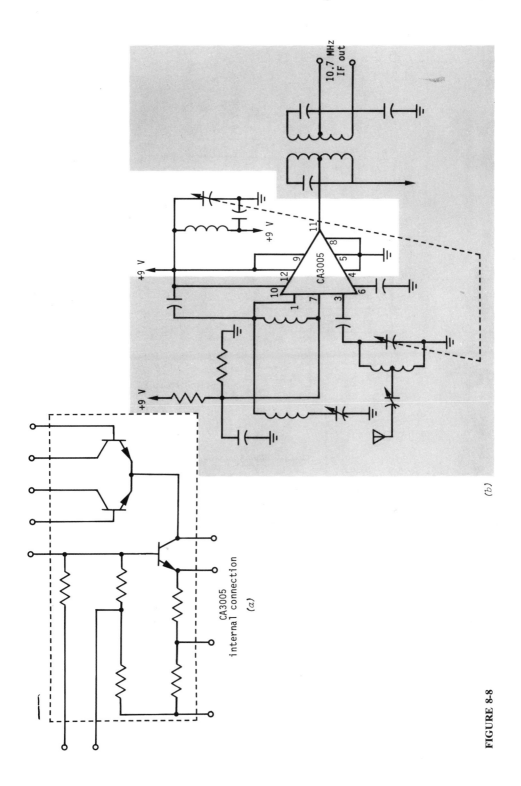

(b)

(a)

CA3005
internal connection

FIGURE 8-8

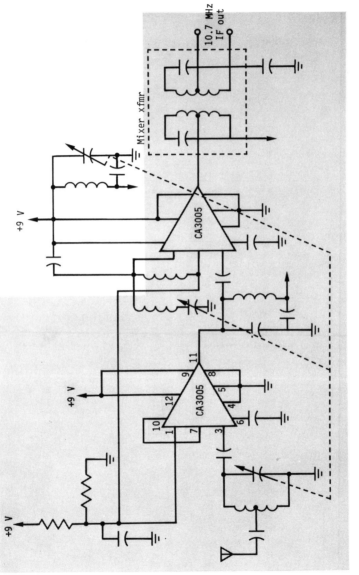

FIGURE 8-8 (*continued*)

(*c*)

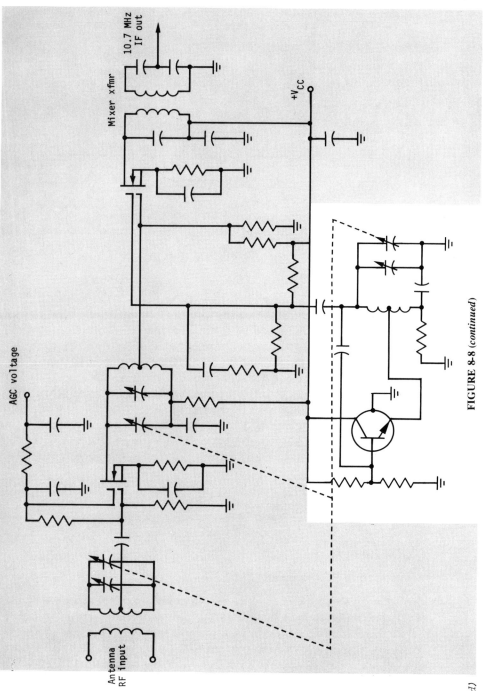

FIGURE 8-8 (continued)

(d)

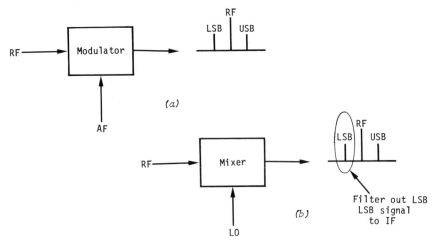

FIGURE 8-9

gain function will be applied to the output IF signal, and treated separately as though the function were achieved with a single transistor, biased to distort, and with the RF signal applied to the base (or gate of an FET) and the LO signal applied to the emitter (or source of an FET). In this case the device gain for the two signals is not equal, and hence requires an increased LO level to offset the lower gain through the emitter (source). Figure 8-10 shows the schematic form of typical mixer circuits.

Mixer modules are often specified with factors that may help you to better understand this most critical circuit function in the superhet receiver. These typical specifications are described below.

1. *Conversion loss* (in dB) is the efficiency of the mixer from its RF input to IF output. An ideal mixer would have 6 dB conversion loss, as the power in *one* sideband is 0.25 (25%) of the modulation power in the RF carrier. The conversion loss will increase if the RF level approaches or exceeds the LO level.

$$\text{Conversion loss} = 10 \log P_{\text{RF(in)}}/P_{\text{IF(out)}}$$

2. *Conversion compression* (or compression point) (in dBm) is the RF signal level that will produce an increased conversion loss (usually 1 dB) above the ideal 6 dB value. In practical mixers the conversion compression typically increases from 0.1 dB, with an RF signal at 10 dB below the LO level, to 1 dB with the RF signal at 6 dB below the LO level. So conversion compression actually specifies the highest practical RF input signal level.

3. *Dynamic range* (in dB) specifies the operating range that the RF signal into the mixer may vary between the limits established by the receiver sensitivity and the mixer conversion compression level.

Dynamic range = conversion compression
  − [input sensitivity (in dBm) + RF gain (in dB)]

The dynamic range of the mixer thus establishes the requirements for the AGC function, which will be reviewed later in this chapter.

4. *Distortion products* (in −dB) specify the level of undesired frequencies in the mixer stage that could convert undesired incoming RF signals to the IF frequency for amplification. This is a complicated sort of specification, as its values will vary depending on the type of mixer circuit and on the levels and frequency ratios of the LO and incoming RF signals. If the mixer circuit distorts by clipping *only* the top or bottom of the input waveform (an unbalanced mixer with nonsymmetrical clipping), there

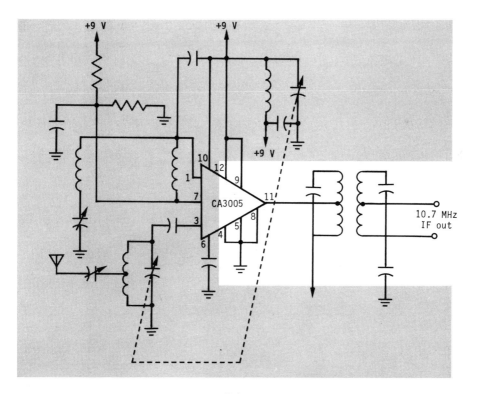

(a)

FIGURE 8-10

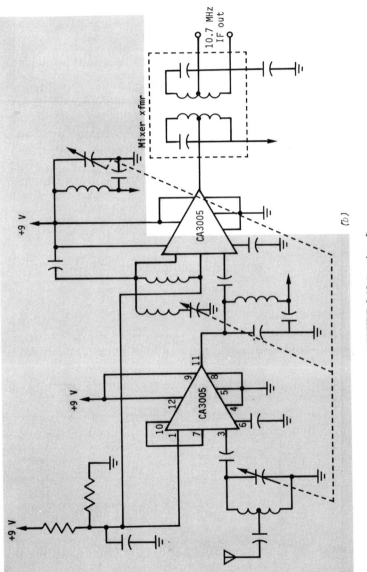

FIGURE 8-10 (continued)

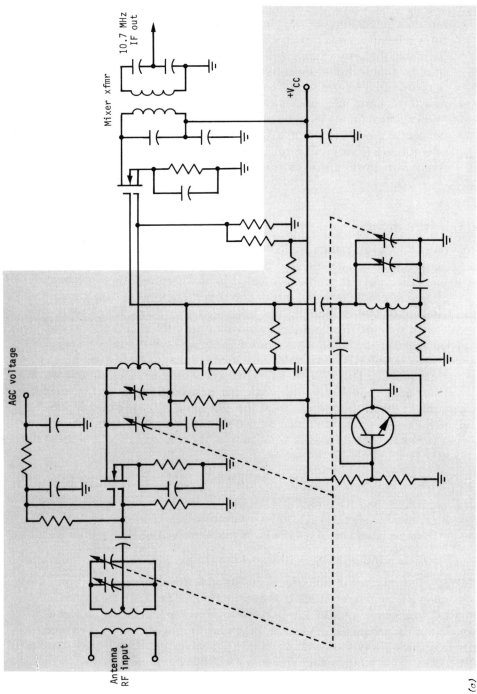

FIGURE 8-10 (*continued*)

(*c*)

will be a high level of second harmonic distortion in its output. If the mixer circuit distorts by clipping *both* top and bottom of the input waveform (a balanced mixer with symmetrical clipping), however, the second harmonic distortion signal will be low, but there will be a high level of third harmonic distortion at its output. These harmonic distortion products are often referred to as *second order* and *third order* for the second and third harmonic, respectively. Practical mixers of the two types (symmetrical and nonsymmetrical) have third- and second-order products between -10 and -20 dB at their conversion compression point of operation (RF signal at 6 to 7 dB below the LO signal level), respectively, with lower distortion levels when the RF signal level is decreased.

Image Frequencies

As you can see, substantial tradeoff problems are associated with the mixer stage in a superhet receiver, due primarily to its required nonlinear operation. In addition to the problems created by the harmonic and intermodulation distortion products in the mixer, the superhet circuit is also susceptible to the reception and IF amplification of *image* frequencies. If you review the means by which the IF signal was obtained as the difference frequency of *two* applied signals—the incoming RF and LO signals—you should observe that a *second* incoming RF signal could also mix with the LO and yield an identical IF frequency. This second RF signal (f_2) would be operating at the IF frequency *above* the LO signal, while the desired RF signal (f_1) would be operating at the IF frequency *below* the LO signal. If, for example, an AM broadcast radio is tuned to receive a 600 kHz RF signal, with the LO operating at 1055 kHz, the difference frequency of the two signals would drive the 455 kHz IF amplifier circuit as:

$$f_{IF} = f_{LO} - f_{RF} = 1055 - 600 = 455 \text{ kHz}$$

If there is a second RF signal at 1510 kHz while the radio remains tuned to 600 kHz, the LO signal will also mix with this *second* frequency, called the *image frequency,* and yield a 455 kHz signal to drive the IF amplifier circuit:

$$f_{IF} = f_{image} - f_{LO} = 1510 - 1055 = 455 \text{ kHz}$$

In fact, the mixer will nonselectively provide the IF frequency from two incoming RF signals: the tuned (desired) RF signal at the IF frequency *below* the LO frequency, and the image (undesired) RF signal at the IF frequency *above* the LO frequency. As such, the image frequency is always twice the IF frequency away from the tuned RF signal in the direction of the LO frequency. That is, the image frequency will be above the tuned RF signal if the LO is above the tuned RF signal (*super*heterodyne); and the image frequency will be below the tuned RF signal if the LO is below the tuned RF signal (*sub*heterodyne).

As you can imagine, image frequencies can create big problems in radio reception. The obvious interference that can result in the standard radio and TV receivers has been somewhat eliminated by station allocation frequencies. From our previous example the allocating agency (U.S. FCC) would never issue station licenses in the same geographical region for 600 kHz and 1510 kHz. In addition, the resonant circuit at the input of the mixer stage is tuned to the frequency of the desired RF signal, so it will also reject the image frequency from the mixer input. In the early days of UHF television, high-level image-frequency signals were not sufficiently rejected by the UHF mixer, however, thus allowing the higher channels to be received when the tuner was set at low channel numbers. Inexpensive radio receivers have poor image-rejection specifications, because they usually do not employ a tuned RF amplifier stage ahead of the mixer. High-quality receivers, however, may often have two stages of RF amplification ahead of the mixer in order to achieve image-rejection specifications in excess of 100 dB, and thereby practically eliminate image-frequency problems.

The use of one, or possibly two, low-noise high-Q RF amplifier stage(s) at the *front end* of a superhet RF receiver, ahead of the mixer stage, provides two important factors for superior performance: low system noise and high system image-frequency rejection (with a tradeoff reduction in dynamic range). High gain in RF amplifiers is not as critical a requirement as are noise-figure specifications and the light resonant circuit loading for high-Q LC circuits as required for image rejection. The higher frequency RF amplifiers, of course, must have very low input capacitance to effectively achieve the light loading for high circuit Q values. In addition to its function as a low-noise and high-Q amplifier, the RF amplifier must provide a variable-gain circuit function to keep the higher level incoming RF signals from overdriving the mixer stage above its conversion compression point. The popular cascode amplifier circuit effectively met these requirements for RF amplifiers in a variety of receiver functions. The more recent use of the dual-gate MOSFET has combined the cascode function into a single electronic device. Figure 8-11 shows typical RF amplifier circuits, including the cascode and dual-gate versions.

Additional Receiver Functions

Many variations of the standard superheterodyne circuit are used in a variety of receiver and electronic system functions. One popular variation is the use of *double conversion* to achieve multiband general coverage and short-wave receiver functions by expanding the *front end* of a basic AM receiver. Figure 8-12 shows the block digram of such a basic AM double-conversion receiver that will tune both the 540 to 1600 kHz broadcast band, and the 6.98 to 8.04 MHz short-wave range. Note that the LO for the short-wave band is a 6.44 MHz fixed-frequency, crystal-stabilized circuit. Note also that the mixer requires no tuned RF amplifier, using only a 7 to 8 MHz bandpass filter to

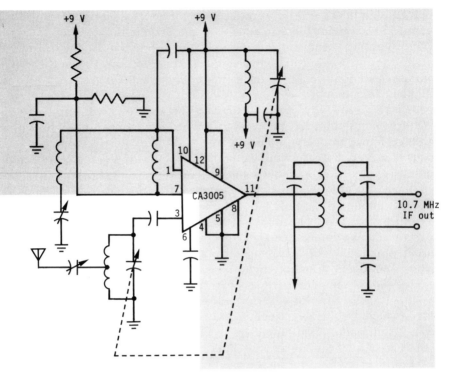

(a)

FIGURE 8-11

reject unwanted RF signals. Low-noise RF stages serve little use in the HF region of the spectrum because atmospheric and galactic noise levels are quite high; and the double-conversion process removes the image frequency from the short-wave tuning range of the mixer. With this arrangement the image frequencies will range from 5.9 to 4.84 MHz, computed as follows:

$$f_{\text{image}} = f_{\text{LO}} - f_{\text{IF}} = 6.44 - 0.54 = 5.90 \text{ MHz}$$
$$f'_{\text{image}} = f_{\text{LO}} - f'_{\text{IF}} = 6.44 - 1.6 = 4.84 \text{ MHz}$$

Although the basic double-conversion approach of Fig. 8-12 is a simple way to expand the tuning range of a receiver, the technique has some obvious limitations. First, the frequency coverage is limited to the narrow range of the lowest tuning frequency, which in the above example was slightly greater than 1 MHz. Second, the range of image frequencies would expand as the coverage of the receiver is extended with additional LO frequencies. A more effective approach used to achieve wide-range multiband receiver functions is with the use of a variable-frequency first LO and fixed-frequency second

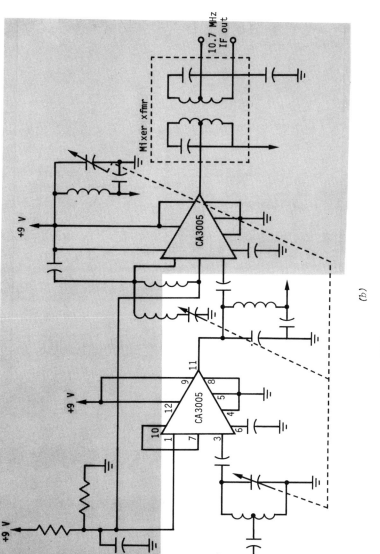

FIGURE 8-11 (continued)

(b)

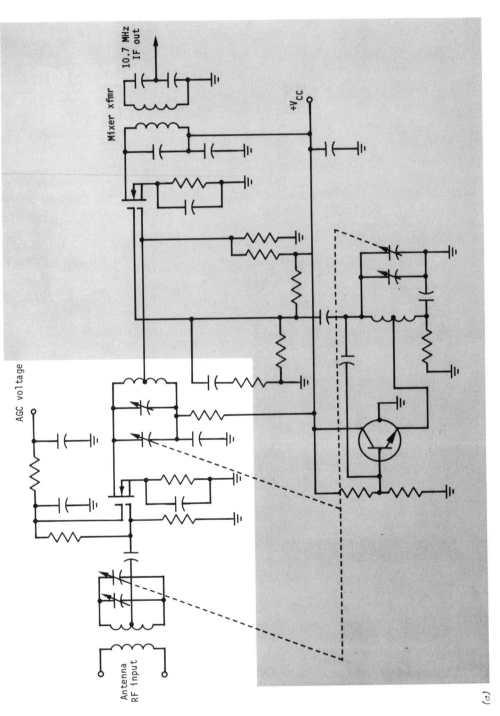

FIGURE 8-11 (*continued*)

(*c*)

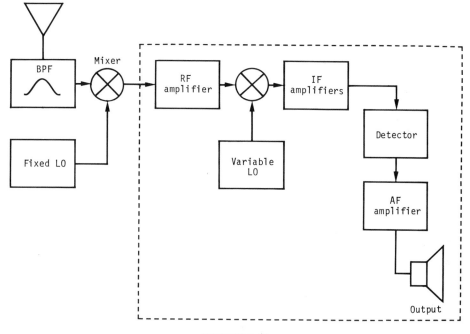

FIGURE 8-12

LO as in the block diagram of Fig. 8-13. With this circuit arrangement, high-*Q* tuned RF amplifier and mixer input circuits are practical, as the same control that varies the LO frequency can also tune the mixer and RF sections of a ganged variable capacitor. For multiband operation this circuit usually employs band-switching coils to operate in conjunction with the same ganged variable capacitor for tuning. For some historic reason multiband receivers of this type have been labeled *communications receivers,* as they provide general coverage of the HF and lower VHF regions of the spectrum, and they have capabilities for receiving and demodulating ICW, or interrupted continuous wave (code), pulse (teletype, etc.), AM and FM (narrowband and wideband) radio communications signals. Most quality communications receivers have RF amplifier stages for wide-range image rejection and low-noise response in the VHF region; and they employ both manual and automatic RF gain controls for optimum sensitivity.

Radio receivers operating in the upper VHF and UHF regions of the spectrum are specialized for the particular communications function they perform. It would be unusual to find the kind of general-purpose communications receiver typical of the lower frequencies used for the reception of VHF, UHF, and SHF signals. Many of these higher frequency receivers also operate in conjunction with a coincident transmitter unit, since most of the higher frequency radio applications require two-way communications. The

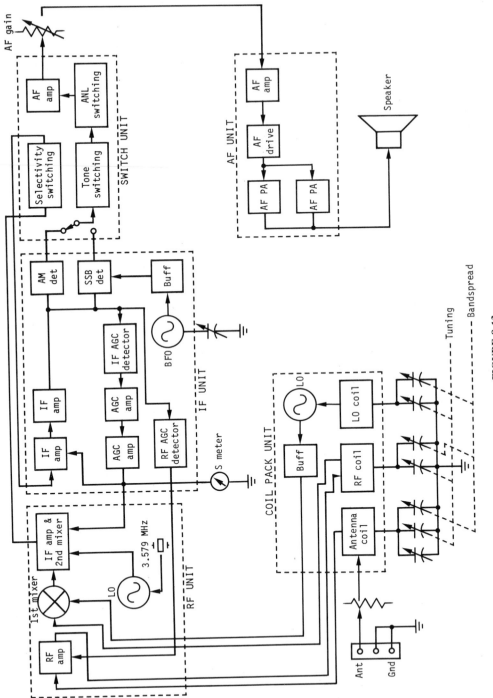

FIGURE 8-13

point to be made here is that regardless of the degree of specialty, cost, or complexity, most of the RF communications receivers employ some variation of the basic superheterodyne function, and they use tuned RF amplifier stages to reduce noise, increase sensitivity, and reject image frequencies. Although there are some unusual receiver designs, these constitute such a small portion of the technology that a detailed review of their characteristics is omitted from this text. Your knowledge of the characteristics of the IF amplifier, used as a TRF receiver with the added mixer and RF stages that provide a heterodyne receiver function, should give you the confidence to understand the operation of most receiver systems in use today, regardless of specialty or complexity.

Sample Problem 2

Given: A poorly designed TV receiver *front end* (tuner) with a high second-order distortion in the mixer stage, and poor image rejection in the tuned circuits of the RF and mixer stages.

Determine: The full complement of possible signal frequencies that can be received and converted for amplification within the IF bandpass of 41 to 47 MHz if the TV set is tuned to U.S. channel 5 at 77.25 MHz and the LO is operating at 123 MHz.

(*a*) The RF bandpass and image frequencies are computed as follows:

$$f_{RF} = f_{LO} - f_{IF} = \begin{cases} 123 - 47 = 76 \text{ MHz} \\ 123 - 41 = 82 \text{ MHz} \end{cases}$$

$$f_{image} = f_{LO} + f_{IF} = \begin{cases} 123 + 41 = 164 \text{ MHz} \\ 123 + 47 = 170 \text{ MHz} \end{cases}$$

(*b*) The LO second-harmonic conversion frequencies are computed as:

$$f_2 = 2f_{LO} - f_{IF} = \begin{cases} (2 \times 123) - 47 = 199 \text{ MHz} \\ (2 \times 123) - 41 = 205 \text{ MHz} \end{cases}$$

$$f_2' = 2f_{LO} + f_{IF} = \begin{cases} (2 \times 123) + 41 = 287 \text{ MHz} \\ (2 \times 123) + 47 = 293 \text{ MHz} \end{cases}$$

(*c*) The possible sources of interference are referenced for strong local RF signals in the frequency ranges of 164–170 MHz, 199–205 MHz, and 287–293 MHz, disclosing the following:

1. U.S. TV channel 11 operates at 198–204 MHz.
2. U.S. Public Safety mobile radio operates in the 162–173 MHz band.
3. U.S. Government radio services operate in the 225–328 MHz band.

The potential for receiver interference is high!

RECEIVER TUNING AND FILTERS

Sample Problem 2 disclosed potential sources of receiver interference caused by poor front-end design in a TV receiver. The high second-order distortion in the mixer stage could possibly be reduced by lowering the levels of the LO or incoming RF signals. The reduction or elimination of potential sources of incoming RF signal interference, however, may best be reduced by employing selective tuning and filter networks in the RF amplifier, mixer, and IF amplifier circuits. Interference problems often result from local signals that are so intense they penetrate the cabinetry of receivers set to tune weak distant RF signals, even though filtering is employed for image rejection. In addition to the obvious receiver enclosure shielding and power line filters, two RF amplifier stages will typically provide in excess of 100 dB of image rejection. If high-quality RF communications receivers are to achieve interference-free reception of extremely weak signals, however, *additional* special filtering must be added to the receiver to further eliminate unwanted signals.

One of the most important techniques used to eliminate predictable sources of interference signals in sensitive receivers is the use of notch filters or "traps" (band-stop filters). If we apply this approach to Sample Problem 1, it would be possible to trap out the image and second-order conversion frequencies from entering the mixer stage, thus eliminating these sources of interference. In addition to trapping incoming RF interference signals, receivers with high-gain IF amplifiers should employ traps to prevent the LO signal, along with its high-level harmonics, from entering the IF circuit.

As with mixers, filters may be fabricated as an integral part of manufactured radio receiver circuits, or they may be employed as subsystem elements in a wider range of receiver and related systems. Filters are available with a wide range of electrical and mechanical characteristics, with either lumped- or distributed-constant designs. The general circuit characteristics of lumped-constant filters are discussed in Chapter 3; the distributed-constant microwave filters are reviewed in Chapter 12. In their application to RF receivers, such filters are characterized with the following specifications:

1. *Function* (lowpass, highpass, bandpass, band-reject, etc.). Most radio receivers employ either lowpass, bandpass, or band-reject filters, with few applications for highpass filters. IF amplifiers often use lowpass

filters, because the IF frequency is usually the lowest radio frequency in the receiver system.

2. *Center-frequency* and *−3 dB bandwidth* (or *passband*) characteristics specify the operating frequency range of the filter.

3. *Insertion loss* (in dB) defines the passband loss affected by the passive filter circuit.

4. *Number of sections* defines the degree of increased selectivity achieved by cascading basic filter sections, as shown in Fig. 8-14.

5. *Stopband attenuation* (in dB) specifies the rejection properties of the filter's frequency vs attenuation characteristics, usually provided in graphical form as shown in Fig. 8-14.

We can here proceed with an example of filter applications in a U.S. CATV (community-antenna or *cable* television) system where adjacent-channel interference of the 6 MHz bandwidth TV signals must be rejected. CATV systems use separate receivers for each TV channel. If channel 3, for example, has a reliable 0 dBmV (1 mV across 75 Ω reference) signal at the receiver, while channel 2, at some distance from the receiver, delivers only a −50 dBmV signal, there will probably be interference from the channel 3 signal in the channel 2 receiver. If the receiver has a typical 60 dB adjacent-channel selectivity, additional filters must be employed ahead of the channel 2 receiver to reduce the channel 3 signal to 50 dB *below* the level of the channel 2 signal at the receiver input. A quick addition of the relative signal levels indicates that a channel 2 bandpass filter with a stopband attenuation

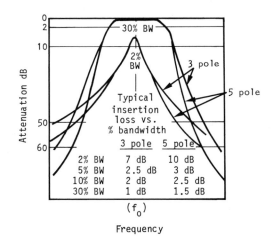

FIGURE 8-14

of 40 dB is required. Note from the bandpass filter curves of Fig. 8-14 that the *8-section* filter will have 40 dB stopband attenuation at the *+1 bandwidth* intersection, which will adequately reduce the adjacent-channel interference as required.

ELECTRONIC TUNING

AFC/AFT

Tuning of the superhet receiver is primarily accomplished by varying the frequency of the LO signal. While most receivers are tuned with multigang variable capacitors that allow simultaneous tuning of RF amplifier and mixer-input resonant circuits, many receivers *also* employ feedback circuits that provide fine-frequency control of the LO to compensate for any slight drift which may occur in the incoming RF signal or LO circuit. This function is called *automatic frequency control* (AFC), or the more popularized TV term *automatic fine tuning* (AFT). The function was absolutely required of the earlier vacuum-tube radio receivers that drifted considerably as the tubes "warmed up." Most modern solid-state receivers have stable LO circuits, but consumer-quality FM and TV receivers still require wide-range AFC to compensate for inaccurate tuning by their users.

Automatic frequency control of receivers is most often achieved with the use of frequency discriminators or phase-locked-loop circuits at the output of the IF amplifier(s), which in turn provide a feedback signal to control a reactance modulator in the LO circuit. You may refer back to the sections of Chapter 7 that detail the circuit operation of *discriminators, PLL's,* and *reactance modulators,* as their functions in AFC circuits are almost identical to those in FM circuits. Figure 8-15 shows a typical AFC circuit in block diagram form. Note that the frequency-discriminator function is a part of most FM receivers, and thus its addition is necessary only in AM and pulse-modulation receivers.

Scanning

With the increased availability of stable wide-range varicap devices, PLL's, and other voltage-controlled oscillator (VCO) IC's, full electronic tuning of superhet receivers is now a most practical achievement. The advantages of electronic tuning are many, including the elimination of mechanical linkage of variable-capacitor shafts to equipment panels, the ability to lock onto and follow a variable-frequency incoming RF signal (used for radio security), and the ability to provide a frequency-sweep function to the receiver (used for scanning receivers and spectrum analyzers). The tradeoff disadvantages to full electronic tuning include either (1) the elimination of tunable RF

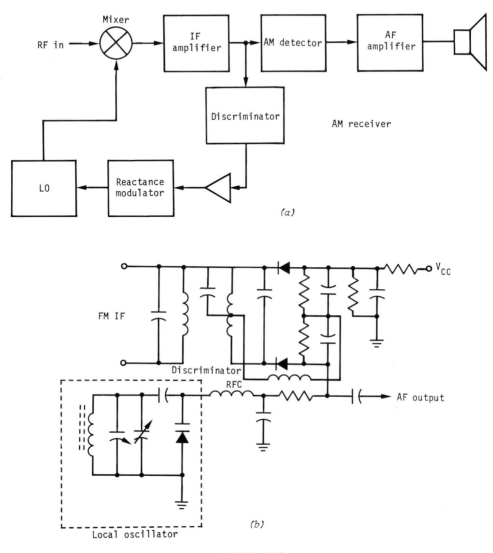

FIGURE 8-15

amplifier and mixer resonant circuits, possibly allowing reception of image frequencies and reducing the Q-rise gain of the circuits, or (2) the requirement for extremely costly dual and triple phase-locked VCO/filter networks to achieve the simultaneous tuning required. Examples of the costly approach can be found in highly specialized microwave, military, and scientific receiver applications. With the effective use of fixed or switchable bandpass filters for RF and mixer-input resonant circuits, however, full electronic tuning is a most practical technique, with frequency control from either

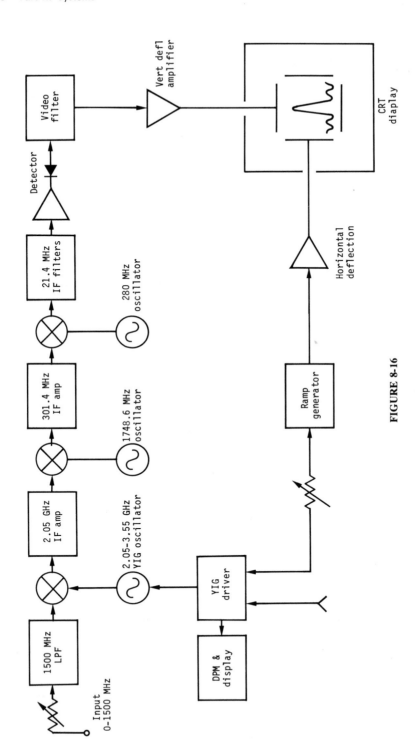

FIGURE 8-16

analog or digital sources. Figure 8-16 shows the block diagram of a spectrum analyzer as an example of a sophisticated frequency-sweeping superhet receiver that uses VCO electronic tuning.

Synthesis

The critical tuning of closely spaced radio channels by a receiver requires an ultrastable LO frequency operating in conjunction with an extremely selective AFC circuit. If the stability of an incoming RF signal is assured, based on usually tight restrictions for transmitting station frequencies, then *digital frequency synthesis* provides a better approach to critical tuning of superhet receivers. Digital synthesis is simply the deployment of digital IC logic arithmetic functions to obtain *any exact* LO frequency from *one* single-fixed-frequency stable oscillator. Such synthesis is achieved by multiplying and dividing the fixed stable frequency by any factors to obtain the desired LO frequency. This approach has been successfully applied to even the popular 40-channel U.S. CB (Citizens Band) radio transceivers, replacing a large number of crystals which would have otherwise been required to switch tune the units.

A simple digitally synthesized LO signal can be obtained with a single-frequency crystal-stabilized oscillator, a phase-locked loop, and two digital frequency divider elements connected as shown in Fig. 8-17. The frequency-standard reference could be a 5 MHz crystal oscillator that drives a ÷1000 digital divider yielding an extremely stable and accurate 5 kHz reference that drives the phase detector of the PLL. The phase detector

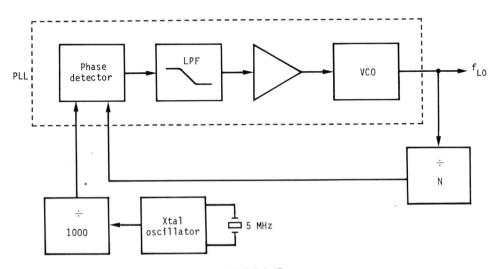

FIGURE 8-17

output is operating at some RF value which will be fixed by the factor *n* of the second digital divider at its output. The feedback signal from the second digital divider must be at the same exact frequency as the 5 kHz reference in order to obtain PLL "lock." If the factor *n* of the second divider is 5401, for example, the VCO of the PLL will lock at 5401 × 5 kHz = 27.005 MHz, which is CB channel 4. When the factor *n* of the second divider is then changed to 5403, the VCO will shift until lock is again achieved at 5403 × 5 kHz = 27.015 MHz, which is CB channel 5. Of course, this example was simplified to yield exact RF channel frequency values. In practical application the reference frequencies and *n* factors would be modified to obtain an LO signal frequency—not the RF channel frequency.

GAIN AND NOISE CONTROL

Throughout this chapter we have repeatedly reviewed the need for critical control of RF gain, for it affects sensitivity, noise, and distortion. Quality electronic communications receivers employ a range of feedback techniques to obtain optimum performance of their circuits and signal processing functions. Included among the important feedback functions are automatic RF and IF gain control, *squelch* and *muting, noise blanking,* and *automatic noise limiting* (ANL).

Automatic Gain Control

Automatic gain control (AGC), historically called *automatic volume control* (AVC), is achieved by feeding back a voltage from the receiver's detector (demodulator) stage, which can electronically control the gain of the first RF or first IF amplifier. The feedback voltage is obtained from the average (peak) dc level of the rectified and filtered IF signal (from the detector). Variable-gain RF amplification in modern receivers is usually achieved with special AGC bipolar transistors and field-effect transistors. Since most FET's are square-law devices, their response is ideally suited for electronic gain control, with the dual-gate MOSFET providing the most effective AGC action. The response curves of Fig. 8-18 show how variable gain can be achieved by changing the transistor's bias points. To achieve AGC action the feedback voltage from the detector simply varies the RF amplifier transistor bias point.

Figure 8-19 shows the AGC feedback to the RF amplifier using a voltage-doubler rectifier connection at the detector stage, and a 3N187 dual-gate MOSFET with its coincident response curves and AGC bias line plotted to show its range of operation. The doubled detector output AGC voltage can range from −1 V dc to −5 V dc, and it is fed to buck the +4 V dc bias on gate 2 of the MOSFET. This AGC circuit connection will thus process threshold-

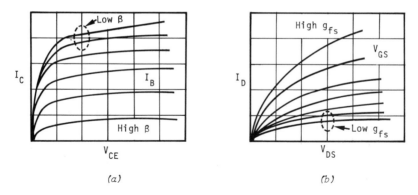

FIGURE 8-18

level signals with an RF amplifier transconductance of about 10 mS, yielding a 0.5 V (peak) signal at the detector; it will compress the transconductance to about 100 μS with a high-level RF input signal that drives the detector to a 2.5 V (peak) value. The AGC feedback voltage thus achieves an RF gain compression of about 100:1 (40 dB). If additional gain compression is required, the first IF amplifier stage can be similar or identical to the RF amplifier; with the AGC feedback voltage applied to both circuits the gain compression can be increased to about 80 dB.

The AGC circuit of Fig. 8-19 is called *simple AGC* because the gain compression occurs in almost direct proportion to any increased signal level at the detector diode. Ideal AGC action would not affect compression, however, until the detector signal was close to its maximum operating level. Such a response could be achieved if the voltage-doubler diodes were biased to prevent rectification until at least a 2 V (peak) diode signal level, as in our previous example, was reached. AGC feedback voltage would then occur as the detector voltage increased from 2.0 to 2.5 V, yielding a 0 to 1.0 V AGC feedback voltage. This technique is called *delayed AGC*, because the AGC action is *delayed* until the detector signal reaches a given level. If we apply delayed AGC to the previous example, the required RF and IF gain compression will not be sufficient. If we combine delayed AGC with an amplifier, however, it is possible to approach the concept of an ideal AGC function. Figure 8-20 shows the delayed and amplified AGC function as it would be applied to the 3N187 MOSFET of our previous example, achieving full gain compression as the detector is driven from 2.25 to 2.50 V (peak).

Squelch and Muting

Squelch and *muting* are somewhat similar convenience features of RF communications receivers. They both eliminate the annoying hiss of amplified

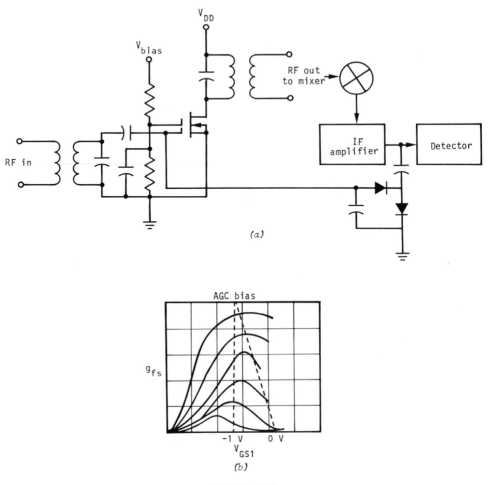

FIGURE 8-19

noise when there is no incoming RF signal at the receiver input. The squelch function is usually applied to voice communications receivers that are set to a particular channel awaiting a message. Muting is usually applied to broadcast receivers to function as they are tuned from station to station across the dial. These functions basically deactivate the audio amplifier section of the receiver when there is zero or some preset value of AGC voltage. With variable muting on an FM broadcast receiver, for example, the control can be set to activate the audio amplifier stage with only strong local stations, thus simplifying the tuning process for nontechnical consumers. Another common technique that is used to obtain squelch in voice communications receivers is with the use of a 5 kHz to 8 kHz filtered audio amplifier in the audio stages of the receiver. Since the audio response of voice communica-

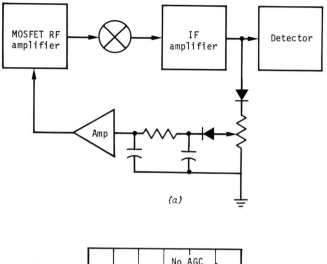

(a)

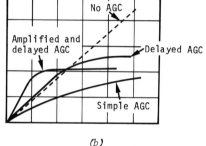

(b)

FIGURE 8-20

tions transmitter modulators is limited to about 3 kHz, the presence of higher frequency noise in the AF section of the receiver indicates that the input signal is not strong enough to develop AGC feedback to "quiet" its sensitivity. Figure 8-21 shows the three approaches to squelch and muting.

Blanking and Automatic Noise Limiting

Noise blanking (NB) and *automatic noise limiting* (ANL) are differently achieved functions that basically reduce the noise level in the presence of a received AM voice signal. Static and ignition noise are particularly prominent in vehicular voice communications equipment in which some form of electronic noise reduction is a necessity. Most of this *noise* or "static" is in the form of impulses or spikes of relatively short duration as compared to the voice-frequency modulation signal that is received. The noise blanker shown in Fig. 8-22 is a simple separate "receiver" that is tuned to an adjacent frequency where no RF station channel exists. Since the static and spikes

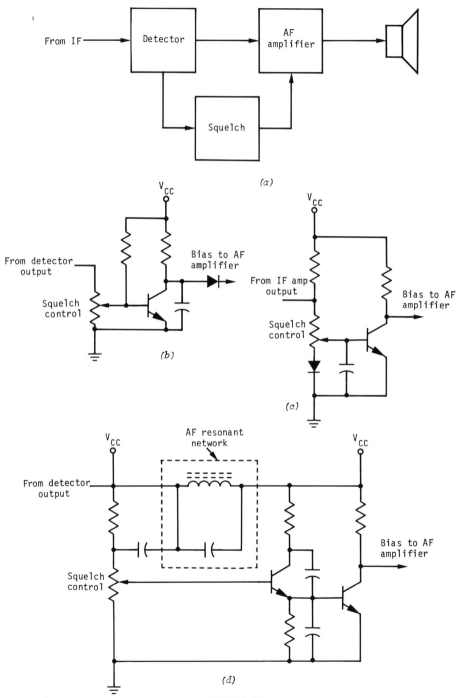

(a)

(b)

(c)

(d)

FIGURE 8-21

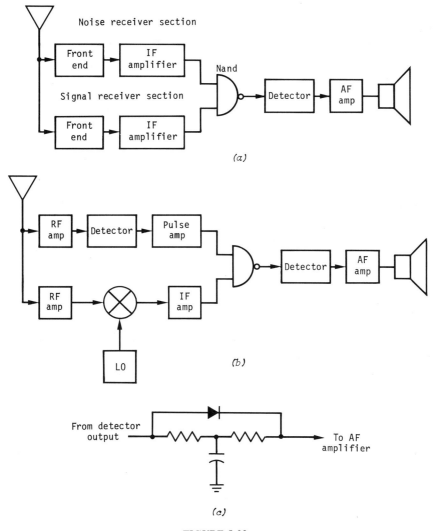

FIGURE 8-22

cover a broad range of frequencies, the NB receiver amplifies and detects this noise as the same noise that is received with the signal in the superhet receiver. The detected static and spikes are amplified and drive a *NAND gate* in the IF amplifier section which momentarily interrupts (blanks out) the modulated IF signal for the instant that the static or spike pulse occurs. When the resulting signal is filtered in the detector output, its waveform is averaged; thus the noise is removed without noticeable deterioration of the signal.

The ANL function is a less effective, but also less costly, technique of static or spike noise reduction. The ANL circuit is basically a diode operating in conjunction with an *RC* T-network at the output of the receiver's detector stage as shown in Fig. 8-22. The *RC* T-network basically filters the detector output signal and establishes a dc voltage level to forward bias the diode. With the diode forward biased, the detected audio modulation signal passes through it to the AF amplifier stage. Since the time constant of the *RC* network is too long to respond to the short-duration noise pulses, however, their increased voltage peaks momentarily turn *off* the diode, interrupting the audio signal for the instant that the static or spike pulse occurs. This function is often called a *series noise gate*.

MEASUREMENTS

The performance of RF communications receivers is generally specified by a series of standard measurement procedures. The minimum instrumentation recommended for making accurate RF measurements includes:

1. A stable, low-distortion, wide-range RF signal generator, operating in the frequency range of the receiver under test, with an accurately calibrated attenuator for reliable output signals from 0.1 μV to at least 0.1 V (rms), and with calibrated 0 to 100% adjustable-index AM, and 0 to 100 kHz adjustable-deviation FM.

2. A stable "audio" frequency signal generator with sine-wave output to at least 1 MHz at levels from 100 μV to at least 3 V (rms).

3. An accurate RF and multifunction electronic voltmeter, with accurate AF and RF frequency response beyond the frequency range of the receiver under test, with accessory AF and RF notch filters for distortion measurements.

4. An accurate frequency counter to cover the full range of RF and modulation signals that are processed by the receiver under test.

5. A range of accessory cables, impedance-matching and load elements, and miscellaneous hardware that may be required to interface the test equipment with the receiver.

Note that the oscilloscope is absent from the above list. In performing RF measurements we basically deal with sine waves, and anyone who knows anything about electronics already knows what a sine wave looks like, so we just don't use expensive high-speed oscilloscopes where RF voltmeters will suffice! Also absent from the list are a noise generator and a spectrum analyzer, which are useful for receiver measurements; but there are receiver

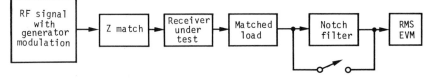

FIGURE 8-23

measurement techniques that eliminate the need for these costly and specialized instruments.

Since we began this chapter with the relationship of noise to receiver
sensitivity, it is appropriate to consider these measurements first, as the
basis of the receiver's ability to amplify and demodulate weak RF signals.
Figure 8-23 shows the typical setup for making SINAD (signal plus noise and
distortion) sensitivity measurements. The procedure requires modulating the
signal generator to full value (100% AM, or full-deviation FM) and "notching
out" the modulation-frequency signal from the receiver's output. The RF
input signal level that yields a 12 dB difference between the unnotched and
notched output levels is the specified sensitivity of the receiver, e.g., 2 μV
for 12 dB SINAD. Note that all low-level receiver measurements of this type
should be carried out in a *screen room* to assure receiver immunity from stray
or unwanted RF signals. For critical measurements you might "pull the
graveyard shift" (work 2 a.m. to 6 a.m.) in order to proceed with minimum
static or other noise pollution.

Noise and Sensitivity

Measurements of FM broadcast receiver sensitivity are made on the basis of
an unmodulated input signal required to achieve a degree of noise quieting.
The measurement setup is similar to that of Fig. 8-23, but with the notch filter
and AF generator removed. The procedure requires the measurement of a
reference noise level with the RF signal generator at its minimum output
level and a 300 Ω resistance across the receiver input. Then the output level
of the signal generator is increased until the output noise level is reduced to
20 or 30 dB below the reference level in the specification, e.g., 2.5 μV for 30
dB quieting. The specification of FM broadcast receiver sensitivity has been
modified to be expressed in power rather than voltage input required to
achieve 30 dB quieting, with the power expressed in dB above a 1.0 fW
(femtowatt = 10^{-15} watt) as dBf. For our above 2.5 μV example the power
sensitivity would be computed as follows:

$$P = v^2/R = (2.5 \times 10^{-6})^2/300 = 2.08 \times 10^{-14} \text{ W}$$
$$P' = 20.8 \text{ fW}$$
$$\text{dBf} = 10 \log P' = 10 \log 20.8 = 13.2 \text{ dBf}$$

The noise figure and signal-to-noise ratio can be obtained from the SINAD and sensitivity measurements setup of Fig. 8-23 by determining the overall power gain of the receiver and reflecting the value of zero-signal output noise back to the receiver input. The comparison of calculated thermal-agitation noise power across the real input resistance, and the output power as reduced by the receiver gain, is the noise factor (or noise figure in dB) of the receiver. The overall noise figure of a given 10 kHz-bandwidth AM communications receiver, for example, can be computed from a measured 1.0 W signal output from a 10 fW signal input, and an 80 mW output noise level with the signal notched out as follows:

$$A_P = P_{out}/P_{in} = 1/10^{-14} = 10^{14}$$
$$P_{nT} = P_{n(out)}/A_P = (8 \times 10^{-2})/10^{14} = 8 \times 10^{-16} \text{ W}$$
$$P'_n = k'B = 4 \times 10^{-21} \times 10^4 = 4 \times 10^{-17}$$
$$NF = 10 \log P_{nT}/P'_n = 10 \log 8 \times 10^{-16}/(4 \times 10^{-17}) = 13 \text{ dB}$$

This noise figure specification is often expressed as:

$$NF = 10 \log [(S/N_{in}) \div (S/N_{out})]$$

so for the above example,

$$NF = 10 \log [(10^{-14}/4 \times 10^{-17}) \div (1/0.08)] = 10 \log 20 = 13 \text{ dB}$$

The signal-to-noise ratio measurement is usually performed at some specified signal input level far above the receiver's sensitivity value. In FM receivers this measurement is typically made at the full limiting level of the IF amplifier section in order to obtain an attractive output S/N value. The procedure requires increasing the RF signal generator to a specified level with full modulation, and then notching out the signal and measuring the level of the remaining noise. For the above AM receiver example the output S/N is simply computed as:

$$S/N = P_{out}/P_{n(out)} = 1/0.08 = 12.5 = 11 \text{ dB}$$

Still using the same basic measurement setup and data as obtained from the previously specified noise and signal level measurements, it is convenient to proceed to AGC measurements. The AGC voltage is measured from its feedback point to the RF and IF amplifier stages as the RF signal generator level is increased with full modulation. The AGC feedback voltage is thus tabulated or plotted against equal increments of RF input level until no change in the AGC voltage is noted and its response has flattened out. A second indirect technique of AGC measurement is the measurement of the change in AF (modulation signal) output that results from raising the input signal level by a specified 60, 70, or 80 dB above the receiver's threshold sensitivity. This second technique specifies AGC compression; e.g., a 70 dB change in input level yields a 10 dB change in AF output and an AGC compression of 60 dB.

Frequency Measurements

The previous group of receiver tests did not use the frequency counter, as the measurements were primarily concerned with response levels. A second group of receiver tests are frequency related, and they may all be completed using the same basic test setup as shown in Fig. 8-24. These receiver measurements include frequency response, tracking, bandpass, selectivity, AFC lock range, image rejection, and distortion products. Most of these measurements are made with a zero-modulation CW RF signal at a relatively high input level; so the output levels may be measured at the detector output of AM receivers or the AGC point of FM receivers. An additional convenience can be realized with a synthesized RF signal generator, thus eliminating the need for a separate frequency counter.

Receiver *frequency response* and *tracking* measurements can be completed simultaneously by plotting the relative output levels and receiver dial setting on a channel-by-channel basis across the operating frequency range of the receiver. Errors in tracking can generally be corrected by adjusting the trimmer and padder capacitors and tuning coil in the receiver's local oscillator circuit—an adjustment that is commonly made while measuring tracking. Well-designed receiver front ends have frequency-response variations of less than 3 dB over their full range of operation. If the response at the high or low end of the tuning range is poor, adjustment of the L/C ratio of the resonant circuits in the RF amplifier and mixer input stages may be required. As with tracking, such adjustment, calibration, and specification of receiver frequency response are generally simultaneous activities.

Selectivity and Image Rejection

Bandpass and *selectivity* are closely related measurements that compare the on-channel and adjacent-channel frequency response of the receiver. The procedure requires tuning the receiver and signal generator to a given RF channel, then varying the frequency of the signal generator to obtain a frequency vs amplitude response. This response can be tabulated or plotted to indicate both bandpass and adjacent-channel response of the receiver, with the level of the adjacent channel, compared to the peak bandpass response,

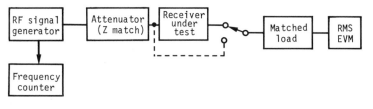

FIGURE 8-24

specified as the selectivity factor (usually in dB). Critical adjustment of the tuned circuits in the IF amplifier section of the receiver will optimize the bandpass and selectivity characteristics of the receiver. This adjustment, along with the tracking and RF tuning adjustments, is called receiver *alignment*.

The adjustment or specification of the AFC lock range is measured by shifting the frequency of the RF signal generator in small increments away from the center frequency of the receiver tuning. If the AFC circuit is operating properly, the LO frequency will shift with the RF signal to convert the incoming signal frequency to the IF frequency. The lock range should be limited to RF signals that vary from the edge of the upper adjacent channel frequency to the edge of the lower adjacent channel frequency, with its value specified in Hz (kHz or MHz).

The rejection of image frequencies and the identification and rejection of other frequencies caused by receiver distortion are factors that are determined by identical measurement procedures, as in the setup of Fig. 8-24. The procedures require establishing a reference receiver output level with the RF signal generator set at the on-channel receiver tuning frequency, and then driving the receiver to its full limiting or maximum AGC voltage level. The frequency of the signal generator is then increased, without altering the receiver tuning, to obtain a frequency vs receiver response at all the mathematically related frequencies that may yield distortion products. The signal generator is also set to twice the IF frequency above (or below) the receiver tuning in order to measure the level of the image frequency signal. The difference in the measured levels is specified as the *image rejection* in dB. In high-quality receivers the image rejection and "spurious signal rejection" may exceed 100 dB.

CONCLUSION

This chapter has reviewed aspects of RF communications receivers as electronic systems that perform a range of signal-processing functions in order to effectively receive, extract, and reproduce modulation signals. As you may have concluded, the radio receiver can be classified as a *complicated arrangement of some basic electronic circuits*. One consolation, however, is that the same basic block-diagram arrangement is used in the full range of communications receivers, from the small portable AM radio to the most sophisticated satellite data receiver. Your thorough knowledge of the superheterodyne circuit with its special functions can be applied to understanding the operation of almost all the RF receiving equipment that you may encounter. Chapter 10 of this text will follow your understanding of modulation, demodulation, and receiver systems with an introductory review of full communications systems which apply the content of this and the two preceding chapters to modern communications practice. The following problems will test your knowledge of radio signals and receiver functions. If you have difficulties, you should refer to this and the preceding chapters, as necessary, to build your knowledge and confidence.

Problems

1. Given: A single-frequency TRF receiver tuned to receive the U.S. NBS station WWV at 15 MHz, made up of five 20 dB RF amplifier modules and a 6 kHz bandpass filter module driving a 50 Ω AM detector and AF amplifier.

Based on 50 Ω impedances on all modules, a specified 10 dB NF for each RF amplifier, and with a requirement for 20 dB RF S/N at the detector input, determine:

- (*a*) Thermal-agitation noise voltage at the 50 Ω receiver input
- (*b*) Equivalent reflected noise resistance at the second RF amplifier input, from the NF and output of the second RF amplifier
- (*c*) Equivalent reflected noise resistance at the first RF amplifier input, from the NF and total equivalent from part *b* above
- (*d*) Approximate total system RF noise figure
- (*e*) Input S/N in dB to obtain 20 dB S/N at the detector input
- (*f*) Receiver sensitivity for 20 dB S/N at the detector input
- (*g*) RF signal voltage incident to the 50 Ω AM detector input

2. Given: A TV receiver with front-end arrangement as shown in Fig. 8-25, with a UHF tuner comprising an LO and 1N82A mixer diode that is to be used for fringe-area reception of U.S. UHF TV channel 60 (747.25 MHz).

Based on a 6 MHz receiver bandpass, 300 Ω UHF tuner input and output impedances, 6 dB mixer conversion loss, and 7 dB mixer noise figure, determine:

- (*a*) Total equivalent noise resistance at the mixer-stage input from its output, NF, and real input (resistive) impedance

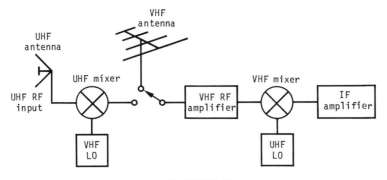

FIGURE 8-25

(*b*)   Total NF of the UHF tuner

(*c*)   UHF tuner sensitivity for a 40 dB *S/N* at its output to the VHF tuner

3.   Given: The TV receiver of Problem 2.

Determine:

(*a*)   The increased UHF tuner sensitivity

(*b*)   The total UHF tuner noise figure, assuming that a low-noise RF amplifier stage with 50 Ω input impedance, 300 Ω output impedance, 6 dB voltage gain, and 2 dB NF was included as part of the UHF tuner circuit

4.   Given: A superheterodyne radio receiver to tune the 4 to 12 MHz short-wave frequencies, with an IF frequency of 455 kHz, high second-harmonic distortion in the mixer, and an RF amplifier and mixer-input response as shown in Fig. 8-26.

Determine:

(*a*)   The operating frequency range of the LO

(*b*)   The change in capacitance (ratio) required to tune the *LC* networks in the LO and RF and mixer stages

(*c*)   The range of image frequencies

(*d*)   The image rejection when the receiver is tuned to 4 MHz and 12 MHz

(*e*)   The frequency ranges of potential reception of unwanted RF signals due to second-order mixing

5.   Given: A 75 Ω mixer module to be used in a broadcast FM tuner, with a 6 dB conversion loss, a conversion-compression point at +20 dBm, and third-order distortion at 20 dB below the compression-point signal.

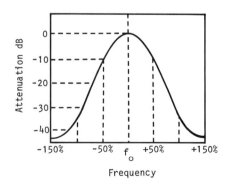

FIGURE 8-26

If the receiver sensitivity yields a 50 μV signal at the mixer input, determine:

(*a*)   Dynamic range of the mixer in dB
(*b*)   The highest practical RF input signal voltage to the mixer
(*c*)   The highest IF output signal voltage from the mixer, based on *b* above
(*d*)   The highest IF output signal voltage from third-order mixing, if the RF and mixer tuning rejection is 40 dB

6.   Given: A simple fixed LO converter to be used to tune U.S. CB radio channels operating from 26.96 to 27.41 MHz, using a basic AM broadcast radio (540 to 1600 kHz) as shown in Fig. 8-27.

Determine:
(*a*)   The fixed LO frequency on the dial for tuning CB channel 20 (27.205 MHz) at 800 kHz and CB channel 1 (26.965 MHz) at 560 kHz
(*b*)   The range of image frequencies that might be received with the converter
(*c*)   The anticipated image rejection achieved by using a simple two-section bandpass filter, as in Fig. 8-14, ahead of the converter input

7.   Given: An FM receiver with a phase-discriminator demodulator, similar to Fig. 7-19 in Chapter 7, yielding an output of 20 mV/kHz.

Determine: The change in varactor-diode capacitance required for AFC control if the LO in the receiver uses a circuit similar to Fig. 7-16*b* in Chapter 7.

8.   Given: A digitally synthesized LO as in Fig. 8-17 to be used in a superhet receiver to tune 20 U.S. Public Service channels in the frequency range of 155.415 to 155.700 MHz, with each channel separated by 15 kHz increments, and a 10.7 MHz IF frequency.

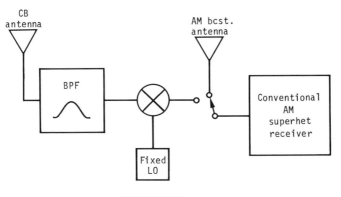

FIGURE 8-27

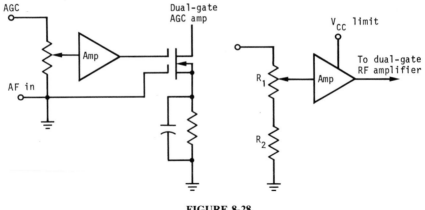

FIGURE 8-28

Determine:
(*a*)   The frequency range of the LO
(*b*)   The highest reference frequency by which all channel LO frequencies are divisible
(*c*)   The required range of *n* factors required of the digital divider for tuning the 22 channels

9.   Given: The delayed AGC and squelch functions in a receiver as in Fig. 8-28, to be used with two 3N187 MOSFET's as first RF and AF amplifiers, respectively.

If the receiver's detector output (peak) voltage can vary from a minimum of 0.3 to a maximum of 1.3 V (refer to the 3N187 response in Fig. 8-19, determine:
(*a*)   The required AGC (dc) amplifier gain, if full-compression delayed AGC is to occur with a 1.1 to 1.3 V variation in the detector output
(*b*)   The amplification required in the squelch amplifier to provide zero squelch for 0.3 V detector signals
(*c*)   The ratio (range) of R_1/R_2 for full squelch of detector signals below 1 V

10.   Given: An AM aircraft VOR (VHF omnirange) receiver to operate in the frequency range of 108 to 118 MHz.

Specify the equipment setup and procedures for performance measurements of:
(*a*)   SINAD sensitivity   (*d*)   Image rejection
(*b*)   Noise figure   (*e*)   AFC lock range
(*c*)   Selectivity   (*f*)   AGC compression

CHAPTER 9

DATA COMMUNICATIONS AND PULSE MODULATION

Although you are probably more familiar with applications of AM and FM
through broadcast radio and television, *digital communication* and *pulse-
modulation* techniques are as common to modern electronic communication
as the more familiar forms. In a strict sense, *digital* or *data* communications
refers to the transmission and reception of *coded* signals that do *not* necessarily modulate an RF carrier, even though they are identified as pulse
modulation. If you interpret modulation as the process of imposing *intelligence* signals on an RF carrier, then data communication and pulse modulation become two *distinct* technical topics. The two topics are combined in
one chapter, however, because they are so combined in engineering practice. This chapter will approach the two topics separately, but will also
show their interrelationship in signal and circuit functions.

The intelligence signals used in broadcast radio are the familiar AF sound
signals that originate from voice or musical sources. In their modulated and
demodulated form such AF signals still retain waveforms that are electrically equivalent to our sense of hearing. The intelligence signals used in
broadcast television, teleprinters, computers, and other digital equipment
are much more complex, however, because they do not have waveforms
that are the simple equivalents of our senses. These signals are transmitted
in coded form, so they must be *encoded* for transmission and *decoded* after
reception, originating and terminating in some form that we, or our computers, can interpret.

This chapter is not *directly* concerned with electronic *systems* (see Chapter 10) or computer circuits, but some basic digital systems concepts are
here reviewed so you can better understand the *functions* of data communications and pulse modulation. It seems that while much popular technical
information is available on AM, FM, and even TV, data, digital, and pulse
techniques have not received sufficient attention from a communications
perspective. This situation is unfortunate, because electronic communications with computer-based systems is becoming a most popular and practical means of getting information from one place to another. You should

thus actively attempt to locate and familiarize yourself with data and pulse-modulation communications equipment as you study this chapter in order to relate theory to real-world practice.

The coding, decoding, and processing of binary (two-level) intelligence signals is the basis of *modern* electronic communications, but the idea of transmitting and receiving coded signals is based on *ancient* communications techniques. The early telegraph, for example, provided for electrical communications using coded (Morse, etc.) signals, similar to those sent by flashing lights, smoke puffs, and jungle drums of yet earlier times. These coded signals were distinguished by their varying rhythms and can be defined in modern terms as pulse-position modulation (PPM) or pulse-width modulation (PWM) signals. Musical rhythms from drums are like encoded PPM signals, and Morse code takes the form of PWM signals. If you sound out the "bump, ba-bump bump" rhythm shown in Fig. 9-1a and translate it to PPM pulses, you should easily distinguish the "dah, di-dah dit" Morse PWM pulses of Fig. 9-1b for the coded letter "C."

Historically, the telegraph and ticker tape preceded the voice telephone, just as radiotelegraphy preceded voice-modulated radio communications. The encoder and decoder of these early coded signals was the human brain of the telegraph operator. Many skilled telegraphers of early times could send and receive Morse code at the rate of speech—40 to 50 five-letter words per minute (W/M). The electromechanical teleprinter (*Teletype* is a trademark brand name) replaced the skill of these early telegraphers and extended both the reliability and speed of electrical communications. Modern electromechanical teleprinters send and receive five-letter words (plus a space) at standard speeds from 60 to 100 W/M, depending on the particular machine and service; and they use a punched-tape or typed-page input/readout format. The units operate with a motor and distributor with *start, stop,* and eight code contacts. A closed contact is called a *mark* or *mark pulse,* and an open contact is called a *space* or *space pulse.* Each pulse time unit is called a *bit.* The combination of spaces and marks yields the associated keyboard code.

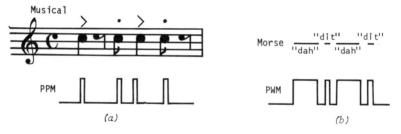

FIGURE 9-1

Both the Morse and electromechanical teleprinter codes are in *serial* form, i.e., the marks and spaces are sequential for each code letter. The code telegraphers were able to separate the letters of the code sequence "by ear," but the teleprinters must employ start and stop pulses at the beginning and end of each pulse-code letter or *character*. The modern American Standard Code for Information Interchange (ASCII) is shown in Fig. 9-2 as used with the ASR-33/35 Teletype as an 11-bit code: *start*, 7-bit *character*, *parity* (error) check, and 2-bit *stop* pulse. When such start/stop pulses are used in the coding of signals, the transmission is called *asynchronous* (nonsynchronous). By contrast, the earlier Morse code transmissions were *synchronous* in function, without start/stop pulses. With synchronous transmission more characters can be transmitted and received within a given sequence of bits.

With the recent advances in computer and digital technology the importance of coded modulation signals has likewise increased to provide for the transmission of digital data. In most cases the digital data signals are converted to serial synchronous or asynchronous pulses in order to be compatible with the previously established standard codes (ASCII, etc.). As with

BIT NUMBERS

$b_7 b_6 b_5$ →	$b_4 b_3 b_2 b_1$	COLUMN → / ROW ↓	$\begin{smallmatrix}0\\0\ 0\end{smallmatrix}$ 0	$\begin{smallmatrix}0\\0\ 1\end{smallmatrix}$ 1	$\begin{smallmatrix}0\\1\ 0\end{smallmatrix}$ 2	$\begin{smallmatrix}0\\1\ 1\end{smallmatrix}$ 3	$\begin{smallmatrix}1\\0\ 0\end{smallmatrix}$ 4	$\begin{smallmatrix}1\\0\ 1\end{smallmatrix}$ 5	$\begin{smallmatrix}1\\1\ 0\end{smallmatrix}$ 6	$\begin{smallmatrix}1\\1\ 1\end{smallmatrix}$ 7
	0 0 0 0	0	NUL	DLE	SP	0	@	P	`	p
	0 0 0 1	1	SOH	DC1	!	1	A	Q	a	q
	0 0 1 0	2	STX	DC2	"	2	B	R	b	r
	0 0 1 1	3	ETX	DC3	#	3	C	S	c	s
	0 1 0 0	4	EOT	DC4	$	4	D	T	d	t
	0 1 0 1	5	ENQ	NAK	%	5	E	U	e	u
	0 1 1 0	6	ACK	SYN	&	6	F	V	f	v
	0 1 1 1	7	BEL	ETB	'	7	G	W	g	w
	1 0 0 0	8	BS	CAN	(	8	H	X	h	x
	1 0 0 1	9	HT	EM	)	9	I	Y	i	y
	1 0 1 0	10	LF	SUB	*	:	J	Z	j	z
	1 0 1 1	11	VT	ESC	+	;	K	[	k	{
	1 1 0 0	12	FF	FS	,	<	L	\	l		
	1 1 0 1	13	CR	GS	-	=	M	]	m	}
	1 1 1 0	14	SO	RS	.	>	N	^	n	~
	1 1 1 1	15	SI	US	/	?	O	_	o	DEL

NUL	Null, or all zeros	FF	Form feed
SOH	Start of heading	CR	Carriage return
STX	Start of text	SO	Shift out
ETX	End of text	SI	Shift in
EOT	End of transmission	DLE	Data link escape
ENQ	Enquiry	DC1	Device control 1
ACK	Acknowledge	DC2	Device control 2
BEL	Bell, or alarm	DC3	Device control 3
BS	Backspace	DC4	Device control 4
HT	Horizontal tabulation	NAK	Negative acknowledge
LF	Line feed	SYN	Synchronous idle
VT	Vertical tabulation		

ETB	End of transmission block
CAN	Cancel
EM	End of medium
SUB	Substitute
ESC	Escape
FS	File separator
GS	Group separator
RS	Record separator
US	Unit separator
SP	Space
DEL	Delete

FIGURE 9-2

the teleprinter pulse signals, most digital data signals are distributed by telephone (telecommunications) systems, so the coding and speed (bit rate or word rate) are limited by the characteristics of the telephone lines and circuits. Even the data communications systems that operate independent of telephone lines use the telecommunications system terminology, however, in order to maintain compatibility with the prevailing telephone/teleprinter technology. As you may conclude, pulse coding is an important aspect of digital data communications, just as it is for the teleprinter and the earlier forms of communications. You should further observe, however, that encoding and decoding are *not* RF modulation processes; so coded "modulation" signals are sent and received in wired direct and AF telephone *carrier* systems just the same as AF voice and music signals are. The *RF* modulation of pulse signals is an *additional* function that is required if the coded pulse or data signals are to be imposed on an RF carrier. The *RF modulation* techniques will be reviewed later in this chapter and also in Chapter 10.

In order to treat encoded pulse signals like the AF voice or music that is sent and received by the telephone systems, the pulses are converted to audio frequencies, with a *mark* at *one* frequency and a *space* at a *second* frequency. As the coded signals shift from mark and space levels, the related audio tones also shift, yielding frequency-shift AF tones as *keyed* by the pulse signals. This technique is appropriately called *frequency-shift keying* (FSK), and it can be considered as a form of audio FM. The electronic device that converts the pulses to FSK tones, and vice versa, is called a *modem* (for *mo*dulator/*dem*odulator). The modems typically transmit and receive FSK signals within the limits of allocated communications channels that are spaced at standard frequency increments. The bandwidth, frequency shift, and spacing of the telegraph or data channels depend on the word speed of the coded input pulses. The word speed is reduced to *bits per second* (based on the coded letter or *character,* plus any additional code bits, as required to yield a 5-character word and a 1-character space), called *baud* after the early Baudot telegraph code. Table 9-1 lists standard telegraph channels, frequency spacing, and audio frequency shift for FSK operation. As you may observe, the modem must employ critical frequency-selective and filter circuits in order to operate reliably within its specified AF range of transmission and reception.

Figure 9-3 shows a representative FSK signal and compares the functional requirements of a modern data communications system with the basic modulation of AF voice or music signals. Although a voice or music signal is most often directly modulated on an RF carrier in broadcast radio, and may pass through a telephone system for "wire" voice communications, the data pulses require two *additional* functions for transmission and reception. The data from a computer or other digital source must be converted first to serial, asynchronous (or synchronous) PPM data pulses of

TABLE 9-1. TELEGRAPH AND MODEM CHANNELS: AUDIO FSK SPACING (ALL VALUES IN Hz)

A. CCITT frequency bands for FSK telegraph channels

A	B	C
M—390	M—3825	M—420
• 420	• 425	• 480
S—450	S—467.5	S—540
M—510	M—552.5	M—660
• 540	• 595	• 720
S—570	S—657.5	S—780
M—630	M—722.5	M—900
• 660	• 765	• 960
S—690	S—807.5	S—1020
M—750	M—892.5	M—1140
• 780	• 935	• 1200
S—810	S—977.5	S—1260
M—870	M—1062.5	
• 900	• 1105	
S—930	S—1147.5	
M—990	M—1232.5	
• 1020	• 1275	
S—1050	S—1317.5	
M—1110		
• 1140		
S—1170		
M—1230		
• 1260		
S—1290		

A—120 Hz channel spacing ± 30 Hz shift
B—170 Hz channel spacing ± 42.5 Hz shift
C—240 Hz channel spacing ± 60 Hz shift
M—Mark
•—Center frequency
S—Space

B. CCITT frequency recommendations

Modem	M	•	S	Freq. Shift
200 baud				
Channel 1	980	1080	1180	±100
Channel 2	1650	1750	1850	
600 baud	1300	1500	1700	±200
1200 baud	1300	1700	2100	±400

C. Bell System Data Phone® low and medium speed (FSK) digital-data transmission frequency assignments

Modem	M	•	S	Freq. Shift
300 baud (or less)— Type 103				
Channel 1	1070	1170	1270	±100
Channel 2	2025	2125	2225	
300–1200 baud— Type 202				
C&D	1200	1700	2200	±400

Note: The frequency tolerance of signals should be held within ± 10 Hz.

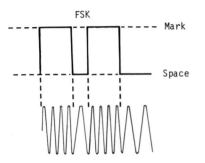

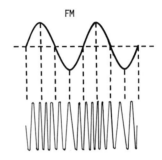

FIGURE 9-3

standard coding. This asynchronous coding (decoding) function is accomplished with a UART (pronounced "u-art"), which is an acronym for *universal asynchronous receiver/transmitter*. (USART's are used when both synchronous and asynchronous capabilities are needed.) UART's are available as LSI (large-scale integration) IC's that convert parallel digital data to an output form equivalent to that obtained from the distributor of a modern teleprinter (ASR-33 Teletype, etc.). The standard coded data pulses are then applied to a modem that converts them to compatible FSK form for application to the telephone system or other "channel," after which additional modulation and processing proceed the same as for voice or music signals. The modulation processes are, of course, reversed for signal and data reception.

The telegraph and data communications techniques discussed thus far in this chapter deal with *coded* signals that *originate* in coded form. Digital communication techniques are *also* used to convert voice, music, and other signals to digital forms and thereby increase the reliability of communications. It is possible to reduce a given signal to a variety of equivalent pulse forms by sampling, as shown in Fig. 9-4*a*. The basic pulse techniques include PAM (pulse-amplitude modulation), PWM (pulse-width modulation), and PPM (pulse-position modulation). These pulse techniques yield coded signals that closely maintain the character of the *original analog* waveform that is sampled. These three pulse forms have little direct advantage over conventional AM or FM approaches, but they provide an efficient means of processing multiple signals in specialized systems applications.

The modulation forms of Fig. 9-4*b*, PCM (pulse-code modulation) and *delta* modulation (DM), are a most efficient *and* reliable means of converting analog signals to pulse-coded signals. The PCM technique converts the sampled signal levels to binary values that correspond to the sampled amplitude of the analog signal. In computer terms the PCM technique functions as a serial A/D (analog to digital) converter for modulation, and a serial D/A (digital to analog) converter for demodulation. The delta modulation technique provides output pulses that correspond to *changes* in the level of an analog signal, with short-duration pulses of positive or negative polarity. Both PCM and DM forms have two distinct advantages over analog signals and PAM/PWM/PPM transmission. First, the sampling rate and timing (position) are fixed, so it is possible to expand the information capacity of a PCM or DM system by time-division multiplex techniques, where sampling of *other* signals can fill the spaces between the *original* pulse samples as shown in Fig. 9-4*c*. (The multiplex techniques and other pulse-system concepts will be reviewed in Chapter 10.) Second, the pulse amplitude in PCM and DM systems is fixed, so *limiting* can be applied to these pulses in order to reduce system noise, just as noise reduction can be achieved with FM systems. So both increased reliability and increased information capacity result when PCM and DM approaches are used for signal processing. Of course, there are the ever-present negative tradeoffs

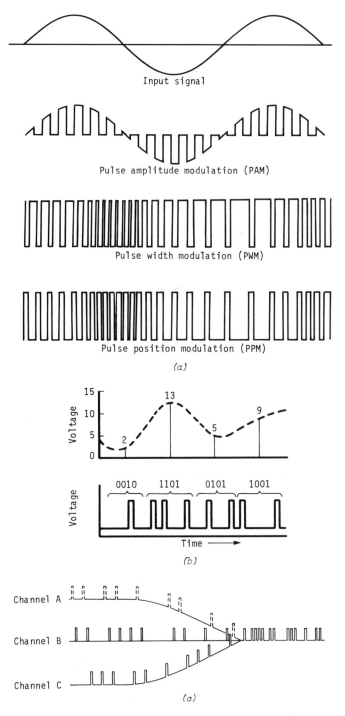

Input signal

Pulse amplitude modulation (PAM)

Pulse width modulation (PWM)

Pulse position modulation (PPM)

(a)

(b)

Channel A

Channel B

Channel C

(c)

FIGURE 9-4

of PCM and DM, including a rather complex system with increased bandwidth requirements.

As you may conclude, pulse modulation and digital communications are rather more complicated topics than the direct AM and FM approaches of Chapters 6 and 7. With the ever-expanding use of computers, along with increasing needs for communications services, the pulse and digital techniques are becoming the *most* important means of modulation. As such, your study of this chapter will round out your knowledge of modulation and demodulation techniques used for electronic communications.

LEARNING OBJECTIVES

Upon completing this chapter, including the problem set, you should be able to:

1. Identify and understand the various modulation forms used in digital communications.
2. Understand the popular data-input codes and their application to communications.
3. Understand data input and output forms, and the circuit functions used for processing and conversion of parallel, serial, synchronous, and asychronous pulses.
4. Understand the operation on modems in the conversion of serial pulses to FSK and PSK signals for application to communications channels.
5. Understand the principles of A/D and D/A conversion and the circuits used to obtain such conversions in communications systems.
6. Understand PCM and other modulation techniques and the circuits used to obtain and decode these pulse forms.
7. Understand how digitally coded signals are modulated and transmitted on RF carriers.
8. Understand and specify measurements of the important digital communications quantities, using frequency, time, and phase domain techniques.

DATA COMMUNICATIONS

In the introductory pages of this chapter, basic signal codes were identified in relation to Morse and teleprinter communications systems. You may recall that these codes were defined as PWM (pulse-width modulation) and PPM (pulse-position modulation), respectively. These binary (two-level, on-off) codes that are used for communications vary either the length of the pulses while keeping the spacing fixed, or the length of the spacing while keeping the pulse length fixed. Although the code *information* may differ, PWM and PPM pulses are electrically similar but out of phase; so changing the polarity of a PWM pulse will convert it to the more common PPM format for signal processing. These coded pulses represent the simplest form of *data communications* signals used for both telegraphic and computer terminal communications. (You should not confuse the binary *level* codes described here with the binary *number* codes described in reference to PCM later in this chapter.)

Codes and Timing

With the *Baudot*-coded teleprinter, for example, the spacing of PPM pulses is based on the 5-level, 6-character speed of the machine. At 60 W/M (words per minute) the start pulse (*space*) and character pulses (*marks* or *spaces*) are typically each 22 ms (milliseconds) in duration; whereas the stop pulse (mark) is about 1.4 times longer than the 22 ms pulses, or 31 ms. Each 22 ms unit of pulse time is called a *bit*; with this particular teleprinter code and speed, the transmission of one 5-level character will require about 7.4 *bits of information* and a total time of about 163 ms. With each *word* comprising 5 characters and a space, it will require approximately 1 second to complete each word, yielding the 60 W/M speed. At 100 W/M a Baudot-coded teleprinter would require 13.5 ms bits and a 19 ms stop pulse.

In data communications the transmission and processing of pulses are based on the highest equivalent frequency, or speed, of the binary signals. This speed is in turn based on the shortest pulse time (bit) for a given code, and is transposed to *bits per second* (b/s), called the *baud rate:* 1 baud = 1 b/s. Baud rates for any given code can be computed based on the following formula:

$$\text{Baud} = (\text{W/M}) \times (\text{C/W}) \times (\text{B/C})/60 \text{ seconds}$$

Baud is the bits per second, W/M is the words per minute, C/W is the characters per word, and B/C is the bits per character as used by a given code. The baud rate for the 60 W/M teleprinter example of the previous paragraph can thus be computed as follows:

$$\text{Baud} = (\text{W/M}) \times (\text{C/W}) \times (\text{B/C})/60 = 60 \times 6 \times 7.4/60 = 44.4 \text{ baud}$$

If you relate this rate to an equivalent frequency of about 45 Hz, you should quickly conclude that it is extremely slow. Obviously our modern data communications systems cannot rely exclusively on electromechanical teleprinters to interface with computers; neither can they rely exclusively on the teleprinter codes to achieve high-speed data transmission.

Data transmission from computer terminals and other digital equipment uses higher word speeds and equivalent baud rates in order to transfer information efficiently. Most of the U.S. equipment is standardized to the

ASCII code, operating from 110 to 19K baud asynchronous, and 2400 to 56K baud synchronous transmission, with only the 110 baud transmission compatible with the electromechanical (Series 33/35) teleprinters at 100 W/M. Given the 8-level ASCII coded signals, with one start bit and one 2-bit stop pulse as shown in Fig. 9-5, the 100 W/M baud rate can be computed as:

Baud = (W/M) × (C/W) × (B/C)/60 = 100 × 6 × 11/60 = 110 baud

The time per bit can be computed as the inverse of the baud rate as:

$(t/\text{bit}) = 1/\text{baud} = 1/110 = 9.09$ ms

Given the synchronous ASCII coded signals at 56K baud, as shown in Fig. 9-5, the W/M rate can be computed as:

(W/M) = baud × 60/[(C/W) × (B/C)] = 56 × 10^3 × 60/(6 × 8)
= 70,000 W/M

The time per bit for the 56K baud rate can be computed as:

$(t/\text{bit}) = 1/\text{baud} = 1/(56 × 10^3) = 17.86$ μs

Figure 9-5 also shows the teleprinter coded pulse times and compares the Baudot (Fig. 9-5*a*) and ASCII (Fig. 9-5*b*) examples in a visual form.

Bandwidth Requirements

Electronic digital and transmission circuits require a minimum bandwidth to *directly* process binary data communications signals based on their baud rate, just as such circuits would require a specified bandwidth to process given AM or FM signals. To preserve the precise, fast rise-time waveform of pulse signals, a theoretically infinite bandwidth would be required for transmission. Fortunately, the detection of digitally coded signals does not require the linearity of AM or FM, because only the presence or absence of a pulse—not the discrete level of the pulse—needs to be determined. A *certain* bandwidth is required, however, to transmit and process degraded pulses in order to assure reliable communications; and this requirement will quickly exceed the capacity of conventional telephone lines for distances greater than a few hundred feet. For this reason there are few direct-wired dc-level data communications systems, except for some ancient slow teleprinter installations and some short-distance computer interconnections.

The practical specific bandwidth requirements for reliable communications vary greatly depending on many empirical system factors, but a theoretical "ballpark" minimum bandwidth value can be obtained from mathematical analysis. A pioneer in the development of PCM (pulse-code modulation), Claude Shannon of Bell Telephone Laboratories developed the theoretical relationship between baud rate, signal-to-noise ratio (*noise*

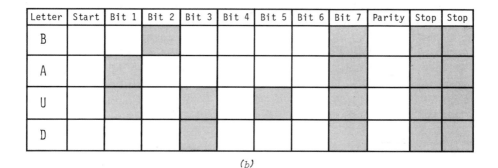

(a)

(b)

FIGURE 9-5

is covered in Chapter 8), and bandwidth for reliable dc-level data communications as:

Baud $= $ BW $\log_2 [1 + (S/N)]$

Bandwidth is in hertz; S/N is the signal-to-noise power, with 30 dB (1000 : 1) the typical minimum figure used in the calculation; and $\log_2$ is the logarithm to the base 2. If you don't have access to extracting $\log_2$ values from your calculator, you can substitute $\log_{10}$ (common log) values and multiply the resulting figures by 3.3:

Baud $= 3.3$ BW $\log_{10} [1 + (S/N)]$

If, for example, a 3 kHz bandwidth transmission circuit was available for dc-level data communications, with a 30 dB minimum S/N, the maximum theoretical baud rate may be computed as follows:

Baud $= 3.3 \times 3 \times 10^3 \times \log 1001 = 29.7$K baud

This computed value is a theoretical limit which is considerably derated in engineering practice, but the calculation provides a basis from which reasonably accurate judgments can be made.

Frequency-Shift Keying

It is far more efficient and reliable to employ the FSK (audio frequency-shift keying) or related modulation techniques, as mentioned in the introductory section of this chapter, rather than dc-level circuits for binary data communications. Such modulation techniques allow the "stacking" (multiplexing) of a number of signals on one transmission circuit. (Multiplex techniques are covered in Chapter 10.) One standard 3 kHz telephone voice channel, for example, can be divided to handle up to 20 or more telegraph (teleprinter) channels, with each channel occupying 120 or 170 Hz for FSK transmission of 60 W/M or 100 W/M, respectively. Because of their higher baud rates and the need for accessible channels, data communications with computers and related digital equipment typically use a telephone *voice* channel for reliable, high-speed FSK modulation of data pulses. Specialized or extremely high-speed data employ dedicated broadband wired channels that are available in increments from 8 kHz to 96 kHz. (The modulation techniques that process digital pulses for FSK and multiplex stacking fall within the broader category of *carrier communications systems,* which is covered in Chapter 10.)

DATA MODEMS

Binary (level) pulse to FSK (or equivalent) conversion is accomplished with the subsystem component called a *data modem.* Most U.S. data modems have been standardized to the characteristics of Bell System designs for data communications. Their early modem products were called *data sets,* and the terminology is still widely used throughout the industry. The popular type 103 modem, for example, provides for two-way FSK data communications up to 300 baud. The 103 uses 1070 Hz for a *space* and 1270 Hz for a *mark* for originating signals, and a 2025 Hz space and 2225 Hz mark for responding signals. The 202-type high-speed modem operates on one direction only on a two-wire circuit, but at data speeds to 1200 baud (600 baud on "voice-quality" lines and circuits). The 202 uses 1200 Hz for a mark and 2200 Hz for a space. International CCITT (*Comité Consultatif International Télégraphique et Téléphonique*) standards for modems use a 980 Hz mark and an 1180 Hz space for origination, and a 1650 Hz mark and an 1850 Hz space for response signals. One-way modems use 1300 Hz for a mark and 2100 Hz for a space. The FSK frequencies for amateur "ham" radio are standardized worldwide for a 2125 Hz mark and a 2975 Hz space. As you may appreciate, a wide range of FSK frequencies is possible within a 3 kHz voice channel, with higher baud rates understandably requiring wider spacing of audio frequencies for marks and spaces.

These modems, which are designed to operate with two-wire and four-wire telephone circuits, are also designated by their *direction* of operation. A simplex system *always* operates in one direction only, with a transmit (*mod*-ulator) unit at one end of a circuit and a receive (*dem*odulator) unit at the other. Such modems may be designated S/O, for "send only" operation, or R/O, for "receive only" operation. Simplex operation can be compared to your experience with conventional broadcast radio and TV transmitters and receivers: they are single-function, one-way devices. A half-duplex system operates in two directions, but in only one direction at a time. Half-duplex units at either end of a circuit may either transmit or receive, provided the corresponding unit is in the opposing mode. This system is analogous to the familiar push-to-talk intercom, where only one conversation can be transmitted or received at a given time. The full-duplex system allows for two-way simultaneous transmission and reception of signals. The common telephone is, of course, a full-duplex voice communications system. The modems discussed here transmit data only—not voice, but a comparison of their functions with the familiar voice equipment should more easily define their operation.

Although most modems are standardized for FSK operation, many other techniques are used for data communications. The Bell types 201 and 301, for example, use phase modulation for voice-grade and broadband circuits, respectively. Such PSK (phase-shift keyed) systems differentially phase shift an audio frequency tone (less than 1 radian) in order to achieve higher baud rate processing. As you may recall from Chapter 7, FM and PM are similar techniques, with phase modulation being the more practical choice for stable signal generation. You can think of FSK as audio FM, and PSK as audio PM, in order to make an equivalent comparison of the two techniques. Another technique, as used with the Bell series 150 and 400 modems, uses the multifrequency "touch-tone" coding system to effect data communications. This system allows simultaneous (*parallel*) transmission of data pulses, but is limited to 12 (or an optional 16) tone combinations at a slow minimum pulse time of 40 ms with a 5 ms space. Figure 9-6 shows the standard touch-tone

	1209 Hz	1336 Hz	1477 Hz	1633 Hz*
697 Hz	1	ABC 2	DEF 3	F U N
770 Hz	GHI 4	JKL 5	MNO 6	SPECIAL FUNCTION E C I T A I L O N S
852 Hz	PRS 7	TUV 8	WXY 9	
941 Hz	*	OPER 0	#	

FIGURE 9-6

frequencies and keyboard. PLL IC's are ideally suited to both the encoding and decoding of touch-tone data, making the technique a practical, low-cost choice for simple, direct communications of low-speed numerical data.

Modem Circuits

Recent advances in IC technology have greatly simplified the circuitry of the more popular FSK modems. Since the FSK signal is essentially FM (or PM), it is ideally detected (decoded) using the PLL-type detector, as described in Chapter 7, or specialized frequency-to-voltage converter devices. Both analog and digital techniques are used to encode and decode FSK signals, with digital synthesis providing more precise FSK encoding frequencies. Figure 9-7 shows the popular A-8402 IC, which can be used as either a voltage-to-frequency (FSK encoder) converter or a frequency-to-voltage (FSK decoder) converter. In the V/F (voltage-to-frequency) mode, the input binary data drives the *integrator* section of the IC, charging the external capacitor C_1. The *trigger* section input of the IC includes a comparator circuit that will consecutively fire the *one-shot* multivibrator and discharge C_1 when the voltage across C_1 rises to some reference value (0.7 V for this device). Since a higher input voltage will charge C_1 in a shorter time

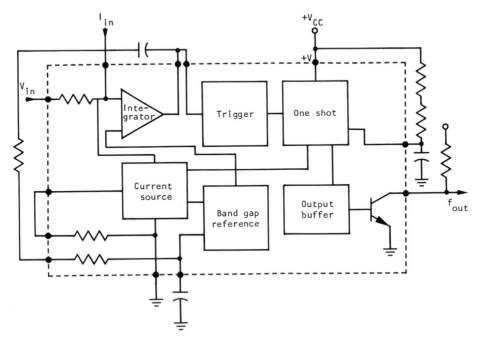

FIGURE 9-7

period, the one-shot will be triggered more rapidly, yielding direct voltage-to-frequency conversion. In the *F/V* (frequency-to-voltage) mode the FSK input signals are applied directly to the *trigger* section comparator input, firing the *one-shot* and *current source* that drives the *integrator* input. The *integrator* output charges the external capacitor C_1 to a voltage proportional to the average current of the source.

Precise *digital synthesis* of the FSK frequencies can be obtained with a crystal oscillator, frequency divider, and ring counter arrangement. Figure 9-8*a* shows an input signal at 16 times the FSK frequency driving the clock input of a CMOS (complementary metal-oxide semiconductor) 74C164, connected as an 8-bit shift register, to develop an approximate sine wave. In operation, summing the ouputs Q_1 through Q_8 and driving the data input of the IC produces the synthesized sine waveform. This digital technique is employed in the MC14412 LSI modem IC shown in Fig. 9-8*b*. This LSI device includes divider and synthesis circuitry for FSK encoding, as well as digital demodulator circuitry for FSK decoding, with the option of all four U.S. or CCITT standard FSK frequencies obtained from the same 1.0 MHz crystal.

As mentioned earlier, FSK encoding and decoding are also practically achieved with the use of linear PLL IC techniques. With this approach, a voltage-controlled oscillator (VCO) can be used as an FSK modulator, and a PLL tone decoder can be used as an FSK demodulator, thus using two versatile IC devices to provide the basic modem circuitry. The XR2206/2207 and XR2211 are typical linear IC's so designed for application in modem

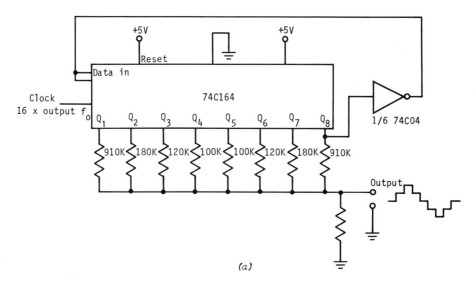

(a)

FIGURE 9-8

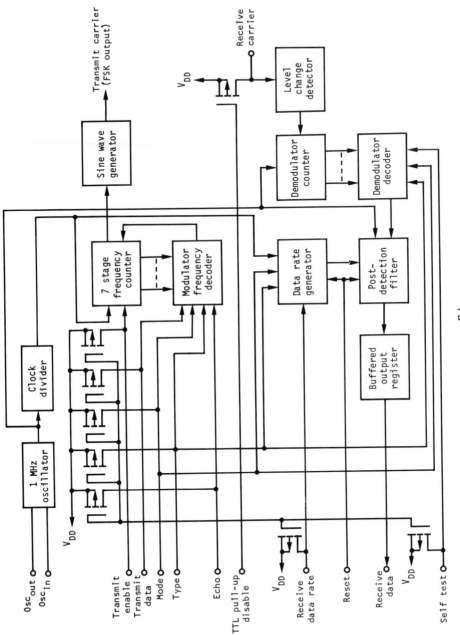

FIGURE 9-8 (continued)

(b)

circuits. One advantage of this linear approach is the availability of low-distortion sine-wave FSK output signals at *any* combination of frequencies below 100 kHz. Figure 9-9*a* shows the XR2206 sine-wave FSK generator with trimmer adjustment of R_6 and R_7, yielding two corresponding mark and space frequencies. The mark and space frequencies can be simply computed as:

$$f_{\text{mark}} = 1/(R_6\,C_3) \quad \text{and} \quad f_{\text{space}} = 1/(R_7\,C_3)$$

Figure 9-9*b* shows the XR2207 triangle-wave FSK generator with trimmer adjustments of R_1 through R_4, yielding two pairs of corresponding mark and space frequencies for optional two-channel applications. In operation the *current switching* section of the IC alternately switches the different timing resistors to the *VCO* input in order to control its output frequency in reference to the binary data input signal. Figure 9-9*c* shows the XR2211 PLL FSK demodulator. The VCO of this IC is set for a frequency midway between the FSK frequencies, e.g., 1170 Hz for the 1070 Hz/1270 Hz originate mode, and 2125 Hz for the 2025 Hz/2225 Hz answer mode. As the incoming FSK signal fluctuates between the *space* and *mark* frequencies, the *phase detector* output will alternately shift from a positive and negative VCO control voltage, which is in turn amplified as decoded data signal pulses by the *FSK comparator* at the output of the IC.

In addition to FSK encoding and decoding circuitry, modems require sharp filter circuits for the reception of FSK signals. Since these signals are often received in degraded form and at low levels, they contain interference spikes and glitches and noise signals that will activate false output pulses, thus introducing errors in the decoded output data. Early modems used heavy and bulky *LC* audio bandpass filters, but thanks to modern technology these have been replaced with high-performance active filters that are available in miniaturized form. These active audio bandpass filters use high-gain op-amp IC's, called *Norton amplifiers* or *current-differentiating amplifiers* (CDA's). Figure 9-10*a* shows an example of an AF state-variable filter that uses LM3900 CDA IC's. In this circuit the output of A_3 drives the inverting input of A_1, making A_1 a highpass unit; while A_2 performs a lowpass function. With the output following A_2, the circuit Q is a function of the feedback resistor R_4.

Interfacing

An additional modem circuit may precede the FSK input filter to provide conditioning of the degraded FSK signals. This conditioning function is often achieved with the use of IC *interface* elements such as differential comparators, analog switches, zero-crossing detectors, or line drivers/receivers.

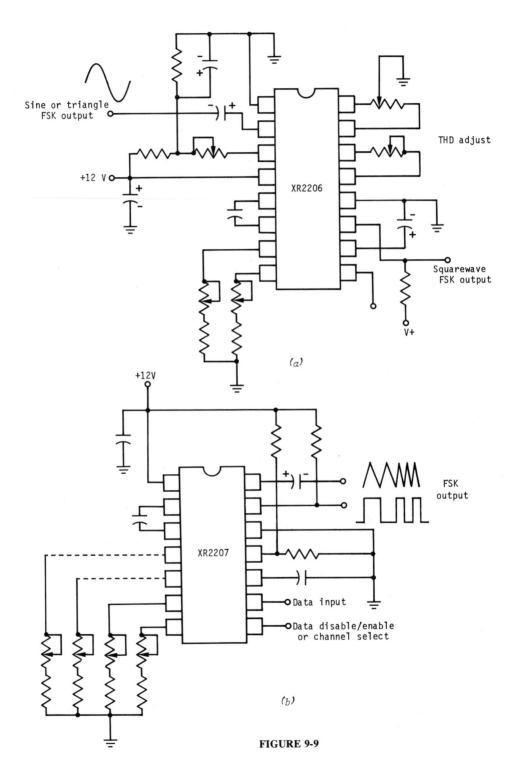

Sine or triangle
FSK output

THD adjust

+12 V

XR2206

Squarewave
FSK output

V+

(a)

+12V

XR2207

FSK
output

Data input

Data disable/enable
or channel select

(b)

FIGURE 9-9

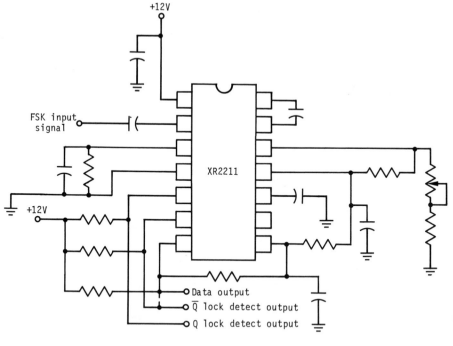

(c)

FIGURE 9-9 (*continued*)

These devices have high CMRR (common-mode rejection ratio) factors, so they will reject noise and interference signals that are common to both leads of balanced, two-wire telephone lines. They will also switch to full output voltage when the input signals reach a given reference value or cross zero, thus converting a degraded, noisy waveform back to a "clean" signal. Figure 9-10b shows a representative circuit and response of this IC interface device as it is used for the reception of FSK signals.

The model *FSK* inputs/outputs are often directly connected to telephone lines through *data access* networks (usually provided by the telephone company) that simulate the loading impedance of a standard telephone for compatibility and toll requirements. This direct connection transfers audio FSK signals to the lines at levels near -10 dBm (0.1 mW across a 600 Ω line $= 245$ mV. Modems are also indirectly connected to telephone systems through devices called *acoustic couplers*. These devices use a cradle to support a standard telephone handpiece, and employ *sound* transducers to send and receive FSK audio tones through the telephone set. Acoustic couplers trade off inconvenience and low data speed for the advantages of portability and electrical isolation from the telephone lines.

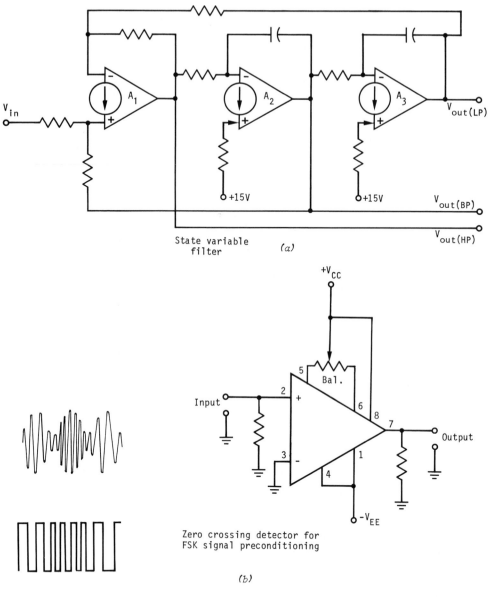

State variable
filter   (a)

Zero crossing detector for
FSK signal preconditioning

(b)

FIGURE 9-10

The modem *data* inputs/outputs connect to digital processing equipment
or teleprinter equipment through standardized interconnection cables. The
majority of such data is in *serial* binary form at levels which have also been
standardized to allow for successful interface with a wide range of communi-

cations equipment. Popular interface standards include those specified by the U.S. EIA (Electronic Industries Association), the U.S. Military (MIL-STD's), the international CCITT, and Western Electric. The most popular standard for U.S. data communications is the EIA RS-232, and the most popular worldwide standard is CCITT. Fortunately there is close compatibility between RS-232, CCITT, and MIL-STD-188 standards for data interconnection, allowing convenient adaptation of equipment inputs and outputs for operation.

Specialized IC devices used for interface connection of data signals between computer and modem equipment are called *line drivers* and *line receivers*. These devices comply with specific interface standards. For the popular EIA RS-232 specification, for example, the 1488 and 1489 IC's are used extensively in line driver/receiver applications. The 1488 accepts logic-level binary input signals, with logic 0 at less than 0.8 V and with logic 1 at greater than 1.9 V. (Modern TTL equipment will typically operate at less than 0.8 V for logic 0 and +4 to +5 V for logic 1.) The 1488 converts the levels of these binary input signals to −5 to −15 V (−10 V nominal) for logic 1 (mark) and +5 to +15 V (+10 V nominal) for logic 0 (space), then drives a 3 kΩ to 7 kΩ (5 kΩ nominal) loaded line with these binary output signals. The 1489 receives the line signals from the 1488, providing the nominal 5 kΩ load impedance for the line and 1488, and converts the RS-232 level signals back to zero and +5 V logic-level signals (−3 to −25 V input levels convert to +4 to +5 V output levels, and +3 to +25 V input levels convert to nominal zero volt output levels). Figure 9-11 shows the schematic and functional application of the 1488 and 1489. The advantages of the bipolar (− and +) binary conversion of these devices include higher signal levels, and more important, "fail-safe" operation, because the absence of a signal (zero volts) will indicate a *disconnect* or malfunction.

As stated earlier, teleprinters generate serially coded data, while computers may also process parallel data. Since data communications systems require serial data, an additional subsystem component may be necessary to interface the computer (or other parallel digital system) to the modem. This subsystem component is the UART identified earlier in this chapter. It functions to convert parallel computer data to serial form and vice versa. Thanks again to new technology, the UART is available as an LSI IC device. The device is *universal* in the sense that it adds start, stop, and parity code bits; and it will handle 5- to 8-bit character codes, including the Baudot, ASCII, and other popular data codes. Figure 9-12 shows a popular AY-5-1013 UART LSI IC. In operation the transmitter section shift register is parallel loaded from the computer and serially clocked out at 16 times the baud rate along with the control bits. The receiver section of the UART clocks-in serial data to its shift register and latches its parallel outputs after extracting the control bits.

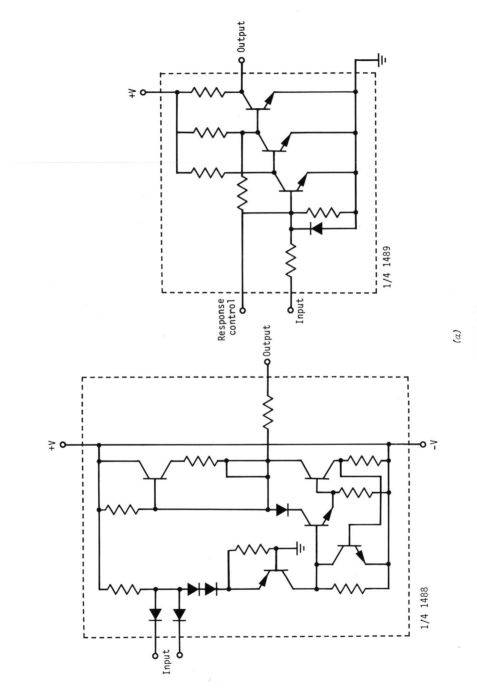

FIGURE 9-11

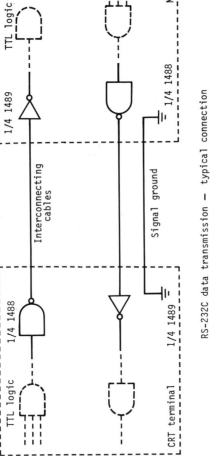

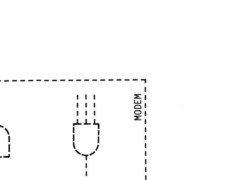

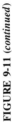

RS-232C data transmission — typical connection

(b)

FIGURE 9-11 (continued)

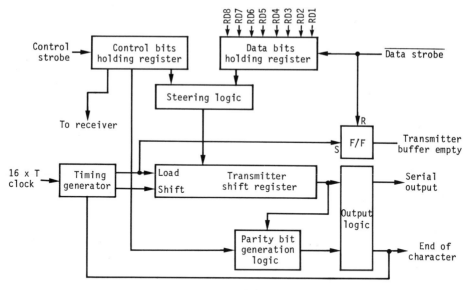

(a)

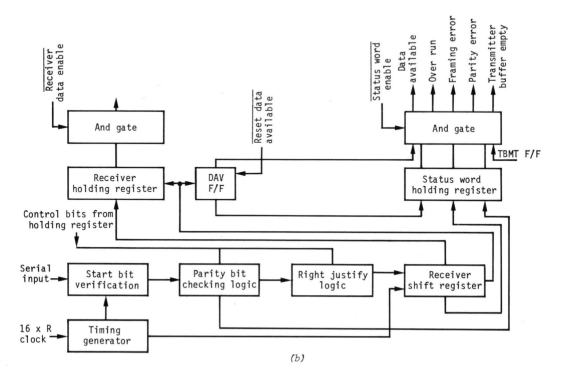

(b)

FIGURE 9-12

PULSE CONVERSION/PULSE-CODE MODULATION

As you should now understand, *data communications* is the technology that defines the processing of signals that *originate* in pulse-coded form from teleprinters, computers, and other digital equipment. This same technology can be used to process audio, video, and other analog-waveform signals if they are *first* converted to pulse-coded form. You may, at first, see no real technical advantage in converting voice signals to pulses, then *back* to analog FSK signals for transmission, when the voice signals can be *directly* transmitted over a telephone circuit. There are problems with random-frequency noise and signal losses on transmission circuits, however, so voice signals may become degraded to the extent that they cannot be recovered through the long-distance amplifying repeaters used in telephone systems. It is possible, though, to reduce noise and recover signals if they are FSK-coded at only *two* frequencies, by using filters and interface devices used in modems (refer to Fig. 9-10). Another important aspect of the voice-to-code-to-FSK conversion for transmission (and the reverse for reception) is its application to *cryptographic* communications, where voice and other analog signals are coded for security purposes.

The process of transmitting coded analog signals can be simplified if the pulses are not reconverted to FSK for transmission. Although it is possible and practical to transmit *pulses* down a transmission line and through transmission circuits, it is not economical to tie up a voice channel (300 Hz to 3 kHz) with just one series of pulses. So along with the conversion of analog signals, multiplex techniques are employed to expand the *number* of signals that can be transmitted in pulse form. These modern pulse-coding/multiplexing techniques are called *pulse-code modulation* (PCM) and *time-division multiplex* (TDM). With present standards, PCM techniques are used to code and decode 24 analog voice channels, and transmit them on telephone circuits that can process only one voice channel in *analog* form. In addition the pulse-coded signals can be reconstructed in repeaters where high equivalent noise levels would make direct voice communications impossible.

THE CODEC

The electronic equipment used for analog-to-pulse conversion is called a *codec*. Codec is an abbreviated conjunction of *co*der/*dec*oder, and the device functions as a reverse modem. Other descriptive names given to equipment and circuitry that perform such coding and decoding are *analog-to-digital* (A/D) and *digital-to-analog* (D/A) converters, often abbreviated ADC and DAC, respectively. Functionally, analog waveforms are digitized in the codec by a process of *sampling* the amplitudes with recurrent short-duration pulses. There is, in fact, a *sampling theorem* advanced by Shannon that mathematically describes how an analog signal can be sampled at a rate of just twice the highest signal frequency, then be fully reconstructed with but negligible distortion. This theory is one that is fully realized in practice, where an 8K baud sampling rate is standard for voice channels extending to 4 kHz. Incidentally, Shannon based his PCM development work in the 1940's on principles advanced by Harry Nyquist in the 1920's. Nyquist advanced a theorem that a given bandwidth can carry pulse signals of half its high-frequency cutoff. Nyquist's relationship was actually the inverse of Shannon's sampling theorem, but in engineering practice the term *Nyquist rate* is often used to specify Shannon's *sampling rate*. Figure 9-13 shows how an analog waveform may be sampled to yield PAM pulses (the first step for PCM coding), PWM pulses, and PPM pulses.

Codec Circuits

PAM coding of analog signals can be achieved with a variety of circuit techniques. One of the most practical basic circuit approaches uses a high-speed IC differential amplifier wired as an analog switch. With this basic circuit, shown in Fig. 9-14a, the current-source transistor in the IC is driven with an analog signal, while one of the differential-pair transistors is switched at the sampling rate. In practice the use of combined differential amplifiers driving one common load will allow sequential sampling of a number of analog signals. This sequential TDM sampling is the basis of efficient PCM operation. Modern codecs use high-speed MOS (metal-oxide semiconductor) and hybrid, multipole IC analog switches to perform the sequential sampling for PAM coding. For specialized applications requiring PWM coding of analog signals, the function is easily achieved with the popular NE555 IC timer. In this application the IC is connected for monostable operation, as shown in Fig. 9-14b, with the analog signal applied to the control-voltage input of the first comparator, and the sampling pulse applied to the input of the second (parallel) comparator. The related PPM (pulse-position modulation) of analog signals can also be achieved with the NE555 connected for astable (free running) operation, as shown in Fig. 9-14c, with the analog signal applied to the control-voltage input the same as for PWM operation.

Pulse-Modulation Detectors

The decoding of PAM pulses is accomplished with a specialized operational amplifier circuit called *sample and hold*. An example of this circuit function is shown in Fig. 9-15, along with the related reconstructed analog waveforms. In operation the PAM pulses are applied to the input of op-amp 1, and amplified to drive the MOSFET (Q_2) through the voltage-variable resistor (JFET Q_1). Q_1 operates as a switch controlled by the amplified sampling pulses from op-amp 2, allowing C_1 to charge to the instantaneous voltage of

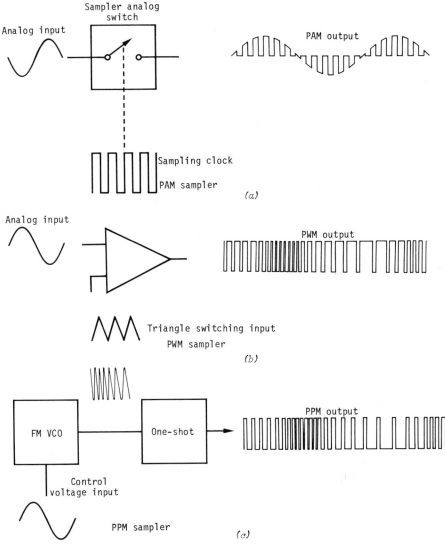

FIGURE 9-13

the applied PAM pulses and *hold* that charge until the application of the next sampling pulse. Careful selection of the capacitor value will assure the required charge and discharge slopes at the high-frequency limit of the decoder.

Envelope detection is a much simpler means of decoding PAM pulses, and the technique can also be used for decoding PWM and PPM pulses. This is

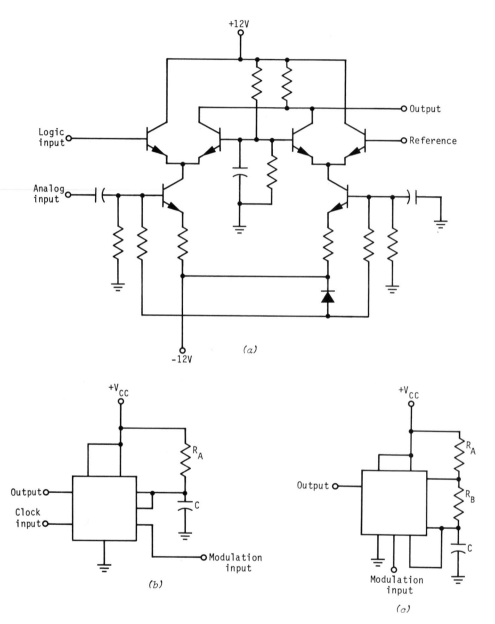

FIGURE 9-14

essentially the same technique used to demodulate AM signals, as described in Chapter 6. The dc-offset PAM waveform of Fig. 9-13 can be compared to the rectified AM envelope signals described in Chapter 6. The envelope decoder is simply a diode detector and an *RC* integrator network (lowpass

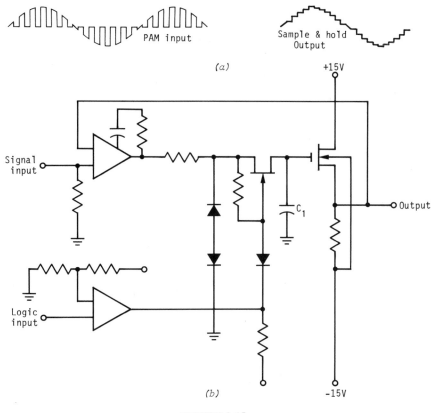

FIGURE 9-15

filter) with a time constant equal to the sampling period of the pulse signals. Often the integrator function is combined with an operational amplifier IC in order to provide amplification and isolation to the decoding function. PPM pulses must be inverted prior to envelope detection to convert their "on-time" to longer-duration pulses and thus *charge* the integrator section of the detector. This inversion process is often accomplished with an IC flip-flop circuit. PWM pulses can be directly converted to analog signals, with the variations in pulse duration directly charging the integrator circuit.

Digital Coding and Decoding

True *digital* coding and decoding of analog signals involves an additional processing function which ultimately converts the sampled signals to binary values. True ADC's have outputs that are actual binary-coded levels. If an analog signal was sampled as in Fig. 9-13, for example, the PAM amplitudes, in volts, would be 0, 10, 14, 10, and 0 for each of the sequential five samples, respectively. These magnitudes have equivalent binary (8-4-2-1) values of 0000, 1010, 1110, 1010, and 0000 for each conversion, respectively. If these binary values are coded in serial pulse form at the ADC output, they can be represented by bursts of short-duration pulses for each conversion, all of equal amplitude and coincident with the equivalent binary numbers. The coding of analog signals for PCM transmission follows this process, with the bursts of short-duration pulses carrying the converted analog information down the transmission line. Figure 9-16 shows this conversion of PAM to equivalent-value binary pulses.

Analog-to-digital conversion of voice-frequency signals is often achieved with an ADC technique called *successive approximation*. Successive-approximation ADC's incorporate a comparator, clocked programmer (sequencer), counter, and digital-to-analog converter (DAC) in their circuitry. This type of ADC provides highly accurate and relatively high-speed signal conversion. In operation the ADC simply compares its input signal to that of the built-in DAC output in sequence (one bit at a time). The simplified block diagram of a 4-bit successive-approximation ADC and its signal processing functions is shown in Fig. 9-17. Observe from this figure that the clock drives the binary counter that inputs to the DAC in sequence, from the *least significant bit* (LSB) to the *most significant bit* (MSB), doubling its output voltage to the comparator with each successive bit. As long as the clock input

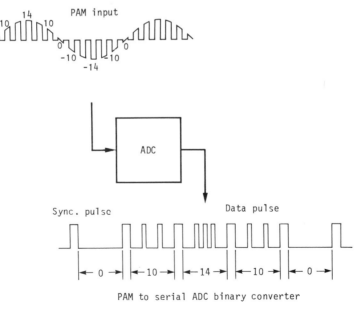

PAM to serial ADC binary converter

FIGURE 9-16

signal exceeds that of the DAC output, the comparator will transfer full-amplitude pulses to the ADC output. When the DAC output level exceeds that of the comparator input signal, the comparator output will become *low* (binary 0), inhibiting the clock. With this function the higher input levels will produce more clock pulses; whereas lower input signals will yield fewer clock pulses. The *binary* (base 2) translation is achieved from the sequentially doubling DAC output voltage that drives the reference input to the comparator.

The serial output of the ADC is a series of pulses—*not* the binary data required for PCM. Binary data is available at the parallel ADC outputs that sequence its internal DAC inputs. ADC's are thus serial analog-input, parallel binary-output devices; and DAC's are parallel binary-input, serial analog-output devices. An additional PCM signal processing function is required to *serialize* the parallel binary data for transmission, and then reconvert the transmitted serial data to parallel form upon reception. The *serializing* function is often achieved with an eight-stage static shift register IC such as the CD4014, connected for parallel-to-serial conversion, as shown in Fig. 9-18*a*. Reconversion of the serial binary data to parallel form at the receiver codec may also be achieved with an IC shift register. Figure 9-18*b* shows an MSI (medium-scale integration) 74C164 device connected for serial-to-parallel conversion.

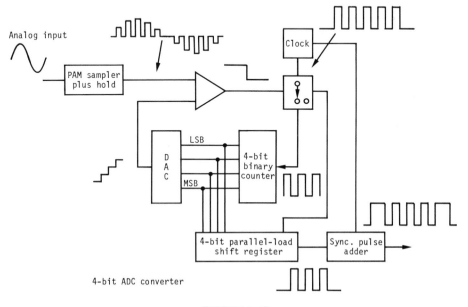

FIGURE 9-17

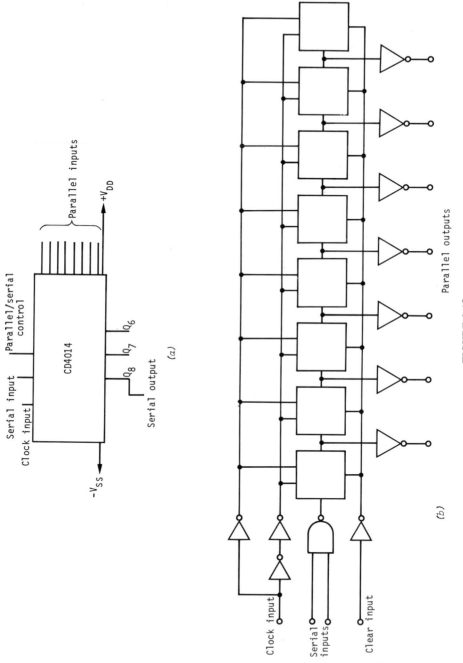

(a)

(b)

FIGURE 9-18

Referring to Fig. 9-17, you should further observe that the successive approximation ADC can function only if the input signal to the comparator remains fixed while the DAC is sequenced from its LSB through MSB levels. To achieve these fixed ADC input steps for PCM conversion, PAM signal levels must first be processed by a *sample and hold* circuit similar to that used for the *direct* demodulation of PAM (see Fig. 9-15). This process is called *quantizing*. An additional factor that will determine the accuracy of the ADC is related to the number of binary sampling bits programmed (sequenced) into its DAC section. The ADC output code is, in fact, a $\log_2$ representation of some analog signal. With the 4-bit example of the previous paragraphs, it is possible to resolve the analog signal at 15 discrete levels: $2^4 = 16$ binary-numbered values. In practice, voice signals are converted to PCM with the use of an 8-bit ADC that uses 1 bit for signaling and 7 bits for binary-code conversion, with a resolution of 127 binary-numbered values ($2^7 = 128$, less the zero value, yields 127).

Time-Division Multiplex

One major advantage of the PCM technique is its ability to sample a *number* of voice or data channels sequentially and transmit them on a single line or channel. The operational specifications for this time-division-multiplex approach have been standardized by the U.S. Bell System and the international CCITT. The Bell System PCM is called the *T1 carrier*. Both the Bell and CCITT T1 or equivalent standards use 8-bit ADC's and an 8K baud sampling rate for each of 24 voice channels of up to 4 kHz bandwidth. Each 24-channel sample is called a *frame*. A single *framing-bit* pulse is added to the code at the end of each frame to assure synchronization. The total equivalent baud rate is thus computed, in megabits per second, as:

Baud = (bits per frame) $\times$ (sampling rate)
     = $[(8 \times 24) + 1] \times 8000 = 1.544$ Mb/s

Figure 9-19 shows the framing and sequential sampling of T1 carrier PCM signals.

A second advantage of PCM transmission is based on the digital demodulation capabilities of modern electronic devices and circuits. Since pulse demodulation is based on detecting *only* the presence or absence of signals—not the *levels* of signals—reliable recovery of degraded or noisy PCM signals is practical where communications would be impossible using other modulation forms. This factor will allow the transmission of 24-channel carrier signals at 1.544 Mb/s using existing telephone circuits that carry only a single analog AF voice signal. The PCM technique is, incidentally, the practical means used for "Picturephone" transmission over telephone circuits. PCM signal conditioning at the demodulator input to the codec, and also in PCM telephone-circuit repeaters, can be achieved with IC *interface*

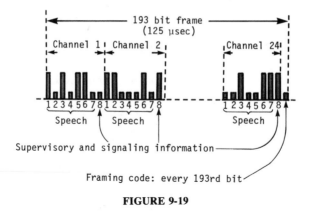

FIGURE 9-19

elements such as are used in data modems. Differential comparators, zero-crossing switches (detectors), or line receivers with high CMRR factors can effectively switch to full output voltage, even with highly degraded and noisy PCM signals. These devices, being typically balanced circuits, cancel signals or interference that appear at both inputs with the same (common mode) waveform. With such a differential input, only signals that appear out of phase at the inputs will be processed. Drivers that generate balanced out-of-phase signals for two-wire transmission act as a source for these coded signals, while the receiver interface elements exclude all but differential signals at their inputs.

PCM Demodulation

After signal recovery, additional demodulation steps are required to separate and reconstruct the analog voice information from PCM pulses. The process begins with the conversion of the serial PCM pulses to parallel form in order to drive an 8-bit DAC. This serial-to-parallel conversion is usually achieved with an 8-bit serial-in, parallel-out shift register, which may be a part of a modern LSI or hybrid IC DAC. Although there are a variety of DAC techniques, the *resistance-ladder network* type is the more popular choice for PCM demodulation. Figure 9-20a shows the functional diagram and DAC output voltage that is achieved by alternate switching of the R-$2R$ resistance-ladder networks within its structure. As you may observe, the R-$2R$ resistance-ladder networks are merely "weighted" voltage dividers that vary the current through (and voltage across) the $2R$-value load resistance. The resulting output voltage is thus proportional to the binary numbers at the LSB to MSB inputs to the DAC, as driven by the parallel outputs of the preceding 8-bit shift register. The DAC output is a series of PAM pulses that contain all 24 coded voice channels. The separation of these

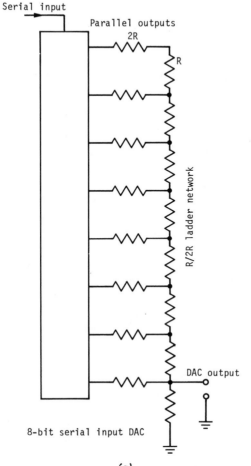

(a)

8-bit serial input DAC

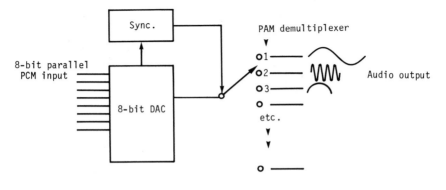

(b)

FIGURE 9-20

channels is achieved with a commutating analog switch, called a *demulti-plexer*. This switch samples the 8-bit PAM pulses in sequence, and in synchronization with the *coding sampler* in the PCM transmitter codec. Figure 9-20*b* shows the functional block diagram of the PCM receiver codec and its related signal separation and reconstruction waveforms.

Synchronization

Synchronization between the PCM transmit and receive codecs must be achieved in order to assure reliable communications using this pulse-modulation technique. Referring back to Fig. 9-19, you may recall that the 193rd data bit is inserted into the PCM transmit code after the time-division multiplexer sequences through each 192-bit, 24-channel *frame*. This 193rd bit, called *the framing bit,* is transmitted alternately high and low (101010 . . . , etc.) to establish a synchronization pattern in the multiplexed PCM signal. The receiver codec contains circuitry that will detect the transmitted framing bits in sequence in order to establish and maintain synchronization. This "sync" function may be achieved with a 193-bit binary counter, flip-flop, and gating circuit, which resets the serial-to-parallel register to zero and switches the demultiplexer to *channel 1* upon detecting the *framing bit*. Should synchronization be lost, the counter circuit would sequentially sample up to the "worst-case" 48 combinations (24 channels with a framing bit of *1* and 24 channels with a framing bit of *0*) to reobtain sync. With each 193-bit frame requiring 125 μs, the worst-case synchronization would require:

$$(48 \text{ channels}) \times (8 \text{ bits/channel}) \times 125 \ \mu s = 48 \text{ ms}$$

So long as this loss of synchronization does not occur frequently, the 48 ms (about 1/20 second) will seldom be noticed in normal voice transmission.

In addition to the popular PCM technique, many analog conversion codes can be applied to the transmission of pulses. Such codes can take many forms for cryptographic, military, general security, and special signal processing requirements. The technical process of sampling and coding used for other transmission codes is similar to that used for PCM, so your understanding of *this* popular modern communications technique should allow you to easily understood other code forms. There is a much greater variety of pulse techniques used in engineering practice.

APPLIED PULSE AND DIGITAL COMMUNICATIONS

Your lack of access to a variety of pulse and digital communications equipment may make the topics of this chapter appear complicated, confusing, and perhaps even uninteresting. After all, to many people *electronic com-*

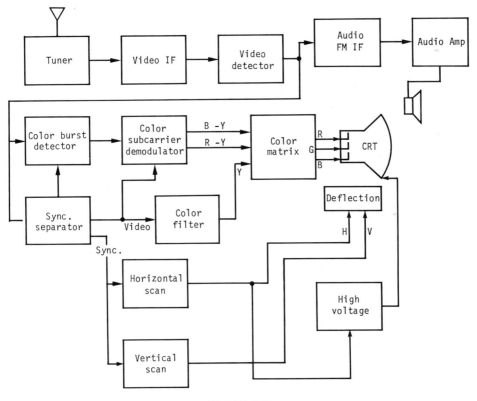

FIGURE 9-21

munications describes the study of radio, television, and related topics—not telephone and computer topics. In fact, however, the pulse and digital techniques described in this chapter are as important to your knowledge of electronic communications as the more familiar RF, AM, and FM principles. The common TV receiver, for example, employs many of the pulse, timing, coding, and multiplex techniques that have been described as used for voice and data communications applications. In order to relate some of the obscure concepts, advanced in the preceding sections of this chapter, to your *direct* experience, the U.S. television system will be used as a familiar application of pulse and data processing circuitry.

The Television Picture

Figure 9-21 is a block diagram of the typical NTSC (U.S. National Television Standards Committee) color TV receiver and its related waveforms. The RF/IF processing sections, through to the video detector, function as an AM VSB (vestigial sideband) superheterodyne receiver. The FM (4.5 MHz) IF and detector section following the video detector processes the audio signals to the loudspeaker. The demodulated analog video signals are amplified and drive all three guns of the picture tube (CRT) in synchronization with the color signals and scanning of the CRT screen. For conventional TV picture reception the video and color signals are analog in form; while the horizontal and vertical sweep signals are a repetitive sawtooth waveform. The *synchronization* of these four signals is achieved with pulse-modulated waveforms, however. The video, color, vertical sweep, and horizontal sweep signals are synchronized during the periods when the CRT electron beam is "blanked" at the edges of the TV picture.

Figure 9-22*a* shows a sequence of transmitted horizontal synchronization pulses that control the timing of the 525 horizontal lines of the TV picture.

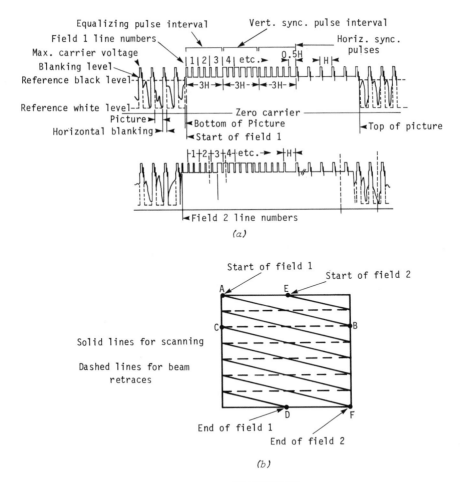

(a)

(b)

FIGURE 9-22

These pulses recur at a time interval of 63.5 μs, corresponding to the 15,750 Hz frequency of the receiver's horizontal sweep circuit that drives the electron beam across the face of the picture tube (CRT). (The horizontal frequency for the NTSC *color* system is actually 15,734.264 Hz.) To complete a full picture *frame* the electron beam is also driven vertically, from the top to the bottom of the CRT screen, *twice* within a time period of 1/30 second (33.33 ms). So the vertical sweep recurs at a rate of 60 times per second. Each vertical sweep, top to bottom, is called a *field,* and it takes two fields to complete a full picture frame. (The vertical frequency for the NTSC *color* system is actually 59.94 Hz.)

The TV *picture* is actually produced by a fast-moving dot on the phosphor face of the CRT. It is only through the *retention* of the instantaneous image by our eyes that we are able to see a continuous 525-horizontal-line display on the screen. The 525 lines are scanned in two fields of 262.5 lines each. The scanning is timed and sequenced with pulses to alternately display even-numbered, then odd-numbered lines with each vertical sweep field. This ingenious technique of interleaving the odd and even lines in each successive field is called *interlace vertical scanning,* and it is employed to reduce flicker in the TV picture. Figure 9-22b shows the interlace scanning of the top portion of the CRT. Observe that the odd-numbered lines (1, 3, 5, etc.) are swept in the first field; and *half* of a line *above* line 1, then lines 2, 4, 6, etc., are swept in the second field. Even though the lines of the second field occur 1/60 of a second after those of the first field, the retention of light by our eyes allows us to distinguish the lines of *both* fields as if they were continuously displayed.

TV Synchronization

Pulse techniques are used to control the timing and interlace scanning of the TV camera, transmitter,* and receiving systems. Although specific circuit practice may vary, we shall here describe the *functional* operation of TV pulse synchronization in relation to the digital circuits identified earlier in this chapter. Figure 9-22 also shows the sequence of transmitted pulses at the end of odd and even frames, and their resulting control of video and vertical interlace scanning. Although there are two 262.5-line fields per 525-line picture frame, video information that fills the TV picture occupies only 241.5 lines per field and 483 lines per frame. The video is *blanked* during the remaining lines that are used for synchronization. Incidentally, each horizontal line is also synchronized with the *horizontal sync pulses,* as shown in Fig.

---

* The use of the word *transmitter* throughout this section of the text refers to the complete transmitting function, including camera chain, modulation, and actual transmission after amplification. You should not confuse this use of the term with the more specific *transmitter* used as a single-function piece of equipment.

9-22a, with video *blanking* of 10.2 μs employed for horizontal *retrace* and synchronization, leaving only 53.3 μs of 63.5 μs for the actual picture (video).

In operation the transmitting TV camera and receiver are synchronized for the vertical sweep of the first (*odd*) field to begin with line 1 at the upper left corner of the CRT. The camera and receiver then continue scanning the odd-numbered lines until line 483 (262 odd-numbered lines) at the bottom of the screen. At the instant that half of line 483 is scanned, the video is blanked, and the transmitted horizontal pulse rate is doubled. Six short (2.5 μs) *preequalizing* pulses, followed by six inverted (29.25 μs) serrated pulses and six additional short (2.5μs) *postequalizing* pulses, are then transmitted at the doubled horizontal pulse rate, with a recurrent time interval of 31.75 μs. This series of pulses is detected in an integrating network that operates like the PWM detector of Fig. 9-15. The detected serrated pulses yield a high output level that drives a comparator-function circuit. This circuit in turn triggers the vertical sweep circuits to return the electron beam to the top of the CRT. The six preequalizing pulses use three horizontal lines at the bottom of the CRT beyond the center of line 483. The vertical retrace that begins at the bottom center of line 489 arrives at the top center of the CRT three lines after (line 495) and continues to the center of line 525, where the blanking pulse is removed from the video.

Since the video returns for half of a line *ahead* of line 1 to begin the second field, the line may be considered as the last half of line 525. The first full line scanned by the second field is line 2. The scanning of even-numbered lines continues through the second (even) field with video until line 482 is scanned. The video is blanked *after* line 482 is completed, and the transmitted horizontal pulse rate is doubled as for the first odd-field sweep. The same sequence of six preequalizing, six serrated, and six postequalizing pulses is then applied to this even-field sweep, but the timing has shifted one-half horizontal-line period as shown in Fig. 9-22. The resulting vertical *retrace* occurs at the *end* of line 488 and arrives at the top *left* of the CRT three lines later (the end of line 494, or beginning of line 496). It then continues to the end of line 524, where the blanking pulse is removed from the video, and the sequence begins again with line 1 of the odd field.

The generation of pulses for precise control of TV synchronization may be realized with sync-pulse generator devices such as the 3262B shown in Fig. 9-23a. The basic clocking frequency for this device is obtained by dividing a 2.0475 MHz input signal by 65 to obtain 31.5 kHz. The 31.5 kHz frequency pulses are, in turn, divided by 2 to obtain the horizontal sync pulses. They are also divided by 525 and 2 to obtain the 60 and 30 Hz field and frame pulses, respectively. The line decoder and frame decoder circuitry provide the timing diagram of Fig. 9-23b. This timing diagram complies with EIA Standard RS-170 for NTSC television transmission. You may think of each horizontal video line, which is presenting analog signals between each of the horizontal sync pulses, as being time-division multiplexed like the multi-channel PCM signals discussed earlier in this chapter. In fact, there is a close functional correlation to the accurate timing of the 483 lines of video information, with many time-division multiplex communication systems. Figure 9-23c shows the TV camera, transmitter, and receiver as a representative time-division multiplex system. The only functional difference you may note in comparing the telephone and television systems is that the telephone switches pulse-coded signals in sequence to separate circuits, whereas the TV switches analog signals in sequence to separate portions of the picture tube.

Color Synchronization

The previous paragraphs described how pulses are used to synchronize and control the video and sweep circuits of the TV receiver. Pulses are also used to synchronize and control the receiver's *color* circuits in order to reproduce accurate color images on the face of the CRT. Although both B & W (black and white) and color TV systems are compatible, there are, as you would expect, a number of differences between the two systems. First, the horizontal and vertical sweep frequencies of the NTSC color system differ slightly from those used for B & W TV in order to accommodate the color signal as a subcarrier within the upper sideband of the VSB video-modulated RF carrier. The color signal frequency was standardized at about 3.58 MHz (actually 3.579545 MHz), and the horizontal sync pulses were reduced slightly to 15,734.264 Hz (from 15,750 Hz for B & W) in order to eliminate the interference of these two signals within the spectrum of *their* sideband response. Figure 9-24a shows the TV-channel spectrum and distribution of video (sync) and color signals. Since the vertical sweep is triggered from a sequence of modified horizontal sync pulses, it too is proportionally reduced to 59.94 Hz (from 60 Hz for B & W).

The color signal is synchronized at the beginning of *each* horizontal line with an 8-cycle *burst* at 3.58 MHz, which phase locks the 3.58 MHz color oscillator circuit within the TV receiver. This oscillator serves to establish a 0° phase-reference signal, against which the phase-modulated color information signals are compared for demodulation. The burst occurs immediately after the horizontal sync pulse, but before removal of the video blanking. Figure 9-24b shows the horizontal *blanking pedestal* as applied for video blanking and horizontal sync. The blanking obviously precedes and lags the sync pulse to allow time for the electron beam to retrace across the CRT. It is during this lag time, called the *black porch,* that this 3.58 MHz burst signal is applied.

The color TV picture is produced on the face of the CRT with a combination of two demodulated information signals: *luminance* and *chrominance.* The luminance signal, specified as the *Y* signal, is equivalent to the AM *video*

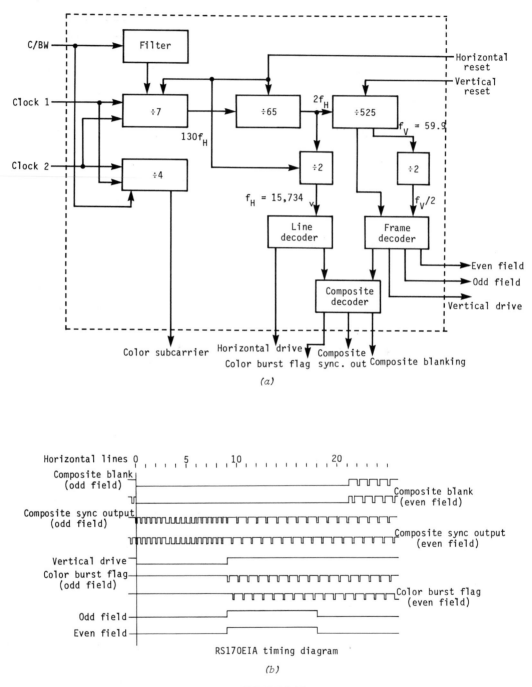

(a)

RS170EIA timing diagram

(b)

FIGURE 9-23

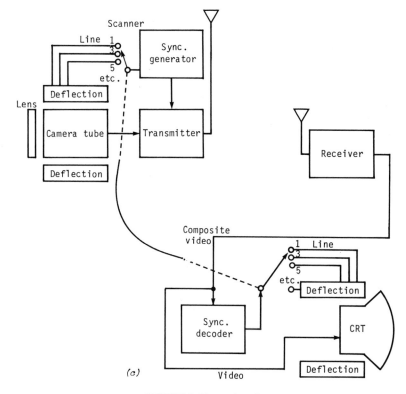

FIGURE 9-23 (*continued*)

signal of B & W systems. The chrominance signal, specified as *chroma,* operates at 3.58 MHz, and is both phase and amplitude modulated in order to carry color and saturation information. Color modulation is achieved in the transmitter by *matrixing* (combining) the three basic color information signals, along with the *Y* signal, from the tricolor TV camera. The resulting matrixed color signals are combined into two values—called *I* and *Q,* for *in-phase* and *quadrature,* respectively—as rectangular coordinate signals which are in turn combined to obtain the polar-form equivalent chrominance signal. By varying the magnitude and sign (+ or −) of the *I* and *Q* signals, they can be combined to yield any proportional magnitude or phase-angle chroma that corresponds to the NTSC vector wheel. Figure 9-25 shows how color signals are combined to obtain chroma, and also shows the related color wheel.

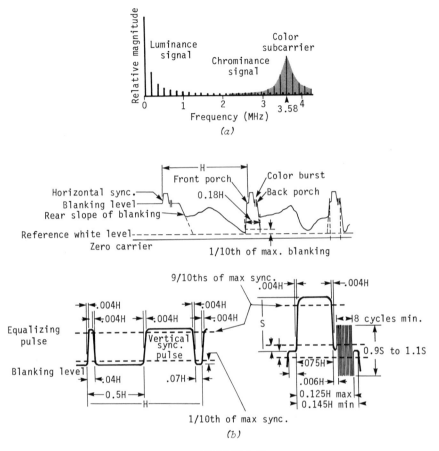

FIGURE 9-24

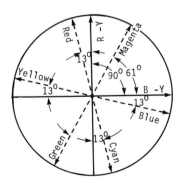

(a)

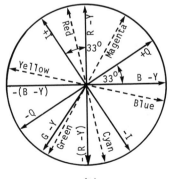

(b)

FIGURE 9-25

Test Signals

In addition to video, sweep, and color synchronizing, pulses are used for TV test signals that are transmitted during the time that the video is blanked at the top of the CRT. You may recall from earlier paragraphs that the video remained blanked after vertical retrace and the postequalizing pulses for a period equal to 12 horizontal lines in each field. Each preequalizing, serrated, and postequalizing pulse uses a period equal to 3 horizontal lines; the video is blanked for a period of 21 lines per field. The standard test signals are applied to lines 18, 19, and 20 of each field. They are appropriately called *vertical interval test signals* (VITS), and include a vertical interval reference signal (VIRS).

The multiburst signal, shown in Fig. 9-26*a*, appears on line 17 of the odd fields, making it line 517 of the 525-line picture. The multiburst frequencies are 500 kHz, 1.25 MHz, 2.0 MHz, 3 MHz, 3.58 MHz, and 4.0 MHz, and they are used as references for video and color response. A color-bar reference signal appears on line 17 of the even fields, and on line 518 of the 525-line picture. This color bar references alternate phases of the 3.58 MHz chroma signal at 12°, 256°, 299°, 119°, 76°, and 192°, for yellow, cyan, green, magenta, red, and blue, respectively. The chroma signals are combined with steps 2 through 7 of an 8-step Y signal, yielding step 1 at 100% luminance without color (white), and step 8 at 7.5% luminance without color (black). Figure 9-26*b* shows this color-bar reference and related chroma and Y values for this test signal. The staircase and *T-bar window* signals shown in Fig. 9-26*c* are transmitted on line 18 of the odd field, and line 519 of the 525-line picture. The T-bar series of sine/square pulses are used for checking the phase delay of signals in the transmitter; the staircase signals are used for checking the gray scale of the Y or video. Line 18 of the even field (line 520) is reserved for test signals originated at the option of the local TV station, and is sometimes used for clocking purposes. The VIRS signals are transmitted on line 19 of both fields (lines 521 and 522). They include a longer 3.58 MHz color burst, and 50% and 7.5% Y (gray) signals as shown in Fig. 9-26*d*. The VIRS signals are used in the transmitter to correct delay and reference signals with each field; the 50% and 7.5% VIRS levels are used in receivers to correct for luminance levels with each field.

The vertical interval test signals must be critically timed and gated in sequence with the sync signals in order to accurately reproduce color images on the face of the CRT. The timing, sequencing, and gating of these signals are achieved with pulse and digital circuitry that employs counters and logic elements. Many TV functions of this type are available in MSI or LSI IC form. The 3262A device of Fig. 9-25 can be connected for color TV synchronization, and its outputs combined with a test-signal generator in order to additionally time, gate, and sequence the test signals. The 3262 uses a 14.31818 MHz clock oscillator to obtain both the slightly reduced sync pulse

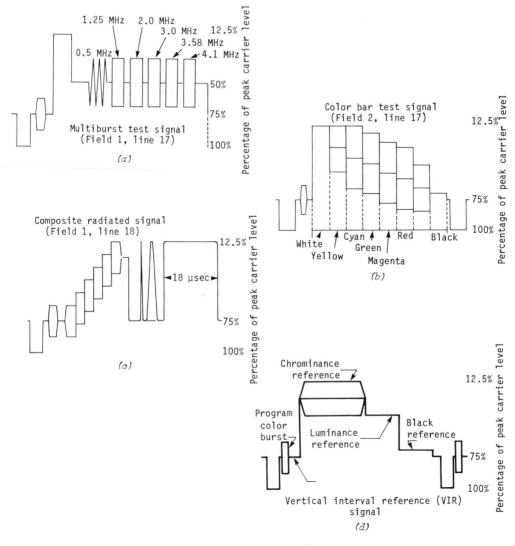

FIGURE 9-26

rates and the 3.579545 MHz color signal. Figure 9-27 shows this multifunction circuit related to the timing diagrams of Fig. 9-23*b*. Note that the counter/divider for the test signals is triggered from the even- and odd-field pulses, which also control and sequence the analog switches to select the appropriate test signals. If you are able to read timing diagrams, and can thus follow this sequence of functions, you should conclude that the *complexity* of sophisticated electronic systems depends on the *number* of separate functions that

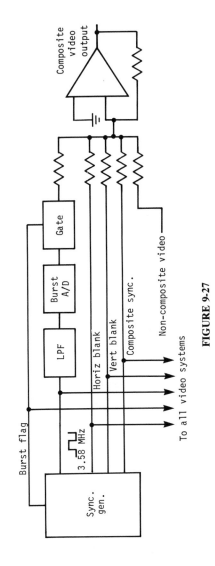

FIGURE 9-27

occur simultaneously. Since the separate functions are, by themselves, rather simple, however, even the most complex system can be simplified by this graphical timing-diagram technique. Incidentally, the idea of the timing diagram is not new—musicians have been using timing diagrams for centuries. A piano score, for example, is simply a timing diagram for two (right-hand and left-hand) stepped analog signals; and a full orchestral score simultaneously relates a greater number of analog signals of some instruments with the pulse signals of the percussion section. With practice you can become quite familiar with even the most complex of electronic systems by using the timing-diagram approach. Perhaps one day you may even wave a baton over your oscilloscope to make your circuits behave.

DIGITAL TELEVISION

The concept of a completely *digital* television system incorporates many of the principles of data, pulse coding, timing, sequencing, and synchronization presented thus far in this chapter. Because of its relative complexity but common familiarity, the TV system is an excellent example for your study of digital communications systems. On this basis your knowledge of a digital color TV system should allow you to readily understand the operation of many less familiar digital communications systems. The color TV system modulates and demodulates pulse and digital signals, directly processes pulses, provides pulse-to-analog, and analog-to-pulse conversion, and uses pulses to time and sequence signals in coded form. In fact, the color TV receiver is probably the most sophisticated mass-produced item that we encounter in our daily lives. In addition to its direct broadcast media application, the TV system finds popular use in security/surveillance, computer terminal, and TV game and other display functions. We will here use the familiar TV game display as a representative example of a dynamic digital communications system.

The TV Game

Figure 9-28 shows the basic block/functional diagrams of three color TV game IC's (57100, 53104, in 1889) and the complete TV game circuit. In operation the 53104 provides *true* and *complement* (upright and inverted) two-phase clock pulses at 1.022727 MHz, as obtained from the same 3.579545 MHz crystal standard used to provide the chroma modulation signal. The clock pulses drive the timing, programming, and sequencing circuits within the 57100 *game-chip* IC, and obtain horizontal and vertical sync by dividing the clock inputs by 65 and 525 ($\times 2$), respectively. The inputs to the 57100 are monostable (one-shot) multivibrators, for left and right channels, that use simple *RC* networks to achieve the synchronized time delays required to position the "paddles" on the CRT display. The output data from the programmer/sequencer addresses the ROM (read-only memory) section of the 57100, which in turn provides binary output data to the *Y* (luminance or video), *I* (in-phase chroma), and *Q* (quadrature chroma) encoder section. The *Y*, *I*, and *Q* decoders of the 57100 are basically 2-bit DAC's that convert the parallel-coded data to serial analog form. The analog *Y*, *I*, and *Q* signals at the output of the 57100 are then applied to the inputs of the 1889 video modulator, which combines the input signals and modulates them onto an RF carrier for transmission to the antenna (RF) input of a standard TV receiver.

The ROM of the 57100 is programmed for the static and dynamic data which determine the particular game and background that are displayed on the face of the CRT. The movement of the "ball," the scoring, and the score display numbers are dynamic, time-varying functions. The data from these dynamic sources address the ROM and result in the combination of motion, color, letters, and numbers that make up the TV game "picture." TV character generator IC's that provide a CRT display of letters and numbers from computer terminals operate basically the same as the 57100, with data from the computer addressing the ROM that delivers output data based on a particular time sequence required to display a given character. The score displays, as well as other moving or changing images, on the face of the CRT are likewise generated in the 57100 based on the programming of the ROM.

Figure 9-29 shows a typical output timing diagram and associated chroma-level combinations for the 57100. Referring to this figure you should observe that the *Y* signal (video) has only four distinct levels: sync, blanking, black, and white. These levels are binary coded at the DAC input as 00, 01, 10, and 11, respectively. You should also observe that the *I* and *Q* signals each have only three distinct levels: +, 0, and − (binary 00, 01, and 11, with 10 reserved for *color burst*); yet they can be combined to produce nine chroma-equivalent colors, plus burst, when the *Y* level is white (11). The magenta chroma signal at 45°, for example, is the polar-form equivalent of the rectangular-form +*I* and +*Q* values, which would, in turn, be coded at each DAC input as 01 and 01, respectively. The orange chroma signal would result from −*I* and +*Q* values that are binary coded as 11 and 01 at the respective DAC inputs. You may recall that the PCM signals described earlier in this chapter required an 8-bit DAC for accurately decoding voice signals to 127 distinct levels. Since the game chip requires only three 2-bit DAC's for both color and video, you should conclude that this is a rather simple technique of displaying images on the face of a CRT. The important factor to keep in mind, however, is that the game-chip system decodes binary data and converts it to time-sequenced analog signals—a functional process that is identical to PCM decoding.

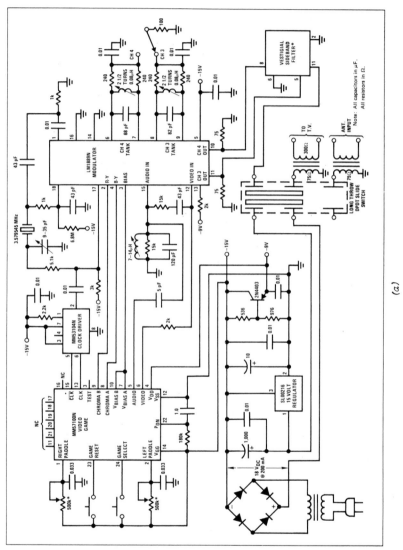

FIGURE 9-28

(a)

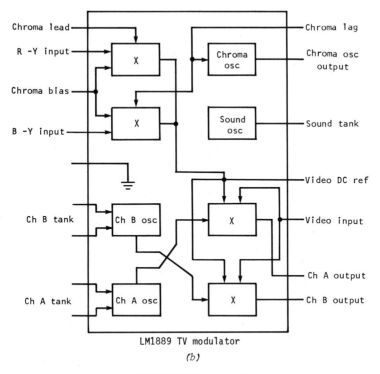

LM1889 TV modulator

(b)

FIGURE 9-28 (*continued*)

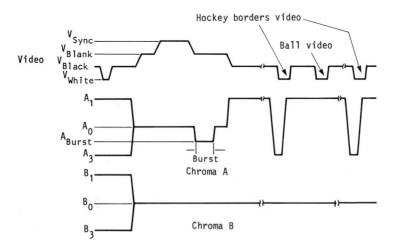

FIGURE 9-29

PCM Video

It is possible and practical, of course, to transmit high-resolution PCM video signals with a system that employs a greater number of bits per binary data sample. The number of bits will translate to the levels of video (Y) and chroma used to resolve the picture image. In addition to the number of *levels* (the 57100 has only four levels), the resolution of the TV image will depend on the *speed* of the data used to define the image. The NTSC system specifies this speed in terms of the video bandwidth for the analog Y signal. The resolution of a TV system, based on a 4.2 MHz Y-signal bandwidth and a 53.3 μs line-sweep period (63.5 μs less the 10.2 μs blanking period per horizontal line), can be computed as:

$$\begin{aligned}
\text{Maximum analog resolution} &= (\text{bandwidth}) \times (\text{period/line}) \\
&= (4 \times 10^6) \times (53.3 \times 10^{-6}) \\
&= 213 \text{ analog levels/line}
\end{aligned}$$

The *period* of *each* of the 213 levels is the inverse of the bandwidth, and it can be computed for this example as:

$$\begin{aligned}
\text{Maximum time per analog level} &= 1/\text{BW} = 1/(4 \times 10^6) \\
&= 0.25 \times 10^{-6} = 0.25 \text{ }\mu\text{s}
\end{aligned}$$

In order to obtain a sufficient range of gray-scale values, blanking, and sync-pulse levels, the Y signal should have a minimum of 32 distinct levels. Figure 9-30 shows representative TV pictures as obtained from 2-bit, 3-bit, 4-bit, and 5-bit binary data for 4, 8, 16, and 32 decoded analog levels, respectively. As you can see in the figure, the 32-level image appears to be continuous tone, because the decoded steps in the analog signal are closely spaced. This NTSC equivalent B & W image can be obtained with 5-bit binary data, sampled and converted at a 4.0 MHz rate. The resulting equivalent baud rate is computed as:

$$\begin{aligned}
\text{Baud} &= (\text{data bits per sample}) \times (\text{equivalent analog bandwidth}) \\
&= 5 \times 4 \times 10^6 = 20 \times 10^6 = 20\text{M baud}
\end{aligned}$$

From this calculation you should conclude that PCM television requires five times the bandwidth of the conventional television system. As noted earlier in this chapter, however, it is possible to reconstruct highly degraded and noisy PCM signals, and thus achieve reliable communications where other techniques would fail. If PCM television communications is required *with* restricted bandwidth limitations, it is always possible to extend the line-sweep period in order to maintain high-resolution images. The reduced bandwidth will yield better signal-to-noise ratios, but the tradeoff disadvantage is that a longer time period will be required to produce a picture image. You should be aware that the high-resolution color TV pictures received from space exploration, as well as your daily weather satellite pictures, use

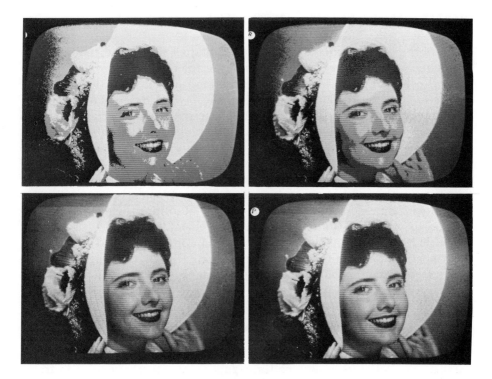

FIGURE 9-30

PCM techniques and thus require a longer period of time to transmit a complete picture. The Bell Picturephone uses a slightly extended line-sweep period, reduced resolution, and fewer horizontal lines to display its picture image, but the system operates at a data rate of only 6.3M baud with transmission over special telephone circuits.

Another advantage afforded by the digital television technique is computer or microprocessor origination of data that can be translated to images. If you compare the simplicity of the 2-bit data required for visual images on the CRT, with the complexity of a TV camera system coupled with lettered graphics, you will appreciate the practicality of TV studio equipment that digitally generates the titles that are keyed onto TV broadcast pictures. Graphic CRT computer terminals are likewise relatively simple, yet they can display extremely complex information in a visual format. The computer terminals that are used in the origination of drawings, PC-board layouts, and IC die patterns for electronics manufacturing are also merely an extension of the same functions used in the TV game chip described earlier.

DIGITAL COMMUNICATION MEASUREMENTS

Much of the measuring equipment and techniques used for digital and data communications are the same as those used for digital and computer electronics technology. Because of their absolute functions, digital circuits do not require the type of calibration, tuning, and "tweaking" that are typical of analog processing equipment. Most digital circuit functions are achieved with IC elements that have two-state (on-off or high-low) outputs, so much of the voltage, impedance, and signal-level measurements of analog functions serve little purpose when dealing with data. Possibly the most important factors in digital and data communications are the timing of pulses and their time relationship with other pulses. These measurements are usually performed with time-domain equipment such as multitrace, delayed-sweep oscilloscopes. Recent additions to the measurement/test equipment technology include a special family of equipment, appropriately called *data-domain* equipment. Data-domain oscilloscopes and LED (light-emitting diode) display testers provide a most efficient approach to measuring binary data and signals.

Test Sets

The basic measurements for data communications include detecting errors in parallel- and serial-form data codes, signal degradation and noise in transmission circuits, cable fault location, and modem performance testing. There is a family of specialized *telecommunications test sets* that are used by this telephone-oriented technology. These units are, for the most part, merely basic test equipment packaged for convenience and adapted for RS-232 or CCITT compatibility. Although specialized testers are available for the measurement of FSK signals and modem performance, these tests can be easily accomplished with the power supply, sine-pulse generator, filter, frequency counter, and oscilloscope. In telecommunications practice the frequency counter and oscilloscope may be replaced with a frequency-selective voltmeter. Modem performance is measured by varying its input first with the power supply and then with the pulse generator at standard levels. The resulting output FSK *tones* are monitored by the frequency counter and displayed on the oscilloscope. High-speed performance can be tested by switching the filter between mark and space frequencies, while observing the alternate signal displayed on the scope. Noise measurements can be made by removing the FSK tones with filters and measuring the remaining noise levels.

Signal degradation and noise can be measured in transmission circuits (lines, etc.) by injecting FSK-equivalent audio *tones* at one end of the transmission circuit, then measuring and evaluating the resulting signals at the other end. The filters can be used to remove the tone signals for noise-level measurements, as noted in the previous paragraph. The critical aspects of such tone measurements include impedance matching of the signal source to the transmission circuits and establishing standardized levels for the tone signal. Most specialized telecommunications test equipment has switchable 135 Ω, 600 Ω, and 900 Ω balanced output impedances, with signal output levels specified in dBm in conjunction with a calibrated attenuator. If conventional test equipment is used for such measurements, provision must be made for standard impedances, signal levels, and adaptation to balanced lines. A calibrated attenuator and matching network, along with a quality audio transformer, can be used to provide these capabilities.

One of the more serious and continuing problem areas with wire and cable transmission systems has to do with cable failure between communications terminals. Although the problems are often due to mechanical breakage or chemical erosion, their locations for repair fall within the realm of electronic system testing. One popular specialized instrument used for cable testing is the *time-domain reflectometer* (TDR). This unit operates somewhat like a *cable radar,* as it transmits a short-duration pulse into a cable and then times and displays the resultant echo signal. An echo signal that is out of phase with the pulse will indicate a short; an in-phase echo will indicate an open in the cable. The time of signal transmission down the cable and back determines the distance from the TDR to the fault. Modern high-resolution TDR's can measure faults at distances to 10 miles with a resolution of 3 feet; and they can resolve faults within an inch at distances to 100 feet.

Data Domain Testing

Since digital systems process information in pulse-time *combinations,* it is important to measure these combinations in order to determine system operation. The *logic-state analyzers* are a family of data-domain instruments used to evaluate the performance of digital signals in combination. Logic-state analyzers are used in communications applications to determine coding errors and the timing and sequencing of serial and parallel data. As you may recall from the early sections of this chapter, data communications originates and terminates as serial data words in coded form. The testing and error detection techniques used with teleprinters and similar apparatus employ a patterned serial input code which is converted to parallel data and drives the input of a *pattern analyzer.* The pattern analyzer is a data-domain test instrument that is programmed to recognize the input code by comparison and to indicate a particular code bit which may be in error.

The testing of complex *parallel* data, such as timing, sequencing, and computing pulses, is most conveniently achieved with *logic analyzer oscilloscopes*. These instruments display multiple-trace timing diagrams, actual binary (and other coded) numbers, and data maps. The logic analyzers typically display 16 channels of timing data in order to analyze digital processing systems. The timing diagrams for the TV game and TV sync generator IC's of Fig. 9-29 were obtained from a logic analyzer display. There are often advantages in displaying parallel digital data as numbers rather than time-base occurrences. Modern logic analyzers have the option of displaying data in *bit* and *word* formats in order to correlate processing functions. Most units operate in binary, octal, and hexadecimal states in order to measure a given computer format. The *mapping* format displays a pattern of data bits which can be quickly recognized or compared to a given code. Because of its reduced space requirement on the face of the CRT, the mapping format can display up to 64K bits of 16 input channels.

Pattern analyzers and logic analyzer oscilloscopes are generally used for digital systems testing. Testing at the *circuit* level is most conveniently achieved with logic probes and logic clips. Although it is possible to measure the state (high or low) of each terminal of a digital IC with a voltmeter, such measurement is inefficient because the precise *level* of a high or low state is not important. Most digital circuit measurements are more conveniently made by sequencing the IC's through their operation while noting the LED indications on a simultaneous parallel-display logic clip. *Logic-clip testers* are also available with programmed drivers to *input* logic levels to digital IC's and compare their outputs to preset values. The *LED-display logic analyzer* combines the functions of many logic-clip testers in order to display multiple-channel data, much like the logic-analyzer oscilloscope binary data display. While the LED-display instruments lack the versatility of the oscil-loscope format, they do provide an inexpensive alternative for the conve-nient and reliable testing of digital circuits and subsystems.

PCM Testing

The testing and evaluation of PCM systems combines the *data* communica-tions and *digital* measurement techniques. Important specialized PCM mea-surements include error detection in data codes, data errors due to signal degradation and noise, and codec performance testing. The family of special-ized *telecommunications test sets,* as mentioned for data communications, includes PCM and TDM test instruments. Although *basic* data and PCM measurements can be achieved with conventional test instruments, effective error detection and isolation in high-speed PCM/TDM systems *does* require specialized instrumentation. The error measurements are typically per-formed with the digital portion of the PCM system, dividing the system into two sections: *codec* and *serial transmission* circuits. The errors due to signal degradation and noise are thus measured in serial form and require the high-speed data transmitter and receiver; whereas the codec functions can be tested with conventional digital and analog instrumentation. Figure 9-31 shows a representative *PCM/TDM error-detecting test set*. The unit includes a high-speed data-pattern sequence generator and receiver, and operates as a

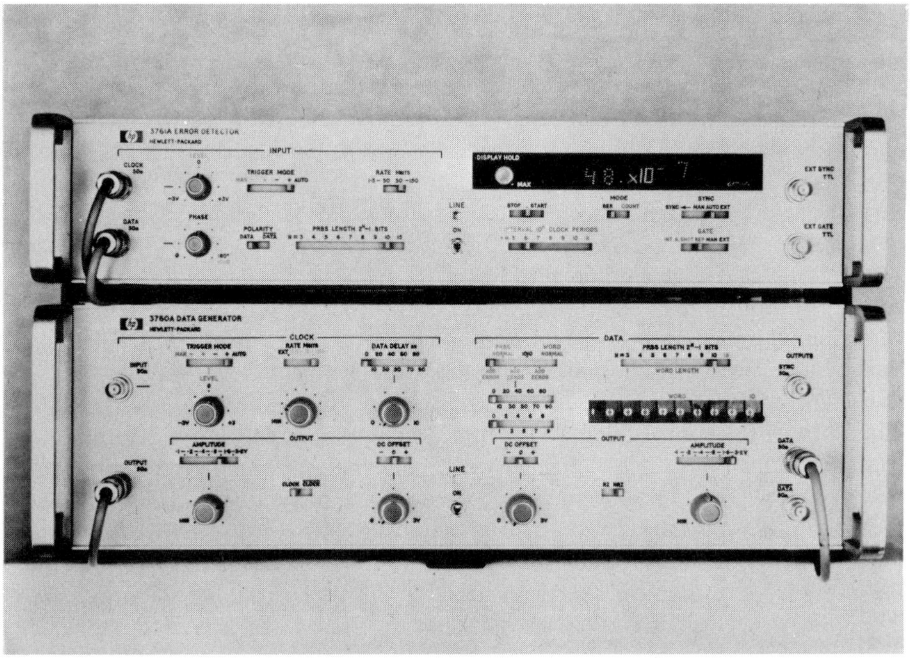

FIGURE 9-31

closed-loop system. In operation the data output from the PCM system under test is received and compared with the data input applied to the system, with errors counted and specified as the *bit error rate* (BER).

The testing of the codec transmitter circuitry can be accomplished with a programmable power supply, audio signal generator, and time-domain/data-domain oscilloscope. In practice a sequence of voltages and audio signals is applied to each input channel of the codec/TDM encoder, with the resulting binary output data observed on the data-domain oscilloscope. The decoding function of the codec receiver may be tested using a patterned PCM serial data input from the PCM/TDM error-detecting test set, or other PCM/TDM programmed data source, and the time-domain oscilloscope. With the serial data programmed for a given sequence of PCM signals, the equivalent analog signals will appear at the codec outputs. Of course, the troubleshooting of the subsystem functions can be accomplished with the same test equipment setups for the isolation of failures or malfunctions.

CONCLUSION

The content of this chapter is concerned with the important aspects of data and digital communications, and with the pulse forms of signals used in these applications. Although much of the equipment used for this technology is not as familiar to you as are common radio and TV receivers, there is no doubt that you will encounter a predominance of such digital equipment if your interests are related to modern electronics. The author contends that people with a strong background in digital technology can compete successfully for employment positions in the communications industry, because their knowledge of modern techniques will more then compensate for any lack of competitive experience with the more traditional aspects of the field. You may have observed that the content of this chapter lacked the calculations for thorough analysis that are typical of the preceding chapters of the text. In fact, little in the way of mathematical analysis is required of this aspect of the technology. The concepts are basically simple, with complexity in the communication of information achieved through coding—not modulation. The concepts of data and digital communications are thus easily understood once you become familiar with their *function* and *application*. The following problems are thus structured to test your knowledge of the *functional* structure of data and digital communications systems.

Problems

1. Given: A standard Baudot-coded teleprinter operating at 75 W/M.

    Identify and determine:
    (*a*)   The baud rate
    (*b*)   The time period of the mark/space, start, and stop bits
    (*c*)   The time period of each 7-level character
    (*d*)   The time period of each 5-letter word
    (*e*)   The required bandwidth for a 30 dB noise figure

    Draw: A time vs amplitude plot for the letters "BS."

2. Given: A 9600 baud, 8-level (7 + parity) synchronous ASCII-coded data signal.

    Identify and determine:
    (*a*)   The time period of the character bits
    (*b*)   The time period of each 8-level character
    (*c*)   The time period of each 5-letter word
    (*d*)   The equivalent code speed in words per minute
    (*e*)   The instruction transmitted with a 111001 code
    (*f*)   The required bandwidth for a 30 dB noise figure

3. Given: An FSK data modem, operating at 1200 baud and a 1200/2200 Hz frequency shift, over a voice-grade telephone circuit.

Determine:
(a) The equivalent FM deviation of the FSK signal (*Hint:* use the center frequency)
(b) The equivalent highest modulation frequency based on the baud rate
(c) The equivalent FM modulation index
(d) The 20 dB and 40 dB bandwidths of the transmission signal

4. Given: An XR2206/2211 modem transmit/receive pair of IC devices (see Figure 9-9) to operate with the characteristics of Problem 3 above.

Determine or specify (in reference to manufacturer's literature):
(a) The values of R_6, R_7, and C_3 required in the XR2206 circuitry to obtain the mark and space frequencies
(b) The values of R_5, C_1, C_F, and C_D required in the XR2211 circuitry to efficiently decode the mark and space frequencies

5. Given: A simplex transmission requirement from an 8-bit parallel data output of a computer to 1488-type IC line drivers, through a cable, to 1489-type line receivers, to an AY-5-1013 UART, and into a data modem for 1200 baud operation.

Identify or specify (in reference to manufacturer's literature):
(a) The typical parallel data levels from the computer into the line driver
(b) The RS-232 levels for mark and space signals
(c) The number and levels of the signals at the input to the UART
(d) The level of the UART output and its required clock-signal rate

6. Given: An 8-bit codec using a successive approximation ADC with DAC.

Draw:
(a) A detailed block diagram of the codec transmit circuitry
(b) A detailed block diagram of the codec receive circuitry
(c) The analog and digital waveforms at each stage of the above block diagrams

7. Given: A time-division multiplex system, using 4-bit ADC's and an 8K baud sampling rate for each of 12 PCM channels.

Determine:
(a) The equivalent baud rate per frame

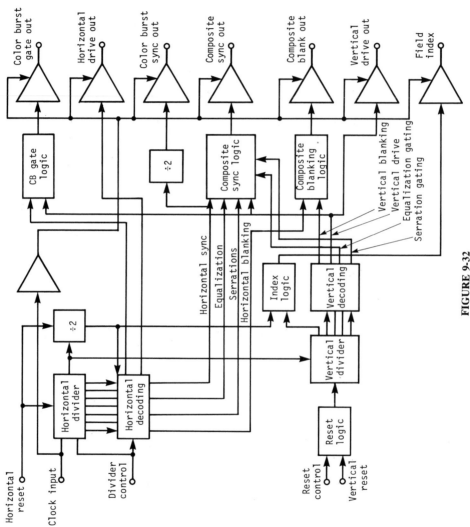

FIGURE 9-32

(*b*)   The bandwidth required for transmission
(*c*)   The equivalent bandwith that would be occupied by the 12 analog
        voice channels before PCM conversion
(*d*)   The location and function of the synchronization framing bit for the
        codec transmit and receive functions with other codecs

8.   Given: An MM5320 TV camera synchronous generator IC, providing
U.S. standard interlaced synchronization signals, with the block diagram of
Figure 9-32.

Determine and graph with a timing diagram:
(*a*)   The horizontal drive output pulses
(*b*)   The vertical drive output pulses
(*c*)   The composite synchronous output, showing equalization and ser-
        rated pulses for both even and odd fields
(*d*)   The color burst gate output pulse in reference to the back porch of
        the horizontal drive pulse

9.   Given: An MM5322 TV color-bar generator IC, providing dot-bar and
color-hue generation signals for the U.S. standard color television system.

Determine and graph with a timing diagram:
(*a*)   The video signal (including the horizontal synchronous pulse) that
        will provide 11 equally spaced vertical lines from top to bottom of
        the screen display
(*b*)   The video signal (including the vertical interval) that will provide
        seven equally spaced horizontal lines across the screen display
(*c*)   The video signal that combines *a* and *b* above for a cross-hatch
        pattern (7 × 11 lines) on the screen
(*d*)   Chroma output for a gated rainbow of white, eight colors at 45°
        increments, and black over the time period of the 11 vertical lines of
        a above

10.   Given: A PCM limited-resolution B & W, 401-line television communications system, using 3-bit binary video data for synchronous tips, blanking,
and a six-value gray scale, and sampled at a 1 MHz rate.

Determine (in reference to PCM voice communications applications,
considering each line as a channel):
(*a*)   The time period of each frame
(*b*)   The number of data bits per frame
(*c*)   The baud rate for the PCM data transmission
(*d*)   The horizontal resolution based on a sweep rate of 12 kHz
        (interlaced)

CHAPTER 10

ELECTRONIC COMMUNICATION:
APPLICATIONS AND SYSTEMS

Not too long ago our most complicated electronic system was a superheterodyne radio. In those days there was little distinction between *radio* and *electronics*—the whole technology was basically divided into *electrical power* and *radio*. This was a time when people who specialized in electronics worked with sound systems or radio systems, which was about all that *electronics* involved. The concept of a "system" was then, and is still, considered as a *combination of parts that provide the functions necessary to completely achieve some process.* In application, a "sound system" required a microphone, preamplifier/equalizer, power amplifier, and loudspeaker, including the interconnecting cables and power supplies required to make the whole "setup" function. The "radio system" required a microphone, AF amplifier, modulator, RF transmitter, transmit and receive antennas, RF receiver/demodulator, AF amplifier, and a loudspeaker, etc.

Based on your previous study of Chapters 6 and 8, you should appreciate that a thorough knowledge of even the rather *basic* AM radio system requires a substantial background in electronic devices, circuits, transmission lines, and antennas. If you now scale the required knowledge to present times, where electronic systems of such variety exist, you should conclude that a lifetime of study will probably be insufficient time to acquire a thorough knowledge of all aspects of the technology. In order to effectively accommodate the wide-ranging diversity in electronics, including electronic communications, people are obliged to specialize in one aspect or another within the field. One such broad specialty area is that of *electronic communication: applications and systems*—the topic of this chapter.

Electronic systems specialists are more often concerned with signals and processing functions than with electronic devices and circuits. On paper they often work with block diagrams rather then schematics; and the resulting "hardware" they most often work with is in the form of assembled racks and modules, rather then active and passive circuit components. The content of Chapter 9 employed many systems concepts in describing the

454

applications of digital and data communications. These descriptions were first necessary in order to define the requirements for the electronic circuitry; but they were greatly influenced by the *limitations* of available electronic circuitry. The systems specialist, as this example indicates, is a sort of "chicken and egg" person—sometimes committed to the *priority* of the system as demanding circuit capabilities that must be developed, and at other times deploying existing or newer circuit capabilities to provide systems that did not previously exist. With all due credit to the other side, the author is somewhat prejudiced in favor of a systems priority: the *need* for an electronic circuit function must *precede* the development and use of the circuit. If the reverse case were argued, it could be likened to the medical researcher whose goal is to develop a cure for no known disease.

The intent of these introductory paragraphs is to place the topics of electronic systems into perspective by identifying some of the major distinctions of the specialty field. We shall now continue to describe the more detailed aspects of communications systems. The technology can actually be categorized from two dimensions: the signal processing *functions,* and the *applications.* You are, of course, familiar with some forms of *modulation* and *demodulation* as examples of signal processing functions; and you may consider *AM* or *FM broadcast transmitters* as examples of systems applications. You may also conclude that systems applications use a variety of systems functions in order to process electronic signals. The rather complex but common color TV receiver, for example, processes modulated RF signals into audio, video, sync, and chroma signals, which in turn reproduce synchronized visual and aural information for our consumption. It achieves this systems function by demodulating AM, FM, pulse, and quadrature signals from the modulated RF carrier. The example of the color TV is here used to further define how the content of this chapter will proceed to categorize *systems.* The color TV receiver is, of course, just one part of a much larger system of communications that may even involve an orbiting satellite. Such *applications* will also be described in the content of this chapter.

LEARNING OBJECTIVES

Upon completing this chapter, including the problem set, you should be able to:

1. Identify the worldwide frequency allocation chart and the radio services specified thereon.
2. Recognize the popular modulation forms, and understand how the bandwidth requirements vary with *information channel* capacity for each form.

3. Understand the techniques for increasing channel information capacity through the use of subcarriers, sideband signals, frequency-division multiplex, and time-division multiplex.
4. Understand the techniques used for two-way and multiple-path electronic communications.
5. Understand wire and cable communications services and their relationship to radio services.
6. Understand the electronic communications systems used in AM, FM, and FM stereo broadcast radio services.
7. Understand the communications systems used for broadcast and other television services.
8. Understand the systems used in two-way fixed, vehicular (mobile), marine, aircraft, and paging/homing radio services.
9. Understand the techniques and operation of radar, radionavigation, and command and control systems are used in commercial, scientific, and military applications.
10. Understand the systems used for space and satellite communications.
11. Understand the diverse techniques used in carrier and radio telecommunications services, including telephone, telegraph, data, and facsimile systems.
12. Apply the basic techniques used in telemetry systems.

RADIO COMMUNICATIONS SERVICES

The radio frequency spectrum is divided and classified to enable international frequency allocation without danger of interference from one country or area to another. The worldwide regulating body for these frequency assignments is the International Telecommunications Union (ITU) of 124 member countries. The related regulating body for the U.S. is tte Federal Communications Commission (FCC). These official agencies have assigned radio communications services in terms of applications as follows:

1. *Fixed:* Radio communication between specified fixed points. Examples are point-to-point high-frequency circuits and microwave links.
2. *Mobile:* Radio communication between stations intended to be used while in motion or during halts at unspecified points, or between such stations and fixed stations.
3. *Aeronautical mobile:* Radio communication between a land station and an aircraft, or between aircraft.
4. *Maritime mobile:* Radio communication between a coast station and a ship, or between ships.
5. *Land mobile:* Radio communication between a base station and a land mobile station, or between land mobile stations (vehicular radio).

6. *Radionavigation:* The determination of position for purposes of navigation by means of the propagation properties of radio waves, including obstruction warning.
7. *Aeronautical radionavigation:* Navigation service for the benefit of aircraft, including VOR (VHF omnirange), tacan (tactical air navigation), radio beacons, ILS (instrument landing system) altimeters, and radars.
8. *Maritime radionavigation:* Navigation service for the benefit of ships, including radio beacons, direction-finding stations, and radars.
9. *Radiolocation:* The determination of position for purposes other than navigation by means of the propagation properties of radio waves, including land and coastal radars and tracking stations.
10. *Broadcasting:* Radio communication intended for direct reception by the general public, including AM and FM radio and television.
11. *Amateur:* Radio communication by persons interested in radio technique without pecuniary interests.
12. *Space:* Radio communication between space stations.
13. *Earth-space:* Radio communication between earth stations and space stations and satellites.
14. *Radioastronomy:* Astronomy based on the reception of radio signals of cosmic origin.
15. *Standard frequency:* Radio transmission of specified frequencies of high precision intended for scientific, technical, and other purposes.

A wide variety of applications fall within the broad categories of the frequency allocations outlined here. These classifications will serve as a basis for identifying the various electronic communications systems that are used for these services.

You should be quite familiar with the more prominent techniques of modulating information signals on RF carriers through your study of the preceding chapters of this book. As you may recall, the text reviewed the ways in which signals could be modified in order to carry information. If you now apply these modulation concepts to the services defined in the frequency allocation chart, you should conclude that the transmission of information is the predominant function of these services. The electronic communications systems used to provide these services thus include two major functions: *modulation* and *RF transmission*. The modulation function includes the modification of signals in amplitude, frequency, phase, and time in the familiar AM, FM, PM, and pulse modulation forms, respectively. These same distinctions are also used to separate signals, with the more prominent techniques applied to frequency and time. The selection of different, simultaneously transmitting radio signals by a receiver, for example, is a means of separating signals by frequency. The processing of digital signals in the PCM codec, as described in the previous chapter, is an example of time-based combination and separation of pulses.

Although the more simple electronic communications systems employ only one or two of the techniques for signal modification/separation, sophisticated systems may employ *all* of the techniques in a variety of combinations. The popular color TV system, parts of which were also described in the previous chapter, is an excellent example of a sophisticated system that employs a variety of these modification/separation techniques in order to process visual and aural information. Referring again to the frequency allocation chart, you should observe that considerable portions of the VHF and UHF regions of the spectrum are occupied by TV broadcast channels. Each channel of the U.S. NTSC standard for television broadcast, in fact, occupies 6 MHz of radio space, enough space in the spectrum for 600 AM broadcast radio channels. The comparison is here made to relate channel bandwidth to the number of signal modification/separation techniques employed. Simple systems occupy relatively little radio space for communication of limited information; whereas the sophisticated systems require greater bandwidth for the communication of extensive information. Since broader bandwidths are more practical to achieve at higher frequencies, the sophisticated systems are typically deployed in the VHF, UHF, and SHF regions of the spectrum. Figure 10-1 compares the receiver (antenna through detector) sections of the AM broadcast and TV receivers, and shows the initial similarity of the systems.

INFORMATION CHANNELS AND BANDWIDTH

The simplest, most primitive form of wire and radio communications— interrupted CW code—requires the least bandwidth of all modulation forms. Since the code is composed of variable pulse-duration periods, a certain range of frequencies, or *bandwidth,* is required to allow at least one full cycle of an ac signal to develop. The modulation and demodulation circuitry in RF systems must also detect this shortest pulse, so the bandwidth required for even this basic modulation form is significant. If a radio telegrapher is transmitting coded signals at rates of 15, 30, and 45 W/M (five-letter words per minute), for example, the shortest "dit" for the Morse-coded letter "E" will probably be about 67, 33, and 22 ms, respectively, based on the equivalent of 60 dits per five-letter word. Since the frequency of a signal can be computed as the inverse of its time period per cycle, so too can the bandwidth be computed as that minimum pulse duration:

$$f = 1/t \quad \text{and} \quad BW = 1/t$$

f is the frequency of a given signal in Hz, BW is the bandwidth of a circuit required to support a signal at a given frequency, and t is the time of the

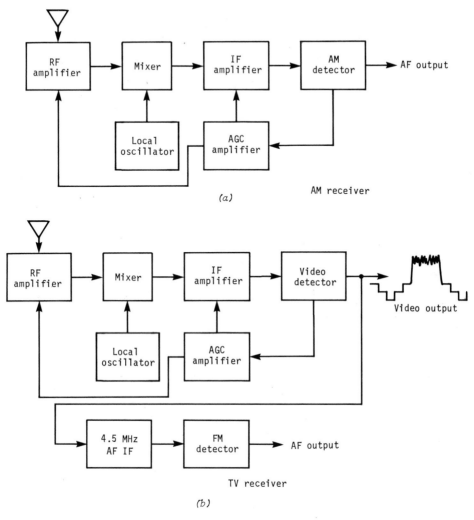

(a)

AM receiver

(b)

TV receiver

FIGURE 10-1

shortest pulse duration in seconds. For the Morse code examples above, the bandwidth requirements can be computed for 15, 30, and 45 W/M, as:

$$BW = 1/t = \begin{cases} 1/0.067 = 15 \text{ Hz} \\ 1/0.033 = 30 \text{ Hz} \\ 1/0.022 = 45 \text{ Hz} \end{cases}$$

There is a direct relationship between the amount of information, words per minute, and the bandwidth required to transmit this information. This bandwidth, which is based on the amount of information transmitted, is specified as an information *channel*. To assure some margin of reliability, practical bandwidth assignments generally allow at least a 50% variation in the pulse duration, making the channel bandwidths for the above examples about 25, 50, and 70 Hz, respectively. Figure 10-2a shows the typical Morse-coded pulses for a five-letter word.

The transmission of Morse-coded signals at a top manual speed of even 50 W/M is slow compared to the human voice which can easily communicate at least 250 W/M via wire or radio transmission. This factor, of course, contributed to the rapid obsolescence of the telegraph caused by the telephone. Transmission at 250 W/M can be obtained by a code machine requiring a channel bandwidth of only about 400 Hz. The superiority of voice signals over coded signals, however, includes additional dimensions of information not available with codes. In addition to the ability to transmit 250 W/M, the voice channel can also transmit *pitch* (voice frequency), *loudness* (voice amplitude), and *timbre* (voice harmonic structure), which add character and emotion to the words. Moreover, the voice channel can, to a limited extent, transmit music. These complications require an expanded channel bandwidth as compared to the scaled-up code channel.

Figure 10-2b graphs the frequency response of voice signals in terms of the power distribution of speech. You may observe that a 20 dB (100:1) variation in power falls within a 3 kHz bandwidth, placing 99% of the voice

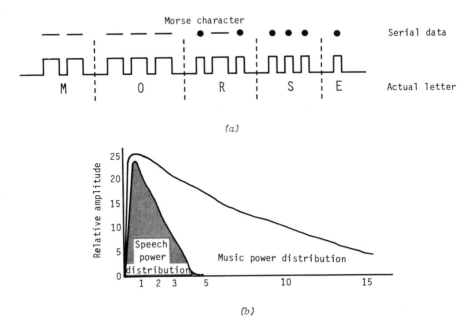

(a)

(b)

FIGURE 10-2

power (information) within this range of frequencies. The standard *voice channel* for communication of speech is thus 3 kHz. The bandwidth requirements for the accurate reproduction of music, being considerably more complicated than speech, extend to at least 15 kHz as shown in the response curves of Fig. 10-2*b*. Based on our example of the NTSC television system, which can reproduce over 200 variations in intensity as the electron beam sweeps across the face of the CRT, the bandwidth requirement extends beyond 4 MHz. This bandwidth was computed in Chapter 9 based on a 53.3 μs line-sweep period and 213 analog levels per line. The total channel bandwidth of the NTSC TV system is actually 6 MHz, because synchronization, color, and audio information are transmitted along with the video (Y) signal.

The generalities of information channel bandwidth requirements reviewed thus far do not consider the additional bandwidth requirements of particular modulation techniques. As you may recall from Chapter 6, amplitude modulation (AM) requires a radio channel bandwidth of twice that of the modulating signals. A voice channel thus amplitude modulating an RF carrier would require a channel bandwidth of 6 kHz instead of 3 kHz. This 2× factor was, of course, eliminated with the application of single sideband (SSB) techniques to modulated signals, again reducing the RF voice-channel requirements to 3 kHz, but at the expense of additional processing circuitry. In the case of frequency modulation, as reviewed in Chapter 7, FM bandwidths are considerably greater than those of other forms. Considering a typical full-deviation modulation index of 5, for example, the comparable 3 kHz voice-channel requirements could extend to nearly 36 kHz for 20 dB (99%) power transmission (see Bessel functions, Chapter 7). There are substantial tradeoff advantages in noise reduction in FM, of course, that make even its greater bandwidth requirements acceptable.

The channel bandwidth requirements for the various forms of data and digital transmission, including FSK and PCM, were thoroughly reviewed in the previous chapter. As you may recall, the bandwidth requirements for direct pulse or data transmission are based on Shannon's relationship for baud rate and noise, which provided 29.7K baud data transmission on a 3 kHz voice channel, with 30 dB S/N. You should also recall, however, that most data transmission is presently converted to FSK by the modem, and is thus actually a form of FM with resulting increased bandwidth requirements. With the example of an FSK teleprinter modem that shifts from 1070 to 1270 Hz for a space and mark, respectively, the deviation is 200 Hz, with a maximum modulating frequency of 300 baud. The related FM (FSK) modulation index is 0.67, and the 20 dB (99%) power transmission (see Bessel functions again, Chapter 7) extends to about 1200 Hz per FSK information channel. These channel capacity factors are, of course, what dictate the maximum baud rate and FSK frequencies for particular modem standards. Table 10-1 summarizes the information channel bandwidth requirements for popular communications signals in both direct cable and modulated forms.

TABLE 10-1. INFORMATION CHANNEL
BANDWIDTH REQUIREMENTS (SUMMARY)

Communications Signal	Bandwidth	
Morse code:		
25 W/M	100	Hz
4-channel TDM, 42.5 baud/channel	850	Hz
Modulated, 1 kHz and 25 W/M	2100	Hz
Amplitude modulation (AM):		
Telephony, 3 kHz AF:		
Double sideband	6	kHz
Single sideband	3	kHz
Broadcast radio	10	kHz
Facsimile	5.45	kHz
U.S. broadcast television	6	MHz
Television relay	13.1	MHz
FSK: 4-channel TDM, 42.5 baud/channel	613	Hz
Frequency modulation (FM):		
Telephony, 3 kHz AF	36	kHz
Broadcast radio	180	kHz
Facsimile	25.45	kHz
Microwave relay:		
60 channels	3.7	MHz
960 channels	16.3	MHz
Television relay	17	MHz
Stereo broadcast radio	300	kHz
Wire:		
Voice-channel telephone	3–14	kHz
Data transmission	Baud rate	

INCREASING CHANNEL CAPACITY

If you compare the bandwidth requirements of particular information channels and their modulated forms with the services identified on the frequency
allocation chart, two factors should become immediately obvious. First, the
services that transmit greater information or complex modulation (e.g., FM)
require a greater bandwidth and are thus more practical for higher frequency
operation. Compare AM and FM broadcast radio again, for example; where
a 10 kHz channel is satisfactory in the 540 to 1600 kHz range for AM
broadcast, the 200 kHz (20 dB) FM broadcast channel is more appropriate
for operation in the 88 to 108 MHz range. Second, the exponentially increasing needs for the electronic transmission of information are continually con-
suming more radio space, leaving few available channels for communications. Even the microwave channels are crowded in most urban areas of the
U.S. One solution to the radio space problem is the sharing of unused frequencies by other services. Vehicular communications, for example, share
portions of the UHF TV band. Another solution involves the deployment of
techniques to increase channel information capacity, such as the addition of
stereo and SCA (subsidiary communications authorization, or "storecast")
signals to the existing FM broadcast channels.

Multiplexing

The two most popular techniques for increasing channel capacity are *frequency-division multiplex* (FDM) and *time-division multiplex* (TDM). As the terms imply, the techniques provide for the combination of many signals to be transmitted within the bandwidth of a given channel. The FDM technique involves the cascading of a number of SSB narrowband channels for transmission within a single wideband radio channel. Figure 10-3*a* diagrams a typical carrier communications system that uses the FDM technique to combine 1800 voice channels for ultimate transmission on a single FM microwave channel. In this process the voice channels are combined on a single carrier by mixing with subcarriers spaced at increments of 4 kHz, from 60 kHz to 108 kHz, to form a basic 12-channel *group,* as shown in Fig. 10-3*b*. With the use of balanced mixers and filters, the carriers and lower sideband frequencies of each of the voice channels are eliminated, leaving only the upper sideband signal within each 4 kHz increment of the group, as shown in Fig. 10-3*c*. The groups can, of course, be combined for yet increased information capacity. In the jargon of the carrier communications industry, these higher capacity combinations are called *supergroup, mastergroup,* and *jumbogroup,* for 60, 600, and 3600 voice channels, respectively. Although this example describes the general structure of a carrier communications application, frequency-division multiplex techniques can be similarly applied to a wide variety of signals and systems.

The technique of time-division multiplex was reviewed in detail in describing the operation of PCM codecs in Chapter 9. You may recall that the advantages of PCM included sequential, synchronized sampling of a number of signals within a specified time frame. Although the sampled signals of the PCM system are in digital form, the sequential sampling process can also be applied to analog signals with equal success. The line structure and video modulation of the television system are a form of TDM. Each of the 525 lines that make up the TV picture is sequenced in the camera to sample the analog levels of the image on the face of the camera tube. These same 525 lines are also sequenced in the TV receiver, producing lines that display demodulated, demultiplexed synchronized analog signals on the face of the picture tube. As with FDM, time-division multiplex techniques can be used in a variety of systems applications in order to increase the information capacity of a given channel.

Multiple Modulation Forms

Channel information capacity can also be increased by combining modulation forms. Again, the popular TV receiver serves as a familiar example of this technique. Synchronization of the TV camera and receiver is achieved by sync pulses that define the line sweep and vertical field of the picture. For the NTSC system the horizontal pulses of 0.5 μs are recurring at a rate of

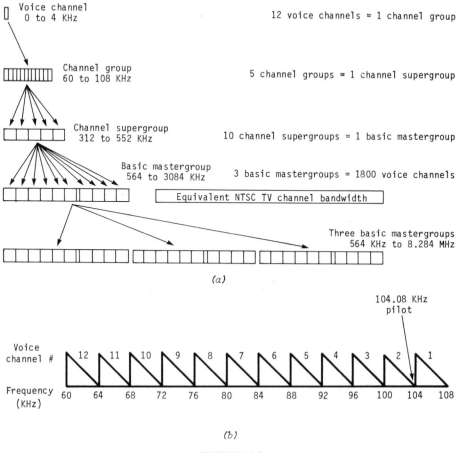

(a)

(b)

FIGURE 10-3

15,750 Hz, and the vertical pulses of 187.5 μs are recurring at a rate of 60 Hz, as detailed in Chapter 9. The video (Y or luminance signal for color transmission) is amplitude modulated with the lower sideband filtered to provide VSB (vestigial sideband) modulation. The color (chrominance) signal is quadrature modulated, defined in Chapter 9 as the polar-form equivalent of the I and Q signals. Finally, the audio signal is a frequency-modulated subcarrier operating at a center frequency 4.5 MHz above that of the RF carrier. Isolation and separation of these four distinct signals that modulate the RF carrier are assured by careful selection of nonharmonically related frequencies and pulse rates for the system. The sync pulses occur during the *blanking* of the video signal at the sides and top and bottom of the picture, and the harmonics of the video signal are interleaved with those of the 3.579545 MHz color signal. The 4.5 MHz audio subcarrier is filtered from the video circuit; while the AM rejection of the FM detector removes any remaining amplitude

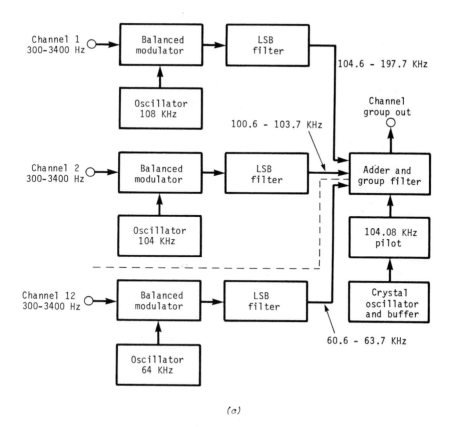

(c)

FIGURE 10-3 (*continued*)

variations from its output. These combined modulation techniques used in the NTSC television system are representative of the much wider range and variety of combined modulation forms that can be deployed in modern electronic communications systems.

Multipath Communications

The content of this chapter has thus far reviewed the techniques for increasing the channel capacity of one-way wire and radio transmission systems. In the sense of *real information* the capacity of two-way communications increases substantially with the benefit of feedback response, just as a two-way telephone conversation communicates more than a one-way telegram. In direct-wired telephone communications, phased transformers are used to attenuate the intense transmitter signal to the receiver in the handset, while allowing the received signal to remain at almost full intensity. This transformer is called a *duplex coil* or *hybrid,* because it provides for simultaneous

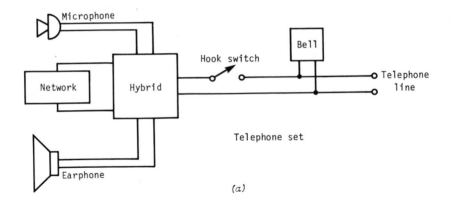

Telephone set

(a)

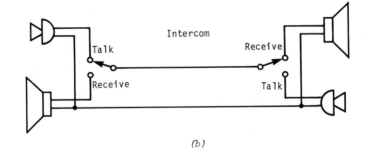

(b)

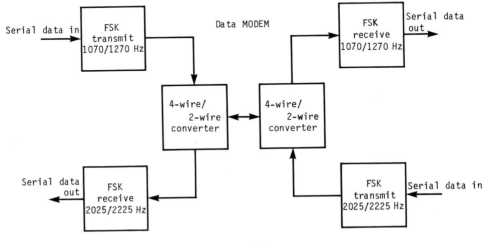

(c)

FIGURE 10-4

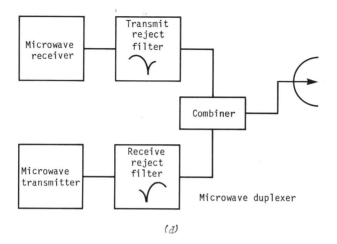

(d)

FIGURE 10-4 (*continued*)

two-way (duplex) transmission of signals on a single pair of wires. Figure 10-4a shows the structure of a conventional telephone set including the duplex coil. Without this coil, telephone sets would operate in half-duplex mode, allowing only one person to speak at a time. Half-duplex operation is most common in so-called *two-way* radio communications systems such as CB (Citizens Band), vehicular, and business transceivers. A more accurate designation for these two-way systems might be "one-way-at-a-time radio," as they operate like the familiar push-to-talk intercom of Fig. 10-4b.

True duplex (two-way) operation of radio systems that allow simultaneous transmission and reception is somewhat more complex than the telephone receiver. As noted in the previous paragraph, however, there are a number of systems where such feedback is absolutely necessary for effective communication of information. One obvious problem with attempts to achieve duplex operation is that intense transmitter power will overload, or burn up, the sensitive receiver input of a transceiver unit. The simplest method of duplex operation is different transmit and receive frequencies for each pair of interacting communications terminals. Such a system is used for duplex operation of modems, for example, with the originating unit set for FSK operation at 1070/1270 Hz, and the responding unit set for operation at 2025/2225 Hz. A similar duplex arrangement is used at the other end of the spectrum for microwave carrier systems which may transmit at the receive frequency of the companion system, with the receive frequencies likewise mated. Figure 10-4c and d shows the modem and microwave system arrangements, respectively.

Duplex operation can be achieved at a single radio frequency for some communications systems that require single-frequency operation. In pulsed radar systems, for example, an extremely high-power, short-duration pulse is transmitted at the same frequency, and through the same antenna system, as its very sensitive receiver section. The duplexer in this case can be a high-power, high-speed switch, a directional coupler or hybrid of some kind, or a ferrite circulator. (The detailed operation of the directional couplers and ferrite circulators is covered in Chapter 12.) These specialized devices functionally allow transmitted signals to travel in one direction only, for example, transmit section to the antenna and not to the receiver input, and received antenna signal to the receiver section only (if necessary).

WIRE AND CABLE COMMUNICATIONS

The frequency allocation chart identified many important electronic communications services used for the transfer and exchange of information. The chart tells only half the story of communications, however, for it lists only those services that *radiate* signals. Although radio serves to provide efficient communications capabilities for particular needs, the more common wire and cable services best meet the needs of personal and business communications. You might better appreciate the value of the traditional telephone system by its continuing success of interconnecting information channels between over a hundred million telephones in the U.S.—not to mention the rest of the world. A comparative radio service, the U.S. Citizens Band, has proven an ineffective choice for widespread personal and business communications. The very factors that make wired services ineffective for mass communication also provide the characteristics that make them ideally suited to the needs of private, secure, reliable, and interference-free transfer of information.

The popular wired communications services include the obvious telephone, teleprinter, computer, and data terminals; the new Picturephone® service and cable television (CATV); and a wide range of specialized private and government business lines that use cable to transmit information from one location to another. Cables can carry wideband and narrowband signals in their original or modulated forms. The general bandwidth requirements for information channels apply to wire and cable, the same as for radio transmission, but with substantial differences in transmission (propagation) characteristics. The details of transmission lines, cables, etc., and the media through which radio signals propagate, are covered in Chapters 11, 12, and 14. The more complex principles and characteristics of transmission are not critical to your understanding of systems operation as it is here presented, so you may consider these systems as using either a twisted pair of wires, or a coaxial cable, for transmission. The emphasis here is on the special *information* transmission properties of cables. In this respect the more prominent advantages of cable transmission, as introduced in the previous paragraph, reduce to *narrowband privacy:* information can be transmitted *only* where wires can be placed.

The bandwidth limitations of cables depend, of course, on their structure. The twisted-wire pairs used with the common telephone are specified for voice-channel service, but they can be successfully used for bandwidths up to and in excess of 200 kHz. Very often people confuse the frequency limitations of telephone *circuits* with *cable limitations*. The switching and other functions of the telephone networks have far narrower bandwidth capabilities than the transmission lines. One of the major problems with the use of twisted-pair transmission lines is their susceptibility to *crosstalk* interference from adjacent wires in bundled form. The advantages of coaxial cable include the shielding necessary to prevent crosstalk at the expense of a more costly structure. The lower uniform capacitance of the coaxial cable also affords greater bandwidth capabilities, so it is the more desirable broadband transmission line. Perhaps the major limitation of cable transmission is the high losses that occur with both distance and frequency. The obvious "*I-R* drop" losses with long cable runs are a function of conductor size within the cable. The additional high frequency losses are due to the capacitance of the cable, and provide a *slope* to its frequency response. Figure 10-5 graphs the attenuation and frequency response of popular paired and coaxial cables and waveguides used in communications applications.

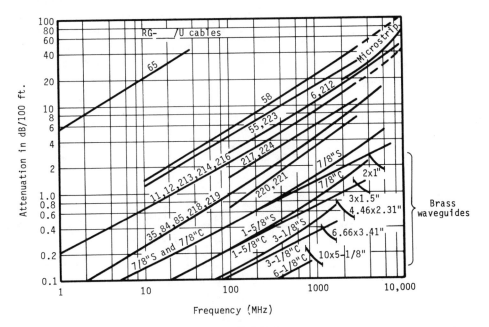

FIGURE 10-5

Given the cable loss characteristics, electronic system elements are employed to compensate for both *I-R* drop and slope in order to maintain signal levels along the cable route. These system elements are called *repeaters*. Special repeaters have been developed for use with two-wire lines for compensation of frequencies to above 1 MHz in order to accommodate and distribute Picturephone service. The repeaters used for coaxial cable transmission systems must have greater bandwidth capabilities to take advantage of the cable characteristics. In addition, most broadband transmission systems use some form of FDM or TDM to exploit the greater information capacity of increased bandwidths. CATV (community antenna, or "cable" TV), for example, can provide practical transmission of 40 television channels, plus FM broadcast, with distribution using coaxial cable and broadband repeaters, operating in the 30 to 300 MHz frequency range with a bandwidth of over 270 MHz. The increased privacy and security of cable transmission, of course, reduce the possibility of interfering with the many radio services that operate with allocations in the 30 to 300 MHz frequency range. Figure 10-6 shows the typical CATV system.

Although wire and cable communications services operate independent of the radio services listed on the frequency allocation chart, most of them still use radio as a part of their total operations. The telephone services, for example, use HF *radiotelephone* to ships at sea, UHF radiotelephone for vehicular applications, microwave point-to-point relay for shorter distance broadband transmission, and satellite relay for long distance, worldwide broadband communications. On the other hand, most radio communications services *also* use wire and cable as a part of their total operations. Many radio and TV broadcast station studio-to-transmitter *links,* for example, lease transmission lines from the local telephone company to transfer their audio and video signals.

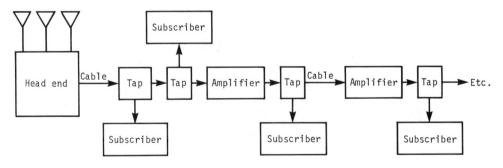

FIGURE 10-6

TELEPHONE SYSTEMS

The basis of successful multiline, private wire communications, such as are used in our telephone systems, is *routing* and *switching*. This sort of basic electrical function might seem a bit mundane for a textbook on electronic communications, but after a little examination the complications of routing and switching millions of separate lines should be quite evident. Setting up and operating a reliable telephone communications system is in no way like installing an intercom in an office complex, even though both systems use wires and switches. The complications of such a large number of interconnections, *however,* require a system every bit as complex as even a multiplex microwave radio or computer system. In the older telephone systems switching is accomplished electromechanically with two types of switches: the *Strowger* and the more modern *crossbar*. These switches are actuated by a conventional rotary telephone dial. The switches, which are actually specialized relays, are arranged in many multiples in order to accommodate the *traffic* into and out of the *exchange* office. The Strowger switch can connect an incoming wire to 100 outgoing wires, but in a step-by-step sequence that ties up all of the connections until the call is completed. The crossbar switch is the more efficient switching array, where a number of points can be accessed and held in a matrix arrangement like a computer *memory,* then released for connecting the dial sequence of another call once the lines are connected. The crossbar switch is thus faster and does not become "tied up" once the connection is made, as does the Strowger switch.

With your knowledge of the capabilities of digital logic circuits, you should conclude that all the electromechanical switches in telephone systems can be efficiently replaced with electronic switches using IC's. Again the concept is easily applied to the simplicity of an intercom, but when it comes to replacing the tremendous amount of existing reliable circuitry in telephone exchanges, you should appreciate the extreme magnitude of the problem. It is for this reason—the inertia of a gigantic *machine*—that solid-state switching is proceeding slowly for the established large telephone services. As a first step in solid-state switching, all rotary-dial telephones are gradually being replaced by *touch-tone*® sets. The touch-tone arrangement provides a standard of 12, with an option of 16, keys that simultaneously generate two audio tones that reference numbers and functions. Figure 10-7*a* shows the typical touch-tone keyboard and its related frequencies. In addition to providing capabilities for rapid switching, the frequency combinations can easily accommodate modems, acoustical couplers, and any number of devices that are awkward or impossible to use with mechanical switching exchanges. Electronic switching, of course, employs the same kind of LSI processor circuits with some sort of memory such as is found in modern computers and related digital equipment. The redundant circuit functions, the programmed

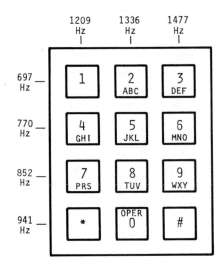

(a)

ESS block diagram

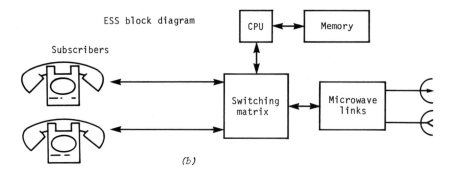

(b)

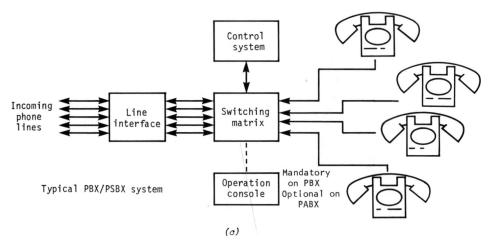

Typical PBX/PSBX system

(c)

FIGURE 10-7

memories, and the large number of input and output combinations make the actual *electronic* operation of an electronically switched telephone exchange almost identical to that of a modern computer system. The functional diagram of these electronic switching techniques is shown in Fig. 10-7*b*.

Although solid-state electronic switching is moving slowly in replacing the electromechanical apparatus in large telephone systems, the technique is widely applied to new and replacement installations in small telephone systems. The most popular small telephone installation is used for private communications in business and industry, where each installation will have either automatic or operator controlled telephone communications. These installations, called *PBX* (private business exchange) and *PABX* (private automatic business exchange), function like small computers, achieving a wide range of communications capabilities within a given business or industrial facility. Figure 10-7*c* shows the block/function diagram of this popular application of a wired communications service.

Beside the switching function, large telephone companies provide an additional *routing* function not duplicated in the PBX installations. The large telephone systems must have convenient and rapid access between local exchanges, and additionally provide long distance access to any exchange around the world. Here again you should consider the immense problems of standardization for access to exchanges where neither the spoken language nor the technology of two distant countries is related. Fortunately, there are international standards, as with the ITU, established by the CCITT (see Chapter 9) for worldwide telephone communications. With these standards in effect, it is possible to dial up number codes that will access the telephone systems of almost any country.

The efficient routing of heavy telephone traffic requires established networks of wire, cable, and radio relays and terminals, as well as utilization of electronic processing to expand the channel capacity of these routes. The U.S. telephone distribution network has both long and shorter distance access routes, so it is possible to transfer signals from one location to another by alternate routes if the more direct routes are occupied. The amount of traffic that any particular route can handle will depend on the information-channel capacity of the transmission technique employed for the particular service. If the signals are to be *carried* on a four-wire, twisted-pair transmission circuit, with a bandwidth capability in excess of 600 kHz, for example, it is possible to transmit and receive one FDM *supergroup* of 60 voice channels with these circuits. A comparable four-cable coaxial-line circuit can transmit and receive one FDM *mastergroup* of 600 voice channels. Like the twisted pairs the coaxial lines can be combined into a large multiline cable to handle greater traffic requirements. This technique that uses FDM, TDM, and a kind of multicable multiplex is defined as *carrier communications,* and the large telephone systems that use these techniques are called *common carrier systems.* Figure 10-8 expands the multiplex grouping techniques of Fig. 10-3 as applied to standard cable and ultimate radio transmission of multiplexed high-capacity information channels.

RF COMMON CARRIER

The radio systems used in carrier communications generally operate in the UHF and SHF region of the spectrum. Those areas on the frequency allocation chart, specified as "CC," are most often used for microwave point-to-point relay and terminal service. Most long-distance carrier communications systems have access to satellite relay capabilities. Satellite communications is fast becoming the most efficient means of transmitting radio signals throughout the world. These satellites typically operate in the lower frequencies of the SHF region and perform as *transponders,* receiving signals at one microwave frequency and returning them to earth at another. Some UHF *tropospheric scatter* radio carrier communications applications were established prior to satellite applications becoming practical. The principles of tropospheric scatter are detailed in Chapter 14 in the section on radio propagation. The technique simply uses the troposphere—an atmospheric layer about 6 miles above the earth's surface—as a reflector of transmitted radio signals in order to reflect them back to some other location on the earth's surface. The function is similar to that achieved with the satellite returning signals to different points on the earth. Satellite communications is far more efficient and reliable than tropospheric scatter, however, so this technique is fast becoming obsolete for carrier communications.

The details of telegraph and data communications techniques were presented in the preceding chapter. Since these signals, along with the PCM signals also described, use the same transmission circuits, they may be multiplexed just like the voice signals described here. A typical satellite ground terminal used for common carrier service, for example, transmits in the 5.9 to 6.4 GHz band and receives in the 3.7 to 4.2 MHz band, with a bandwidth of 40 MHz for FM RF and IF signals. The related satellite transponders determine the total information handling capabilities of the system. The current WESTAR (Western Union) satellite, for example, is capable of processing 1200 voice channels (of 4 kHz each), or data at 50 Mb/s, or 16 PCM channels at 1.544 MHz (1.544 Mb/s). This satellite is equipped with 10 transponders (plus two backup units) to process the signals of a number of earth stations simultaneously. INTELSAT (International Telecommunications Satellite) V, now in service, has 50 such transponders to handle the increased traffic requirements for satellite communications. Since dollar comparisons are often a popular index of the practicality of a system or technique, they can be applied here to indicate the relative success of the INTELSAT program. INTELSAT I, launched in 1965, processed 240 voice channels with two transponders at an average yearly cost per voice channel

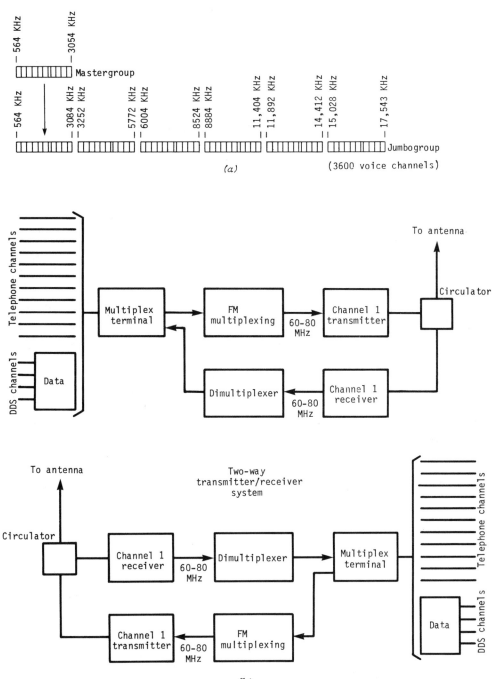

(a)

(3600 voice channels)

Two-way
transmitter/receiver
system

(b)

FIGURE 10-8

of $23,000. The INTELSAT V, launched in 1978, processes 60,000 voice channels with 50 transponders at an average yearly cost of *less* than just $60 per voice channel.

FACSIMILE

In addition to voice, data, and digital information, we also use direct wire, carrier, and radio communications circuits/systems for the transmission of pictures and other high-resolution graphics. Not to be confused with television, this technique is called *facsimile*, perhaps more familiar to you as the *wirephoto* transmission system used to distribute pictures to newspapers. Unlike television, the reproduction of pictures on a facsimile machine is a slow process, but the information can be distributed through existing telephone systems to any private line. Facsimile, or "fax" for short, is thus a narrowband, slow-scan technique that typically yields a paper (hard) copy of a picture or document. Complex facsimile systems can transmit full gray-scale black-and-white and color images, as well as line copy; whereas simpler fax machines are limited to line-copy reproduction only. The basic structure of fax machines includes synchronized rotating drums with some form of line scanner in the transmitter and line *writer* (printer) in the receiver.

In addition to newspaper wirephoto applications, fax machines are used for the distribution of weather charts, "mug" shots for police departments, and even complete newspaper printing. One such system, used by the *Wall Street Journal,* remotely proofs the New York daily financial newspaper in California for U.S. west coast same-day printing and distribution. In relation to bandwidth, the fax transmission for Xerox-quality copies of an 8½ by 11-inch document, for example, takes about 6 minutes using a single voice-channel telephone circuit. The same document can be transmitted and reproduced in 30 seconds with broadband transmission circuits, and in 1.4 seconds with the 1 MHz T1 PCM circuit used for Picturephone transmission. Although most fax systems operate into common carrier systems, some use direct radio transmission techniques. With radio transmission you may conclude that facsimile and television are almost identical. The distinct differences include a tradeoff of motion for resolution. Facsimile is capable of transmitting high-resolution still pictures with about 1000 lines per inch; but the TV system can resolve only about 200 lines across the full width of one horizontal scanning line (about 25 lines per inch on a 19-inch CRT).

Figure 10-9 shows the block/function diagram of a typical facsimile system, along with representative fax transceivers that are used for the transmission of documents. The TV receiver facsimile for home use is now in trial production. The receive-only unit will deliver hardcopy as transmitted on a subcarrier of the broadband television signal. CATV systems are also anticipating the provision of such fax services. One of the major obstacles to current widespread public use of fax is the cost of the

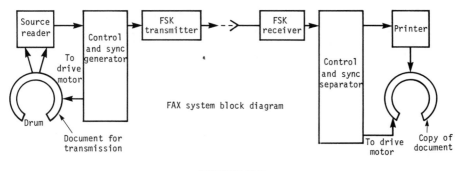

FAX system block diagram

FIGURE 10-9

mechanical, drum-scanning (printing) apparatus—not the electronics. The business applications of the system are now reaching cost-effectiveness, however, and fax machines will soon be as common as office copy machines. A data modem connected into a processor can actuate the telephone, input data to the computer, provide teleprinter hardcopy or CRT display, and decode and process signals for facsimile copy. Such office systems now in use are costly; but with improved scanner technology, they will actually reduce the costs of some services that now require hand processing. Within a few years, considering the rising costs of mail service and delivery, you may anticipate fax transmission for less than the cost of mail.

EMISSION STANDARDS

Most broadcast radio services use the AM or FM techniques advanced in Chapters 6, 7, and 8. You should thus be very familiar with the equipment and principles employed in these services after completing those chapters. The range of modulation standards and associated information-channel bandwidths as designated by the ITU and FCC can be found in government frequency allocation charts and emission tables. In the terminology of emission tables, the familiar AM broadcast radio is generally termed *telephony,* and designated A3. Its further detailed designation is *sound broadcasting* of speech and music, 8A3, where the modulating signal may vary between 4 and 10 kHz and yield an 8 to 20 kHz bandwidth depending on available local radio space. The familiar FM broadcast radio service is also generally termed *telephony,* but designated F3, with further detailed designation as 180F3, indicating the 180 kHz bandwidth required for the information channel.

In addition to the familiar "radio" broadcast services, the tables specify emission standards for telegraphy, facsimile, television, and radar, covering all the radio services in current use. The modulation techniques for these services are generally categorized as *CW* and *AM, FM,* and *pulse,* with the majority of radio services using some form of AM or FM for the communication of information. A careful review of the emission tables should disclose many important aspects of communications services, including channel bandwidth and the characteristics of the modulation signals imposed on the RF carrier. Using audio-frequency (analog) facsimile (wirephoto), designated 5.45A4, for example, you should obtain enough information to understand its operation; that is, with a drum diameter of 70 mm, rotating at 60 r/min, a resolution of 5 lines/mm will require a channel bandwidth of 5.45 kHz. The microwave relay service used for multiplexed telephone voice channels, as discussed in previous sections of this chapter, is designated as *composite transmission,* 15300F9, for up to 960 channel (80 group) transmission with an FM *pilot* carrier. The specifications indicate a deviation of 4.136 MHz, a composite modulating frequency of 4.028 MHz, and a resulting 16.32 MHz bandwidth requirement.

Stereo FM Broadcast Standards

U.S. stereophonic FM broadcasting, designated as *composite transmission,* 300F9, is a popular form of broadcast radio. Although it is often called *stereo multiplex,* its operation is unique to composite service and should not be confused with the FDM techniques used in voice-channel carrier communications. The specifications in the emission table indicate the same 75 kHz deviation as with conventional FM broadcast (180F3). To achieve compatibility, FM broadcast stations simultaneously transmit the same program material in either 180F3 or 300F9 for reception by both conventional *and* stereo receivers. Figure 10-10 shows the spectrum of the *modulation* signal used for stereo FM broadcast, the block diagram of a transmitter stereo coding arrangement, and the block diagram of an FM receiver stereo decoding section.

Referring to Fig. 10-10 you should first dispense with the conventional FM broadcast signal, as modulated by the *combined L* (left channel) and *R* (right channel) signals. These combined signals, identified as $(L + R)$, modulate the FM transmitter in the conventional way described in Chapter 7. For stereo transmission the *R* audio signal is first inverted before its combination with the *L* signal, yielding an output identified as $(L - R)$. (The minus sign refers to *phase* cancellation—not *subtraction* in a mathematical sense.) The $(L - R)$ signal then feeds a 38 kHz balanced modulator, and provides a DSB-SC output of two sideband signals that extend to 15 kHz above and below the 38 kHz stereo carrier. These two sideband signals also modulate the FM transmitter. The 38 kHz stereo carrier is obtained by doubling a 19 kHz CW pilot signal, which also modulates the FM transmitter as a reference, to allow both stereo and *monophonic* signals to be processed without receiver adjustment. The resulting spectrum of this composite modulating signal is shown in Fig. 10-10*a*, extending from 50 Hz to 53 kHz. In addition to the two-channel stereo, some FM broadcast stations also use the full 75 kHz modulation frequency limit of specification 300F9 by adding an additional signal to their transmission. This signal, called *SCA* (subsidiary communications authorization) or *storecast,* is separate background-music programming material simultaneously transmitted to leased receivers in commercial establishments. The 7.5 kHz (maximum) AF SCA signals are combined with a 67 kHz balanced modulator, producing the sideband structure from 59.5 to 74.5 kHz. In practice, those FM stations that also provide SCA service must reduce the deviation, and consequent quality, of their stereo signal in order to prevent interference of the signals with different program material.

The reception and decoding of FM stereo broadcast signals is achieved with the circuit functions shown in Fig. 10-10c. The output of the FM detector first yields the spectrum of Fig. 10-10a. The $(L + R)$ signal is routed through a lowpass filter, and then through a combining network to left- and right-channel audio amplifiers. The combining network may be a combination of simple summing circuits or switched envelope detectors. When the FM broadcast signal is monophonic, the $(L + R)$ audio signals are the only input and outputs, driving both L and R audio amplifiers with the same signals. When the $(L - R)$ signal is present, the two signals input to the combining network as $(L + R) + (L - R)$ and $(L + R) - (L - R)$, yielding outputs of $2L$ and $2R$, respectively. The receiver automatically demodulates the stereo sideband signals with a 38 kHz product detector driven by the doubled 19 kHz pilot signal. If there is thus no transmitted pilot signal, the receiver operates in monophonic mode. The techniques used for composite FM stereo signal processing are also used in a wide range of electronic communications systems. You should be able to apply these functions, as necessary, to the more complex composite systems designated in the emission table.

Television Broadcast Standards

Broadcast television also uses a form of composite modulation of the RF carrier signal. The U.S. NTSC television system uses a combination of A5 and F3 modulation for picture and sound information, respectively. It is thus designated 5750A5C/250F3 for a 5750 kHz AM "visual" bandwidth and a 250 kHz FM "aural" bandwidth. As you may recall from the preceding chapter, the television picture information required a video bandwidth of 4 MHz for a resolution of about 214 line-pairs (within the 53.3 μs picture portion of the horizontal sweep.) If this information signal is conventionally amplitude modulated, a channel bandwidth in excess of 8 MHz would be required for the picture (video), without even accommodating the sound signal. Since the bandwidth reduction techniques of PCM were not yet developed when television standards were first established, other methods were pursued to reduce the excessive bandwidth requirements of video transmission. The SSB technique was investigated, but its difficulty in tuning

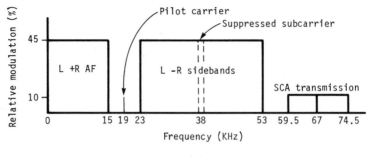

(a)

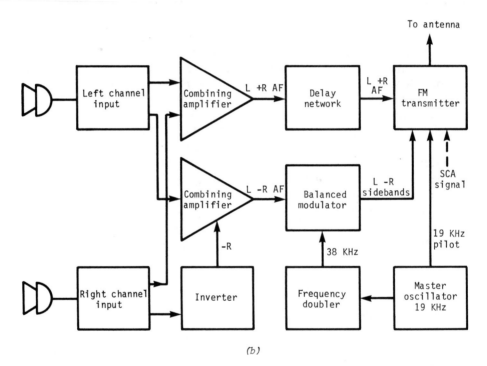

(b)

FIGURE 10-10

makes the modulation form inappropriate for popular television service. The tradeoff compromise, *vestigial sideband* modulation, provides the most practical analog alternative for the broadband, high-resolution requirements of television systems. Figure 10-11 shows the spectrum of the *modulation* signal used for NTSC broadcast television, plus the bandpass characteristics of the compensating video amplifier in the receiver, along with representative functional block diagrams of the transmitting and receiving systems.

Referring to Fig. 10-11a, observe that the modulated RF signal is actually a DSB-AM signal to 750 kHz above and below the carrier frequency, with

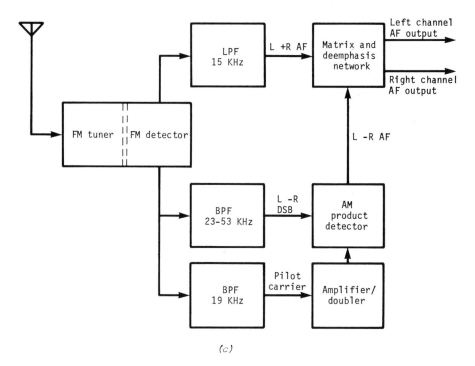

(c)

FIGURE 10-10 (*continued*)

the lower sideband signal rapidly attenuated within the next 500 kHz. The upper sideband signal is not attenuated, however, until it extends to beyond the 4 MHz required for the 200+ line-pair resolution of the video. The upper sideband signal is rapidly attenuated within the space of 500 kHz above the 4 MHz response, at which frequency a second carrier, that of the sound signal, is inserted. To prevent interference and achieve convenient separation of the sound and picture signals, the sound signal carrier is frequency modulated, with a deviation of 25 kHz. The sideband structure of this FM carrier may extend to 250 kHz (including guard bands), thus the designation 250F3. The bandpass frequency-response characteristics of the receiver video amplifier, shown in Fig. 10-11*b*, compensate for the increased amplitude of the DSB-AM portion of the modulated signal with attenuation of those frequencies below 1.25 MHz.

Table 10-2 compares the popular television systems used throughout the world. You should note the predominance of the 625-line, 50 field/second standard, with the older British 405-line system and the French 819-line system representing the extremes of resolution. The AM video bandwidth of the 405-line system is only 3 MHz; and the AM video bandwidth of the 819-line system is 10 MHz. You should also note that these same two sys-

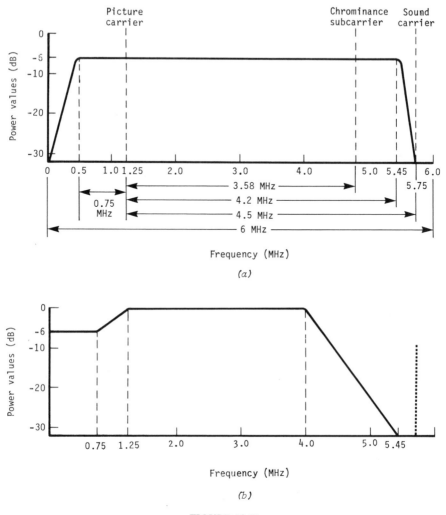

FIGURE 10-11

tems are the only ones that use AM (A3) audio modulation; the remaining systems all use FM (F3) modulation.

There are three predominant color standards for broadcast television: NTSC, PAL, and SECAM. The quadrature modulation I and Q signals used with the NTSC system, as detailed in Chapter 9, are really a part of the AM video modulation, with an 8-cycle reference burst transmitted at the beginning of each horizontal line to synchronize the color demodulator in the receiver. In the PAL (phase alternation line) system the phase of the color subcarrier is changed from each horizontal sweep line to the next. Both a color burst signal and a line switching signal are transmitted with this system in order to alternately switch the phase and synchronize the color demodulator of the receivers. In the SECAM (sequential with memory) system the color signal is transmitted as an FM subcarrier of the AM video signal. Like the PAL, in the SECAM system the color signal modulates each horizontal line alternately, requiring a switching signal at the beginning of each line to synchronize the color demodulator of the receivers.

The transmission of television signals via satellite is increasing with our technical capabilities in deploying this newer communications technology. There are two major applications for television via satellite: direct transmission of signals and long distance relay of signals. Satellite relay functions for television signals are similar to the relay functions for voice and other signals as detailed earlier in this chapter. The direct transmission of television signals generally employs a synchronous satellite that beams its signals over a portion of the earth's surface. This technique is now used for television reception in sparsely populated areas where local stations and transmitters are impractical.

RADIOTELEPHONY

In addition to the familiar broadcast services, emission tables specify transmission characteristics for AM and FM telephony. These classifications of radio transmission are used primarily for fixed and mobile, vehicular, marine, and aircraft communications services where wire or cable communications is impractical. The modulation forms specified include DSB-AM, designated 6A3; DSB-SC, designated 6A3B; SSB, designated 3A3A; and FM, designated 36F3. The services are intended for voice transmission, with the information-channel bandwidth based on a maximum modulation signal frequency of 3 kHz. The minimum bandwidth of 3 kHz is thus achieved with SSB modulation; the maximum bandwidth of 36 kHz is specified for the 15 kHz deviation of a commercial telephony transmitter. These radio services generally operate in two-way simplex, half-duplex, or full-duplex modes, and at a wide range of frequencies allocated for such service through the VHF and UHF regions of the spectrum. These frequencies are generally designated in the 30 to 50 MHz, 150 to 174 MHz, and 450 to 470 MHz bands in the frequency allocation chart. Table 10-3 identifies these frequencies in greater detail for the particular service; categories include *public safety, industrial/business, amateur, aviation, marine,* and *weather* communications.

Public safety communications services include radio transmission by police and fire departments, hospitals, doctors and ambulance firms, and other similar functionaries that protect life and property. Industrial and business communications services include radio transmission to and from mobile telephones, public and private transportation units, and individual walkie-talkie and paging units. For the most part these services use FM transceiver

TABLE 10-2. INTERNATIONAL TELEVISION STANDARDS

A.

	A	M	N	B	C	G	H	I	D,K	L	E
Lines/frame	405	525	625	625	625	625	625	625	625	625	819
Fields/second	50	60	50	50	50	50	50	50	50	50	50
Interlace	2/1	2/1	2/1	2/1	2/1	2/1	2/1	2/1	2/1	2/1	2/1
Frames/second	25	30	25	25	25	25	25	25	25	25	25
Lines/second	10 125	15 750	15 625	15 625	15 625	15 625	15 625	15 625	15 625	15 625	20 475
Aspect ratio[1]	4/3	4/3	4/3	4/3	4/3	4/3	4/3	4/3	4/3	4/3	4/3
Video band (MHz)	3	4.2	4.2	5	5	5	5	5.5	6	6	10
RF band (MHz)	5	6	6	7	7	8	8	8	8	8	14
Visual polarity[2]	+	−	−	−	+	−	−	−	−	+	+
Sound modulation	A3	F3	F3	F3	A3	F3	F3	F3	F3	F3	A3
Preemphasis in microseconds	—	75	75	50	50	50	50	50	50	—	—
Deviation (kHz)	—	25	25	50	50	50	50	50	50	—	—

[1] In all systems the scanning sequence is from left to right and top to bottom.
[2] All visual carriers are amplitude modulated. Positive polarity indicates that an increase in light intensity causes an increase in radiated power. Negative polarity (as used in the U.S.—Standard M) means that a decrease in light intensity causes an increase in radiated power.

B.

Country	Standard Used	Country	Standard Used	Country	Standard Used
Argentina	N	Israel	B,G	Peru	M
Australia	B	Italy	B,G	Poland	D,K
Austria	B,G	Japan	M	Portugal	B,G
Belgium	C,B,H	Korea	M	Rhodesia	B,G
Bulgaria	D,K	Luxembourg	C,L	Romania	D,K
Canada	M	Mexico	M	Saudi Arabia	B
Czechoslovakia	D,K	Monaco	E,L	Spain	B,G
Denmark	B,G	Morocco	B,H	Sweden	B,G
Finland	B,G	Netherlands Antilles	M	Switzerland	B,G
France	E,L	New Zealand	B	The Netherlands	B,G
Hungary	D,K	Nigeria	B,I	United Kingdom	A,I
India	B	Norway	B,G	U.S.A.	M
Iran	B,G	Pakistan	B	U.S.S.R.	D,K
Ireland	A,I	Panama	M	West Germany	B,G

C. Frequencies Relative to the Center Frequency of the RF (video) Carrier

System	Location of Audio Carrier (MHz)	Overall Bandwidth of Channel (MHz)	Bandwidth of Main Sideband (MHz)	Bandwidth of Vestigial Sideband (MHz)
A	-3.7	5	3.2 LSB	0.75 USB
M,N	+4.5	6	4.2 USB	0.75 LSB
B,G	+5.6	7	5 USB	0.75 LSB
C	+5.7	7	5 USB	0.75 LSB
H	+5.7	8	5 USB	1.3 LSB
I	+6	8	5.5 USB	1.3 LSB
D,K	+6.8	8	6 USB	0.75 LSB
L	+6.8	8	6 USB	1.2 LSB
E	+12.2	14	10 USB	2 LSB

TABLE 10-3.   POPULAR RADIOTELEPHONY
SERVICES

Service	Frequency Range (MHz)
Air Traffic Control	118–136
Amateur	28–29.7
	50–54
	144–148
	420–450
Citizens Band	26.965–27.405
Civil Air Patrol	26.62
Industrial	29.7–30.82
	47.69–49.6
	152.8–153.8
	173
Land Transportation, etc.	150–152.8
	157.45–161.57
Manufacturers	72–76
Maritime	156.25–157.45
Public Safety (Police, Fire, etc.)	30.86–47.69
Public Safety	153.8–156.25
	161.8–172.4
Public Safety, Industrial, Land Transportation, etc.	451–470

units in base-station and portable form, and often employ fixed-station repeaters throughout a given area to assure successful communications with
the lower power portable units in service. The channel bandwidth restrictions for this type of FM equipment include a 3 kHz voice frequency and a
maximum deviation of 5 kHz, yielding a maximum channel bandwidth of 20
kHz. This service is designated *narrowband FM* by the FCC.

Amateur radio communications are allocated to a wide range of frequency
bands throughout the HF, VHF, UHF, and even the SHF regions of the
spectrum. Because of the noncommercial, noncritical nature of amateur
communications, there are limited restrictions on the operation of varied
modulation forms within these frequencies. The radio amateur may elect to
employ a wide range of modulation forms within some frequency segment of
the band designated for such service. Radio amateur licensing is technically
restrictive, meeting the objectives for use by those individuals who have a
technical interest in radio communications. As such, amateur radio service
employs most all modulation forms, from A0 to A5 and F0 to F5, including
television, the use of repeaters, and even satellite relay.

The widespread use of radio for aviation communications includes voice
transmission as well as navigational assistance. The navigational aspects of
aviation will be reviewed in a later section of this chapter. Most aviation

voice communications use the VHF band from 118 to 136 MHz. In contrast to the public safety and industrial/business radio, which use narrowband FM, aviation radio uses simplex one-way and two-way AM transmission for communication with aircraft. There are 360 channels in the 18 MHz-wide aviation band. Even though the transmission is specified 6A3, occupying 6 kHz of bandwidth, the channels are spaced at 50 kHz increments to assure reliable, interference-free communications. Most of the services within this range of frequencies are intended for ground-to-air/air-to-ground transmission for purposes of radiolocation and traffic control, including distress and search-and-rescue frequencies.

Marine radio services include communications and navigational aids. As with aviation, the navigational aspects of this service will also be reviewed in a later section of this chapter. Marine communications include both ship-to-ship and ship-to-shore services operating in the 156 to 162 MHz portion of the VHF band. There are 88 designated channels within this band, spaced at 25 kHz increments. The use of this band is typically confined to shorter distance, two-way FM transmission. The older long distance marine-radio service operated in the HF "short-wave" region of the spectrum, with conventional AM transmission, at frequencies distributed through the 3 to 30 MHz range (see the frequency allocation chart). The recent deployment of the MARISAT series of satellites is rapidly replacing the older long distance maritime radio service, however, with more reliable, wideband SHF communications systems for long distance voice, data, telegraph, facsimile, and television service.

Weather service radio transmission provides a most important source of information for transportation and other aspects of government and industry. Such weather information is available from the U.S. National Weather Service (NWS) stations, for example, which use narrowband FM transmission in the 162 MHz VHF band. More detailed weather information is available through the transmission of weather maps by facsimile techniques. The orbiting weather satellite *Nimbus,* for example, regularly transmits weather charts that show the movements of atmospheric conditions. The facsimile transmission of these signals is in the UHF 402 and 1690 MHz bands. In addition to the television techniques used to obtain weather information for this transmission, the Nimbus and other sources of weather reports use data obtained by radar and telemetry sensors. These devices are reviewed in the following sections of this chapter.

TELEMETRY, COMMAND, AND CONTROL

Telemetry is simply a technique of remote metering. In the previous example of the weather satellite, infrared (and other) sensors are included to obtain information about the temperature of large areas of the earth's surface, simultaneous with the movement of cloud patterns and air masses. The output from the infrared sensor is sampled, and the resulting data is modulated on the RF carrier, along with the TV (facsimile) picture information, for transmission to an earth receiving station. This rather elementary example of telemetry is the basis, of course, of remote metering. If you now consider the requirements for extensive information about the launch of a rocket and its progress through space, transmitted and monitored by computers and dozens of technicians at a *ground-control* station, you should appreciate the importance of telemetry systems to space technology.

Telemetry system applications are not confined to space technology, since the dimensions and capabilities of these systems can be applied to a wide variety of areas. At present, one of the foremost uses of telemetry communications is in the transportation of oil and natural gas. Many miles of pipeline throughout the world are used to transfer this critical fuel from field to processing plant. Many sensors are placed along these pipelines to determine the temperature, pressure, flow, and other critical factors required to keep the gas and oil flowing at optimum rates. The remote metering of data dictates the control of many interrelated pumping stations throughout the system, so most pipeline operations use broadband microwave communications to transmit both telemetry and control signals to a central, computer-controlled operations facility. The array of sensors along the pipeline is continuously sampled, using time-division multiplex techniques, with the accumulated data transmitted by frequency-division multiplex microwave systems.

Other popular applications of telemetry include the familiar weather balloons that carry *radiosonde* transmitters operating in the 403 MHz band. As these balloons ascend through the atmosphere, they continuously transmit FM telemetry signals from which temperature, pressure, humidity, and other meteorological data are obtained. The military applications of telemetry are widespread, including the transmission of signals for monitoring the critical movements of guided missiles and other military vehicles. Possibly the ultimate telemetry system may soon be deployed in the reading of our home and business electric, gas, and water utility meters, alleviating the tedious and costly manual meter reading process. The technique can deploy telephone lines to transmit data to a central computer from which billing and consumption information can be obtained.

In space and military technology the terms *command* and *control* apply to the transmission of signals to a remote vehicle or location. In many instances command and control functions follow the reception and analysis of telemetry signals from sensors that indicate a given response. In the case of the oil pipeline, data indicating low pressure and flow may require a response in increased pumping pressure at various stations along its route. Similar, but much more rapid, responses to telemetry signals are required in guiding a space vehicle, rocket, or missile in order to keep it on course. Most systems of this type are *closed loop*, i.e., the command signals are automatically returned in response to a given telemetry signal, with the entire process controlled by a central computer.

Figure 10-12 shows a simplified electronic command and control system which responds to preprogrammed information and telemetry signals that indicate the condition of a remote vehicle. The *command* may, for example, specify that the vehicle change its position, thus actuating displacement apparatus. As the vehicle is changing, telemetry signals are transmitted indicating the degree of shift. The telemetry signals are compared to preprogrammed data in the computer, which in turn transmits control data back to the vehicle in order to obtain its final position. The whole process is sort of an electronic *blindman's buff*. "Find the object" corresponds to the *command*. Your observation of the movements of the blindfolded player corresponds to the *telemetry* information. Your response, "You're getting warmer" or "You're getting colder," effects a *control* function for the player who is to find an object that he cannot see.

RADIONAVIGATION AND RADAR

Radionavigation, as defined in an early section of this chapter, refers to two categories of signals: *beacons* and *radars*. Fixed beacon transmitters can beam signals to receivers in land, air, and sea vehicles in order to indicate their location, direction, and progress toward a destination. Such beacon signals are used extensively in guiding shipping and aircraft. The instrument

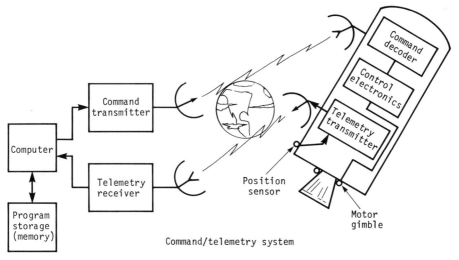

Command/telemetry system

FIGURE 10-12

landing system (ILS), for example, provides lateral and vertical guidance to aircraft during landing, as well as distance data from a landing point. The VHF omnirange (VOR) and UHF tactical air navigation (tacan) are beacon signals from which aircraft receivers can obtain bearing and location information while in flight. Marine radio beacons, such as those of the maritime radio service, are placed at coastal and island locations to benefit shipping. One major distinction between the marine and aircraft beacons is frequency, with lower frequency transmission more practical for shipping. These lower frequency, 190 to 410 kHz beacons are used by ship operators to determine the bearing, location, and movement of their vessels. The basic operation of these marine and aircraft beacon systems is relatively simple, requiring only direction-finding antennas coupled to sensitive radio receivers.

The far more complicated, but tremendously more valuable, navigation/identification/location system is *radar* (*ra*dio *d*etection *a*nd *r*anging). Unlike the beacon navigation system, where transmitters and receivers are located at fixed *and* moving points, respectively, the radar transmitter and receiver are at a *single* fixed or moving location. Although a ship with a beacon receiver, for example, can only determine its location with respect to beacon transmitters whose locations are known, a ship equipped with radar can determine its location with respect to any body of land or other ships in its area of coverage. Radars essentially provide us with *vision* far beyond the limitations of our eyes, and they can provide us with information about bearing, speed, size, and other factors not now obtainable by other techniques.

The radar technique applies the principles of radio wave reflection in order to obtain navigation/identification information. With both the transmitter and receiver at the same location, the system transmits an intense short burst (pulse) of RF waves, rapidly switching to reception mode to receive the reflected *echo* of the initial transmitted signal. Since the velocity of the radio waves is constant (the approximate speed of light), the time required to receive the echo signal can be converted to distance for *ranging* measurements. With the simultaneously synchronized moving antenna and display scope (or plotter), the *direction* and *elevation* of the return echo signal can also be determined. These directional data are obtained by comparing the time of the transmit and echo signals with the movements of the antenna/display units. Since the technique deploys short-duration bursts of RF signals and highly directive antennas, most radar applications use frequencies in the microwave region of the spectrum, above 1000 MHz. The technical details of the microwave components, transmission lines, and antennas used in radar systems are covered in Chapters 11 through 14. Figure 10-13 shows the basic block/function diagrams of the radar systems and compares typical display techniques.

In addition to location and ranging of stationary *targets,* a pulse radar can also identify the size of targets, as well as their speed and direction if they are moving. A larger target at a given distance and direction, for example, will reflect a more intense echo than a smaller target, and will indicate a higher amplitude level on the receiver display. Thus the sensitivity of a radar receiver can be calibrated against distance to indicate relative target size. Reducing the receiver sensitivity results in only the larger areas being displayed. Moving targets can also be characterized by noting their location with successive echo signals; but they can also be accurately characterized with a simpler technique that uses the doppler effect to obtain a frequency shift in the echo signal. An increasing frequency will indicate a target moving in the direction of the radar receiver; a decreasing frequency will indicate movement away from the radar. The *rate* of frequency shift can be translated to speed at a given distance.

Radar systems can be categorized by technique and function. The three basic radar techniques are *CW, FM,* and *pulse.* CW (continuous wave) radars are simple systems used primarily for high-directivity, low-power applications such as vehicular speed detectors used by police for traffic control. These CW radars use the doppler effect for obtaining frequency-shift data from rapidly moving targets at shorter distances from the receiver. Since both the transmitter and receiver of the CW radar are active, isolation devices must be used to prevent the intense transmitter power from desensitizing the receiver input. FM (frequency modulation) radars extend the speed-determining capability of the CW radar by providing a differential frequency shift in the transmitter that can be compared to the frequency of the echo. The FM technique can also be used to measure the *distance* of a fixed (or slowly moving) target by continuously varying the transmitter frequency and comparing the frequency of the echo signal to that of the transmitter at the time that the echo returns. The delay time of the echo signal, as obtained from the rate of transmitter frequency shift, can be directly converted to time.

The pulsed radar system of Fig. 10-13, described earlier, is the most popular and versatile technique for determining the location, distance, and movement of targets. The terms *pulse* and *pulsed* here apply to a pulse-modulation signal applied to the transmitter, which, in turn, transmits a short burst of RF waves. The inverted form of the pulse-modulation signal actuates the input of the receiver section of the radar, blocking its sensitive input from the high-power burst of transmitted RF energy. The simplest form of pulse radar is used in applications such as altimeters in aircraft, measuring the return time from echoes of bursts of RF directed to the earth's surface in order to determine their altitude. The complex forms of pulsed radar systems deploy the varied forms of moving antennas and displays, as well as the frequency-shift doppler and FM techniques. These complex systems can distinguish fixed and moving targets, in addition to determining their size, location, and speed.

The more popular radar functions include the *navigation* applications for shipping and aircraft; fixed and mobile *search* radars for surveillance; the *fire-control* radars for military weapons systems; and other specialized applications, including weather radars, police radars, and simple doppler intruder alarm systems. Each function requires radar systems with special capabilities appropriate to the needs of the application, with the complex military systems representing the most sophisticated deployment of the technology. Radar systems are also used in transponder functions to identify target echo

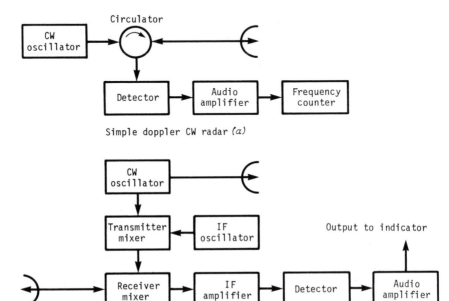

Simple doppler CW radar (a)

CW doppler radar with IF amplification (b)

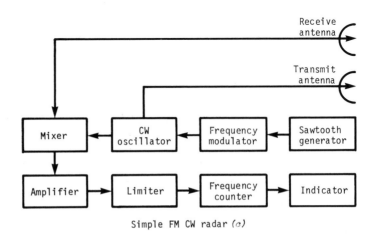

Simple FM CW radar (c)

FIGURE 10-13

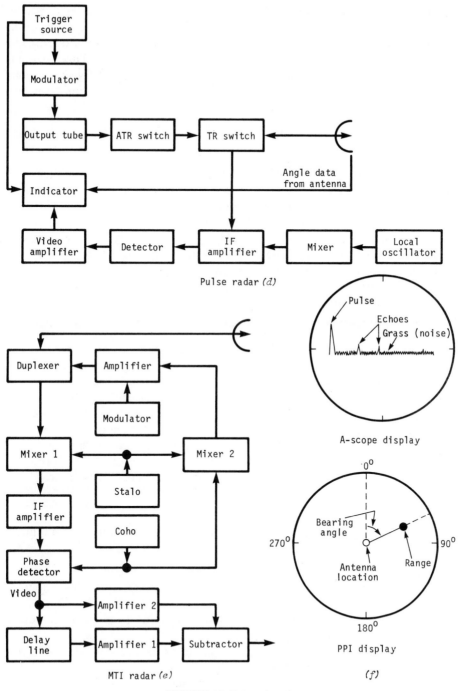

Pulse radar *(d)*

A-scope display

PPI display

(f)

MTI radar *(e)*

FIGURE 10-13 *(continued)*

signals. An aircraft receiving a coded pulse radar signal will return a response signal in the direction of the transmitting radar, for example, with the use of a radar transponder. So, in addition to receiving an echo, the originating radar will receive a coded signal that identifies the aircraft. This system is used extensively in both commercial aviation and military defense systems, where it is called *IFF radar,* an abbreviation for "*i*dentification: *f*riend or *f*oe."

CONCLUSION

The content of this chapter is concerned with the broad aspects of electronic systems' principles and applications. Your review of this information should provide you with an exposure and appreciation of the ever-widening diversity of electronic communications. The content does not provide in-depth coverage of any particular aspect of the technology, however, as these specialized topics are best presented in books entirely devoted to a specialized system. In reviewing this chapter you may find that some particular aspect of the technology is of greater interest than others. You should pursue this interest through reading, exposure, and experimentation with that specialty area. For the most part, employment, advancement, and success in electronics technology require in-depth knowledge and skills with a particular specialty field within the broad technology. Your choice of a specialty that combines your strong interest with a dynamic aspect of the technology will help to assure future success and rewards. Much of the dynamic growth in electronic communications is in the higher frequency areas, including microwaves. As such, it is appropriate for you to have knowledge of the transmission, microwave, and antenna topics of the last part of this text: Chapters 11 through 14. It is recommended that you withhold your decisions about specialization until after your completion of Chapters 11 through 14, followed by a second review of this chapter.

Problems

1. Given: A shipping company requirement for specification of communications services and frequencies used on their ships.

Determine:
- (*a*) The HF maritime mobile communications frequencies allocated below 30 MHz
- (*b*) The VHF maritime mobile communications frequencies allocated from 30 to 300 MHz
- (*c*) The UHF maritime mobile communications frequencies allocated from 300 to 3000 MHz
- (*d*) The SHF maritime mobile communications frequencies allocated from 3 to 30 GHz

(e) The maritime mobile satellite communications frequencies
(f) The special characteristics and functions of the HF, VHF, UHF, and SHF maritime mobile communications equipment
(g) The distinction between the UHF maritime mobile and UHF maritime mobile satellite communications equipment

2. Given: A shipping company requirement for specification of radionavigation services and frequencies used on their ships.

Determine:
(a) The characteristics and function of maritime radionavigation services in the 300 kHz frequency region
(b) The characteristics and function of maritime radionavigation services in the 3 GHz frequency region
(c) The frequencies allocated for maritime radionavigation satellite services

3. Given: A taxicab company starting operations and requiring specification of two-way radio-communications equipment for office and vehicular installation.

Determine:
(a) The frequencies available for their communications needs
(b) The specific recommended frequency of operation for this service
(c) The type of emission (modulation and bandwidth, etc.) specified for this service
(d) The RF transmitter power limitation for this service
(e) The technique (simplex, half-duplex, or full-duplex, etc.) used for this communications service

4. Given: The WESTAR communications satellite with the following signal processing capabilities in each of 10 operational 36 MHz-bandwidth transponders (there are two additional backup spare transponders, for a total of 12):

1. One color television channel with audio, or
2. 1200 voice channels, or
3. Data at a 50M baud rate, or
4. TDM/PCM: 16 channels at 1.544M baud, or
5. TDM/PCM: 400 channels at 64k baud, or
6. TDM/PCM: 600 channels at 40k baud

The transponders process FM signals, including 23,300F9 and 17,000F9 composite TV; or two 17,000F9, or one 3400F9 composite FDM voice channels; or 50,000F9Y composite FSK data, or that equivalent in composite TDM/PCM digital.

Determine:
(*a*)   The transponder earth-to-space and space-to-earth frequencies
(*b*)   The typical receiving antenna aperture for this communications service
(*c*)   The bandwidth requirements for each of the processed signals, identified as 1 through 6 above
(*d*)   The total bandwidth occupied by the 12 transponders in the two frequency bands (earth-to-space and space-to-earth)
(*e*)   The reasons why two-frequency duplex operation is used, and the reasons why the earth-to-space and space-to-earth frequencies are separated into bands approximately 2 GHz apart

5.   Given: The U.S. Standard NTSC color television system.

Define and describe:
(*a*)   The modulation techniques for the video, audio, color, and synchronization signals, combined to form the composite VSB modulated RF carrier
(*b*)   How the four modulation signals are each combined and extracted without interfering with each other
(*c*)   The relative advantages and disadvantages of operating with vestigial sideband modulation, rather than full DSB-AM, or FM for the transmission of the television signals

6.   Given: A microwave radio system intended to perform duplex, two-way relay service in the 6.625 to 6.875 GHz frequency band, handling one mastergroup of 600 voice channels at 564 to 3164 kHz.

Determine:
(*a*)   The bandwidth of the SSB baseband mastergroup of 600 voice channels, and the bandwidth of each voice channel
(*b*)   The deviation of the FM microwave carrier signal with a given modulation index of 4.0
(*c*)   The 40 dB bandwidth of the transmitted FM microwave signal
(*d*)   The emission designation for the service (_____ F9)
(*e*)   The percentage of the 6.625 to 6.875 GHz band occupied by this transmitted signal

7.   Given: A CATV system that operates from 54 to 300 MHz using the conventional U.S. TV broadcast frequencies of channels 2 through 13 (maintaining the channel 4-5 and 6-7 gaps) and the entire FM broadcast radio band.

Determine:
(*a*)   The number of additional midband TV channels for cable transmission between the FM band and TV channel 7

(b)   The number of additional superband TV channels for cable transmission between TV channel 13 and 300 MHz

(c)   The total possible number of TV channels available for transmission by the cable service

(d)   The additional number of 300F9 FM stereo music channels that may also be transmitted in the TV channel 4-5 gap

8.   Given: A facsimile system which is to operate an acoustically coupled 1200 baud FSK data modem, capable of 50 line-pairs per inch (2 lines/mm) resolution, with an 8-inch circumference drum, and which is used to transmit 8- by 10-inch typewritten copy over standard telephone circuits.

Determine:

(a)   The total line resolution per revolution of the drum

(b)   The required drum rotation in revolutions per second and revolutions per minute

(c)   The number of revolutions required to scan the 10-inch length of printed copy

(d)   The length of time required to scan the length of printed copy, and the consequent telephone circuit time

(e)   The length of time required to scan a 150 line/inch high-resolution halftone lithograph with the same 8- by 10-inch dimensions

9.   Given: A small-boat, 9.445 GHz pulse radar system with a pulse length of 1 μs and a pulse repetition rate of 1000 pps.

Determine:

(a)   The minimum and maximum range limits, in meters, of the system (at 3×10^8 m/s)

(b)   The minimum receiver IF bandwidth based on the pulse length of the radar signal

(c)   The rate of antenna rotation, in r/min, for 360° coverage based on a 1° target resolution (one pulse/degree of rotation)

(d)   The actual *size* of the target resolution, in meters, at the minimum and maximum range limits of the system

PART III TRANSMISSION

CHAPTER 11

TRANSMISSION LINES AND NETWORKS

Radio frequency signals are transmitted from one point to another in the form of changing voltage and current. The signals are transmitted through wires, waveguides, or the air and space. If the distance over which the signals travel is great enough to yield a phase shift because of the transit time, the characteristics of the wires or space have an electrical effect on the signals, and the signals thus become a kind of electrical transmission network. Would you believe that air, or more accurately *free space,* has an impedance of 377 Ω? You have of course measured a length of supposed 300 Ω flat TV lead-in wire with an ohmmeter and found it to measure infinitely high resistance. The TV lead-in was not measured as 300 Ω because no signal was applied to it. The fact is that all wires, materials, space, and structures do exhibit electrical properties when electrical or electromagnetic signals are applied to them. It is the purpose of this chapter to review the principles of these electrical properties as applied to RF transmission lines and networks.

LEARNING OBJECTIVES

Upon completing this chapter, including the problem set, you should be able to:

1. Understand the concepts of lumped and distributed constants in high-frequency components and transmission lines.

2. Understand the concepts of signal propagation along a two-conductor transmission line.

3. Understand and compute voltage, current, and power transfer relationships.

4. Understand the concepts of reflected voltage, current, and power.

5. Understand the relationship of incident and reflected signals, standing waves, and wavelengths.

6. Understand and compute the power transfer characteristics of VSWR, reflection coefficient, mismatch loss, and return loss.

7. Understand the structure and applications of the Smith chart.

8. Use the Smith chart to determine transmission line and network characteristics.

9. Understand the concepts and application of impedance-matching techniques.

10. Compute component and network requirements for impedance matching used both lumped- and distributed-constant components.

ELECTRICAL PROPERTIES

If you recall your results of network and circuit calculations from the early chapters of this book, you should agree with the following:

1. The required values of L and C circuit components become smaller as the frequency of the RF signal is increased.

2. The circuit mounting and package inductance and capacitance may exceed the requirements for small-value L and C components at high frequencies where wavelengths are short.

3. Longer wiring paths will increase both the inductance (due to *longer* conductivity *paths*) and the capacitance (due to *increased* conductivity *area*) to externally affect the L and C properties of separately wired components and networks.

4. The interconnection of high-frequency circuits and systems cannot be efficiently accomplished using techniques where lead paths are short, because the wiring inductance and capacitance may exceed that of the circuit demands.

From these statements you may conclude that lower frequency circuit techniques will not necessarily work for transferring high-frequency signals from one point to another.

Constants

The type of network and circuit analysis discussed in the early chapters of this book dealt with what is called *lumped constants*. Loosely defined, a lumped constant refers to the property of a circuit component that can be placed at a given point in a circuit, i.e., a lump of resistance in series with a lump of inductance, etc. The study of basic electronics confines itself to dealing with lumped constants of R, C, and L. If lumped-constant techniques are to be employed at higher frequencies, where wavelengths are shorter, the circuitry, mounting, and package structure must, of course, shrink in size in order to keep the values of L and C appropriately small. Keep in mind that a 1-inch (25 mm) length of No. 20 gauge hookup wire has an inductance of about 20 nH. Thus, if you are working with frequencies and circuits that require component values of 20 nH, you cannot use hookup wire and leads of components that even approach 10% of the required inductance value—the leads cannot be any longer than 1/10 of an inch (2.5 mm).

In this age of microminiature integrated and hybrid circuits, short lead lengths and microscopic components are practical, but the wiring (*a*) into and out of IC's, (*b*) to and from the components that cannot be miniaturized, and (*c*) to and from systems and antennas requires a *different approach* to electronic circuits. The approach that *is* suitable for transferring high-frequency signals uses what is called *distributed constants*. In contrast to lumped constants, the distributed type of constant refers to properties of R, L, and C which are *not* placed at a given point in a circuit, but which are distributed throughout the circuit. Using a conventional coil as an example here, we can compare the lumped- and distributed-constant interpretation of its properties. Figure 11-1*a* shows the familiar lumped-constant schematic form that sums the resistance distributed along the length of the coil and places it in series with the coil, as though there were an actual resistor connected in series with the coil's inductance. A more accurate distributed-constant rep-

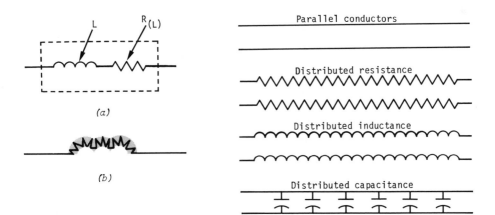

FIGURE 11-1

resentation of the coil's actual resistance and inductance would require the unusual drawing as shown in Figure 11-1*b*, however. While you probably will not encounter drawings like that of Figure 11-1*b*, which represents distributed constants, you should still interpret the electrical properties of components as taking that form when signal wavelengths are short.

If we place two 3 ft (0.91 m) lengths of No. 20 gauge hookup wire, equally spaced at 1 inch (25 mm) in parallel, the wire would have the following approximate distributed constants: $R = 0.03$ Ω, $L = 1.5$ μH, and $C = 6$ pF; or more properly stated, $R = 0.01$ Ω/ft (0.0328 Ω/m), $L = 500$ nH/ft (1.64 μH/m), and $C = 2$ pF/ft (6.57 pF/m). Increasing or decreasing the length of the wire would proportionally increase or decrease the constants. If we carefully connect one end of the wires to a signal generator and terminate the other end with an appropriate load resistor, an increasing applied frequency would affect *only* the distributed resistance along the wires that is due to *skin effect*. The capacitance and inductance are constants that do not change with frequency. If the frequency of the signal is then increased to yield about three wavelengths over the 3 ft (0.91 m) length of the wires, we could theoretically obtain an *instantaneous* lumped-constant equivalent circuit that represents the transmission line as shown in Fig. 11-2. The instantaneous voltage peaks provide a "maximum charge separated by a dielectric, yielding an electric field," which is a definition for *capacitance*. The peaks also provide a difference in voltage *along* the wire, causing a flow of current limited by the series resistance, and developing a magnetic field around the wire as a function of its inductance. Since the flow of current is at maximum midway between the (+) and (−) voltage peaks, the magnetism and inductance are also at maximum at these points.

To obtain the instantaneous equivalent circuit, we had to theoretically stop the movement of the signal. Since RF signals do *not* stop, however, except perhaps in our imagination, the equivalent circuit only represents the effects of the signal waves on the parallel wires at *one instant* in time. In the *next* instant the phase of the signal at the generator will advance and the waves will move further along the wires, with a given wave eventually moving to the load and its power dissipated in the load resistance. In fact, a more accurate circuit picture for your imagination would be the two wires, the waves, and the lumped components all moving along the wires. By the process described the signal generator actually *launches* waves into and along the two-wire circuit, and the *circuit* transmits the waves to the load. As such, the two-wire circuit with uniform distributed constants becomes a *transmission line*.

Characteristic Impedance

For the signal generator to launch waves into the transmission line, i.e., actually transmit *power* into the line, the generator must see the line as

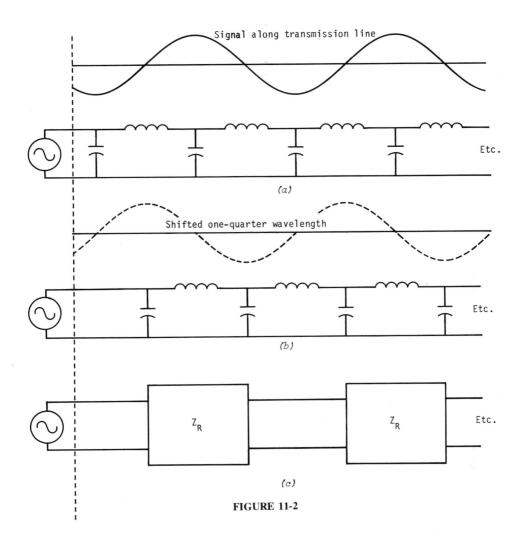

Signal along transmission line

(a)

Shifted one-quarter wavelength

(b)

Z_R Z_R Etc.

(c)

FIGURE 11-2

having a finite impedance value. If, for example, the internal impedance of the signal generator is 300 Ω, the maximum power will be transferred into a transmission line with an impedance value also of 300 Ω. Transmission lines do, in fact, have finite impedance values that are a function of their series and parallel constants. You might even consider the transmission line as a series of T or π matching-network sections, with each network maintaining its input and output impedance at the value of the generator and load impedance in order to assure maximum power transfer through the networks. The instantaneous network impedance that is maintained along with the length of a transmission line is called its *characteristic impedance*, and it is designated Z_0. The characteristic impedance of a line can be computed as the ratio of signal voltage to signal current and series constants to shunt constants along the line:

$$Z_0 = V/I = \sqrt{(R +jX)/(G +jB)}$$

$R +jX$ is the series resistance and inductive reactance distributed along the line, and $G +jB$ is the shunt conductance and capacitive susceptance distributed along the line. If the resistance and conductance are proportionally small, making the resistive (heat) losses insignificant, the complex ratio simplifies to

$$Z_0 = \sqrt{L/C}$$

L is the series inductance along the line, and C is the shunt capacitance along the line. Referring back to our earlier example of two parallel lengths of No. 20 gauge hookup wire acting as a transmission line, the characteristic impedance would be computed as:

$$Z_0 = \sqrt{L/C} = \sqrt{500 \text{ nH}/2 \text{ pF}} = \sqrt{25 \times 10^4} = 500 \ \Omega$$

For the convenience of computation with certain circuit conditions the parallel equivalent, *characteristic admittance* (Y_0), may be used. Y_0 is, of course, the inverse of Z_0 $(Y_0 = 1/Z_0)$, so the characteristic admittance of our hookup wire example would be:

$$Y_0 = 1/Z_0 = 1/500 = 2 \times 10^{-3} = 2 \text{ mS}$$

Propagation Constant

In engineering practice the term used to describe the launching and transmission of a signal is *propagation constant*. The propagation constant is a measure of the *loss* and *phase shift* imposed on the signal as it passes through the distributed equivalent networks along the transmission line. As such, the propagation constant is complex, and thus identified in two parts—real and imaginary—just as a complex impedance is identified in two parts: R (real) and $\pm jX$ (imaginary). The propagation constant is designated by the Greek letter γ (small *gamma*). Its real part, *attenuation constant,* which indicates loss, is designated by the Greek letter α (small alpha); while its imaginary part, *phase constant,* which indicates proportional phase shift, is designated by the Greek letter β (small beta). The propagation constant can be computed as the product of the series constants to shunt constants along the transmission line:

$$\gamma = \alpha +j\beta = \sqrt{(R +jX)(G +jB)} = \sqrt{ZY}$$

The attenuation constant value (α) is measured in *nepers* per unit length, where nepers is from natural (Naperian) logarithm. This value can be converted to the more acceptable decibels per unit length by multiplying the neper value by the constant 8.69. The phase constant value (β) is measured in *radians* per unit length, and can be compared to the basic 2π radians = $360° = 1$ wavelength. Thus, $\beta = 2\pi$ per wavelength.

If we proceed to calculate the propagation constant for our example of two parallel lengths of No. 20 gauge hookup wire, the real-part losses would be too small to provide significant data. If the line were theoretically lengthened to 1000 ft (305 m), however, the distributed constant values would become significant. We may assume for this example that the high-frequency skin effect increased the distributed series resistance by a factor of 10. The values for our 1000 ft (305 m) theoretical transmission line are now: $R = 100\ \Omega$, $L = 500\ \mu H$, and $C = 2$ nF. The propagation constant would be computed as follows (assuming that an operating frequency of 1000 MHz yields the wavelength of about 1 ft (0.305 m) as in the example of Figure 11-2):*

i. First, the reactance and susceptance are found for a frequency of 1000 MHz:

$$X = \omega L = 6.28 \times 10^9 \times 5 \times 10^{-4} = 3.14 \times 10^6$$
$$B = \omega C = 6.28 \times 10^9 \times 2 \times 10^{-9} = 12.56$$

ii. Then, the constants are computed:

$$\gamma = \sqrt{(100 + j3.14\ M\Omega)(0 + j12.56)}$$
$$= \sqrt{(3.14 \times 10^6 \angle 89.99°)(12.56 \angle 90°)}$$
$$= 6.28 \times 10^3 \angle 89.99° = 0.1 + j6283/10000\ \text{ft} = \alpha + j\beta$$

iii. Then, the attenuation (α) of 1 neper is converted to dB:

$$0.1 \times 8.7 = 0.87\ \text{dB}/1000\ \text{ft}$$

iv. Finally, the phase constant is compared to the number of wavelengths (each wavelength is 1 ft long):

$$6.28 \times 10^3\ \text{rad}/1000\ \text{ft} = 6.28\ \text{rad/ft}$$

Although the above procedures may require a high-capability calculator, and the resultant values may be very small, they are nonetheless important to the analysis of transmission lines and distributed-constant networks.

The phase constant of the transmission line should not be confused with the velocity of the signal wave as it moves along the line, even though the two are related. The *free-space velocity* of electromagnetic radiation, current, and waves in wires is nearly the speed of light. This velocity is a fixed value, designated v_o (sometimes v_c or c) and generally specified in convenient units:

---

* All calculation steps must be carried in the calculator, as slight errors will accrue significant values of α.

1.86 × 10⁵ mi/s   9.84 × 10⁸ ft/s
3 × 10⁸ m/s   1.18 × 10¹⁰ in./s
3.28 × 10⁸ yd/s   3 × 10¹⁰ cm/s

The wavelength of a signal is computed by dividing its velocity by its frequency. Wavelength is identified by the Greek letter λ (small lambda). The free-space wavelength, in feet, of the 1000 MHz signal is computed as:

$$\lambda = v_o/f = (9.84 \times 10^8)/(1 \times 10^9) = 0.984 \text{ ft}$$

The wavelength at 1000 MHz of our previous example of parallel hookup wires was computed to be 1.0 ft, however, with 6.28×10^3 rad/1000 ft length. So we have *two* wavelengths to consider: one in free space (0.984 ft) and one along the transmission line (1.0 ft).

Phase Velocity

On the basis that we cannot change the absolute speed of light, we really cannot change the velocity of waves or radiation. We *can,* however, advance and retard phase shift of signals and waves with capacitors and inductors. The phase-constant (β) aspect of propagation describes this phenomenon. The velocity associated with the rate at which the phase changes is called the *phase velocity*. Phase velocity is not necessarily equal to free-space velocity or the speed of propagation. It may be computed by dividing ω ($2\pi f$) by the phase constant (β). The phase velocity, in feet, of our previous example at 1000 MHz is computed as:

$$v' = \omega/\beta = 6.28 \times 10^9/6.28 \text{ rad} \cong 1 \times 10^9 \text{ ft/s (actually } 0.999 \times 10^9 \text{ ft/s)}$$

This value of phase velocity exceeds the speed of light; but since no mass or real energy is involved with the concept of phase, it is possible to obtain such values. If we increase the capacitance of the transmission line of our example from 2 pF/ft to 3 pF/ft, keeping all other values the same, the propagation constant would increase as follows. (Refer to the earlier calculations of γ for comparison.)

i. First, the increased susceptance is computed as:

$$B = \omega C = 6.28 \times 10^9 \times 3 \times 10^{-9} = 18.8 \text{ S/1000 ft}$$

ii. Then the increased propagation constant is computed:

$$\gamma = \sqrt{ZY} = \sqrt{(3.14 \times 10^6 \angle 89.99°)(18.8 \angle 90°)}$$
$$= 7.69 \times 10^3 \angle 89.99° = 0.1 + j7695/1000 \text{ ft} = \alpha + j\beta$$

iii. Then the attenuation would increase to

0.1 (nepers) × 8.7 ≅ 1 dB

and the phase constant of 7.69 rad/ft would yield a wavelength of

$$\lambda = 2\pi/\beta = 6.28/7.69 = 0.816 \text{ ft}$$

and a phase velocity of

$$v' = \omega/\beta = 6.28 \times 10^9/7.69 = 8.16 \times 10^8 \text{ ft/s}$$

Now, the value of phase velocity is less than the speed of light, and the wavelength is proportionally shorter. Rather than repeat the tedium of the propagation calculations, a simpler approach can be used to determine phase velocity as a function of increased distributed capacitance: the ratio of two phase velocities varies as the square root of the capacitance or inductance ratios. Applying this relationship to our transmission line example, we can compute the new phase velocity as follows:

$$v' = v_1/\sqrt{C_2/C_1} = 1 \times 10^9/\sqrt{3 \text{ pF}/2 \text{ pF}}$$
$$= 8.16 \times 10^8 \text{ ft/s}$$

Wavelength is determined by phase velocity, which is in turn dependent on the propagation constant, so the wavelengths for a given frequency signal can vary in transmission lines from their value in air (or free space). Wavelengths are typically longer than their free-space values in waveguides, and shorter in two-conductor and coaxial cables. One analogy you may consider as representative of the phase velocity concept, and how phase velocity can differ from free-space velocity (the speed of light, which is a fixed constant), is that of an automobile sliding across an ice pond. Let the rotation of the driven rear tires represent the phase velocity, and the motion of the car and front (free-wheeling) tires represent the free-space velocity—the *real* motion of the car. Driving the car in low gear without applying the brakes will result in (a) the driven rear tires slipping on the ice with higher engine RPM, yielding a higher phase velocity, and (b) slipping (skidding) on the ice with lower engine RPM, yielding a lower phase velocity; yet the car and free-wheeling front tires will advance at an approximate fixed velocity. With this analogy, *wavelength* becomes the lineal distance that the car moves along the ice per each 360° rotation of the tires.

All transmission lines, including the familiar flat TV twin-lead, twisted-pair telephone lines, coaxial cables, waveguides, and even long-distance power lines, are analyzed in terms of their distributed constants: L, C, R and G, Z_0, γ. These are the basic principles that describe how a transmission line can support and transfer a signal wave from one point to another. The decision to employ or neglect these distributed constants is made on the basis of the electrical distance the transmission line represents. If a 10% change is considered significant, it represents about 6° of a sine wave. Most often, decisions are made in favor of employing distributed constants when the length of the line exceeds a phase shift of about 6° or 0.017 λ. Placing this distance in the perspective of frequency yields about 20 inches (50 cm) at 10 MHz, scaled down to 1 inch (25 mm) at 200 MHz. It is a common misconception to relegate such analysis exclusively to the microwave frequencies.

POWER TRANSFER

The *power* relationships of signal transfer are analyzed using the more familiar basic electrical principles of Ohm's, Kirchhoff's, and Watt's laws. If you consider that transmission lines have impedance (Z_0), power can be transferred from a signal source into the line. Since the transmission line is primarily reactive, but has insignificant initial phase shift of signal voltage, it can be considered to be a *lossless* resistance, and basic electrical computations may thus be employed. Once the signal power is on the transmission line, it is transferred along the line to a load. When the signal encounters the load at the end of the line, the same basic electrical computations are again employed to determine power transfer from the line to the load. Figure 11-3 shows the transmission circuit and the equivalent relationships for basic electrical analysis.

The analysis technique for computing RF power transfer characteristics from the signal source to the line, and from the line to the load, may be summarized as follows (consider that all impedances are resistive, i.e., have 0° phase shift):

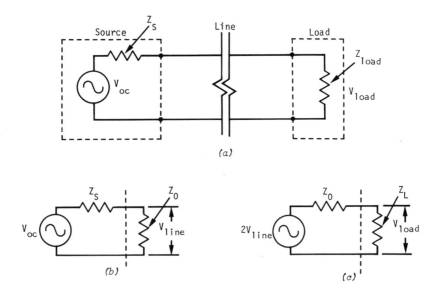

FIGURE 11-3

1. Determine the open-circuit voltage of the source as twice its terminal voltage when the source and load impedance are equal.

2. Consider that the source impedance is in series with the Z_0 of the transmission line, and compute the voltage, current, and power in the line as:

$$V_{line} = V_{oc}Z_0/(Z_s + Z_0)$$
$$I_{line} = V_{line}/Z_o = V_{oc}/(Z_s + Z_0)$$
$$P_{line} = V_{line}I_{line}$$

3. Now bring the signal to the load and consider the transmission line as a source, with its open-circuit voltage twice that of V_{line}, and compute the voltage, current, and power in the load as:

$$V_{load} = 2V_{line}Z_{load}/(Z_{load} + Z_0)$$
$$I_{load} = V_{load}/Z_{load} = 2V_{line}/(Z_{load} + Z_0)$$
$$P_{load} = V_{load}I_{load}$$

Table 11-1 shows the resulting voltage, current, and power from the application of the above procedures to a 50 Ω signal source, with 10 V_{oc}, and various values of Z_0. Table 11-2 shows the resulting current and power from the application of the above procedures to the values of Z_0, and corresponding V_{line} values, and a 50 Ω Z_{load}. You should note, in observing the results of the

TABLE 11-1.   POWER TRANSFER—SOURCE TO LINE
($Z_{source} = 50 \ \Omega$; $V_{oc} = 10$ V)

Z_0	V_{line} (V)	I_{line} (mA)	P_{line} (mW)
0	0	200	0
10	1.67	167	277
25	3.33	133	444
50	5	100	500
100	6.67	67	444
250	8.33	33	277
∞	10	0	0

TABLE 11-2.   POWER TRANSFER—SOURCE TO LOAD
($Z_{load} = 50 \ \Omega$; V_{line} from Z_0 values of Table 11-1)

Z_0	$2V_{line}$ (V)	V_{load} (V)	I_{load} (mA)	P_{load} (mW)
10	3.33	2.78	56	154
25	6.67	4.45	89	396
50	10	5	100	500
100	13.1	4.45	89	396
250	16.67	2.78	56	154

tabular calculations, that maximum power is transferred to the load *only* when the impedances of all three elements—source, line, and load—are matched. If one or more of the elements are not matched, maximum power transfer does *not* occur, and some of the available power is left in the signal source. If you refer to Table 11-1 and use the Z_0 of 100 Ω as an example, you should note that 444 of the available 500 mW were transferred, with 56 mW not leaving the generator.

Reflected Power

Table 11-3 has voltage, current, and power values identical to those in Table 11-1, but these values were obtained for a fixed transmission line Z_0 of 50 Ω and varying values of Z_{load}. You may observe the same 444 of the available 500 mW transferred to the 100 Ω load; but a unique condition exists here, since 500 mW is already on the transmission line. The transmission line is lossless and it cannot dissipate power, so the 56 mW is carried *back* along the line to the generator. Let us review the entire process from source-to-line-to-load-to-line-to-source:

1. The source and line impedance are matched, so 500 mW is transferred into the 50 Ω line.
2. The line carries the 500 mW to the load.
3. The 100 Ω mismatched load dissipates only 444 of the available 500 mW and *reflects* the remaining 56 mW.
4. The line carries the remaining 56 mW back to the source.
5. The source and line impedance are matched, so the reflected 56 mW is transferred back into the source.
6. The reflected 56 mW adds to the initial 500 mW dissipated in the source, bringing the total source power dissipation to 556 mW, for an output of 500 mW, thus decreasing the source efficiency.

If you apply the above sequence to higher load mismatch conditions, you should conclude (1) that the generator efficiency is ultimately reduced by

TABLE 11-3. POWER TRANSFER—LINE TO LOAD
($Z_0 = Z_{\text{line}} = 50 \ \Omega; V_{\text{line}} = 5$ V; $V_{\text{oc}} = 10$ V)

Z_{load}	V_{load} (V)	I_{load} (mA)	P_{load} (mW)	P_{refl} (mW)
0	0	200	0	500
10	1.67	167	277	223
25	3.33	133	444	56
50	5	100	500	0
100	6.67	67	444	56
250	8.33	33	277	223
∞	10	0	0	500

mismatched load impedances, and (2) that when the Z_0 of the line is equal to the source impedance, the generator operates as if no transmission line exists, with the load directly connected to the source. You should also conclude that the transmission line is a two-way element, carrying signals simultaneously from and back to the signal source.

RF power is carried on a transmission line in the form of voltage and current. Since the *perfect* line dissipates zero power, the voltage and current are 90° out of phase with each other and, verified by Watt's law: $VI \cos \angle \theta$. If $\angle \theta$ is 90°, $\cos \angle 90° = 0$; so the power is always zero regardless of the voltage and current values. The losses in series R and shunt G of the propagation constant yield phase angles in excess of 89° in the most lossy practical transmission lines, with less than 1% power dissipation per wavelength. So for all practical purposes the voltage and current can be considered to be out of phase by 90° on all transmission lines.

Line and Load Power

In engineering practice the line voltage, current, and power *from* the generator are called *incident,* as they are incident to the load. The line voltage, current, and power flowing *back to* the generator are called *reflected,* as they are reflected from the load. The incident voltage and current on the line are 90° out of phase, as are the reflected voltage and current. When the incident voltage encounters a resistive load, it develops an in-phase current, dissipating real power in the load. If an open circuit terminates the line, the incident current goes to zero and the incident voltage reflects in-phase, resulting in an open-circuit voltage of twice the incident value. If a short circuit terminates the line, the incident voltage goes to zero and the incident current reflects in-phase, resulting in a short-circuit current of twice the incident value. You can verify these relationships by applying step 3 of the three-step summary for RF power transfer analysis given at the beginning of this section. If you consider a 50 Ω Z_0 transmission line with an incident voltage of 5 V and an incident current of 100 mA (5 V/50 Ω = 100 mA), terminated with an open circuit, you can compute the (open-circuit) load voltage as:

$$V_L = 2V_{inc}Z_L/(Z_L + Z_0) = 2 \times 5 \times \infty/(\infty + 50) = 10 \text{ V}$$

If the same line is terminated with a short circuit, you can compute the (short-circuit) load current as:

$$I_L = 2V_{inc}/(Z_L + Z_0) = 2 \times 5/(0 + 50) = 200 \text{ mA}$$

From the above procedures you should note that while transmission line concepts are specialized, their analysis techniques employ *basic* electrical laws that relate impedance, voltage, current, and power. Some of the terminology and approaches may seem obscure, but you should realize that it is just "old-hat" theory dressed up.

If you refer back to Table 11-3 for the Z_0 of 50 Ω, with Z_L values of 25, 50, and 100 Ω, you should note that the transferred and reflected power are the same where Z_L is $Z_0/2$, and where Z_L is $2Z_0$, even though the load voltage and current values are *not* the same. You should further observe that Z_L values less than Z_0 yield higher load currents and lower load voltages than the matched condition where Z_L and Z_0 are equal. Since the reflected power is the same in both mismatched cases, the reflected voltages and currents must also be the same in both cases; yet the *load* voltage and current are *not* the same in both cases.

To account for this variation in load voltage and current, you might review the open-circuit and short-circuit conditions ($Z_L = \infty$ and 0, respectively) of Table 11-3 and the calculations of the previous paragraph. In both extreme cases, 100% of the incident power is reflected, with $2V_{inc}$ and zero I_{inc} as the load conditions when $Z_L = \infty$, and with zero V_{inc} and $2I_{inc}$ as the load conditions when $Z_L = 0$. If 100% of the incident power is reflected by the open or shorted load conditions, so *also* are 100% of the voltage and current. In the open-circuit case we have 200% of the incident voltage and zero current; whereas in the short-circuit case we have 200% of the incident current and zero voltage.

From the above observations you should agree with the following conclusions about incident, reflected, and load signals:

1. Reflected voltage *adds* to incident voltage when the load impedance exceeds Z_0.
2. Reflected voltage *cancels* incident voltage when the load impedance is less than Z_0.
3. Reflected current *adds* to incident current when the load impedance is less than Z_0.
4. Reflected current *cancels* incident current when the load impedance exceeds Z_0.

We can now proceed to the formulas that describe the related incident, reflected, and load signals, and that provide a basis for the calculation of these values.

$$\text{If } Z_L > Z_0: \quad V_L = V_{inc} + V_{refl}; \quad V_{refl} = V_L - V_{inc}$$
$$I_L = I_{inc} - I_{refl}; \quad I_{refl} = I_{inc} - I_L$$
$$\text{If } Z_L < Z_0: \quad V_L = V_{inc} - V_{refl}; \quad V_{refl} = V_{inc} - V_L$$
$$I_L = I_{inc} + I_{refl}; \quad I_{refl} = I_L - I_{inc}$$

Sample Problem 1

Given: A 50 Ω Z_0 transmission line with a V_{inc} of 2 V, terminated with a 150 Ω resistive load.

Determine: I_{inc}, P_{inc}, V_L, I_L, P_L, V_{refl}, I_{refl}, and P_{refl}.

(a)   The incident current is first computed as:

$$I_{inc} = V_{inc}/Z_0 = 2/50 = 40 \text{ mA}$$

(b)   Then the incident power is:

$$P_{inc} = V_{inc}I_{inc} = 2 \times 40 = 80 \text{ mW}$$

(c)   Then the load characteristics are computed:

$$V_L = 2V_{inc}Z_L/(Z_L + Z_0)$$
$$= 2 \times 2 \times 150/(150 + 50) = 3 \text{ V}$$
$$I_L = V_L/Z_L = 3/150 = 20 \text{ mA}$$
$$P_L = V_LI_L = 3 \times 20 = 60 \text{ mW}$$

(d)   Finally, the reflected values are computed:

$$V_{refl} = V_L - V_{inc} = 3 - 2 = 1 \text{ V}$$
$$I_{refl} = I_{inc} - I_L = 40 - 20 = 20 \text{ mA}$$
$$P_{refl} = V_{refl}I_{refl} = 1 \times 20 = 20 \text{ mW}$$
$$= P_{inc} - P_L = 80 - 60 = 20 \text{ mW}$$

REFLECTIONS AND STANDING WAVES

Now that you know what happens to transmission-line signals as they encounter mismatched loads, we will proceed to review what happens to these signals as they flow back along the line toward the generator. First, you should note that both the incident and reflected signals are continuous, so both incident and reflected power are carried simultaneously (in both directions) on the transmission line. The incident and reflected signals have the same phase velocity; but since they are moving in opposite directions, the phase relationships of the two signals are continuously changing. In one instance the two signals are in phase ($\angle\phi = 0°$), and in another instance the two signals are out of phase ($\angle\phi = 180°$). Between these two extremes the phase relationships vary (0° to 180° and 180° to 360° or 0°); and the resulting signal intensity of their interaction varies *also*.

Second, you should note from the previous problem that the initial phase relationship of incident and reflected signals is established at (or by) the load terminating the transmission line. When the incident and reflected signals *add*, they are in phase ($\angle\phi = 0°$); and when they *cancel*, they are out of phase ($\angle\phi = 180°$). Signal voltages are in phase at the load when the resistive Z_L exceeds Z_0, and they are out of phase when Z_L is less than Z_0. Signal currents, on the other hand, are in phase at the load when the resistive Z_L is less than Z_0, and they are out of phase when Z_L exceeds Z_0.

Signals along the Line

Let us proceed with the example of our previous problem, with a resistive Z_L of 150 Ω and a Z_0 of 50 Ω. The incident and reflected voltage signals were in phase at the load, so their values *added* for a total V_L of 3 volts. If we now move back along the transmission line a distance of one quarter-wavelength (90°), the incident signal will be 90° earlier ($-90°$) than its phase at the load; and the reflected signal, at this same quarter-wavelength point, will be 90° past ($+90°$) its phase at the load. With respect to each other, the incident and reflected signals are $-90°$ and $+90°$, or 180° out of phase at this quarter-wavelength point; so the incident and reflected voltages *cancel* for a total $V_{\lambda/4} = 2$ V $-$ 1 V $=$ 1 volt. If we again proceed an additional quarter-wavelength back along the transmission line, for a total distance of half a wavelength from the load, the incident signal will be 180° earlier than its phase at the load; and the reflected signal, at the same half-wavelength point, will be 180° past its phase at the load. With respect to each other, the incident and reflected signals are now $-180°$ and $+180°$, or 360° or 0°, and in phase.

If we continue to move back along the transmission line, in quarter-wavelength steps, the phase conditions will repeat in the manner outlined above. At this point you should conclude that the phase relationships of incident and reflected signals *reverse* every quarter-wavelength along the transmission line. The resultant voltages (3 volts and 1 volt in our example) at the quarter-wavelength intervals along the transmission line are stationary values, and form a series of stationary waves along the line. This stationary wave pattern repeats all the way back to the generator, and it is called *standing waves*. Figure 11-4 shows the standing-wave pattern that results from incident signals interacting with reflected signals.

If we repeat the same phase analysis for the incident and reflected currents of our problem example, we can determine that the currents *cancel* at the load for an I_L value of 20 mA; and that they *add* at the quarter-wavelength point for 40 mA $+$ 20 mA $=$ 60 mA. This pattern—20 mA, 60 mA, 20 mA, 60 mA, etc.—repeats all the way back to the generator. You should observe that the incident and reflected *currents* are in phase at the quarter-wavelength points where the incident and reflected *voltages* are out of phase, and vice versa. At the first quarter-wavelength back from the load, the voltage of our problem example is 1 volt and the current is 60 mA. A quick Ohm's law calculation will show that the impedance at this quarter-wavelength point is no longer the 50 Ω Z_0 of the transmission line, but:

$$Z_{\lambda/4} = (V_{\text{inc}} - V_{\text{refl}})/(I_{\text{inc}} + I_{\text{refl}}) = 1/60 \text{ mA} = 16.7 \ \Omega$$

At the second quarter-wavelength (half-wavelength, total) back from the

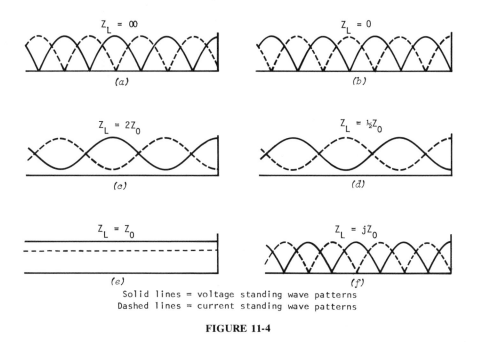

Solid lines = voltage standing wave patterns
Dashed lines = current standing wave patterns

FIGURE 11-4

load, the voltage is 3 volts and the current is 20 mA; so the impedance at the half-wavelength point is:

$$Z_{\lambda/2} = (V_{\text{inc}} + V_{\text{refl}})/(I_{\text{inc}} - I_{\text{refl}}) = 3/20 \text{ mA} = 150 \text{ }\Omega$$

Impedance Relationships

You should observe here that the characteristic impedance of the transmission line cannot be maintained if incident and reflected signals interact to produce standing waves on the line. You should observe also that the phase relationships of the incident and reflected signals are identical at the load and the half-wavelength points back along the line. So too are the impedances the same at the load and the half-wavelength points along the line. Another relationship you may note is that the Z_L/Z_0 ratio of $3:1$ yields the maximum voltage and impedance at the half-wavelength points along the line; and that this $3:1$ ratio is inverted at the odd quarter-wavelength ($^1/_4$, $^3/_4$, $^5/_4$, etc.) points along the line, yielding minimum voltages and impedances Z_0/Z_L of $3:1$. Finally, you should note that the Z_L/Z_0 and inverted Z_0/Z_L relationship is the same ratio as the minimum-to-maximum voltages and currents in the standing-wave pattern along the transmission line.

Power transfer and impedance relationships can be identified in the standing-wave patterns along transmission lines. This technique of standing-wave analysis is used extensively in engineering practice, where measurements of the maximum and minimum voltages along the standing-wave pattern are easily obtained. The ratio of these voltages is called *voltage standing-wave ratio* (VSWR). If only the VSWR and Z_0 of a transmission system are known, many other important circuit characteristics can be calculated, including:

1. The impedance of resistive loads
2. The minimum and maximum impedance, and the relative minimum and maximum voltage and current along the standing-wave pattern on the transmission line
3. The relative load voltage and current, the percent voltage reflected from the mismatched load
4. The percent power transferred to and reflected from the mismatched load

On the basis of these interrelationships, VSWR and impedance are two of the most important values found in the specifications of higher frequency RF components and systems.

At this point you should be able to relate the load and line impedances if VSWR and Z_0 are known values. You may also relate the maximum and minimum voltages of the standing-wave pattern, and transpose the formula to obtain load and reflected voltages if the incident voltage is known, as follows:

$$V_{max} = V_{inc} + V_{refl}$$
$$V_{min} = V_{inc} - V_{refl}$$
$$\text{VSWR} = V_{max}/V_{min} = (V_{inc} + V_{refl})/(V_{inc} - V_{refl})$$

and thus by transposition,

$$V_{refl} = V_{inc}(\text{VSWR} - 1)/(\text{VSWR} + 1)$$

Sample Problem 2

Given: A 300 Ω transmission line terminated in a resistive load, with a measured VSWR of 6, an incident voltage of 50 mV, and a voltage minimum at 5 half-wavelengths back from the load.

Determine: The resistive load impedance, the minimum and maximum line impedance and voltage of the standing-wave pattern along the line, the reflected voltage, and the power transferred to and reflected from the load.

(a)   First, the load resistance is determined to be less than Z_0 because a minimum in the standing-wave pattern is at a half-wavelength interval from the load. The resistance can therefore be computed as:

$$R_L = Z_0/\text{VSWR} = 300/6 = 50 \ \Omega$$

This value is also the minimum line impedance.

(b) The maximum line impedance inverts the above ratio and can be computed as:

$$Z_{\max} = Z_0 \ \text{VSWR} = 300 \times 6 = 1800 \ \Omega$$

(c) Now the reflected voltage is computed as:

$$V_{\text{refl}} = V_{\text{inc}}(\text{VSWR} - 1)/(\text{VSWR} + 1) = 50 \times 5/7 = 35.7 \ \text{mV}$$

(d) Then $V_{\max}$ and $V_{\min}$ are computed:

$$V_{\max} = V_{\text{inc}} + V_{\text{refl}} = 50 + 35.7 = 85.7 \ \text{mV}$$
$$V_{\min} = V_{\text{inc}} - V_{\text{refl}} = 50 - 35.7 = 14.3 \ \text{mV}$$

(e) The incident, load, and reflected powers are then computed as:

$$P_{\text{inc}} = (V_{\text{inc}})^2/Z_0 = (50 \ \text{mV})^2/300 = 8.33 \ \mu\text{W}$$
$$P_L = V_L{}^2/R_L = (14.3 \ \text{mV})^2/50 = 4.1 \ \mu\text{W}$$
$$P_{\text{refl}} = P_{\text{inc}} - P_L = 8.33 - 4.1 = 4.23 \ \mu\text{W}$$

The important relationships of the above problem include the identification of load characteristics through the standing-wave pattern on the line, and the incident and reflected signal ratios through the VSWR value.

SPECIFICATIONS

The specifications of RF systems and components generally include Z_0, VSWR, and a *maximum* power rating. In order to use these ratings and calculate power transfer without assuming a specific incident voltage, it is practical to work with *ratios* (or percent) of incident and reflected signals. The accepted terminology for this approach includes *reflection coefficients, mismatch loss,* and *return loss.* The reflection coefficient is the ratio of reflected-to-incident voltage, current, or power, and is identified with the Greek letter ρ (small rho). Since impedance values are fixed (generally Z_0), power ratios are the *square* of the voltage or current ratios:

$$\rho = V_{\text{refl}}/V_{\text{inc}} = I_{\text{refl}}/I_{\text{inc}} = (\text{VSWR} - 1)/(\text{VSWR} + 1)$$
$$\rho^2 = P_{\text{refl}}/P_{\text{inc}}$$

Based on the incident voltage, current, and power of 1.0 (100%), the load voltage, current, and power are:

$$V_L = V_{\text{inc}}(1 \pm \rho)$$
$$I_L = I_{\text{inc}}(1 \pm \rho)$$
$$P_L = P_{\text{inc}}(1 - \rho^2)$$

Mismatch loss is specified in dB, and is defined as the *power ratio* below (less than) 100% power transfer:

Mismatch loss $= 10 \log (1 - \rho^2)$

There is no commonly accepted symbol for mismatch loss even though it is a commonly used term. Small dB values of mismatch loss result from higher power transfer and closer impedance matching.

Return loss is also specified in dB, and it is defined as the *voltage ratio* above (greater than) 100% voltage *reflection:*

Return loss $= 20 \log 1/\rho$

There is also no commonly accepted symbol for return loss, although it too is a commonly used term. Return loss is often misused in place of VSWR where convenient measurement techniques yield return loss values in dB. Large dB values of return loss result from higher power transfer and closer impedance matching. VSWR values can be obtained from mismatch loss and return loss, through calculations of voltage or current reflection coefficients, in two steps as follows:

$$\rho = \sqrt{1 - (\log^{-1} (\text{mismatch loss dB}/10))} = 1/\log^{-1} (\text{return loss dB}/20)$$

and then

$$\text{VSWR} = (1 + \rho)/(1 - \rho)$$

Sample Problem 3

Given: A transmission line with a measured return loss of 6 dB.

Determine: The voltage reflection coefficient, the percent power transferred to the load, the mismatch loss, and the VSWR.

(a) First, the voltage reflection coefficient is computed from the return loss as:

$$\rho = 1/\log^{-1} (\text{dB}/20) = 1/\log^{-1} (6/20) = 0.5$$

(b) The power transferred to the load is then computed:

$$P_L = P_{\text{inc}} - P_{\text{refl}}$$

If $P_{\text{inc}} = 1.0$ and $P_{\text{refl}}/P_{\text{inc}} = \rho^2$,

$$\text{Percent } P_L = (1 - \rho^2) \, 100\% = (1 - 0.25) \times 100\% = 75\%$$

(c) Then, the mismatch loss is computed:

$$10 \log (1 - \rho^2) = 10 \log (0.75) = 1.25 \text{ dB}$$

(d) Finally, the VSWR is computed:

$$\text{VSWR} = (1 + \rho)/(1 - \rho) = (1 + 0.5)/(1 - 0.5) = 3$$

SMITH CHART ANALYSIS

Up to this point we have analyzed *only* the maximum and minimum impedance and voltage points along the standing-wave patterns on the transmission line, and we related these values to *only* resistive load impedances.
These limitations were purposely applied to simplify the mathematics and
direct your attention to the important principles that relate to standing waves
on transmission lines. In fact, very few RF load and circuit elements are
purely resistive over a broader frequency range of operation. The mathematics required for the analysis of reactive and complex elements, however, is
so painfully inconvenient that graphical techniques have been adopted to
replace the math. The best-known and most widely used graph for
transmission-line calculations is the Smith chart. It is used in solving a wide
range of network, circuit, and transmission problems encountered at higher
radio frequencies.

Structure

The Smith chart (Figure 11-5) is a circular graph on which resistance, reactance, and complex (rectangular-form) impedance can be plotted. Observe
that the chart combines circles of R (resistance) values with arcs of $+jX$

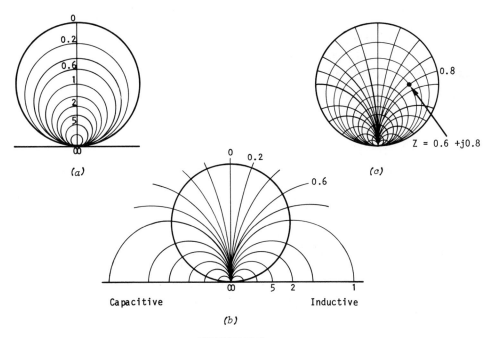

FIGURE 11-5

(inductive reactance) and $-jX$ (capacitive reactance) values. The circles and arcs are spaced logarithmically above and below a 1.0 (100%) value to accommodate the widest range of circuit values. The 1.0 center values on the chart most often represent the characteristic impedance of the transmission line, or a nominal circuit impedance value. A complex impedance would be plotted on the chart at the intersection of the R-value circle and the $\pm jX$-value arc. Figure 11-5c shows an impedance plot of $R + jX = 0.6 \pm j0.8$ Ω.

The complete detailed Smith chart is shown in Figure 11-6. The perimeter of this chart is divided in phase angles in degrees ($\pm 180°$) and phase shift in wavelengths (± 0.50 λ). A line extending from the exact center of the chart, through the impedance plot of $0.6 + j0.8$, will intersect the perimeter at $90°$ and 0.125 λ, indicating that the incident and reflected voltages are $90°$ out of phase, that there is 0.125 λ to the nearest minimum point of the standing-wave pattern along the line, and that this minimum is in the direction of the load. VSWR values are plotted on the Smith chart where the R circle that equals the VSWR crosses the $\pm jX = 0$ line. You should recall that VSWR $= R_L/Z_0$, so with Z_0 at the exact center of the chart ($Z_0 = 1.0 \pm j0$), the VSWR and R_L would be plotted at the same point.

The reference identification printed along the $\pm jX = 0$ line, which reads "Resistance component (R/Z_0)," and the other references on the chart—(G/Y_0), $(-jX/Z_0)$, $(-jB/Y_0)$, $(+jX/Z_0)$, and $(+jB/Y_0)$—relate to the concept of *normalization*, not VSWR. Before any actual impedances can be plotted on the Smith chart, they must be normalized by dividing their actual values by Z_0. A complex load impedance of $20 - j30$ Ω terminating a 50 Ω Z_0 transmission line, for example, would be normalized to $0.4 - j0.6$, and the Z_0 would be normalized to 1.0, then these normalized values would be plotted. (Normalization allows universal use of the chart with any transmission line or circuit impedance value, thus eliminating a need for Smith charts that are calibrated for specific impedance values.)

The VSWR Circle

If you now proceed to draw a circle with a compass, placing the center point of the radius at the exact center of the Smith chart ($R = 1.0$ and $\pm jX = 0$), and rotate a full circle through the VSWR plot and back again, you will have *all* the incremental values for the standing-wave pattern. If you refer to Figure 11-7 where a VSWR $= 3$ circle is drawn, and find the point at the top of the chart where the VSWR circle intersects the $\pm jX = 0$ line, you should read an R value of $Z_0/3 = 0.33$. This is where the impedance has *inverted* from $3Z_0$ being the maximum impedance, to $Z_0/3$ being the minimum impedance. A line from the center of the chart, through this minimum impedance of $0.33 \pm j0$, will intersect the perimeter of the chart at a point indicating that the incident and reflected voltages are $180°$ out of phase, and that there are either zero or 0.50 wavelengths to the minimum impedance points along the standing-wave pattern on the line. If you now rotate around the Smith chart

IMPEDANCE OR ADMITTANCE COORDINATES

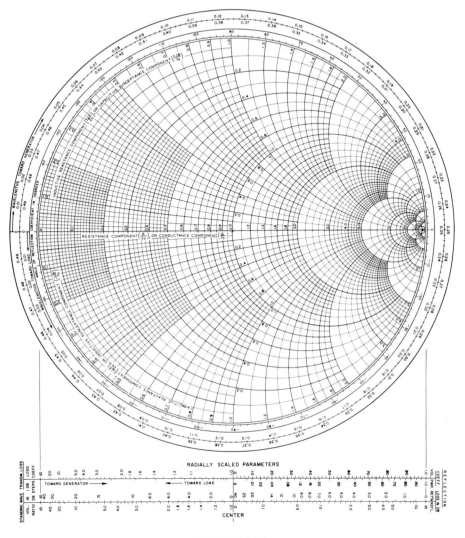

FIGURE 11-6

in equal increments, each point will provide phase, distance, and impedance data from which an accurate plot of a standing-wave pattern can be obtained. In Fig. 11-7 we used 45° increments and tabulated the data in Table 11-4, which was in turn used to plot the standing-wave pattern in the figure.

The point of this exercise is to show that the Smith chart can be used to exactly identify *all* the impedances at *all* the points along the standing-wave pattern. We can apply this capability with the Smith chart to many practical problems, including the identification of complex load and input impedances and the matching of mismatched elements.

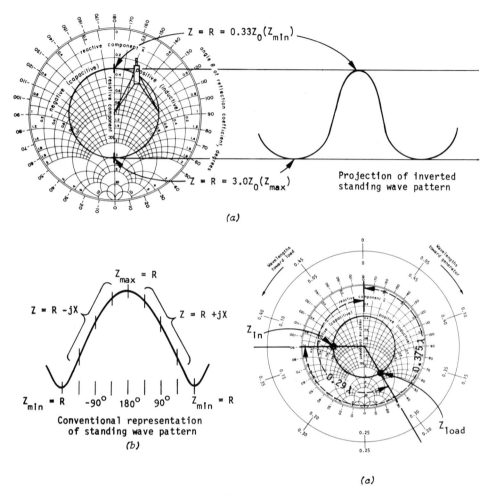

$$Z = R = 0.33Z_0 (Z_{min})$$

$$Z = R = 3.0Z_0 (Z_{max})$$

Projection of inverted
standing wave pattern

(a)

$Z_{max} = R$

$Z = R -jX$ $Z = R +jX$

$Z_{min} = R$ $-90°$ $180°$ $90°$ $Z_{min} = R$

Conventional representation
of standing wave pattern

(b)

Z_{in}

Z_{load}

(c)

FIGURE 11-7

Sample Problem 4

Given: A 100 Ω transmission line with a measured VSWR of 3, a minimum point in the standing-wave pattern at 0.29 λ from the load, and with the overall length of the line at 0.665 λ from the load.

Determine: The normalized and actual load and input impedances.

(a) First, the VSWR circle of 3 is drawn on the Smith chart.

(b) The distance 0.29 λ toward the load is plotted along the chart perimeter.

(c) A line is drawn from this point on the perimeter through the VSWR circle and to the center of the chart. The intersection of this line with the VSWR circle is observed and read as 2.0 +j1.3, which is the normalized load impedance.

TABLE 11-4. TABULAR VALUES FOR STANDING-WAVE PATTERN (VSWR = 3.0; Z_0 = 50 Ω)

Angle of Reflection Coefficient (degrees)	Rectangular-Form Impedance	Polar-Form Impedance	Unnormalized Impedance (Ω)	Computed Line Voltage (%)
0	$3 \pm j0$	$3\angle 0°$	$150\angle 0°$	100
45	$1.38 \pm j1.3$	$1.9\angle 44°$	$95\angle 44°$	80
90	$0.6 + j0.8$	$1\angle 53°$	$50\angle 53°$	58
135	$0.38 + j0.36$	$0.52\angle 44°$	$26\angle 44°$	42
180	$0.33 \pm j0$	$0.33\angle 0°$	$17\angle 0°$	33
-135	$0.38 - j0.36$	$0.52\angle -44°$	$26\angle -44°$	42
-90	$0.6 - j0.8$	$1\angle -53°$	$50\angle -53°$	58
-45	$1.38 - j1.3$	$1.9\angle -44°$	$95\angle -44°$	80
0	$3 \pm j0$	$3\angle 0°$	$150\angle 0°$	100

(d) The distance of 0.29 λ is subtracted from the total of 0.665 λ, leaving 0.375 λ. This distance of 0.375 λ toward the generator is plotted along the perimeter of the chart.

(e) A line is drawn from this second point on the perimeter, through the VSWR circle and to the center of the chart. The intersection of this line with the VSWR circle is observed and read as 0.6–j0.8, which is the normalized input impedance.

(f) Finally, the actual impedances are calculated from the normalized values as $100(2.0 + j1.3) = 200 + j130$ Ω and $100(0.6 - j0.8) = 60 - 80$ Ω.

With a little practice you should be able to execute the procedures of the above problem in a minute or so, while the mathematical approach would probably take 10 minutes or longer, fill two pages of paper, and provide many opportunities for error. You can actually use the Smith chart to graph and analyze every important aspect of impedance, VSWR, and power transfer in RF transmission lines and networks. Even the basic electrical networks of the early chapters in this book can be analyzed with the chart, but in that case you probably will not find it as convenient a problem solver as your electronic scientific calculator. In engineering practice the chart is almost exclusively used for high-frequency work.

One convenient and often overlooked use of the chart is for the quick conversion of rectangular-form impedance to equivalent rectangular-form admittance (and vice versa). This is accomplished by simply plotting a point halfway around the chart (180°) to the opposite side of the rectangular-form impedance (or admittance). The example of Fig. 11-8a shows an impedance plot of $1.4 + j1.3$ converted to an equivalent admittance of $0.38 - j0.36$. Of course, all values obtained from the chart are in normalized form and must be converted to *actual R, X, G,* and *B* values.

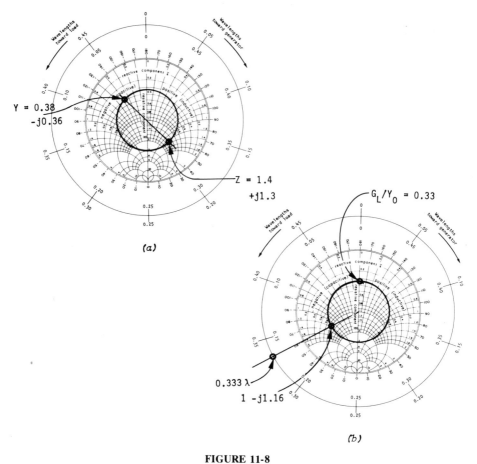

Y = 0.38
-j0.36

Z = 1.4
+j1.3

(a)

G_L/Y_0 = 0.33

0.333 λ

1 -j1.16

(b)

FIGURE 11-8

Admittance

The entire Smith chart can be used as an admittance chart for parallel-network analysis. The exact center of the chart must be converted to a normalized admittance value for such use, with the associated ±j arcs connected to susceptance values, and the circles converted from resistance to conductance values.

Sample Problem 5

Given: A 50 MHz signal with a wavelength of 6 meters on a 50 Ω transmission line that is terminated in a 150 Ω resistive load. There is a point on the line back from the load where a capacitor can be placed in *parallel* to cancel the mismatched load.

Determine: The value of the capacitor.

(a)   First, the equivalent admittances of Z_0 and R_L are computed:

$Y_0 = 1/Z_0 = 1/50 = 20$ mS   and $G_L = 1/R_L = 1/150 = 6.7$ mS

(b)   The center of the Smith chart is normalized to the Y_0 of 20 mS, and G_L is plotted as $G_L/Y_0 = 6.7/20 = 0.33 \pm j0$ on the chart.

(c)   A VSWR circle is drawn (VSWR = 3) through the $0.33 \pm j0$ and rotated toward the generator (clockwise), past the maximum admittance, and to a point that crosses the $G = 1.00$ circle on the $-jB$ side of the chart. At this point the admittance is read as $1.0 - j1.16$.

(d)   A line is drawn from the center of the chart through this admittance point, and to the perimeter of the chart where the wavelength is read as 0.333 λ.

(e)   The distance from the load is determined as $0.333 \times 6 = 2$ meters, and this is the location where the matching capacitor will be connected. The capacitive susceptance will add to the line admittance at that point:

$(0 + j1.16) + (1.0 - j1.16) = 1.0 \pm j0$

(f)   The *actual* value of the normalized susceptance is determined as:

20 mS $\times 1.16 = 23.2$ mS

(g)   Finally, the capacitance for the actual susceptance value is computed at 50 MHz:

$C = B/\omega = 23.2 \times 10^{-3}/(6.28 \times 50 \times 10^6) = 73.9$ pF

The procedures of the above problem were used to cancel a mismatched impedance using transmission-line analysis techniques. The capacitor reduced the VSWR of 3 to a value of 1, resulting in a matched impedance at (and only at) 50 MHz. The admittance approach was used because it was practical to wire the matching capacitor in parallel, *across* the transmission line. Placing the capacitor in series would require cutting the line at yet a different point, and then using the impedance function of the Smith chart.

Radially Scaled Parameters

In addition to the *direct* plots, the complete Smith chart includes incremental values of VSWR, attenuation, reflection coefficient, mismatch loss, and return loss scaled at one side of the circle graph and in parallel with the $\pm jX = 0$ line. These incremental values are identified as *radially scaled parameters*. (See Fig. 11-9.) This author prefers to use the Smith chart with the $\pm jX = 0$ line vertical and the radially scaled parameters to the left. Others prefer to rotate the chart so that the radially scaled parameters appear at the bottom,

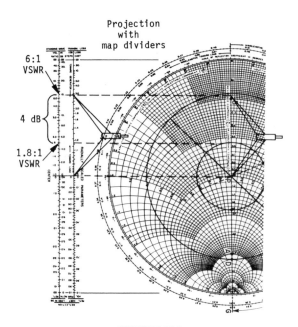

FIGURE 11-9

with the $\pm jX = 0$ line horizontally oriented. With either orientation you will have to rotate the chart, or your head, to read the marked values scaled there. On the *reflection* side are scales marked COEF./VOL. and PWR, which refer to the voltage (ρ) and power (ρ^2) reflection coefficients. The scales marked LOSS IN dB/RETN. REFL. refer to the return loss and mismatch loss, respectively. These values correspond to those along the $\pm jX = 0$ line from the center to the perimeter of the chart. Unless the chart is used with a T-square or parallel rule, it is difficult to directly transfer data from the $\pm jX = 0$ line to the radially scaled parameters. The writer prefers to use a pair of drafting or map dividers to first scale the distance on the circle graph and then transfer the length to the radially scaled parameters.

At the opposite end of the radially scaled parameters are the *standing wave* and *transmission loss* scales. The scales marked as VOL. RATIO and dB are values of VSWR. The scale marked 1 dB STEPS, under transmission loss, represents 1 dB increments of attenuation along the transmission line, which corresponds to the direction that the VSWR circle is rotated on the chart: toward the generator or toward the load. The scale marked LOSS COEF., under transmission loss, represents the *standing wave loss factor* and accounts for losses due to increased maximum voltages and currents, which in turn result from increased VSWR values. This loss factor is computed as VSWR loss = $(1 + \text{VSWR}^2)/2\ \text{VSWR}$. The two transmission loss scales are primarily used with lossy lines or networks.

Sample Problem 6

Given: A lossy 450 Ω transmission line section to be used as a 6 dB attenuator and that is exactly one half-wavelength long and is terminated by a resistive load of 100 Ω.

Using the radially scaled parameters of the Smith chart, determine: The values of transmission loss, VSWR reflection coefficient, mismatch loss, and return loss all referenced to the input, and then find the new input impedance.

(a)   First, the chart is normalized to 450 Ω and the load R_L is plotted on the chart as $100/450 = 0.22 \pm j0$. This value is then transferred to the transmission loss radial scale, 1 dB STEPS, intersecting the scale at about 2 dB in from the edge.

(b)   Then six increments are marked off *toward the generator* to account for the 6 dB of attenuation, and the dividers are readjusted to this new equivalent distance from the *center*, and moved to each radial scale to read the following values: loss coefficient = 1.05, VSWR = 1.4 and 28 dB, $\rho = 0.16$ and $\rho^2 = 0.026$, mismatch loss = 1 dB, and return loss = 16 dB.

(c)   The equivalent values are transferred from the radial scales to the circle chart, yielding an input impedance of $0.7 \pm j0$.

(d)   Finally, the actual impedance is computed as:

$$Z_0 Z_{norm} = 450 \times 0.7 = 315 \pm j0 \ \Omega$$

You should observe the reduced VSWR at the input to the attenuator which resulted from what is called *two-way return loss*, where *both* the incident signal to the 100 Ω load and the reflected signal from the load were reduced by the 6 dB attenuation. The total loss of 6 dB, a power ratio of 3.98, must also be corrected by the loss coefficient of 1.05, bringing the total to $1.05 \times 3.98 = 4.18 = 6.2$ dB.

Reactive Stubs

If the end of a transmission line is shorted or open, it is plotted on the Smith chart at the point where the perimeter of the chart intersects the $\pm jX = 0$ line. As the length of the shorted or open line is extended, the point on the chart will rotate clockwise along the perimeter, intersecting reactive values from $\pm jX = 0$ to $\pm jX = \infty$. By this method it is possible to obtain high-frequency network elements using shorted or open sections of transmission lines. If, for example, a capacitive reactance of 200 Ω is required for a high-frequency circuit, it could be obtained using an open section of 50 Ω transmission line, and then plotted on the Smith chart as a normalized value of $0 - j4.0$. The distance from the open to the reactive point is 0.039λ as shown in Fig. 11-10. If you now relate this reactance aspect of the chart to basic electrical circuits and networks, you should see that lumped-constant networks are possible to construct with transmission-line elements when higher frequency signals make standard capacitors and coils impractically small.

Sample Problem 7

Given: A high-frequency RF network attached to a transmission line of 300 Ω Z_0, and that requires a lumped capacitor and coil of 300 Ω reactance each to make a resonant circuit at 600 MHz.

Determine: The inductance and capacitance values, as well as the shortest electrical length for the two elements using the Smith chart considering that

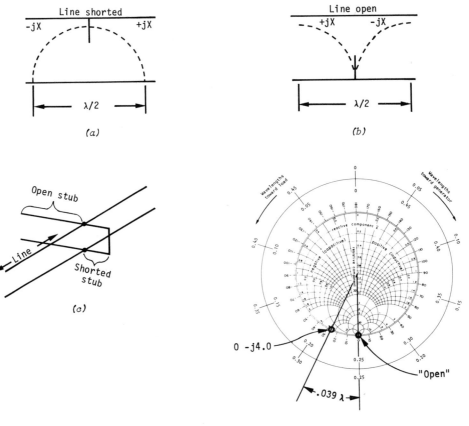

FIGURE 11-10

they are constructed of open and shorted sections of 300 Ω line, as shown in Fig. 11-10.

(*a*)   First, the inductance and capacitance are computed as:

$$C = 1/\omega X = 1/(6.28 \times 6 \times 10^8 \times 300) = 0.89 \text{ pF}$$
$$L = X/\omega = 300/(6.28 \times 6 \times 10^8) = 80 \text{ nH}$$

(*b*)   Since both component values are too small for standard parts, the Smith chart is normalized to 300 Ω and the inductor and capacitor are plotted as reactances of $+j1$ and $-j1$, respectively, at the perimeter of the chart.

(*c*)   The corresponding wavelengths are then plotted at 0.125 λ for $+jX = 1.0$ from a short, and 0.125 λ for $-jX = 1.0$ from an open.

IMPEDANCE MATCHING

The importance of impedance matching of RF networks and circuits was reviewed in the early chapters of this text. Impedance matching is perhaps even more important when using distributed-constant elements at the higher radio frequencies, where VSWR, power transfer, and efficiency are more critical factors. Although impedance-matching *principles* are the same with distributed constants as they are for the lower frequency lumped-constant networks, the techniques differ considerably when employing distributed-constant elements. The matching techniques that we will deal with are based on the usually accepted assumption that the generator source impedance is resistive and relatively well matched to the transmission line. The major concern of engineering practice in this respect is to optimize all elements to the Z_0 of the transmission line, rather than using the line and load in combination to obtain a conjugate system match, as may be done at lower frequencies.

Stub Matching

You have already been introduced to *one* concept of matching through Sample Problem 5 which showed how the Smith chart is used to transform impedances to the $R = 1.0$ or $G = 1.0$ circle. Any complex mismatched impedance with a real part (R or G) of 1.0 (normalized) can be matched without loss of RF power by employing a reactive or susceptive element of opposite sign ($\pm jX$) in series or ($\pm jB$) in parallel. The advantages of employing distributed-constant techniques include the use of the transmission line as a transformer. Even though a complex load impedance may not have its real R or G equal to Z_0 or Y_0, a length of line can easily transform the complex impedance to $R \pm jX$, or the complex admittance to $G \pm jB$. The choice of employing Z, R, and X values is determined by the physical practicality of cutting the transmission line and inserting series components. This approach is impractical at the higher frequencies, because the length of the series component may be a significant part of the signal wavelength. Where the series approach is impractical, Y, G, and B values are employed in conjunction with the admittance parameters on the Smith chart. Very often the reactive or susceptive element used for matching is an open or shorted section of line. These matching sections are connected to the main transmission line and serve to add a lumped reactance or susceptance where they are attached. Popular names applied to these matching sections include *matching stubs, reactive stubs,* and *tuning stubs.*

The same concepts of reactive matching can be employed without transformation; but if the R or G is not equal to Z_0 or Y_0, there will be a power loss in exchange (as a tradeoff) for the advantages of low VSWR. The technique is used when power loss is not critical or where impedance transformation is not practical; it requires that a given impedance or admittance value be combined with component values in order to bring the net normalized impedance or admittance to $1.0 \pm j0$. If, for example, an element with a normalized impedance of $0.6 \pm j0.8$ terminated a transmission line, a resistance of 0.4 and a capacitive reactance of $-j0.8$ added in series with the load would provide matching by bringing the effective load impedance to $1.0 \pm j0$. The disadvantage of this technique is that the load voltage and power are reduced because of the losses in the series resistance or parallel resistance required to bring the total load resistance to a normalized value of 1.0.

Quarter-Wave Transformers

Another popular impedance-matching technique employs a quarter-wave transformer. Although the first technique introduced required that a given load impedance be transformed to a value intersecting the $R = 1.0$ or $G = 1.0$ circle, this quarter-wave technique requires transforming the load impedance to a value intersecting the $\pm jX = 0$ or $\pm jB = 0$ line on the Smith chart. This transformation is achieved by rotating the VSWR circle from the plotted point of the load on the chart clockwise (toward the generator) to intersect the $\pm jX = 0$ line. From this zero reactance point the two impedance values, Z_1 (the $R \pm j0$ transformed from the load) and Z_0, are matched with a quarter-wavelength section of line that has an intermediate value of impedance. This intermediate value (Z') is computed as $Z' = \sqrt{Z_1 Z_0}$, and it is valid both in normalized and actual ohmic values. If, for example, a mismatched load that produced a VSWR of 3 on a 50 Ω transmission line were transformed to an R_L/Z_0 of 3 on the chart, and an actual value of $3Z_0 = 3 \times 50 = 150$ Ω, the impedance of the quarter-wave matching section would be computed as $Z' = \sqrt{Z_1 Z_0} = \sqrt{150 \times 50} = 86.6$ Ω, and normalized as $Z'_n = Z'/Z_0 = 86.6/50 = 1.73$. Figure 11-11c shows the matching effect on the Smith chart as achieved with the quarter-wave transformer.

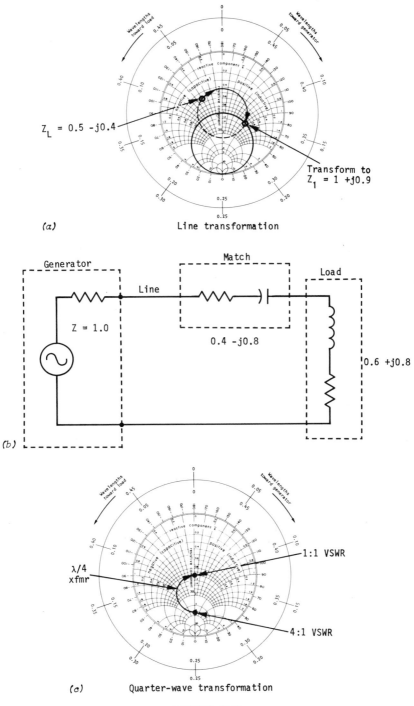

(a)   Line transformation

(b)

(c)   Quarter-wave transformation

FIGURE 11-11

Broadband Matching

All three of the impedance-matching procedures thus far outlined have one major drawback: they provide matching over only a narrow range of frequencies. Of course, frequency response and matching problems could be neglected if all electronic components were purely resistive and had impedances at standardized values; but even straight pieces of wire become reactive at high frequencies! One popular approach to broadening the frequency response of matching networks is the use of multiple quarter-wavelength transformers, providing a gradual transition from Z_1 to Z_0 through a series of intermediate Z' values.

If we extend the example used in the previous quarter-wave transformer problem, from one matching section to three sections, the matching will take place over a distance of three quarter-wavelengths. The intermediate impedances can then be computed, using Z' as the second transformer section; so the first matching section impedance is $Z'' = \sqrt{Z_1 Z'} = \sqrt{150 \times 86.6} = 114\ \Omega$, and then the third matching section impedance is $Z''' = \sqrt{Z' Z_0} = \sqrt{86.6 \times 50} = 66\ \Omega$. Figure 11-12a shows this *three-step transition*.

If we extend the concept of multistep quarter-wavelength transformers to five or seven steps, the intermediate impedance values become more closely spaced. The 150 Ω to 50 Ω transformer, of seven steps, for example, would have intermediate impedance values of 131, 114, 100, 87, 75, 66, and 58 Ω, respectively; and the total length of the transition would be 1.75 λ. Figure 11-12b shows the graphical results of connecting the average values of the intermediate impedances to form a uniform logarithmic taper of impedance values. *Taper transitions* of this type are a most popular means of impedance matching over a broader range of frequencies. Like the quarter-wave and multistep transitions, however, the taper transition (or *taper transformer*) must still be placed at a point along the transmission line where the load impedance has been transformed to intersect the $\pm jX = 0$ line on the Smith chart.

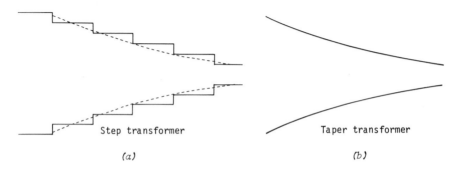

Step transformer   Taper transformer

(a)   (b)

FIGURE 11-12

CONCLUSION

This chapter has been concerned with the applied theoretical principles upon which practical high-frequency circuits and transmission systems operate. Your knowledge of these concepts, coupled with your ability to use the Smith chart effectively, will provide the necessary foundations for further study of practical transmission lines, distributed-constant circuits, microwave devices, and antennas. It is suggested that you equip yourself with additional Smith charts and an accurate compass/divider set or a Smith chart slide rule to use throughout the remainder of this text, and wherever you encounter high-frequency circuits and transmission lines. The problems that follow will serve to test your working knowledge of transmission line principles and to measure your achievement of the learning objectives specified at the beginning of this chapter. A further measure of your success with these objectives will be your ability to work with the applied topics of the next chapters. If you have difficulties, review this chapter as necessary until you can reduce the advanced concepts to the fundamentals presented here.

Problems

1. Given: A 100 ft length of transmission line with distributed constants of 60 Ω series resistance, 10 μH series inductance, 2 nF shunt capacitance, and negligible shunt conductance.

For operation at 500 MHz, determine:
- (*a*) Characteristic impedance (Z_0)
- (*b*) Propagation constant (γ)
- (*c*) Attenuation in dB/100 ft
- (*d*) Phase velocity
- (*e*) Wavelength

2. Given: A 150 Ω, 20 V open-circuit signal source driving a 50 Ω transmission line that is terminated with a 25 Ω resistive load.

Determine:
- (*a*) Incident voltage
- (*b*) Incident current
- (*c*) Incident power
- (*d*) Load voltage
- (*e*) Load current
- (*f*) Load power
- (*g*) Reflected voltage
- (*h*) Reflected current
- (*i*) Reflected power

3. Given: A 100 Ω transmission line with 2 V incident, terminated with a resistive load yielding a VSWR of 4, and with a maximum voltage point 2.50 wavelengths from the load.

Determine:
- (*a*) Load impedance
- (*b*) Maximum line impedance

(*c*)   Minimum line impedance

(*d*)   Reflected voltage

(*e*)   Reflected power

(*f*)   Load voltage

(*g*)   Load power

(*h*)   Maximum line voltage

(*i*)   Minimum line voltage

4.   Given: A 300 Ω transmission line terminated with a 100 Ω resistive load.

Determine:

(*a*)   VSWR

(*b*)   Voltage reflection coefficient

(*c*)   Power reflection coefficient

(*d*)   Percent power transferred to the load

(*e*)   Mismatch loss

(*f*)   Return loss

5.   Given: A 50 Ω transmission line with an incident power of 100 mW and a measured return loss of 12 dB.

Determine:

(*a*)   Voltage reflection coefficient

(*b*)   Maximum line voltage

(*c*)   VSWR

(*d*)   Maximum line impedance

(*e*)   Minimum line impedance

(*f*)   Load power

6.   Given: A complex load impedance of 80 ±j60 Ω terminating a 50 Ω transmission line.

With the aid of the Smith chart, determine:

(*a*)   Normalized load impedance

(*b*)   VSWR

(*c*)   Wavelength from the load to the nearest maximum impedance

(*d*)   Wavelength from the load to the nearest minimum impedance

(*e*)   Normalized minimum impedance

(*f*)   Actual impedance at 0.30 λ from the load

(*g*)   Actual impedance at 5.50 λ from the load

7.   Given: A 100 Ω transmission line with a minimum line impedance of 25 Ω at a distance of 0.10 λ from the load.

With the aid of the Smith chart, determine:

(*a*)   VSWR

(*b*)   Normalized load impedance

(*c*)   Angle of the voltage reflection coefficient

(*d*)   Normalized impedance 0.30 λ from the load

(*e*)   Wavelengths from the load where the VSWR circle intersects the ±jX = 0 line at maximum impedance

(*f*)   Wavelengths from the load where the VSWR circle intersects the 1.0 −jX line

8. Given: A 300 Ω transmission line with 4 dB of attenuation over its length, and terminated with a 75 Ω resistive load.

With the aid of the radial scales on the Smith chart, determine:

(a) Load VSWR
(b) Input VSWR
(c) Transmission loss coefficient
(d) Voltage reflection coefficient
(e) Power reflection coefficient
(f) Mismatch loss
(g) Return loss

9. Given: A mismatched load on a 600 Ω transmission line yielding a minimum line impedance at 0.20 λ from the load and a VSWR of 3, to be transformed to $1 + jX$ for matching with a series capacitor.

For a frequency of 100 MHz, determine:

(a) Normalized load impedance
(b) Actual ohmic load impedance
(c) Transformed normalized impedance value ($1 + jX$)
(d) Reactance of the matching capacitor
(e) Capacitance value of the matching capacitor
(f) Wavelength from the load where the capacitor is to be inserted into the line

10. Given: A mismatched load yielding a minimum transmission line impedance 0.15 λ from the load and a VSWR of 2.5, to be transformed to $1 + jB$ for matching with a parallel-connected shorted stub.

For a frequency of 3 GHz and a line wavelength of 10 cm, determine:

(a) Normalized load admittance
(b) Transformed normalized admittance value ($1 + jB$)
(c) Distance in centimeters from the load where the stub is to be connected
(d) The susceptance of the stub terminals
(e) The length of the stub in centimeters

11. Given: A normalized load impedance of $0.6 - j0.8$ to be matched to a 300 Ω Z_0 transmission line using a three-section, quarter-wave step transition.

For a signal wavelength of 30 inches, determine:

(a) VSWR
(b) Distance from the load to the first quarter-wave step in inches
(c) Actual impedance of the second quarter-wave section
(d) Actual impedance of the third quarter-wave section
(e) Actual impedance of the first quarter-wave section
(f) Overall length of the transition in inches

CHAPTER 12

TEM AND WAVEGUIDE TRANSMISSION LINES AND COMPONENTS

In Chapter 11 you reviewed the theoretical aspects of the transmission line, distributed constants, and RF power transfer. Now theory alone is fine, but it cannot build anything. Technology requires the application of theory to hardware, and that is the basis of this chapter.

Practical transmission lines and elements are required to transfer RF signals from point to point in electronic circuits and systems. These practical lines evolve from two distinct principles of operation: transverse electromagnetic (TEM), or either transverse electric (TE) or transverse magnetic (TM). The common two-wire, coaxial, and strip lines are categorized as TEM lines; the less common, but important, waveguides are categorized as TE or RM lines. These terms and categories will be defined and explained in detail later in this chapter. The requirement to employ such transmission lines rather than just "wire up" circuits and systems depends almost entirely on the wavelength of the RF signal and the distance that the signal is to travel, as elaborated on in Chapter 11. Any RF signal that is to be transferred a distance greater than 1% of its wavelength should employ some type of transmission line. Since most signals *do* move distances greater than 1% of their wavelength in RF systems, transmission lines are indispensable system elements.

If you carry the "1% of wavelength" argument to the higher frequency electronic circuits, even the components, networks, and active devices must be built in the form of transmission lines for operation. The stripline transistor package and cutaway coaxial attenuator are examples of high-frequency components. The physical form that these elements take allows them to function electrically in high-frequency circuits. This chapter will review the structures required of *R, L,* and *C* components and networks for such operation at high frequencies.

In addition to transmission-line packages for *R, L,* and *C* components, networks, and active electronic devices, there is a family of special distributed-constant elements that are unique to very short wavelength RF circuits and systems. These elements are often identified as *microwave* devices, and the frequencies at which the devices are commonly used are likewise identified as *microwave* frequencies. As such, the term *microwaves*—beyond its direct translation of "short waves"—is not easily defined. Microwave circuits and systems are generally those that employ the special family of distributed-constant microwave devices. In practice, the frequencies at which microwave devices become practical extend from below 800 MHz to above 100 GHz. Since waveguide is the more practical transmission line at these higher microwave frequencies, *waveguide* devices are, by definition, also *microwave* devices. So by definition the components and techniques covered in this chapter will also include microwave devices.

LEARNING OBJECTIVES

Upon completing this chapter, including the problem set, you should be able to:

1. Understand the concepts of TEM transmission and be able to compute Z_0, L, C and the dimensions for two-wire, coaxial, and strip transmission lines.
2. Relate practical TEM transmission-line characteristics to the theoretical principles of Z_0 and propagation constant advanced in Chapter 11.
3. Understand the operation and construction techniques of R, L, and C components in transmission-line form.
4. Compute the characteristics and dimensions for the fabrication of simple electrical networks in transmission-line form.
5. Have knowledge of popular TEM transmission lines, RF connector pairs, and special (including microwave) elements.
6. Understand the concepts of TE-TM propagation in waveguide lines and elements, and compute dimensions for waveguide and the characteristics of waveguide signals.
7. Understand the operation and construction techniques for achieving R, L, and C components in waveguide form.
8. Compute the characteristics and dimensions for the fabrication of simple electrical networks in waveguide form.
9. Have knowledge of microwave waveguide components used in communications systems.
10. Have knowledge of and apply high-frequency measurement techniques to waveguide and microwave components and signals.

TEM LINES AND ELEMENTS

When you think of a transmission line you probably visualize a flexible cable that connects a communications system to an antenna. Such transmission lines are in the form of either two-wire or coaxial cables, and they are available in a range of popular Z_0 values to match antenna or system requirements. The popular flat 300 Ω "TV twin lead" represents the typical two-wire form of line; whereas its counterpart 75 Ω "TV cable" represents the typical coaxial form of line. Each form has its representative advantages and tradeoff disadvantages, but the coaxial lines are by far the most popular choice for RF transmission.

TEM is the accepted abbreviation for *transverse electromagnetic*, which is technically defined as "an electrical conductor (or conductors) with signal current flowing along its length, and with electric and magnetic fields *transverse* (at right angles) to the current flow and to each other." This rather abstract definition might be better understood if you refer to the vector drawing of Figure 12-1, where the heavy-line vector represents the direction of alternating RF current, and the light solid and dashed-line vectors represent the alternating direction and magnitude of the electric and magnetic fields, respectively, that result from the RF conductor current. From a more practical point of view, the operation of the TEM line depends on a conductivity path (conductor), and the dielectric constant and permeability of the material that surrounds the conductor and thus supports the transverse (90°) electric and magnetic fields.

Most classical engineering books on transmission lines describe TEM concepts based on the mathematical wave theory of J. C. Maxwell. Maxwell's equations define what we describe above with words and pictures, but the mathematics is so abstract that it is easier to visualize TEM concepts rather

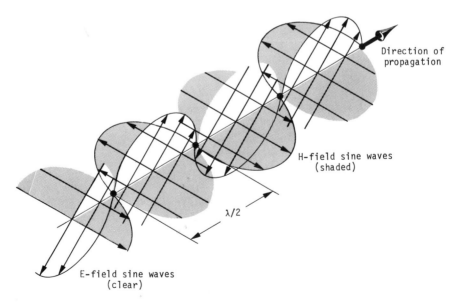

Direction of propagation

H-field sine waves (shaded)

λ/2

E-field sine waves (clear)

FIGURE 12-1

than to compute field forces. While TEM field forces are abstract and difficult to compute, their *effects* are most practical to apply to transmission lines; that is, the *permeability* of the medium surrounding the conductor directly affects the series inductance of the TEM line, while the *dielectric constant* of the medium directly affects the shunt capacitance of the line. Armed with this information you should be able to correlate the electrical properties of the materials used to fabricate transmission lines with the principles of signal propagation presented in Chapter 11. Incidentally, you should be aware that the permeability of common insulating materials used in the structure of practical transmission lines (including air and other gases) is very close to 1.0, thus magnetic effects due to permeability are most often neglected in the computation of transmission line characteristics.

Two-Wire Lines

The concepts of distributed constants were developed in Chapter 11 using the two-wire line as an example. If you apply the theoretical concepts to practical two-wire lines, you should relate the characteristic impedance to the diameters and spacing of the two conductors. The smaller conductor diameters yield increased inductance; whereas the greater the spacing between the conductors, the lesser the capacitance. Since the characteristic impedance (Z_0) is directly dependent upon the L/C ratio of the conductors, a wide range of impedance values can be obtained by varying these two dimensions. A third variable in the structure of practical lines is the dielectric constant, designated by the Greek ϵ (small epsilon), of the insulating material that separates the two conductors. Most popular dielectrics used in the structure of transmission lines and elements fall within the range of 1.5 to 3.0. The characteristics of two-wire lines may be computed using the following formulas:

$$L = \begin{cases} 0.28 \log (d/r) & \text{(in } \mu\text{H/ft)} \\ 0.92 \log (d/r) & \text{(in } \mu\text{H/m)} \end{cases}$$

$$C = \begin{cases} 3.68 \ \epsilon/\log (d/r) & \text{(in pF/ft)} \\ 12.08 \ \epsilon/\log (d/r) & \text{(in } \mu\text{H/m)} \end{cases}$$

$$Z_0 = (276/\sqrt{\epsilon}) \log (d/r) \text{ (in ohms)}$$

where r is the radius of one conductor, d is the center-to-center distance between the conductors, and ϵ is the dielectric constant of the insulating material between the conductors.

The characteristics of an unknown two-wire cable can be determined by measuring the conductors and spacing, and then identifying the dielectric material. This information can then be applied to the formulas of the previous paragraph to calculate L, C, and Z_0. Assume, for example, that you have a two-wire polyfoam dielectric cable with AWG 22 conductors that are spaced at a distance of 0.3 inch. After obtaining the wire diameter from Table 12-1 and the dielectric constant from Table 12-2, you can perform the calculations as follows:

$$L = 0.28 \log (d/r) = 0.28 \log (0.3/0.0125) = 0.39 \ \mu\text{H/ft}$$
$$C = 3.68\epsilon/\log (d/r) = 3.68 \times 1.56/\log (0.3/0.0125) = 4.16 \ \text{pF/ft}$$
$$Z_0 = (276/\sqrt{\epsilon}) \log (d/r) = (276/\sqrt{1.56}) \log (0.3/0.0125)$$
$$= 305 \ \Omega = \sqrt{L/C} = \sqrt{(3.9 \times 10^{-7})/(4.16 \times 10^{-12})} = 305 \ \Omega$$

Reversing the above problem in order to obtain dimensions from given electrical requirements will require reversing the mathematical formulas. This approach is by far more practical if you are to build transmission-line elements to meet electrical specifications. The formulas will transpose as follows:

$$(d/r) = \log^{-1} (L/0.28) \ \text{in.}$$
$$(d/r) = \log^{-1} (3.68\epsilon/C) \ \text{in.}$$
$$(d/r) = \log^{-1} (Z_0 \sqrt{\epsilon}/276)$$

TABLE 12-1.   WIRE DIAMETER AND RESISTANCE (COPPER)

Gauge (AWG or B&S)	Diameter (in.)	Resistance (Ω/1000 ft)	Gauge (AWG or B&S)	Diameter (in.)	Resistance (Ω/1000 ft)
0000	.4600	.04901	22	.02535	16.14
000	.4096	.06180	23	.02257	20.36
00	.3648	.07793	24	.02010	25.67
0	.3249	.09827	25	.01790	32.37
1	.2893	.1239	26	.01594	40.81
2	.2576	.1563	27	.01420	51.47
3	.2294	.1970	28	.01264	64.90
4	.2043	.2485	29	.01126	81.83
5	.1819	.3133	30	.01003	103.2
6	.1620	.3951	31	.008928	130.1
7	.1443	.4982	32	.007950	164.1
8	.1285	.6282	33	.007080	206.9
9	.1144	.7921	34	.006305	260.9
10	.1019	.9989	35	.005615	329.0
11	.09074	1.260	36	.005000	414.8
12	.08081	1.588	37	.004453	523.1
13	.07196	2.003	38	.003965	659.6
14	.06408	2.525	39	.003531	831.8
15	.05707	3.184	40	.003145	1049.
16	.05082	4.016	41	.00280	1323.
17	.04526	5.064	42	.00249	1673.
18	.04030	6.385	43	.00222	2104.
19	.03589	8.051	44	.00197	2672.
20	.03196	10.15	45	.00176	3348.
21	.02846	12.80	46	.00157	4207.

TABLE 12-2. DIELECTRIC CONSTANTS OF
COMMON MATERIALS ($T = 25°C, f = 100$ MHz)

Material	Dielectric Constant
Aluminum oxide	8.8
Titanium dioxide	100
Silicon dioxide (glass)	3.78
Epoxy resin	3.35
Polychlorotrifluorethylene	2.32
Polyethylene	2.25
Cellular polyethylene (polyfoam)	1.64
Polypropylene	2.25
Cellular polypropylene	1.56
Polystyrene	2.5
Polytetrafluorethylene (Teflon)	2.1
Polyvinylchloride (PVC)	2.7
Butyl rubber	2.35
Neoprene rubber	4.5
Balsawood	1.3
Douglas fir wood	1.9
Mica, glass, titanium dioxide	9.0
Soil, sandy, dry	2.5
Ice, pure	3.5
Water (distilled)	78

In their original and transposed forms the formulas for two-wire lines may be applied to a wide variety of practical problems. If, for example, two 300 Ω antennas are to be stacked for parallel operation from a center-fed transmission line, using ¼-inch aluminum tubing as both electrical and mechanical connectors as in Figure 12-2, the spacing of the tubing can be quickly computed as:

$$d = r \log^{-1} (Z_0 \sqrt{\epsilon}/276) = 0.125 \log^{-1} (300 \times 1.0/276)$$
$$= 1.53 \text{ in. (center to center)}$$

A second example is here presented from which you may observe some important characteristics of transmission-line applications. Let us assume that some person figured to save money by using two-wire *zip cord* (lamp cord) to run an intercom system over a distance of 1000 ft. Typical zip cord uses AWG 18 (0.040 in.) conductors spaced at 0.100 in. centers with vinyl (PVC) insulation. Since the cable is to be used for audio frequencies with wavelengths far in excess of the 1000 ft distance, we need not be concerned with the Z_0 of the cable. The cable does have resistance, inductance, and capacitance, however, and these factors may affect the operation of the system. So, while the cable resistance can be obtained from Table 12-1, L and C must be computed as follows:

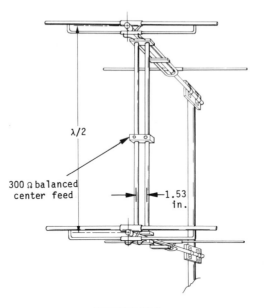

λ/2

300 Ω balanced
center feed

1.53
in.

FIGURE 12-2

$L = 0.28 \log (d/r) = 0.28 \log (0.1/0.02) = 0.2 \ \mu\text{H/ft}$
$\quad = 200 \ \mu\text{H/1000 ft}$
$C = 3.68\epsilon/\log (d/r) = 3.68 \times 2.7/\log (0.1/0.02)$
$\quad = 14.2 \ \text{pF/ft} = 14.2 \ \text{nF/1000 ft}$

Given the input impedance (resistance) of the intercom receiver as 10 kΩ, the transmission line is treated as a simple series-parallel network, with 200 μH in series with the parallel combination of 14.2 nF and 10 kΩ. Thorough analysis of the simple network will require identification of resonance and the −3 dB (corner) frequency point, along with its frequency response characteristics as follows:

$f_R = 1/2\pi \sqrt{LC} = 1/(6.28 \sqrt{2 \times 10^{-4} \times 14 \times 10^{-9}})$
$\quad = 95 \ \text{kHz}$
$X_C = X_L = \omega L = 6.28 \times 95 \times 10^3 \times 2 \times 10^{-4} = 120 \ \Omega$

Next, the −3 dB frequency is computed, where X_C equals the 10 Ω input resistance of the intercom:

$f = 1/2\pi CX = 1/(6.28 \times 14 \times 10^{-9} \times 10^4) = 1.14 \ \text{kHz}$

Thus, the frequency response of the cable, due to capacitive losses, will have a 6 dB/octave (20 dB/decade) slope above the −3 dB frequency of 1.14 kHz.

The zip cord/intercom example of the previous paragraph may be somewhat ridiculous, but it points out many factors that you should be aware of

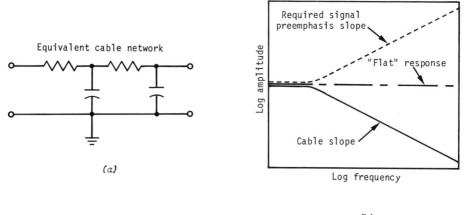

Equivalent cable network

(a)

Log amplitude

Log frequency

Required signal preemphasis slope

"Flat" response

Cable slope

(b)

FIGURE 12-3

when employing cables to transfer signals. Although it might occur to you that the use of transformers in the circuit would lower the 10 kΩ impedance enough to obtain flat frequency response throughout the audio range of the intercom, there will still be *some* frequency where the cable will begin its 6 dB/octave slope. Since cables are used to transfer RF as well as AF signals, the frequency response of the cable is an important factor to consider in communications applications. Most communications signals that are to be transmitted by cable (transmission line) employ preemphasis in order to counteract the capacitive cable losses and obtain a flat frequency response at the receiver. Figure 12-3 shows the equivalent cable network and frequency response with required preemphasis to "flatten the response." Perhaps the other factor that you should have observed from the problem of the previous paragraph is that only technically qualified people should be allowed to specify transmission lines.

Telephone systems are the largest users of two-wire transmission lines for communications. Their standard line is called a *twisted pair*, in which, even in multiple-circuit cables, wires are paired off and twisted to minimize crosstalk and interference. At telephone voice frequencies the line capacitance will significantly affect the transmission of signals with the -6 dB/octave frequency response as noted in the previous paragraph. Typical twisted-pair AWG 24 telephone cables have a capacitance of about 30 pF/ft, an inductance of about 80 nH/ft, and a resistance of about 25 Ω/1000 ft, hence long cable runs to accumulate high capacitance values along with resistive losses. (The standard impedance of telephone systems is 600 Ω—not the 10 kΩ of the intercom example in the previous paragraph.)

In addition to the amplitude vs frequency compensation, *phase compensation* is also applied to prevent distortion of the higher voice frequencies. This kind of phase shift is called *delay distortion;* that is, the phase shift of the higher frequency signals caused by the cable capacitance is greater than for the lower frequency signals. Phase compensation is achieved with loading coils that provide the opposing inductive reactance required to cancel the cable capacitance at the higher frequencies. Phase and frequency (amplitude) compensation are important for telephone circuits that use cables to transmit and receive pulse and digital signals, because the rise and fall times of the pulse signals are lengthened by the cable capacitance and are equivalent to frequencies far above the 3 kHz voice-frequency limit of telephone systems.

Two-wire lines are occasionally used for the transmission of higher frequency RF signals. They are especially useful as antenna feeds or matching elements because their spacing and length can be easily adjusted. A simple 960 MHz, quarter-wave, 300 Ω to 75 Ω transformer, for example, can be specified for construction using AWG 10 wires as follows (see Figure 12-4):

i. $\lambda = v_0/f = 1.18 \times 10^{10}/960 \times 10^6 = 12.3$ in.
   $\lambda/4 = 12.3/4 = 3.08$ in.
ii. $Z' = \sqrt{Z_1 Z_0} = \sqrt{300 \times 125} = 193\Omega$
iii. $d' = r \log^{-1} (Z_0'/276) = .05 \log^{-1} (193/276) = 0.25$ in.
   $d_1 = r \log^{-1} (Z_1/276) = .05 \log^{-1} (300/276) = 0.61$ in.
   $d_0 = r \log^{-1} (Z_0/276) = .05 \log^{-1} (125/276) = 0.14$ in.

The spacing of the transformer-section conductors would thus be 1.22 in. for 300 Ω, to a 0.35 in. quarter-wave matching section, and then to 0.187 in. for 75 Ω. From this example you should conclude that it is quite easy to convert the theoretical concepts advanced in Chapter 11 to actual practice using two-wire lines.

Coaxial Lines

The most popular RF transmission line is the coaxial cable where the outer conductor acts as an RF shield to a concentric (coaxial) center conductor. *Coax* has the advantages of a completely shielded system that should neither radiate nor "pick up" stray signals; but with the grounded shield, it is an *unbalanced* transmission line (having a single "hot" conductor), which may be disadvantageous in some applications. Coaxial cables are available in a wide range of electrical and mechanical configurations, and they have a full complement of associated interconnecting hardware as standard manufacturers' catalog items. Flexible coaxial cables are common, but there are also varying forms of specialized rigid and semirigid coaxial lines that are used in a wide variety of RF systems applications. The more popular flexible cables are listed under the military designations by which they were originally

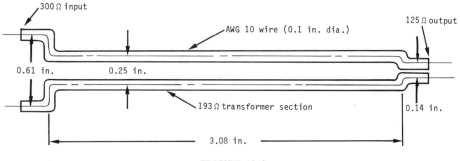

FIGURE 12-4

identified. These designations identify the cables by an RG-*n*/U, with the *n* identifying the particular cable type. Many mating connectors and coupling devices for the cables also have military designations UG-*n*/U, with the *n* identifying the particular connector type. RG cables have braided-wire outer conductors, but less critical (and less costly) cable types may use aluminum foil and strands of bare wire for the outer conductor. Semirigid coax lines often employ outer conductors of solid soft-annealed copper or soft-alloy aluminum. The popular dielectric materials used in flexible and semirigid coax include solid polyethylene, cellular, or foam polyethylene and Teflon. With such a wide variety of cable types and associated hardware available for application, you should review the manufacturers' catalogs to assure your familiarity with the popular coaxial transmission lines and couplings.

As with two-wire transmission lines, the diameters and spacing of the coaxial conductors and the dielectric constant are the three factors that determine the series inductance, shunt capacitance, and characteristic impedance values of coaxial transmission lines and cables. The characteristics of coaxial lines may be computed using the following formulas:

$$L = 0.14 \log (D/d) \qquad \text{(in } \mu\text{H/ft)}$$
$$C = 7.35\epsilon/\log (D/d) \qquad \text{(in pF/ft)}$$
$$Z_0 = (138/\sqrt{\epsilon}) \log (D/d) \qquad \text{(in ohms)}$$

D is the inner diameter of the outer conductor, *d* is the outer diameter of the inner conductor, and ϵ is the dielectric constant of the insulating material (dielectric) separating the conductors. If you are required to obtain physical dimensions from given electrical requirements in order to build coaxial elements to meet electrical specifications, you may apply the formulas in their transposed form as follows:

$$(D/d) = \log^{-1} (L/0.14)$$
$$(D/d) = \log^{-1} (7.35\epsilon/C)$$
$$(D/d) = \log^{-1} (Z_0 \sqrt{\epsilon}/138)$$

The formulas for coaxial lines may be applied to a wide variety of practical problems. Although coaxial cables are available in mass-produced form and are somewhat impractical to fabricate, the formulas may be applied to obtain electrical characteristics from the measured dimensions of unknown or unmarked cables. If, for example, an unknown cable has an IDOC (inner diameter of the outer conductor) of 0.5 in., an ODIC (outer diameter of the inner conductor) of 0.083 in., and a polyethylene dielectric ($\epsilon = 2.25$), the characteristic impedance and shunt capacitance can be quickly computed as:

$$Z_0 = (138/\sqrt{\epsilon}) \log (D/d) = (138/1.5) \log (0.5/0.083) = 72 \ \Omega$$
$$C = 7.35\epsilon/\log (D/d) = 7.35 \times 2.25/\log (0.5/0.083) = 21.2 \text{ pF/ft}$$

There are many special coaxial structures, components, and circuit elements that are specified and fabricated by engineers and technicians that do require the determination of physical characteristics. If, for example, a coaxial form of a 960 MHz, quarter-wave, 125 Ω to 50 Ω transformer, similar to our earlier two-wire example, were to be fabricated by altering the diameter of the inner conductor only (IDOC = 2 in.), the element can be specified for construction using a polystyrene dielectric ($\epsilon = 2.5$) structure as follows:

i. $\lambda = v_0/f \sqrt{\epsilon} = 1.18 \times 10^{10}/(960 \times 10^6 \times \sqrt{2.5}) = 7.77$ in.
$\lambda/4 = 7.77/4 = 1.94$ in.
$Z' = 80 \ \Omega$

ii. $d' = D/[\log^{-1} (Z_0 \sqrt{\epsilon}/138)] = 2/[\log^{-1} (80 \times \sqrt{2.5}/138)] = 0.243$ in.
$d_1 = 2/[\log^{-1} (125 \times \sqrt{2.5}/138)] = 0.074$ in.
$d_0 = 2/[\log^{-1} (50 \times \sqrt{2.5}/138)] = 0.535$ in.

Thus, the center conductor would be a stepped section of three diameters: 0.074 in. for 125 Ω, 0.243 in. for the quarter-wave matching section, and 0.535 in. for 50 Ω. Figure 12-5 shows three similar quarter-wave matching transformer sections using the two-wire and coaxial techniques as in the previous examples, and the most practical *stripline* technique that will be next introduced.

Striplines

Although the strip transmission line technique is somewhat impractical for transferring signals over longer distances, the approach is ideally suited to the fabrication of higher frequency RF networks and components. You can, for example, easily build a stripline component using the same layout and fabrication procedures that you would use in building a printed-circuit (PC) board. The stripline additionally affords most of the shielding advantages of coax without the physical constraints of the cylindrical coax structure. Modern electronic devices, such as transistors and diodes, are also available in stripline packages for direct insertion into strip circuits. An additional advan-

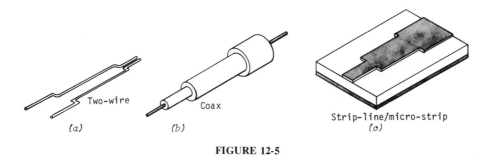

FIGURE 12-5

tage of stripline networks and components is that they may employ a wide variety of slab-shaped dielectric materials that may extend their electrical properties. One such popular material is alumina ceramic, which has a dielectric constant of 9 and which compresses the RF signal wavelength and component size by a factor of 3. By strict definition, *stripline* refers to a flat strip conductor sandwiched between *two* ground planes, while *microstrip* refers to a flat strip conductor above only *one* ground plane. Microstrip affords nearly the shielding effect achieved with two ground planes, but has the advantage of full access for the mounting of electronic components and other circuit elements without altering the structure of the transmission line. With the increasing popularity of stripline elements, a full range of standard coaxial RF connectors is available to adapt and interconnect the elements with flexible coaxial cable. Figure 12-6 shows typical stripline and microstrip structures and their properties.

The series inductance, shunt capacitance, and characteristic impedance of strip and microstrip lines are determined by four factors: width of the center strip, thickness of the center strip, distance between the center strip and ground plane(s) (the thickness of the dielectric), and the dielectric constant. As such, the electrical characteristics may vary greatly, especially with the wide range of available dielectric materials. To reduce the variables you may consider that nearly all stripline structures are fabricated using 1 oz (per ft^2) copper-clad circuit board. Since the 0.0014 in. thickness of the 1 oz. copper is a relatively insignificant proportion of the total thickness of most dielectrics, the thickness factor may be neglected in most stripline calculations. Even with this simplification, however, the calculation of the line characteristics is awkward because the electric fields within the line structure are nonuniform (nonsymmetrical).

The major variable of the stripline structure is its capacitance. As you may observe from Figure 12-6, the lines are built somewhat like a capacitor—two (or three) plates separated by a dielectric; but the ground planes extend beyond the width of the center strips, yielding an additional capacitance, called *fringing-field capacitance* (C_f). The value of C_f depends primarily on

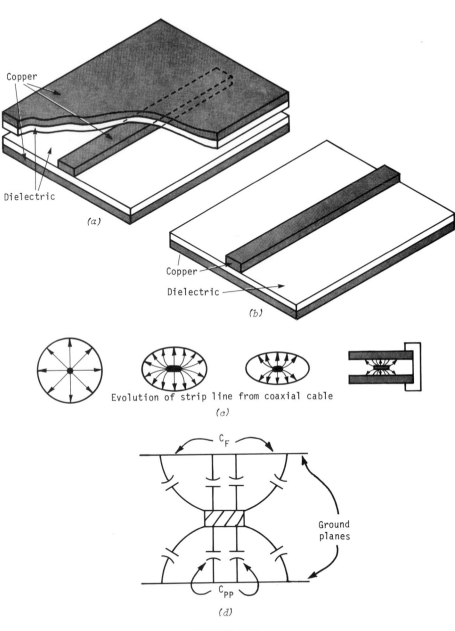

Copper

Dielectric

(a)

Copper

Dielectric

(b)

Evolution of strip line from coaxial cable

(c)

C_F

Ground
planes

C_{pp}

(d)

FIGURE 12-6

the thickness of the dielectric and does not change significantly as the width of the center strip is varied. The *parallel-plate capacitance* (C_{pp}), however, varies directly with the width of the center strip as well as the thickness of the dielectric. As such, the total stripline capacitance is the sum of the C_f and C_{pp} values, with C_f becoming less significant as the width of the center strip increases. The capacitance values for striplines may be computed, in pF/in., as:

$$C_T = 2C_{pp} + 4C_f$$
$$C_f = 1.6 \times 10^{-4}\ \epsilon/t$$
$$C_{pp} = 0.225\ W\epsilon/t$$

C_{pp} is the parallel-plate capacitance from the center strip to *one* ground plane, C_f is the fringing-field capacitance from *one* end of the center strip to *one* ground plane, W is the width of the center strip, t is the thickness of the dielectric between the center strip and *one* ground plane, and ϵ is the dielectric constant. As you may conclude, the capacitance values for microstrip structures are about half that of the striplines:

$$C_T = C_{pp} + 2C_f$$

If, for example, you were to compute the total capacitance for two widths of a stripline structure on $^1/_{16}$ in., 1 oz copper-clad circuit board ($\epsilon = 2.5$), and with center strips of 0.200 in. and 0.020 in., respectively, you might proceed as follows (values in pF/in.):

$$C_{pp1} = 0.225\ W\epsilon/t = 0.225 \times 0.2 \times 2.5/0.063 = 1.8$$
$$C_{pp2} = 0.225 \times 0.02 \times 2.5/0.063 = 0.18$$
$$C_f = 1.6 \times 10^{-4}\ \epsilon/t = 1.6 \times 10^{-4} \times 2.5/0.063$$
$$= 6.35 \times 10^{-3}$$
$$C_{T1} = 2C_{pp} + 4C_f = (2 \times 1.8) + (4 \times 6.35 \times 10^{-3})$$
$$= 3.6 + 0.025 = 3.625$$
$$C_{T2} = (2 \times 0.18) + (4 \times 6.35 \times 10^{-3}) = 0.36 + 0.025$$
$$= 0.385$$

If you closely observe the results of the above computations, you should conclude that the fringing-field capacitance significantly affects the total capacitance of narrow center strips, but C_f becomes insignificant with wide center strips. You may neglect the effects of fringing-field capacitance when the width of the stripline center strip exceeds the thickness of the dielectric.

The characteristic impedance of strip transmission lines can be computed using the following formula if the width of the center strip exceeds the thickness of the dielectric:

$$Z_0 = 40/(\sqrt{\epsilon}\ \log 2.5\ (W/t))$$

W is the width of the center strip, t is the thickness of the dielectric between the center strip and *one* ground plane, and ϵ is the dielectric constant. If the width of the center strip is less than the dielectric thickness, however, the impedance will be affected by the fringing-field capacitance and the above formula must then be modified to account for the significant value of C_f as follows (for W/t values from 1.0 to 0.2):

$$Z_0' = 140/(\sqrt{\epsilon} \log 25 \ (W/t))$$

The above formulas do not exactly apply to microstrip lines, because the fields above (air) and below (dielectric) the center strip are not uniform. The extent to which the Z_0 of microstrip lines vary from the calculations using the above formulas depends on the dielectric and shielding effects of the microstrip enclosure, hence it is difficult, if not impossible, to accurately calculate microstrip impedance. You can apply the above stripline formulas to obtain an approximation of microstrip impedance, however, by increasing the calculated Z_0 for comparable stripline values by about 50%.

To gain some perspective on the typical values of Z_0, we will apply the above formulas to our previous example of 0.200 in. and 0.020 in. strips on $^1/_{16}$ in., 1 oz copper-clad board as follows:

$$Z_{01} = 40/[\sqrt{\epsilon} \log 2.5 \ (W/t)] = 40/[\sqrt{2.5} \log 2.5 \times (0.2/0.063)]$$
$$= 27 \ \Omega$$
$$Z_{02} = 140/[\sqrt{\epsilon} \log 25 \ (W/t)] = 140/[\sqrt{2.5} \log 25 \times (0.02/0.063)]$$
$$= 93 \ \Omega$$

The values for the comparable microstrip elements with the same width and dielectric can be approximated as:

$$Z_{0(\text{microstrip})} = 1.5 Z_{0(\text{stripline})} = \begin{cases} 1.5 \times 28 \ \Omega = 42 \ \Omega \\ 1.5 \times 98 \ \Omega = 147 \ \Omega \end{cases}$$

The inductance values for strip and microstrip lines are most conveniently calculated from known values of characteristic impedance and capacitance, using the transposed form of the $Z_0 = \sqrt{L/C}$ formula:

$$L \ (\text{in henrys/in.}) = Z_0^2 C$$

If we apply the results of our previously computed examples, we can compute the inductance values for the 0.200 in. and 0.020 in. strips as follows:

$$L_1 = Z_0^2 C = 27^2 \times 3.625 \times 10^{-12} = 2.6 \ \text{nH/in.}$$
$$L_2 = Z_0^2 C = 93^2 \times 0.385 \times 10^{-12} = 3.3 \ \text{nH/in.}$$

The stripline formulas can be transposed in order to determine the width of the center strip based on circuit requirements for given L, C, or Z_0 values. When you need to determine the requirements for building circuits, the board thickness and dielectric constant are fixed values, thus the single variable that you can adjust is the *width* of the center strip. The formulas are here transposed to obtain values of W based on given circuit-board materials (fixed t and ϵ) and electrical specifications:

When $W/t > 1$,

$$W = \begin{cases} 0.4t \ \log^{-1} 40/(Z_0 \sqrt{\epsilon}) \\ 2C_{pp}t/0.255\epsilon = C_T t/0.225 \ \epsilon \end{cases}$$

When W/t is from 0.2 to 1.0,

$$W = \begin{cases} 0.04t \ \log^{-1} 140/(Z_0 \sqrt{\epsilon}) \\ [C_T - 6.4(\epsilon/t)]/0.225 \ \epsilon \end{cases}$$

If, for example, you need to determine the width of the center strip in a stripline structure that will provide a Z_0 of 50 Ω using the $^1/_{16}$ in., 1 oz copper-clad circuit board ($\epsilon = 2.5$), you may proceed as follows:

$$W = 0.4t \ \log^{-1} 40/(Z_0 \sqrt{8}) = 0.4 \times 0.063 \ \log^{-1} [(40/50)x \sqrt{2.5}] = 0.08 \ \text{in.}$$

The first form of the transposed formula will be valid in this example, as 0.08 in. is greater than 0.063 in. If the computed value of W was less than t in the above example, the calculation would have to be repeated using the second form of the transposed formula for values of W less than t. Specified values of capacitance can be computed using the other forms of the transposed formulas above, making first a sample calculation with the formula using C_{pp} alone to determine if W is greater or less than t.

Inductance values for stripline and microstrip structures do not change much, because the capacitance increases at a greater rate than Z_0 decreases with increasing W/t ratios. When increased series inductance values are required, W/t ratios of less than 1.0 are usually employed, with a tradeoff increase in Z_0. This higher Z_0 value must, in turn, be matched to other line sections using the Smith chart in order to establish the correct length and width of the center strip to achieve the required inductance.

TEM RF CIRCUITS

All TEM RF circuits, including the two-wire, coaxial, stripline, and micro-strip structures, may employ some variation of their physical structure to achieve inductance and/or capacitance values as required for circuit functions. The decision to use line structures or separate electrical components, such as chip capacitors, small coils, or ferrite elements, depends on a number of factors, including frequency of operation, component availability and cost, and fabrication capabilities and techniques.

Without a doubt, printed-circuit technology has advanced to become the most practical approach to achieving circuit structures and functions. At frequencies above 100 MHz, small inductance values of less than 100 nH are practical and easily achieved with spiral or zigzag center-strip conductors on stripline or microstrip structures. RF capacitance, however, is most often achieved using separate components, because the dielectrics of most micro-strip and stripline structures are lossy. There are special, and of course very expensive, RF dielectric materials that will allow for the direct fabrication of capacitance in the center-strip structure. Typical dielectric *substrate* materials of this type are alumina and beryllia ceramics, silicon dioxide, and even sapphire. The higher RF and microwave-frequency circuits of this type are called *thick film* and *thin film* based on the fabrication technique: plated strip conductors or etched (on monolithic substrate) strip conductors. Figure 12-7 is a typical microstrip schematic diagram and structure for a 2 GHz thin-film RF amplifier circuit.

Although two-wire and coaxial lines are mostly used for "transmission," and strip and microstrip lines are mostly used for RF circuit structures, there are many instances where coax or two-wire structures are necessary or more practical for particular circuits. Higher power circuits, for example, require larger and less lossy structures than are practical with printed-circuit, thick-film, or thin-film techniques. There are also many RF circuits that require adjustment or tuning by moving conductors—functions that are most practical with two-wire or coax elements. Most RF interconnection systems and couplings are standardized on the use of coax elements, and there are a

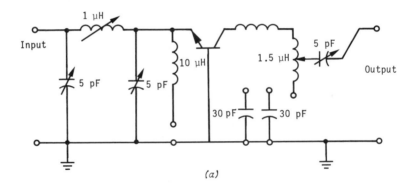

(a)

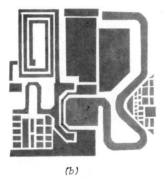

(b)

FIGURE 12-7

number of RF circuit functions that are conveniently packaged in coax connectorlike structures.

In building, repairing, or specifying RF circuits and techniques, the engineer and technician must be aware of the range and availability of off-the-shelf RF components. In addition to the specialized RF transistors and chips, there is a wide range of capacitors, ferrite and inductive elements, special resistive materials, and families of connector systems. Your knowledge of RF circuits and techniques will be greatly enhanced if you become familiar with the products for modern circuit technology. Your serious review and response to the advertisements in current technical journals will provide you with both important knowledge and an ever-expanding technical reference library.

CIRCUIT ANALYSIS

In this portion of the chapter we will apply the principles of network analysis using the Smith chart, as introduced in Chapter 11, to the fabrication and characteristics of circuit elements that employ TEM lines and structures. These approaches are used primarily for electronic communications circuits and systems that operate in the VHF and UHF portions of the radio spectrum (30 to 3000 MHz). As noted earlier, high frequency RF circuits may employ a combination of lumped and distributed elements, depending on the physical size of the circuit and components as compared to the wavelength of the signal. As you may learn from your review of off-the-shelf RF component catalogs, physically small capacitors, and transistor and diode chips, are available for application as lumped elements in RF circuits. The higher frequency signal paths and inductors, however, are distributed circuit elements. With the use of the Smith chart you can determine *where* in the circuit path lumped elements are to be placed, as well as the *electrical value* of the lumped (or distributed) element that is required for circuit operation. As such, Smith chart analysis is often a necessary part of building, adjusting, and modifying RF circuits.

Figure 12-8a shows a typical RF amplifier circuit which is to operate at 500 MHz, for example, on a microstrip structure (ϵ = 2.5, t = 0.050 in., 1 oz copper). The transistor of the example is specified as having a *normalized* output impedance of 1.5 +j2.0 and input impedance of 3.0 −j1.6, and it is to operate in a system normalized to 50 Ω (50 Ω = 1.0 ±j0). For operation the output section of the circuit will be matched with a chip capacitor that has a reactance of −j2.0 to cancel the inductive reactance of +j2.0; then the remaining R = 1.5 value will be transformed with a quarter-wave matching section. The input section of the circuit will be transformed with a 50 Ω line to a point where $Z = R + jX = 1.0 + jX$; and a capacitor will be placed at this point along the line to cancel the +jX value.

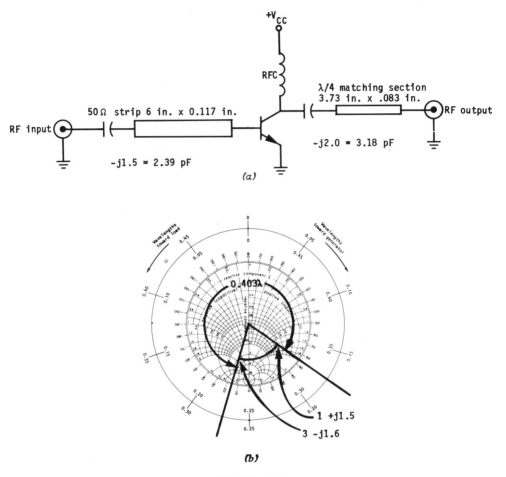

FIGURE 12-8

Sample Problem 1

Given: The transistor circuit of Fig. 12-8*b*, with the characteristics described in the previous paragraph.

Determine: The transmission-line dimensions, and the position and value of chip-capacitor elements in order to obtain a 50 Ω input and output impedance, with minimum VSWR, for the circuit.

(*a*)   The width of the center strip for the 50 Ω Z_0 is computed as follows:

$$Z_{0/(\text{microstrip})} = 1.5 Z_{0/(\text{stripline})}$$

Thus,

$$Z_{0(s)} = Z_{0(m)}/1.5 = 50/1.5 = 33\ \Omega$$

and

$$W = 0.4t \ \log^{-1} 40/(Z_0 \sqrt{\epsilon})$$
$$= 0.4 \times 0.050 \ \log^{-1} [40/(33 \sqrt{2.5})] = 0.117 \ \text{in.}$$

where the subscripts (s) and (m) stand for stripline and microstrip, respectively.

(b) The length, Z', and width of the center strip for the matching section are computed as follows:

$$\lambda = v_0/(f\sqrt{\epsilon}) = 1.18 \times 10^{10}/(500 \times 10^6 \times \sqrt{2.5}) = 14.9 \ \text{in.}$$
$$\lambda/4 = 14.9/4 = 3.73 \ \text{in.}$$

Then,

$$Z' = \sqrt{Z_1 Z_0} = \sqrt{(1.5 \times 50)} \times 50 = 61.24 \ \Omega$$
$$Z_{0(s)} = Z_{0(m)}/1.5 = 61.24/1.5 = 40.82 \ \Omega$$
$$W = 0.4t \ \log^{-1} 40/(Z_0 \sqrt{\epsilon})$$
$$= 0.4 \times 0.050 \ \log^{-1} [40/(40.8 \sqrt{2.5})] = 0.083 \ \text{in.}$$

(c) The length of center strip required to transform to $Z = 1.0 + jX$, and the value of X are plotted as in the Smith chart diagram of Figure 12-8b. Transposing,

$$3.0 - j1.6 \ \text{at} \ 0.273 \ \lambda \quad \text{becomes} \quad 1.0 + j1.5 \ \text{at} \ 0.176 \ \lambda$$

The length around the Smith chart accumulates to

$$0.5 \ \lambda - (0.273 \ \lambda - 0.176 \ \lambda) = 0.403 \ \lambda$$

and, in inches, to

$$0.403 \times 14.9 = 6.0 \ \text{in.}$$

(d) The values of the required chip capacitors of $-j2.0$ and $-j1.5$, respectively, to be placed in *series* with the center strip, are computed for operation at 500 MHz as follows:

$$C = 1/\omega X = 1/[(6.28 \times 5 \times 10^8 \times (2.0 \times 50)] = 3.18 \ \text{pF}$$
$$C' = 1/[(6.28 \times 5 \times 10^8 \times (1.5 \times 50)] = 4.24 \ \text{pF}$$

MICROWAVE DEVICES

Specialized TEM microwave devices include *dividers,* couplers, tuners, filters, and a variety of TEM structures that operate exclusively with distributed-constant functions. Devices of this type generally operate at frequencies between 500 MHz and 18 GHz and are constructed using coax or stripline techniques. Figure 12-9 shows two types of TEM microwave filters of stripline construction. The first (Figure 12-9*a*) is a straightforward arrangement of *L* values, as achieved with the narrow center-strip sections, whereas the *C* values are achieved by adding side conductors radiating from the center strip at quarter-wave spacing. The second filter (Fig. 12-9*b*) is made up of adjacent quarter-wave sections of line, each making a resonant network at the resonant frequency of the *L* and *C* values. The RF signal is inductively coupled from one section to the next as each adjacent strip operates as a transformer.

Couplers

The directional coupler is a popular microwave device. It is used in a wide variety of power splitting, combining, and sampling functions in microwave systems. The coupler has one input terminal and three output terminals: a main-arm output, a forward sampling output, and a reflected (reverse) sampling output. The sampling output terminals are specified to provide a signal at some dB level below the input signal. This specification is called the *coupling factor*. Popular couplers are available with coupling factors of 3, 6, 10, and 20 dB. The 20 dB coupler is an ideal transmitted *power monitor,* since 1% (-20 dB) of the input signal power will appear at the forward sampling output terminal, with only about 1% loss in the signal passing through the main-arm output of the coupler. The 10 dB coupler is an ideal *reflectometer,* which, through sampling forward and reverse (reflected) signals, will provide a measure of reflection coefficient from which VSWR can be computed:

$$\text{Reflection coefficient} = \rho = V_{\text{refl}}/V_{\text{inc}}$$
$$\text{VSWR} = (1 + \rho)/(1 - \rho)$$

The 3 dB coupler is often used as a power divider or combiner with isolation advantages so that impedance mismatches from circuits driving the sampling terminals do not greatly affect the main-arm signal. Figure 12-10 shows the stripline structure for a typical directional coupler and multiplexer. The coupling (transformer) section is phased so that signals entering port 1 and out port 3 will produce an in-phase signal at port 2, the forward sampling output. The signals at port 4 will be out of phase with the input, so cancellation will occur, resulting in virtually zero signal at this terminal. A proportion of any signal *into* port 3, however, will be sampled in phase at port 4; so any reflected signal from a mismatched load on port 3 can be conveniently measured in proportion to the sampled forward signals at port 2. The *directivity* specification of a coupler identifies the extent to which the sampling ports will isolate signals moving in opposite directions, i.e., the ratio of port 1 signal that appears at port 3 as compared to the signal at port 2. A 20 dB isolation is considered satisfactory for most coupler sampling applications.

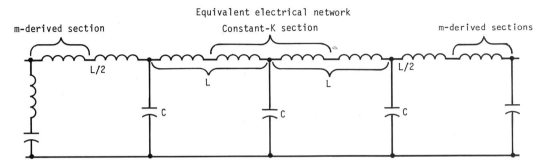

m-derived section

Equivalent electrical network
Constant-K section

m-derived sections

L/2   L   L   L/2

C   C   C

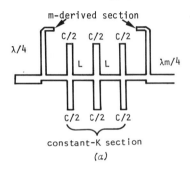

Filter layout

m-derived section

C/2   C/2   C/2

λ/4   L   L   λm/4

C/2   C/2   C/2

constant-K section

(a)

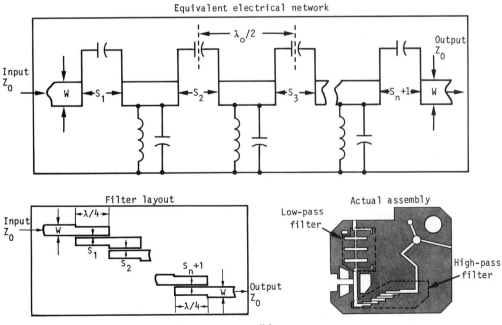

Equivalent electrical network

$\lambda_o/2$

Input Z_0

W $\leftarrow S_1 \rightarrow$ $\leftarrow S_2 \rightarrow$ $\leftarrow S_3 \rightarrow$ $\leftarrow S_n+1 \rightarrow$ W

Output Z_0

Filter layout

Input Z_0

λ/4

W

S_1

S_2

S_n+1

W

Output Z_0

λ/4

Actual assembly

Low-pass filter

High-pass filter

(b)

FIGURE 12-9

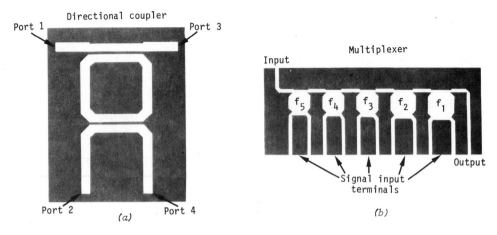

FIGURE 12-10

Combiners

Unlike directional couplers, power dividers/combiners do not provide the isolation of strip (transformer) coupling. They are direct-coupled devices and should thus operate in systems with closely matched impedances. Power dividers/combiners are less costly and less critical devices, however, and they operate at much broader bandwidths than couplers. Power dividers/combiners operate like parallel resistive circuit elements, where each of two parallel 100 Ω elements, for example, will divide (or combine) the power in a 50 Ω system, while maintaining a fixed total impedance.

Figure 12-11 shows coaxial stub tuners and a line stretcher. These are among the most popular TEM microwave tuning devices. The stub tuners are basically a convenient adjustable means of impedance matching with the use of shorted (stub) sections of transmission line. As you may observe, the rigid, air-dielectric coax structure is ideally suited for this purpose, using a movable cylindrical conductive plug to shorten or lengthen the stub section as may be required to obtain a given susceptance value. The stubs in the double-stub tuners are spaced at $\frac{3}{8}$ λ, while those in the triple-stub tuners are spaced at λ/4 per section. The line stretcher of Figure 12-11c is constructed of telescoping sections of rigid, air-dielectric coax. In impedance-matching functions the line stretcher is adjusted to change the phase of incident and reflected signals for minimum mismatch and low VSWR.

Connectors

The most common, and possibly most critical, TEM microwave component is the coaxial connector pair. Coax connectors are popularly available for operation to 18 GHz, where their physical length may represent nearly one

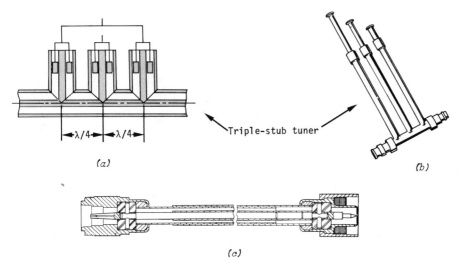

(a)

Triple-stub tuner

(b)

(c)

FIGURE 12-11

wavelength of the signal that they transmit. You should thus appreciate the fact that such ''precision RF connectors'' (as their name implies) must be carefully assembled and installed in order to assure the proper functioning of RF systems. At the VHF and lower UHF frequencies the BNC and UHF connector families are widely used without much regard to their precise assembly and installation. Even at these lower frequencies, however, connectors can cause major problems in otherwise perfectly operating communications systems. It is recommended that you obtain the detailed assembly, installation, handling, and applications information from RF connector manufacturers, and diligently apply their recommendations. Table 12-3 summarizes the physical and electrical properties of popular RF and microwave connector pairs. Figure 12-12 shows a variety of these connectors.

Attenuators and Loads

Resistive microwave products are available in many forms for application in TEM networks. The most popular uses for microwave resistors are in attenuator and load (termination) applications. As such, the resistors are shaped in rod and disk form for use in coax structures, and chip and pill form for use in stripline structures. The resistive materials are metal film or carbon film deposited on a dielectric substrate. They are available off-the-shelf and are specified in resistance values between 5 Ω and 500 Ω, and in power dissipation ratings of from 0.5 to 10 W. Deposited resistive-film substrate cards are also available for circuit breadboard or custom trimming requirements. These cards are generally specified in *ohms per square,* indicating the

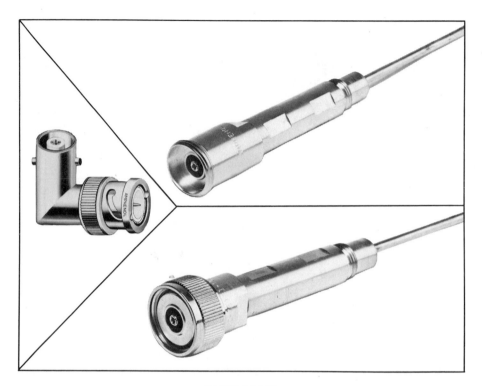

FIGURE 12-12

resistance value of any size square shape of the material. Thus, high resistance values would result from long narrow strips (many squares in series), whereas low resistance values would result from short wide strips (many squares in parallel). Off-the-shelf ohms-per-square values between 25 Ω and 750 Ω are typical for film-resistance cards.

TABLE 12-3.   DESIGNATIONS AND PROPERTIES OF RF/MICROWAVE CONNECTOR PAIRS

	Dimensions (in.)		Nominal	Equivalent Military Numbers*			
Type	Dia.	Length	Impedance	Cable Male	Cable Female	Panel Male	Panel Female
UHF	$^3/_4$	$1^1/_2$	50	PL-259	—		SO-239
BNC	1	$^9/_{16}$	50	UG/88	UG/89	UG/1104	UG/290
TNC	$1^1/_8$	$^9/_{16}$	50	M39012/ 26001	M39012/ 27001	M39012/ 28001	M39012/ 29001
N	$^3/_4$	$1^1/_2$	50	UG/21	UG/23	—	UG/22
SMA	$^5/_{15}$	$^1/_2$	50	Complies with Mil-C-39012 SMA			
APC	$^3/_4$	$1^1/_2$	50	Precision connectors for test equipment			

* These number designations are only representative of a wide variety of cable connectors available. Refer to connector manufacturers' catalogs for greater detail and availability.

WAVEGUIDE

The idea that RF electrical signals could be transmitted along a solid hollow pipe was initially baffling, because the pipe is a continuous conductor which presumably would short out the signal. This waveguide effect is one of the phenomenal applications of radio and electronics that we experience. In general, people do not encounter many examples of waveguide transmission unless they specifically search them out, because waveguide is a specialized and extremely expensive means of transmitting RF signals from one point to another. The technology is on the verge of a breakthrough, however, with the use of fiberoptic (optically conductive) waveguides that work on much the same principle as metal (electrically conductive) waveguides.* Although the existing metal-waveguide technology is a specialized microwave transmission technique, it is nonetheless an important technology for the electronic communications industry, where point-to-point microwave and satellite communications systems are commonplace. In fact, waveguide is the transmission line capable of lowest loss, lowest noise, and highest power; and it is the only practical transmission line to use for the many microwave systems that operate above 18 GHz.

As contrasted to the TEM line, waveguide does *not* operate on the same *current-path-originated* transverse electric and magnetic fields that is the basis for the two-wire, coax, and strip transmission lines. In waveguide it is the energy in the *field* that *originates* the current flow on the inner conductive walls of the hollow pipe. Since either electric-field or magnetic-field forces can affect this current flow, waveguide is specified as either a TE (transverse-electric) or TM (transverse-magnetic) line. Although the theoretical concepts of waveguide are most abstract, they can be explained, and hopefully understood, in terms of the visualization technique as suggested earlier in this chapter for TEM concepts. We will proceed in the following paragraphs to develop this visual understanding of waveguide wave propagation, building upon your knowledge of the previous materials in Chapters 11 and 12.

Waveguides, as their name properly suggests, guide radiated electromagnetic waves along the inside of hollow metal tubes, with a voltage signal across the dielectric (normally air) within a tube, and the current signal

---

* Fiberoptic transmission lines are the newest practical means of transmitting communications signals from one place to another. At the date of this writing the lines are primarily used for shorter distance data communications, where their immunity to electrical interference drastically reduces errors in transmission. The real breakthrough in the use of fiberoptic lines will occur when both frequency-division and time-division multiplexing apparatus become a practical reality for driving and receiving signals at optical frequencies. Theoretically, fiberoptic lines are nothing more than very small cylindrical waveguides that operate at optical rather than microwave frequencies. Unfortunately, the waveguide terminology of this chapter cannot be applied to fiberoptics: even though the concepts are consistent, the terms are not. Your thorough understanding of waveguide principles and techniques, however, should serve as a necessary basis for your future study of fiberoptics as the technology becomes significant.

along the inner conductive surface (walls) of a tube. As such, waveguides guide and transmit the energy in electromagnetic waves in much the same manner that flumes or aqueducts can transmit the energy of water waves: standing waves exist from sidewall to sidewall, while wave motion exists along the length of the tube. The most popular standardized form for waveguides is the rectangular structure, where (due to the historic standards) the *outside* width dimension is twice the *outside* height dimension, and the inside dimensions are slightly greater than a 2 : 1 ratio. In this shape waveguide the electric-field signal is across the exact center of the wide top and bottom walls, but the magnetic field results from current flow across the conductive surfaces (top, sidewalls, and bottom). The electric field is most often associated with the top and bottom walls of the waveguide structure, whereas the magnetic field is associated with the sidewalls of the waveguide. As such, the orientation of the field forces is in two planes, with the vertical identified as the *E* (for electric) *plane* and the horizontal identified as the *H* (for magnetic) *plane*. You can easily remember the orientation of the waveguide planes by thinking about how it would bend: *E* for the *easy* way and *H* for the *hard* way. Figure 12-13 depicts this rectangular waveguide with a visual representation of the electric- and magnetic-field signals, along with some less-popular waveguide structures.

Critical Dimensions

You can visualize the rectangular waveguide structure and signals of Fig. 12-13*a* as a simple two-wire line with a series of quarter-wavelength ($\lambda/4$) stubs radiating along the signal path. As such, the wide dimension of the waveguide is equal to one-half wavelength ($\lambda/2$) of the signal, and the shorted stubs at that distance transform from 0 Ω to ∞ Ω across the $\lambda/4$ distance to the center of the waveguide structure; so the theoretical two-wire signal path does not "see" the shorts. You should deduce that waveguide operating this way cannot support a signal at frequencies where the wavelengths are longer than twice the wide dimension of the waveguide; making the waveguide a sort of highpass filter. This frequency and wavelength are identified as waveguide *cutoff*, and may be computed, based on the wide dimension of the structure, as:

$$\lambda_c = 2a \quad \text{and} \quad f_c = v_o/\lambda_c = v_o/2a$$

where λ_c is the cutoff wavelength, a is the wide (inside) dimension of the waveguide structure, f_c is the cutoff frequency, and v_o is the velocity of radiation based on the *air* dielectric ($\epsilon = 1.0$).

A most popular waveguide, specified by the EIA as WR90, or RG-52/U by the military, has inside dimensions of 0.400 in. × 0.900 in., with a wall thickness of 0.050 in. The cutoff wavelength and frequency for this waveguide can be computed as follows:

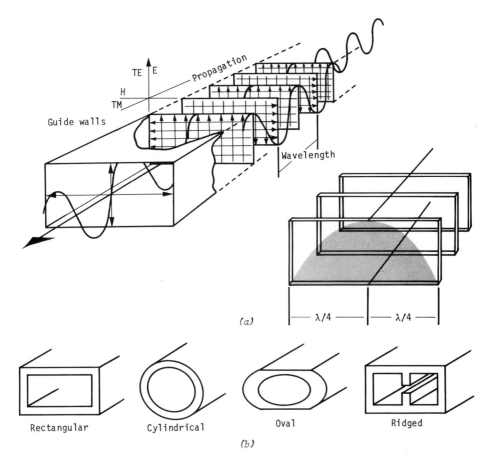

FIGURE 12-13

$$\lambda_c = 2a = 2 \times 0.900 = 1.8 \text{ in.}$$
$$f_c = v_o/2a = 1.18 \times 10^{10}/1.8 = 6.56 \times 10^9 = 6.56 \text{ GHz}$$

The conductive sidewalls of the waveguide will thus short out signals with frequencies and wavelengths *below* (or sometimes called *beyond*) cutoff, which for this example are signal frequencies below 6.56 GHz.

The operation of waveguide at frequencies above cutoff is a bit more complicated to understand. First of all, the $\lambda/4$ distance from the center to sidewalls of the waveguide must be maintained if the theoretical two-wire signal path is not to see the shorted stubs; so there must be $\lambda/2$ of the signal across the waveguide regardless of frequency. It is difficult to change frequency without changing wavelength in the process, but this phenomenon *does* in fact occur in waveguide. As the frequency of a waveguide signal is increased above cutoff, the *phase velocity* (refer to Chapter 11 for review if necessary) of the signal also increases in order to maintain the required $\lambda/2$ across the wide dimension of the waveguide. As a result of this increased phase velocity, the velocity of the signal moving *along* the waveguide is proportionally reduced, and the resulting waveguide wavelength is increased. The velocity of the signal along the waveguide is called *group velocity*, as it describes the particular "group" of waves that move together down the length of the waveguide. The phase velocity, group velocity, and resulting waveguide wavelength may be conveniently computed as follows:

$$\angle\theta = \cos^{-1}(\lambda_o/\lambda_c) = \cos^{-1}[v_o/(f_o\lambda_c)]$$
$$v_p = v_o/\sin\theta; \qquad v_g = v_o\sin\theta$$
$$\lambda_g = \lambda_o/\sin\theta = v_o/(f_o\sin\theta)$$

where λ_o, f_o, and v_o are the free-space, air-dielectric ($\epsilon = 1.0$) operating wavelength, operating frequency, and velocity of radiation, respectively; v_p is the phase velocity, v_g is the group velocity, and λ_g is the waveguide wavelength of a signal operating at f_o.

The phase angle ($\angle\theta$) identified in the above formulas defines the way in which the wavelength above cutoff is increased with frequency in order to maintain the fixed dimension of twice the width of the waveguide. You know, of course, how to get a longer slice of French bread than the diameter of the loaf: you slice the bread at an angle. Water waves are likewise shortest when sliced at right angles to their flow, and they appear longer when sliced on an angle. Surfers, for example, get a much longer ride by crossing a wave diagonally. If you can visualize the cutting of the water waves as being similar to the way that waveguide signal waves are incident to the direction of signal flow within the waveguide, you should conclude that it is possible to maintain a fixed $\lambda/2$ across the waveguide even with an increasing frequency and wavelength down the length of the waveguide. The angle that these waves are incident to the waveguide is specified as $\angle\theta$ in the formulas of the above paragraph.

If we operate the WR90 waveguide of our previous example to guide a 10.0 GHz RF signal, we can compute the characteristics of the signal within the waveguide as follows:

$$\angle\theta = \cos^{-1}[v_o/(f_o\lambda_c)] = \cos^{-1}[1.18\times10^{10}/(10\times10^9\times1.8) = 49°$$
$$v_p = v_o/\sin\theta = 1.18\times10^{10}/\sin 49° = 1.56\times10^{10} \text{ in./s}$$
$$\lambda_g = v_o/(f_o\sin\theta) = 1.18\times10^{10}/(10\times10^9\times\sin 49°) = 1.56 \text{ in.}$$
$$\lambda_o = v_o/f_o = 1.18\times10^{10}/(10\times10^9) = 1.18 \text{ in. (free space)}$$

Waveguide Modes

If the frequency of the signal within the waveguide is increased to greater than twice the initial cutoff frequency, a second form of signal transmission may be possible within the same waveguide structure, just as if the waveguide were made up of two pairs of adjacent two-wire lines with shorter $\lambda/4$ stubs for the higher frequency signal. For WR90 waveguide this *second* wave pattern could exist at frequencies above 13.11 GHz. If the frequency of the signal is increased still higher, the short dimension (0.400 in. for WR90) of the waveguide can likewise reach some cutoff frequency for yet another possible form of transmission within the waveguide. Figure 12-14 shows a number of different possible forms of waveguide transmission. The form that the signal wave takes as it is transmitted through the waveguide is called the *waveguide mode*. You should conclude that higher given frequency signals within a particular-size waveguide may be transmitted in a greater number of modes as the signal frequency increases.

Modes in waveguide are designated based on the wave pattern within the waveguide. TE modes identify those wave patterns where the electric field is oriented at right angles to the direction of signal flow down the waveguide. The wave patterns, or modes, that would result from the previously described examples in Figure 12-13 are TE modes. TM modes identify those wave patterns where the magnetic field is oriented at right angles to the direction of signal flow down the waveguide. TM modes are seldom used in waveguide transmission or components, as they are more difficult to launch and maintain in the popular waveguide structures.

The modes in rectangular waveguides are additionally identified by two subscripts: m and n, for example, TE_{mn}. The subscript m specifies the number of half-wavelengths ($\lambda/2$'s) across the large dimension of the waveguide; the subscript n specifies the number of half-wavelengths across the small dimension of the waveguide. The waveguide mode for our example of Figure 12-13 is the TE_{10} mode (pronounced, "tee-eee one-oh," not "tee-eee ten"), because there is *one* electric-field half-wavelength across the large dimension of the waveguide, and *no* electric-field half-wavelengths across the small dimension of the waveguide. For cylindrical waveguides the subscript m specifies the number of full wavelengths around the circumference, and the subscript n specifies the number of $\lambda/2$'s across the radius (from center to edge) of the waveguide.

Dominant Mode

The most popular mode for rectangular waveguide is the TE_{10} mode. With a slightly greater than $2:1$ size ratio (0.900 × 0.400 in. for WR90, as an example), the TE_{10} mode will operate at the lowest frequency for a particular size waveguide. The TE_{10} mode will also exist exclusively, without the possible excitation of other modes, over a greater proportional frequency range than any other mode. The TE_{10} mode is thus called the *dominant mode* for rectangular waveguide. The dominant-mode frequency range for standard rectangular waveguides extends from f_c to $2f_c$, while the most practical region of operation within this range is from $1.2f_c$ to $1.9f_c$. The popular rectangular

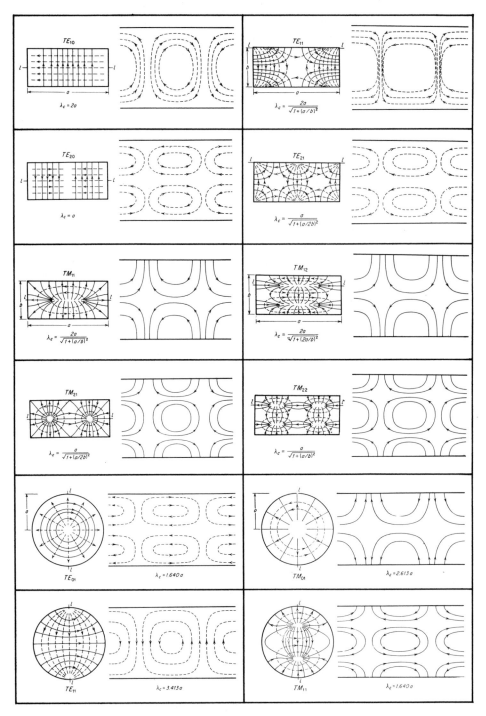

FIGURE 12-14

waveguide sizes, frequency ranges for TE_{10}, and designations are listed in Table 12-4. Table 12-5 lists the older and current frequency-band designations below 18 GHz. You should observe that the frequency limits for a given band were established by the waveguide TE_{10} operating range, and that these band designations are maintained to describe a particular waveguide size. The current designations have been adopted to reflect frequency usage—not waveguide frequency limits—where modern coaxial components and transmission lines are more predominant.

The dominant mode for cylindrical waveguide is designated TE_{11}, while the wave pattern within the waveguide and operation is similar to the TE_{10} mode for rectangular waveguide. Because of their circular orientation, modes within cylindrical waveguide are determined by Bessel function (refer to Chapter 7 for review if necessary) relationships, with the TE modes at the peak values of the Bessel curves and the TM modes at the zero-crossing values of the curves. Although the TE_{11} mode may be used for transmission, other modes are more popular. The TE_{01} and TM_{01} modes, for example, are symmetrical and would not be affected by rotation of the waveguide; the TE_{01} has the lowest sidewall conductivity, making it the lowest loss line for long distance transmission.

Special cylindrical waveguide structures, such as rotary joints for radar antennas and resonant cavities and frequency meters, also use other modes appropriate to the special function served by the cylindrical waveguide. Table 12-6 lists cutoff wavelengths for the more popular cylindrical waveguide modes. The d (inside) dimensions refer to the diameter of the cylindrical waveguide structure. As you may observe, the operating frequency range for any particular mode is far less for cylindrical waveguide than for rectangular waveguide. Since higher order modes are more predominant with cylindrical waveguide, special structures must be provided in the waveguide to assure excitation of only one mode. These structures are called *mode filters*.

If, for example, a section of cylindrical waveguide were to be used as a rotary joint, operating in TM_{01} mode over a frequency range from 10.0 to 12.0 GHz, its diameter and operating characteristics would be computed as follows:

    i. For TM_{01}, $\lambda_c = 1.306d$, where $d = \lambda_c/1.306$ and $\lambda_c = v_o/f_c$.
           Thus, $d = v_o/(1.306f_c) = 1.18 \times 10^{10}/(1.306 \times 10 \times 10^9)$
                $= 0.904$ in.
   ii. For TE_{21}, $\lambda_c = 1.028d$
           Thus, $f_c = v_o/\lambda_c = v_o/(1.028d) = 1.18 \times 10^{10}/(1.028 \times 0.904)$
                $= 12.7$ GHz

From the above calculations you should observe that the next higher order mode is above 12 GHz and will not be excited, so only the dominant TE_{11}

TABLE 12-4.   REFERENCE TABLE OF RIGID RECTANGULAR
WAVEGUIDE DATA AND FITTINGS

EIA WG Designation WR( )	Recommended Operating Range for TE$_{10}$ Mode		Cutoff for TE$_{10}$ Mode		Range in 2λ/λ$_c$	Range in λ$_g$/λ	Theoretical cw power rating lowest to highest frequency (MW)	Theoretical attenuation lowest to highest frequency (dB/100 ft)
	Frequency (GHz)	Wavelength (cm)	Frequency (GHz)	Wavelength (cm)				
2300	0.32–0.49	93.68–61.18	0.256	116.84	1.60–1.05	1.68–1.17	153.0–212.0	.051–.031
2100	0.35–0.53	85.65–56.56	0.281	106.68	1.62–1.06	1.68–1.18	120.0–173.0	.054–.034
1800	0.41–0.625	73.11–47.96	0.328	91.44	1.60–1.05	1.67–1.18	93.4–131.9	.056–.038
1500	0.49–0.75	61.18–39.97	0.393	76.20	1.61–1.05	1.62–1.17	67.6–93.3	.069–.050
1150	0.64–0.96	46.84–31.23	0.513	58.42	1.60–1.07	1.82–1.18	35.0–53.8	.128–.075
975	0.75–1.12	39.95–26.76	0.605	49.53	1.61–1.08	1.70–1.19	27.0–38.5	.137–.095
770	0.96–1.45	31.23–20.67	0.766	39.12	1.60–1.06	1.66–1.18	17.2–24.1	.201–.136
650	1.12–1.70	26.76–17.63	0.908	33.02	1.62–1.07	1.70–1.18	11.9–17.2	.317–.212 / .269–.178
510	1.45–2.20	20.67–13.62	1.157	25.91	1.60–1.05	1.67–1.18	7.5–10.7	
430	1.70–2.60	17.63–11.53	1.372	21.84	1.61–1.06	1.70–1.18	5.2–7.5	.588–.385 / .501–.330
340	2.20–3.30	13.63–9.08	1.736	17.27	1.58–1.05	1.78–1.22	3.1–4.5	.877–.572 / .751–.492
284	2.60–3.95	11.53–7.59	2.078	14.43	1.60–1.05	1.67–1.17	2.2–3.2	1.102–.752 / .940–.641
229	3.30–4.90	9.08–6.12	2.577	11.63	1.56–1.05	1.62–1.17	1.6–2.2	
187	3.95–5.85	7.59–5.12	3.152	9.510	1.60–1.08	1.67–1.19	1.4–2.0	2.08–1.44 / 1.77–1.12
159	4.90–7.05	6.12–4.25	3.711	8.078	1.51–1.05	1.52–1.19	0.79–1.0	
137	5.85–8.20	5.12–3.66	4.301	6.970	1.47–1.05	1.48–1.17	0.56–0.71	2.87–2.30 / 2.45–1.94
112	7.05–10.00	4.25–2.99	5.259	5.700	1.49–1.05	1.51–1.17	0.35–0.46	4.12–3.21 / 3.50–2.74
90	8.20–12.40	3.66–2.42	6.557	4.572	1.60–1.06	1.68–1.18	0.20–0.29	6.45–4.48 / 5.49–3.83
75	10.00–15.00	2.99–2.00	7.868	3.810	1.57–1.05	1.64–1.17	0.17–0.23	
62	12.4–18.00	2.42–1.66	9.486	3.160	1.53–1.05	1.55–1.18	0.12–0.16	9.51–8.31 / 6.14–5.36
51	15.00–22.00	2.00–1.36	11.574	2.590	1.54–1.05	1.58–1.18	0.080–0.107	
42	18.00–26.50	1.66–1.13	14.047	2.134	1.56–1.06	1.60–1.18	0.043–0.058	20.7–14.8 / 17.6–12.6 / 13.3–9.5
34	22.00–33.00	1.36–0.91	17.328	1.730	1.57–1.05	1.62–1.18	0.034–0.048	
28	26.50–40.00	1.13–0.75	21.081	1.422	1.59–1.05	1.65–1.17	0.022–0.031	— — / 21.9–15.0
22	33.00–50.00	0.91–0.60	26.342	1.138	1.60–1.05	1.67–1.17	0.014–0.020	— — / 31.0–20.9
19	40.00–60.00	0.75–0.50	31.357	0.956	1.57–1.05	1.63–1.16	0.011–0.015	
15	50.00–75.00	0.60–0.40	39.863	0.752	1.60–1.06	1.67–1.17	0.0063–0.0090	— — / 52.9–39.1
12	60.00–90.00	0.50–0.33	48.350	0.620	1.61–1.06	1.68–1.18	0.0042–0.0060	— — / 93.3–52.2
10	75.00–110.00	0.40–0.27	59.010	0.508	1.57–1.06	1.61–1.18	0.0030–0.0041	
8	90.00–140.00	0.333–0.214	73.840	.406	1.64–1.05	1.75–1.17	0.0018–0.0026	152–99
7	110.00–170.00	0.272–0.176	90.840	.330	1.64–1.06	1.77–1.18	0.0012–0.0017	163–137
5	140.00–220.00	0.214–0.136	115.750	.259	1.65–1.05	1.78–1.17	0.00071–0.00107	308–193
4	170.00–260.00	0.176–0.115	137.520	.218	1.61–1.05	1.69–1.17	0.00052–0.00075	384–254
3	220.00–325.00	0.136–0.092	173.280	.173	1.57–1.06	1.62–1.18	0.00035–0.00047	512–348

TABLE 12-4   (*continued*)

Material Alloy	JAN WG Designation RG ( )/U	JAN Flange Designation Choke UG ( )/ U	Cover UG ( )/ U	EIA WG Designation WR ( )	Inside	Tol.	Outside	Tol.	Wall Thickness Nominal
Alum.				2300	23.000–11.500	±.020	23.250–11.750	±.020	0.125
Alum.				2100	21.000–10.500	±.020	21.250–10.750	±.020	0.125
Alum.	201			1800	18.000–9.000	±.020	18.250–9.250	±.020	0.125
Alum.	202			1500	15.000–7.500	±.015	15.250–7.750	±.015	0.125
Alum.	203			1150	11.500–5.750	±.015	11.750–6.000	±.015	0.125
Alum.	204			975	9.750–4.875	±.010	10.000–5.125	±.010	0.125
Alum.	205			770	7.700–3.850	±.005	7.950–4.100	±.005	0.125
Brass	69		417A	650	6.500–3.250	±.005	6.660–3.410	±.005	0.080
Alum.	103		418A						
				510	5.100–2.550	±.005	5.260–2.710	±.005	0.080
Brass	104		435A	430	4.300–2.150	±.005	4.460–2.310	±.005	0.080
Alum.	105		437A						
Brass	112		553	340	3.400–1.700	±.005	3.560–1.860	±.005	0.080
Alum.	113		554						
Brass	48	54A	53	284	2.840–1.340	±.005	3.000–1.500	±.005	0.080
Alum.	75	585	584						
				229	2.290–1.145	±.005	2.418–1.273	±.005	0.064
Brass	49	148B	149A	187	1.872–0.872	±.005	2.000–1.000	±.005	0.064
Alum.	95	406A	407						
				159	1.590–0.795	±.004	1.718–0.923	±.004	0.064
Brass	50	343A	344	137	1.372–0.622	±.004	1.500–0.750	±.004	0.064
Alum.	106	440A	441						
Brass	51	52A	51	112	1.122–0.497	±.004	1.250–0.625	±.004	0.064
Alum.	68	137A	138						
Brass	52	40A	39	90	0.900–0.400	±.003	1.000–0.500	±.003	0.050
Alum.	67	136A	135						
				75	0.750–0.375	±.003	0.850–0.475	±.003	0.050
Brass	91	541	419	62	0.622–0.311	±.0025	0.702–0.391	±.003	0.040
Alum.	—	—	—						
Silver	107								
				51	0.510–0.255	±.0025	0.590–0.335	±.003	0.040
Brass	53	596	595	42	0.420–0.170	±.0020	0.500–0.250	±.003	0.040
Alum.	121	598	597						
Silver	66	—	—						
				34	0.340–0.170	±.0020	0.420–0.250	±.003	0.040
Brass	—	660	599						
Alum.	—	—	—	28	0.280–0.140	±.0015	0.360–0.220	±.002	0.040
Silver	96	—							
Brass	—		383	22	0.224–0.112	±.0010	0.304–0.192	±.002	0.040
Silver	97		—						
				19	0.188–0.094	±.0010	0.268–0.174	±.002	0.040
Brass	—		385	15	0.148–0.074	±.0010	0.228–0.154	±.002	0.040
Silver	98		—						
Brass	—		387	12	0.122–0.061	±.0005	0.202–0.141	±.002	0.040
Silver	99		—						
				10	0.100–0.050	±.0005	0.180–0.130	±.002	0.040
Silver	138	—	—	8	0.080–0.040	±.0003	0.156 DIA	±.001	—
Silver	136	—	—	7	0.065–0.0325	±.00025	0.156 DIA	±.001	—
Silver	135	—	—	5	0.051–0.0255	±.00025	0.156 DIA	±.001	—
Silver	137	—	—	4	0.043–0.0215	±.00020	0.156 DIA	±.001	—
Silver	139	—	—	3	0.034–0.0170	±.00020	0.156 DIA	±.001	—

TABLE 12-5. MICROWAVE FREQUENCY BAND DESIGNATIONS

| TE$_{10}$ mode (GHz) | | EIA | Designation | | | | | |
Frequency	Cutoff freq.	WR ( )	FXR	Hewlett-Packard	Microwave Assoc.	NARDA	Sperry Microline	New Military
0.320–0.490	0.256	2300	...	...	...	2300	...	...
0.350–0.530	0.281	2100	J	...	...	2100	...	...
0.410–0.625	0.328	1800	...	...	...	1800	...	...
0.490–0.750	0.393	1500	...	...	...	1500	...	...
0.640–0.960	0.513	1150	...	...	...	1150	...	C
0.750–1.120	0.605	975	...	...	...	975	...	...
0.950–1.500	0.766	770	...	...	...	770	...	...
1.120–1.700*	**0.908**	**650**	**L**	...	**L**	**L**	**L**	**D**
1.450–2.200	1.158	510	...	...	...	510	...	...
1.700–2.600	**1.375**	**430**	**R**	...	...	**LS**	...	**E**
2.200–3.300	1.737	340	...	...	...	340	...	F
2.600–3.950	**2.080**	**284**	**S**	**S**	**S**	**S**	**S**	...
3.300–4.900	2.579	229	...	...	...	229	...	G
3.950–5.850	**3.155**	**187**	**H**	**G**	**C**	**C**	**C**	...
4.900–7.050	3.714	159	...	C	...	159	...	...
5.850–8.200	**4.285**	**137**	**C**	**J**	**XB**	**XN**	**G**	...

			W	H	XL	XB	H	H
7.050–10.00	5.260	112	X	X	XS	X	X	H
8.200–12.40	6.560	90	:	M	:	75	:	I
10.00–15.00	7.873	75	Y	P	KU	KU	U	J
12.40–18.00	9.490	62	:	N	:	51	:	:
15.00–22.00	11.578	51	K	K	K	K	K	:
18.00–26.50	14.080	42	:	:	:	34	:	:
22.00–33.00	17.368	34	U	R	KA	V	V	K
26.50–40.00	21.100	28	Q	:	Q	Q	:	:
33.00–50.00	26.350	22	:	:	:	19	:	:
40.00–60.00	31.410	19	M	V	V	M	:	:
50.00–75.00	39.900	15	E	:	E	E	:	:
60.00–90.00	48.400	12	:	:	:	10	:	:
75.00–110.00	59.050	10	F	:	:	N	:	:
90.00–140.00	73.840	:	:	:	:	RG136	:	:
110.00–170.00	90.845	:	G	:	:	A	:	:
140.00–220.00	115.750	:	:	:	:	RG137	:	:
170.00–260.00	137.520	:	:	:	:	R	:	:
220.00–325.00	173.290	:	:	:	:	:	:	:

* Boldface type designates the more popular bands.

TABLE 12-6. CUTOFF WAVE-LENGTHS FOR CYLINDRICAL WAVEGUIDE MODES

Mode	Cutoff Wavelength (d = I.D.)
TE_{01}, TM_{11}	$0.82d$
TE_{21}	$1.02d$
TM_{01}	$1.31d$
TE_{11}	$1.71d$

mode must be filtered from the structure to assure TM_{01} operation. For this example, then, it would be possible to shift the waveguide cutoff frequency down to about 9.7 GHz, with the f_c for TE_{21} at 12.3 GHz, in order to center the 10 to 12 GHz operating range.

Throughout this chapter, reference has been made to an 18 GHz range for coax systems, with waveguide being required above this frequency. Actually the variables of power-handling requirements, noise, and losses may exclude the use of coax down to *even* the 1 to 2 GHz frequency band. You might consider that coax losses can be reduced by increasing the size of the cable or rigid line, but herein lies the problem: Coax lines will support signal modes and begin to operate as waveguides if the wavelength of the signal becomes small compared to the dielectric space between the inner and outer conductors of the line. For this reason, higher frequency signals require smaller coax lines to prevent what is called *moding*. Smaller coax lines, however, have greater losses, higher noise, and lower power-handling ability, so there are tradeoffs in the selection and use of coax or waveguide at frequencies below 18 GHz. The general formula for computing the "safe" high frequency limit for a given coax line is:

$$f_h = v_0/[\pi (D + d) \sqrt{\epsilon}]$$

f_h is the practical high-frequency limit for a given coax line to assure that moding will not occur; D and d are the IDOC and ODIC of the coax line; and ϵ is the dielectric constant.

Waveguide *L*, *C*, and *Z*₀

If you refer back to the first description of dominant-mode (TE_{10}) rectangular waveguide that began with the two-wire line and $\lambda/4$ stubs as in Fig. 12-19, you should conclude that the b (height) dimension of rectangular waveguide influences its impedance and L/C ratio. A smaller b dimension would thus yield a greater capacitance and lower characteristic impedance for the TE_{10} mode. You should recall, however, that the angle of signal flow down the waveguide varies with frequency and also changes phase and group velocities within the waveguide, yielding electric- and magnetic-field forces that also vary with frequency. Since the capacitance and inductance within the waveguide result from the field forces, these values (C and L) *and* the resulting characteristic impedance (Z_0) also change with frequency. For TE modes the impedance decreases with frequency, and for TM modes the impedance increases with frequency. The impedance for rectangular waveguide modes can be computed based on the angle of signal flow down the waveguide, as was earlier used to determine phase velocity, group velocity, and waveguide wavelength. For TE modes in rectangular waveguide the impedance is computed as:

$$Z_0 = 377 \ (b/a)/\sin \theta = 377 \ b/(a \ \sin \theta)$$

where

$$\angle \theta = \cos^{-1} [v_0/(f_o \lambda_c)]$$

For TM modes in rectangular waveguide the impedance is computed as:

$$Z_0 = 377 \ (b/a) \ \sin \theta$$

As the formulas indicate, the waveguide Z_0 changes with frequency; but since waveguide transmission lines and elements within a given band work in conjunction with each other, impedance mismatches do not occur.

On the same basis that Z_0 changes with frequency across a given waveguide band, so do capacitance and inductance. The absolute capacitance or inductance value is thus not useful to consider in waveguide. What *is* important, however, is the *relative excess* capacitance or inductance at a particular frequency of operation. Here, as with impedances, *normalized* values are most useful; thus a normalized capacitance of 1.5, for example, indicates an excess capacitance of 50% of whatever the line capacitance happens to be at a particular frequency of operation. If, for example, the *b* dimension of rectangular waveguide, operating in TE_{10} at a frequency of 10.0 GHz, is reduced to 0.200 in. at some point along a section of WR90, the increased capacitance will be the major factor that changes the impedance at that point. The formula for characteristic impedance from the previous paragraph shows a direct relationship of Z_0 to b; so we can conveniently compute the two values of Z_0 and Z_0' as follows:

$$\angle \theta = \cos^{-1} [v_0/(f_o \lambda_c)] = \cos^{-1} [1.18 \times 10^{10}/(10 \times 10^9 \times 1.8)] = 49°$$
$$Z_0 = 377b/(a \ \sin \theta) = 377 \times 0.4/(0.9 \sin 49°) = 222 \ \Omega$$
$$Z_0' = 377 \times 0.2/(0.9 \times \sin 49°) = 111 \ \Omega$$

Since the increased capacitance reduced the characteristic impedance by a factor of 2, and the impedance varies as the inverse square root of the capacitance, the normalized capacitance value can be computed as the square of the inverse impedance ratios as follows:

$Z_0 = \sqrt{(L/C)}$, thus $C = L/Z_0^2$

If L remains relatively constant,

$C(\text{normalized}) = (Z_0/Z_0')^2$

For this example,

$C_n = (222/111)^2 = 4$

Now, as a second example, we can reduce the a dimension of the WR90 waveguide at a particular point along its length to 0.600 in. This change will basically reduce the length of the stubs and effect a current flow across the waveguide, resulting in increased inductance. To determine the normalized increase in inductance, we can proceed with the impedance calculations of the previous paragraph as follows:

$Z_0'' = 377 \times 0.4/(0.6 \times \sin 49°) = 333 \ \Omega$

Since the increased inductance yields an increased Z_0, and since Z_0 varies as the square root of the inductance, the normalized inductance value can be computed as the square of the impedance ratios as follows:

$L = Z_0^2 C$

If C remains relatively constant,

$L(\text{normalized}) = (Z_{0n}/Z_0)^2$

For this example,

$L_n = (333/222)^2 = 2.25$

The calculations for normalized capacitance and inductance values of the two previous paragraphs can be simplified, for in each formula only *one* quantity (a or b) is changed to yield a normalized (proportional) C or L value. The simplified formulas are:

$C_n = (b/b')^2$ and $L_n = (a/a'')^2$

b' is the reduced b dimension of the waveguide, and a'' is the reduced a dimension of the waveguide. In practice such reductions in waveguide dimensions are achieved by partially sawing the waveguide and inserting a conductive plate that is soldered in place. These plates in the waveguide are appropriately called *inductive* or *capacitive irises*. If both increased capacitance *and* inductance are required, both the b and a dimensions of the waveguide are reduced with a plate called a *waveguide window*. One major problem with the mechanically critical small waveguide is that calculations or fabrication procedures may produce an undesired error in the depth of the iris into the waveguide structure. An alternative, and more popular, approach is with the use of *tuning screws* placed in the center of the waveguide

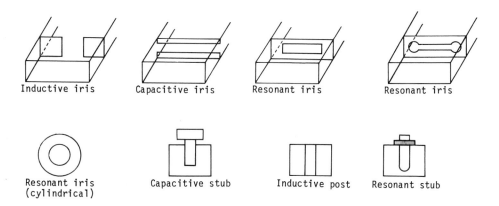

| Inductive iris | Capacitive iris | Resonant iris | Resonant iris |

| Resonant iris (cylindrical) | Capacitive stub | Inductive post | Resonant stub |

FIGURE 12-15

to effect a variable capacitance. More precise inductance values can be achieved with the use of inductive *posts* across the waveguide structure. These posts allow for fine adjustments of normalized inductance by using successively larger diameter conductive rods inserted in place through pre- cisely drilled holes in the waveguide. Figure 12-15 shows some representa- tive waveguide structures that achieve increased inductance and capacitance as may be required for particular circuit functions.

WAVEGUIDE CIRCUIT ANALYSIS

In this portion of the chapter we will apply the principles of network analysis using the Smith chart, as we did earlier for TEM lines. In this instance, however, we will analyze the fabrication and characteristics of electronic devices placed in waveguide structures. These approaches are used where waveguide elements and lines are required in the SHF and EHF portions of the radio spectrum (3 to 300 GHz). As you can observe, the higher frequency waveguide structures require increasingly smaller electronic lumped compo- nents to operate as amplifiers or detectors, etc. These lumped components are combined with the distributed elements in the waveguide structure, and thus require impedance-matching techniques to make them operate prop- erly. There are few, if any, off-the-shelf electrical components to achieve matching within the waveguide structure, as such techniques require the mechanical modification of the waveguide. Perhaps the tuning screw is the closest off-the-shelf device that resembles an electrical waveguide com- ponent.

Figure 12-17a shows a typical waveguide detector mount with the diode element centered in the waveguide exactly λ/4 from the shorted end. The

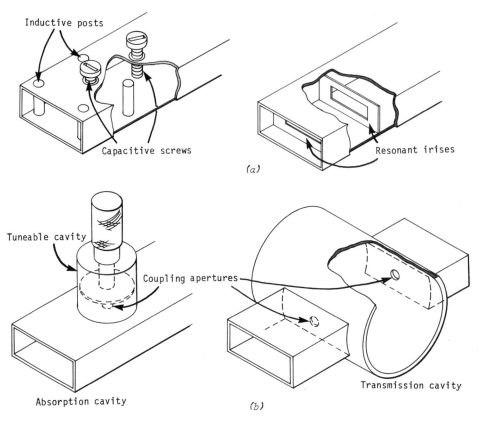

Inductive posts

Capacitive screws

Resonant irises

(a)

Tuneable cavity

Coupling apertures

Absorption cavity

Transmission cavity

(b)

FIGURE 12-16

end of the waveguide thus becomes a quarter-wave stub and does not consequently load the diode. As with most electronic components, the impedance of the diode will vary with frequency, so it must be impedance matched to the waveguide circuit in which it operates. Components of this type placed in a waveguide structure may be matched with values of L or C. (A combination of L and C is required to properly match resonant elements.) The approach used for matching is to transform the diode impedance, with the use of the Smith chart, to a point along the waveguide ahead of the diode, where a parallel capacitance (susceptance) will cancel an equivalent admittance of $Y_n = 1.0 - jB$.

Sample Problem 2

Given: The waveguide diode-detector structure, as in Fig. 12-17a, operating in WR90 at 10.5 GHz, with the diode impedance of $643 - j\,343\ \Omega$ and placement at the center of the waveguide $\lambda/4$ from the shorted end.

Determine: The position of the diode from the end of the waveguide, and the position and depth of a capacitive iris placed along the waveguide in front of the diode in order to effect a matched impedance and minimum VSWR.

(a) The $\lambda/4$ distance within the waveguide is computed as follows:

$$\angle\,\theta = \cos^{-1}\left[v_o/(f_o\lambda_c)\right] = \cos^{-1}\left[1.18 \times 10^{10}\,(10.5 \times 10^9 \times 1.8)\right]$$
$$= 51.4°$$
$$\lambda_g = v_o/(f_o \sin\theta) = 1.18 \times 10^{10}/(10.5 \times 10^9 \times \sin 51.4)$$
$$= 1.44 \text{ in.}$$
$$\lambda_g/4 = 1.44/4 = 0.36 \text{ in.}$$

(b) The waveguide impedance and normalized diode impedance are computed as follows:

$$Z_0 = 377\,b/(a \sin\theta) = 377 \times 0.4/(0.9 \sin 51.4°) = 214\ \Omega$$
$$Z_n = (R/Z_0) \pm j(X/Z_0) = (643/214) - j(343/214) = 3.0 - j1.6$$

(c) The equivalent diode admittance, the distance required to transform Y_n to $1.0 - jB$, and the normalized value of B are plotted in accordance with the Smith chart as follows:

$$Z_n = 3.0 - j1.6$$

will transpose to $Y_n = 0.26 + j0.104$ at $0.024\ \lambda$

which will then transpose to

$$Y_n = 1.0 - j1.5 \text{ at } 0.324\ \lambda$$

The length around the Smith chart accumulates to

$$0.324 - 0.024 = 0.300\ \lambda$$

and, in inches, to

$$0.300\ \lambda_g = 0.3 \times 1.44 = 0.432 \text{ in.}$$

(d) The value of required susceptance (normalized capacitance) is obtained with an iris that reduces the b dimension to the following calculated height:

$$C_n = B_n = (b/b')^2$$

Thus

$$b' = b/\sqrt{C_n} = 0.4/\sqrt{1.5} = 0.327 \text{ in.}$$

The depth of the iris is

$$0.400 - 0.327 = 0.073 \text{ in.}$$

WAVEGUIDE COMPONENTS

The term *component,* as applied to lower frequency lumped electrical and electronic parts, does not necessarily apply to waveguide parts. Most often the term *waveguide component* refers to a complete functional network in a waveguide structure, e.g., a directional coupler, a resonant cavity, a detector mount, etc. While many circuit functions may be achieved in waveguide form using diodes and transistors, proper circuit operation cannot be assured without the addition of specialized waveguide elements. Waveguide systems or subsystems used in electronic communications are composed of different waveguide elements that achieve equivalent signal processing functions that are more easily achieved with lower frequency TEM elements. The waveguide equivalent of even a simple branching (parallel) circuit requires a complicated, and expensive, piece of hardware. For these reasons *systems* engineers and technicians do not often fabricate waveguide components as they would conventional electronic circuitry. Rather, they purchase waveguide components from specialty manufacturers and interconnect them to achieve systems operation. As such, waveguide components are designed to "bolt together" to make up a communications system.

Filters and Cavities

The more popular waveguide components include filters, resonant cavities, attenuators, detector mounts, transitions, loads, tuners, switches, couplers, and power dividers. These components have little resemblance to their TEM counterparts; they are both electrically and physically specialized. Based on the earlier waveguide example in Sample Problem 2, you should observe that the right combination of iris structures in waveguide can achieve a range of filter functions. The most popular technique is to space inductive irises along the waveguide at slightly less than $\lambda/2$ and place a capacitive tuning screw between the irises. Multiple-section, higher Q filters are made up of a series of these arrangements. The same filter functions can be achieved with inductive posts and tuning screws. Waveguide filter manufacturers have collected extensive empirical data on post and screw sizes and placement to achieve the high-Q filters required for communications systems. Figure 12-16a shows typical waveguide filters.

Waveguide resonant cavities are closely related to filters in function, but they differ physically and electrically from the iris and post and screw filters of Fig. 12-16. If you refer back to the section on waveguide mode designations, you should recall the two subscripts (m and n) that describe the wave pattern within the waveguide. If dominant-mode rectangular waveguide were closed (shorted) at both ends of a half-wavelength (λ_g) section, it would be resonant at the cutoff frequency of the waveguide, and its resonant frequency could be adjusted by lengthening the section. Resonant cavities are thus

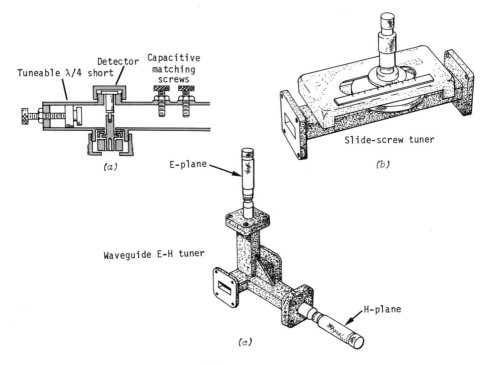

Tuneable λ/4 short

Detector

Capacitive matching screws

(a)

E-plane

Slide-screw tuner

(b)

Waveguide E-H tuner

H-plane

(c)

FIGURE 12-17

short waveguide sections with a critical length. The cavity length is identified in terms of the particular mode of its operation with the addition of a third subscript, p. A dominant-mode rectangular cavity would be designated $TE_{mnp} = TE_{101}$, with the p indicating that *one* electrical half-wavelength of the resonant-frequency signal occupies the length of the cavity. Cylindrical waveguide cavities are likewise designated, with the dominant mode being the TE_{111}. The special low-loss TE_{011} mode in cylindrical waveguide yields resonant cavities with Q's in the range of an unbelievable 20,000! The cylindrical structure is also most appropriate for tunable cavities, with a movable plunger (plug) on a screw thread for fine adjustment of resonance. The cavities are available in two general circuit functions: transmission and absorption. The transmission cavities act as bandpass filters, and the absorption cavities act as bandstop filters and serve wavemeter functions. The RF signals can be coupled into or out of cavity structures with irises, windows, or conductive coupling (current) loops. As with lower frequency resonant circuits, the higher coupling coefficients yield greater resonant loading and lower Q's. Figure 12-16b shows representative waveguide cavities.

Waveguide tuners are elements that combine the L and C functions of irises, windows, or posts and screws for impedance matching, with the cavity adjustments for resonance or peak response at a particular frequency.

You may recall that with the detector mount of Fig. 12-17a there was exactly $\lambda/4$ between the diode and the shorted end of the waveguide. A sliding or movable shorting block placed in a longer section of waveguide behind the diode will make the detector mount tunable, so it would be possible to reflect the required $+j1.5$ susceptance value to the diode without the necessity of an iris ahead of the diode. Tuning of the same detector mount could *also* be accomplished with screws instead of an iris ahead of the diode. In laboratory practice, slide-screw tuners or multistub (capacitive screw) tuners are typical components used for prototype work. In production, the tuner's positions and settings are transferred to screw adjustments that become a part of a product's structure. *E-H* tuners are sections of waveguide with movable shorting blocks that extend from the top wall (*E*-plane) and sidewall (*H*-plane) of a waveguide structure. As such, *E-H* tuners affect a wide range of inductive and capacitive values into the waveguide. Figure 12-17b and c show waveguide tuners.

Couplers and Other Components

Waveguide directional couplers and power dividers are functionally identical to their TEM counterpart devices. They are available with a similar range of coupling factors and are used for the same power division and sampling purposes that were earlier described for TEM couplers and dividers. Waveguide couplers differ physically and structurally, however, as coupling, from the main guide section to the sampling guide section is accomplished by connecting two common-wall waveguide sections with aperture holes that function as coupling windows, and transferring a portion of the input signal to a sampling terminal. One major advantage of broad-wall waveguide couplers is extremely high directivity (in excess of 60 dB for narrowband devices) for accurate sampling in reflectometer functions. Waveguide power dividers are simply T or Y structures of waveguide, in the *E* or *H* planes, with the necessary matching screws, posts, or irises to prevent reflections. As with the TEM power dividers, there is no isolation, so they should operate only in systems with closely matched impedances.

Waveguide attenuators use resistive-film materials deposited on thin dielectric substrates. As with the TEM film resistors, these elements are available in card form and specified in *ohms per square*. The resistive cards are often available in precut sizes to fit within standard waveguide structures. When the resistive-film cards are placed in the electric field within the waveguide structure, I^2R losses occur as a result of the signal flow through the distributed resistance along the length of the cards. The attenuation factor will vary with the resistance and length of the card, and the power-handling capability of the attenuator will vary with the length of the card.

Variable attenuator structures are generally achieved by (1) sliding the resistive card into the electric field, from the inside edge to the center of the waveguide; (2) inserting the resistive card at varying distances into a slot cut into the top wall of the waveguide; and (3) rotating the resistive card in a TE_{11} mode cylindrical waveguide section, from 0° to 90°, with the orientation of the electric field within the waveguide.

Waveguide loads are also resistive elements, often made of the same resistive-card materials that are used for attenuators. Of course the highest distributed resistance values are required for load functions, where the signal should be attenuated by at least 100 dB. Higher power resistive loads of distributed-resistance ceramic materials are available in appropriate wedge or pyramid shapes, and in sizes to fit within standard waveguide structures.

A large group of specialized microwave waveguide components uses a combination of electronic components and special materials that function with electronic components. Such components include attenuators, mixers, oscillators, amplifiers, phase shifters, circulators, and other elements. Another large group of specialized components, including duplexers, diplexers, switches, and other elements, is used in communications systems. Some of these microwave components will be covered in Chapter 13; others have been omitted as being more appropriate for study by microwave specialists. The ever-changing technology will doubtless make many such specialized microwave components obsolete, so your familiarity with those here described should serve only as a basis for your further study.

MICROWAVE MEASUREMENTS

Many of the traditional measurements that are used at lower frequencies cannot be applied to microwave devices, components, and circuits. Some of the measurement limitations are based on practicality; other exclusions are based on the fact that distributed-constant networks are not measured using the same techniques that are applied to lumped-constant networks. The most important microwave measurement categories include (1) absolute and relative power, (2) frequency and wavelength, and (3) VSWR and impedance. Although the physical appearance of the apparatus used to make these measurements differs considerably in TEM and waveguide forms, and also from manufacturer to manufacturer, the functional measurement setups can be reduced to a few general block-diagram forms. It is recommended that you expand your library with catalogs and applications notes on microwave measurements and measuring equipment in order to maintain your familiarity with industrial practice. If you relate the current hardware and practice to the following block-diagram descriptions, you should easily understand and be able to effectively apply microwave measurements.

Power

Most measurements of microwave power rely on the heating effect of RF energy. For low-power measurements, thermal elements are placed in the path of the RF signals, and the extent to which they are heated is compared to a standard, with their temperature converted to conveniently read out in watts (milliwatts or microwatts) on the face of a panel meter or display. For more specialized high-power measurements the RF signals are applied to heat-circulating liquids, with the same comparison to standards and conversion to indicate power in watts. The more popular low-power measurements use thermistor or thermocouple elements mounted in either a coax or waveguide detectorlike power-sensor structure, and which operate in conjunction with a calibrated power meter. The power meter is a precise comparison bridge and amplifier and may be individually calibrated with a given thermal element in the power sensor. These devices typically measure absolute microwave power in dBm or milliwatts.

Many measurements of microwave power, including gain, loss, and frequency-response level, do not require readings of absolute power levels. For such purposes a more convenient *relative* power measurement technique is employed using a combination of diode detector elements and calibrated attenuators in conjunction with a precision audio or video amplifier. This technique allows relative power measurements to be made in conjunction with a sweep-signal source and oscilloscope or a spectrum analyzer. The detectors are generally used in coax structures below 18 GHz in order to obtain broadband response. At frequencies above 18 GHz they are placed in waveguide structures and have frequency limitations based on the practical operating range of a given-size waveguide. For most accurate relative power measurements the signal level incident upon the diode is held within a narrow range with the use of an accurately calibrated RF attenuator ahead of the diode. Most microwave detector diodes are operated in the square-law region (near the origin) of their response curve, so their output current is directly proportional to the level of applied RF power. Most amplifiers, panel meters, digital and CRT displays can thus be directly calibrated in linear power or log dB ratios when they are used in conjunction with microwave detectors in making relative power measurements.

Frequency

Microwave frequencies are generally measured with a frequency counter that establishes its time base on the higher harmonics of a frequency standard. Two techniques are used to establish accurate counter time bases: heterodyne and transfer. At UHF frequencies below 3 GHz the microwave signal is heterodyned (mixed) with a stable, phase-locked signal within the counter. The difference of the two signals is fed to the counter, which converts its numerical display to the microwave frequency (instead of the difference frequency that it is *really* counting). At the higher microwave frequencies the microwave signal is zero beat with a given harmonic-number signal from a stable signal source, called a *transfer oscillator*. The counter basically counts the oscillator frequency and then *transfers* this value to an equivalent microwave frequency (again, not the frequency that it is really counting) at its numerical display. This transfer is achieved by multiplying the transfer oscillator frequency times the harmonic number required to zero beat with the microwave signal. Although both techniques seem cumbersome, they are extremely precise.

Microwave frequencies are also measured indirectly with the use of resonant-cavity frequency meters. These components, in fact, measure the *wavelength* of microwave signals—not their frequency. These are most often absorption-type cavities that couple only a small amount of RF energy from a circuit at resonance, while they are effectively out of the circuit when they are tuned off resonance. As mentioned earlier under the components section, such cavities have extremely high Q values, so they can provide the required resolution to precisely measure the frequency (wavelength) of microwave signals. They are also extremely convenient to use, as they remain physically connected to a circuit, and require only the tuning of a dial to provide a quick and accurate measurement of frequency.

Direct measurement of waveguide wavelength can best be accomplished with a waveguide slotted section that is used for VSWR and impedance measurements. Here a detector probes the standing-wave pattern, and the wavelength is obtained by doubling the distance between adjacent null points in the standing-wave pattern along the waveguide. The measurement of wavelength in coaxial and TEM structures is accomplished indirectly through the measurement of the dielectric constant (velocity factor). This measurement is accomplished with a coax-T device terminated in a detector, which essentially probes the standing-wave pattern. However, rather than moving the probe, as with the waveguide measurement, the frequency of the applied signal is varied to obtain two adjacent null points at the detector. The dielectric constant is then computed using the formula:

$$\epsilon = 2f_1/(f_2 - f_1)$$

The frequencies f_1 and f_2 indicate a null at the detector.

VSWR and Impedance

Perhaps the most important microwave measurements are for VSWR and impedance values. The traditional techniques of VSWR and impedance measurements used a slotted line to measure the standing-wave pattern and compare it to a reference (open or short) position in order to transfer the measured information to a Smith chart. Except in cases where actual circuit analysis is performed, however, simple VSWR measurements are sufficient to specify or calibrate the performance of a microwave device. With the availability of broadband directional couplers and sweeping signal sources, VSWR is most often indirectly measured using reflectometer procedures. The most sophisticated of these is a transmission/reflection test setup that displays both responses on the face of a CRT. The ultimate microwave impedance measuring system is the network analyzer that provides a direct Smith chart readout of the CRT face. The network analyzer will provide measurements of gain, attenuation, reflection coefficient, VSWR, and impedance over broadband frequencies to 40 GHz (in TEM below 18 GHz, and in waveguide above 12 GHz). The network analyzer basically samples transmission and reflection signals and compares their amplitude and phase characteristics.

CONCLUSION

This chapter has been concerned with applying the theoretical principles of RF transmission lines, as advanced in Chapter 11, to the practical hardware and circuits of the real world of RF and microwave electronics. You should now appreciate the extreme complexity of microwave electronics, realizing of course that the content of this chapter merely scratched the surface of the wide range of transmission lines, *passive* components, and techniques that are a most important part of modern RF/electronic communications systems. The following chapter will review the more popular *active* microwave devices, circuit elements, and techniques that provide for the origination, amplification, and processing of higher radio-frequency and microwave signals. Chapter 14 will deal with antennas that rely almost entirely on field forces and transmission principles. Your success with the topics of Chapters 13 and 14 will be greater if you are now comfortable with Chapters 11 and 12. If you have doubts or difficulties, carefully review these two chapters as necessary to build your level of confidence. The problems that follow will test your ability to apply theory to practice, and measure your achievement of the *learning objectives* specified at the beginning of this chapter.

Problems

1.   Given: A pulse signal, with an equivalent bandwidth of 100 kHz, at the 600 Ω output of a modem to be transferred a distance of 5 miles with a twisted-pair transmission line (Belden No. 8795).

If the line is specified with 0.026 in. conductors and 0.016-in.-thickness polyethylene insulation, determine:

(a) Total inductance
(b) Total capacitance
(c) Characteristic impedance
(d) Wavelength at 100 kHz (remember ϵ)
(e) Line matching transformer requirements
(f) -3 dB frequency
(g) Cable loss at 100 kHz (in dB)

2. Given: Two 300 Ω antennas for operation at 450 MHz to be stacked as in Fig. 12-2 using plated brass rod, with each $\lambda/4$ matching section to transform to the 300 Ω transmission-line terminal midway between the two antennas.

If the antenna terminals are spaced at 2.2 in., determine:
(a) Length of the stacking rods
(b) Impedance that each $\lambda/4$ section *transposes* (transforms) to the center *line* terminals
(c) The characteristic impedance of each matching section
(d) The diameter of the plated brass rod

3. Given: A 75 Ω CATV system using an RG-6/U coax feed cable, operating from 50 to 300 MHz over a distance of 1.0 mile.

From catalog information, determine:
(a) IDOC and ODIC of the cable
(b) Dielectric constant of the cable
(c) Phase constant of the signal within the cable
(d) Cable wavelength for U.S. TV channel 6
(e) Cable capacitance
(f) Cable inductance
(g) Cable "slope" in dB/octave
(h) Total cable loss at 300 MHz (in dB)

4. Given: A coaxial impedance-matching balun as in Fig. 12-18 to be adapted for coupling an RG-58/U cable with an "F" connector to a 960 MHz, 72 Ω dipole antenna.

Using a 2 in. ID decoupling sleeve and a $5/8$ in. OD, $9/16$ in. ID copper tubing for the outer conductor of the matching section, determine:
(a) Lengths of the center conductor, outer conductor, and decoupling sleeve
(b) Z_0 of the $\lambda/4$ matching section
(c) Diameter of the center conductor
(d) Spacing of the antenna elements

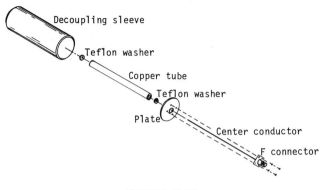

FIGURE 12-18

5.   Given: The microstrip circuit of Fig. 12-19 for a 200 to 400 MHz broadband RF amplifier with conductor dimensions as noted. (Dielectric: $\epsilon = 2.2$, $t = .050$ in.)

Determine:
- (*a*)   Characteristic impedance of the center strip
- (*b*)   Capacitance of the center strip
- (*c*)   Inductance of the center strip
- (*d*)   Function of each center-strip section, 1 through 7.

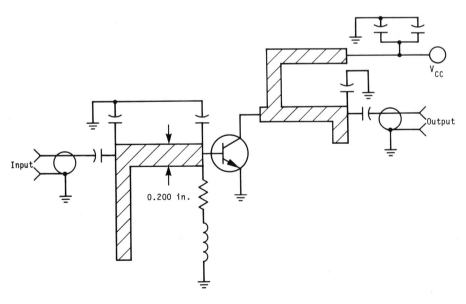

FIGURE 12-19

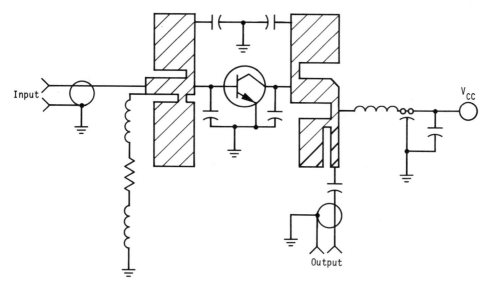

FIGURE 12-20

6. Given: The microstrip circuit of Fig. 12-20 for a 400 MHz RF power amplifier, with the transistor input inpedance at $12 - j18\ \Omega$ and an output impedance of $20 + j10\ \Omega$, to be matched with open (shunt susceptance) stubs located as in Fig. 12-32.

Determine:
(a) The width and length of the 50 Ω center strips
(b) The length of the 50 Ω open stubs

7. Given: A section of rectangular waveguide with inside dimensions 0.420×0.170 in. for operation in TE_{10} at $f_o = 1.5f_c$.

Determine:
(a) λ_c (d) Waveguide wavelength
(b) f_c (e) Characteristic impedance
(c) Phase velocity

8. Given: The waveguide detector structure of Fig. 12-16a, operating in WR62 at 15.0 GHz, and with a slotted-line measurement indicating a VSWR of 4.0 and a null at 0.25 in. toward the generator from the null short measured with the diode removed (remember to shift $\lambda/4$ for the diode-to-short distance).

With the aid of the Smith chart, determine:
(a) Normalized diode impedance
(b) Normalized diode admittance

(*c*)   Required capacitive susceptance to match the transformed diode admittance to the waveguide Z_0 (1.0)

(*d*)   Distance ahead of the diode where the capacitance is to be placed

(*e*)   Approximate depth of a capacitive tuning screw required to effect a match

9.   Given: A TEM directional coupler used in a reflectometer setup, with 100 mW input power, 1.0 mW out of the forward sampling terminal, and 247 μW out of the reverse sampling terminal.

For the measurement of a test device, determine:

(*a*)   Coupling factor

(*b*)   Main-arm output power

(*c*)   Voltage reflection coefficient of the test device

(*d*)   VSWR of the test device

10.   Given: The waveguide structure of Fig. 12-21.

Identify the function of each of the following components: (*a*), (*b*), (*c*), (*d*), (*e*), (*f*), and (*g*).

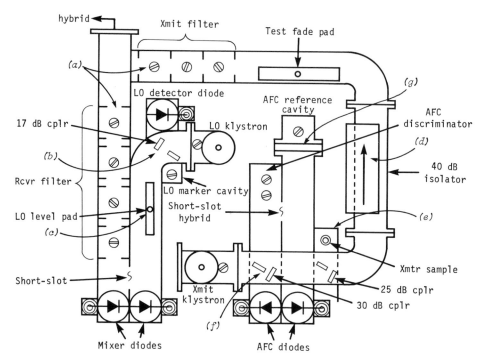

FIGURE 12-21

CHAPTER 13

ACTIVE MICROWAVE DEVICES

A biased argument is advanced, tongue in cheek, by those people who work in the microwave field, claiming that "more than 99% of RF is microwave technology." This specious claim cannot really be denied, but a little explanation should quickly bring the statement into perspective. If you statistically accept the range of microwave frequencies to extend from about 800 MHz to 100 GHz (100,000 MHz), then all the frequencies from dc to 800 MHz would be only 800/100,000th, or 0.8%, of the total range of frequencies available for electronic communications. We should all know, however, that the greater proportion of electronic communications takes place at frequencies below 800 MHz, thus discounting the microwavers' claim.

Most of the recent and future growth of electronic communications, however, is specified for the microwave frequencies and above. Among the reasons for this inevitable dominance of microwaves for communications is the fact that all the lower frequency RF allocations are used up—there is no radio space available for new and expanding technology below the microwave region. If you are looking to the future, you should realize that many of the circuit and systems functions used at lower frequencies will be required for microwave communications, thus making microwaves the area "where the action is."

The point to be made here is that a specialized range of active electronic devices is required to operate in conjunction with the passive network hardware defined in Chapter 12 in order to achieve systems functions at microwave frequencies. Though substantial abstract theory is involved, electronics is still a hardware-oriented technology, and it is primarily the hardware aspects of microwave components and systems that make them unique. The block diagram of a superheterodyne *microwave* radio is, for example, much the same as the block diagram for an FM broadcast-band radio. The microwave hardware (passive circuit elements and active devices) provides the technology that allows us to achieve communications in the UHF, SHF, and EHF regions of the radio spectrum. It is the purpose of this chapter to review the theoretical and practical aspects of these active microwave devices.

By way of introduction, active microwave devices can be grouped into three categories: vacuum tube, solid state, and ferrite. These active devices function as signal sources, amplifiers, mixers, detectors, and network elements. Although solid-state devices have replaced almost all but the highest power vacuum tubes at the lower radio frequencies, there are specialized microwave tubes that as yet have no solid-state counterparts. Some reliable microwave tubes are often still preferred for systems applications even though equivalent solid-state devices have been developed. We are now experiencing great advances in solid-state microwave technology, anticipating that vacuum tubes will gradually be phased out of the microwave field as they were at lower frequencies. At the present time, however, the tubes are still important, so they will be here covered in detail.

Historically, microwave tubes and solid-state detector and mixer diodes were developed in conjunction with radar systems in the early 1940's. The *klystron* was one of the first practical tubes used for amplifier and oscillator applications; and the point-contact silicon diode, types 1N21 and 1N23, was among the first practical microwave mixer and detector devices. It is interesting that while 40-year old AM radios are considered antiques, there are still many klystrons and 1N21/23's in active operation today. The early high-power microwave oscillator tube was the *magnetron,* so called because it required intense permanent magnets for operation. Although it was invented in 1921, the magnetron was little used until it was advanced for radar service in the early 1940's. This tube is still used today in radar, microwave ovens, and industrial applications requiring high-power microwave energy. Since oscillators must also amplify signals, it was to be expected that a high-power amplifier tube would be developed along with the magnetron for microwave applications. This somewhat obscure magnetron-type amplifier tube is called a *crossed-field amplifier* (CFA).

A more recently developed (1950's) tube family that has replaced many klystron and CFA devices is the *traveling-wave tube* (TWT). The TWT has the advantage of wide frequency response without the necessity of mechanical tuning that was required of klystrons and magnetrons. The oscillator version of the TWT is the *backward-wave oscillator* (BWO), which also operates over a wide frequency range without the necessity of mechanical tuning.

The application of simple 1N21/23 series microwave diodes actually predated the invention of the transistor by many years. Although practical microwave transistors have only recently (1970's) achieved popularity in systems applications, microwave diodes of many types have been developed and successfully employed for a wide range of microwave functions, including attenuation, switching, frequency multiplication, tuning, modulation, and signal generation, as well as basic mixing and detection. Among the popular microwave diode types are the *point-contact, pin, varactor, step-recovery, tunnel, impatt, trapatt, Gunn, hot-carrier,* and *Schottky-barrier*. In a general sense, attenuation, switching, and amplitude modulation are achieved by varying the conductivity of diodes placed in series or shunt with a transmission line. Frequency modulation and multiplication are achieved by varying the reactance and linearity of a diode placed in shunt with a transmission line or tuned network. Direct signal generation is achieved with avalanche diodes, in a tuned network, that function as simple relaxation oscillators.

Microwave transistors are presently available in bipolar and field-effect types to achieve amplification and oscillation functions. These devices are presently limited in *power, frequency response,* and *noise figure,* making them most useful for lower power and lower frequency microwave applications. The technology is gradually progressing to improve transistor performance, however, and we can expect advances in these three important areas in the future. You should note that each improvement in solid-state technology displaces the traditional vacuum-tube device and circuitry, making the same microwave function available in small, lightweight, and lower power-consumption packages. At this time, reliable cost-competitive transistors are abundantly available for operation below 18 GHz. The bipolar transistors are more suitable for operation below 4 GHz, whereas the GaAs FET's (gallium-arsenide field-effect transistors, called *gas fets*) are more practical for operation from 2 to 18 GHz. The present practical power limitations for microwave transistors are typically about 10 W at 1 HGz and 1 W at 18 GHz. Table 13-1 shows the specifications of typical small-signal and power transistors.

The limitations of solid-state devices are important to know as they serve to define the point where vacuum-tube devices are still the better choice for microwave amplification and oscillation. Where vacuum tubes are required, however, the costs of microwave systems increase substantially. If we compare the typical 10 GHz, 100 mW signal sources of klystron and

TABLE 13-1. TYPICAL SPECIFICATIONS OF MICROWAVE SMALL-SIGNAL AND POWER TRANSISTORS

Type of Transistor	Current (mA)	Forward Ratio	Power Gain, A_P (dB)	Noise Figure (dB)	Power, P_{out} (dBm)
	I_{DSS}	g_{fs}	*At 10 GHz*	*At 10 GHz*	
GaAs FET	40	40 μS	10	4	15
	I_C	h_{FE}	*At 2 GHz*	*At 2 GHz*	
Small-signal bipolar	6	150	15	2	10
Power bipolar	100	60	6	10	30

Gunn-diode designs, the advantages of the solid-state approach should be obvious. To begin with, the typical power requirements for the klystron would include a 6 V, 3 A heater supply, a 500 V, 40 mA anode supply, and a 300 V, 2 mA repeller supply, for a total power consumption of 38 W from a bulky, high-voltage integrated power supply unit. The comparable power requirements for the Gunn-diode source would be only 10 V and 400 mA, for a total power consumption of 4 W from a simple, low-voltage power supply.

The point to be made here is that aside from the size and weight of the actual microwave devices, the power supplies and other electronic circuitry are larger and heavier. If the costs of the two approaches are compared, the Gunn-diode system may be less than 20% of the cost of a vacuum-tube system. Where high frequency, high power, ultralow noise, or a combination of these factors, are required for microwave signal processing, however, the additional high costs must be an accepted part of vacuum-tube applications.

Chapter 12 reviewed a number of microwave devices, including power dividers, couplers, attenuators, tuners, and filters, that operate on transmission-line principles. There is a closely related family of microwave devices that operate within transmission structures, but use the magnetic properties of materials to process microwave signals. These devices are identified as microwave *ferrite components,* and they provide special functions, including isolation, phase shifting, polarization, and filtering. Microwave ferrites are materials composed of iron oxide that is combined with other materials, such as aluminum, cobalt, manganese, nickel, yttrium, and zinc. Microwave signals will propagate ferrites, because the material has dielectric properties that can approximate those of common transmission lines and other microwave elements. The reason that ferrites are categorized with *active* devices is because of their *electronic* behavior in the presence of the microwave signal that propagates the material. In operation, ferrite devices are biased with an applied magnetic field that imposes a pattern to the spinning electrons in the atomic structure of the material. It is this pattern that magnetizes the ferrite. The electromagnetic microwave signals passing through the ferrite interact with this magnetization; thus the propagation of the signal is affected, thereby achieving the special processing functions identified earlier in this paragraph.

LEARNING OBJECTIVES

Upon completing this chapter, including the problem set, you should be able to:

1. Have knowledge of the operational characteristics and power supply requirements for klystrons, magnetrons, CFA's, TWT's, and BWO's.

2. Understand *S*-parameters as they are applied to active microwave devices, and apply the Smith chart to *S*-parameter analysis.

3. Have knowledge of the operational characteristics and bias requirements of bipolar and GaAs field-effect transistors as used in microwave circuits.

4. Understand and compute circuit conditions and component values required to apply transistors as amplifiers and oscillators at microwave frequencies.

5. Have knowledge of the operational characteristics of varactor and step-recovery diodes as used in microwave signal sources and amplifiers.

6. Have knowledge of the operational characteristics of point-contact, hot-carrier, and Schottky-barrier diodes as used in microwave mixers and detectors.

7. Have knowledge of the operational characteristics of tunnel, Gunn, impatt, and trapatt diodes as used in microwave signal sources and amplifiers.

8. Have knowledge of the operational characteristics of *pin* and small-signal diodes as used in microwave attenuators and switches.

9. Understand the operation and applications of microwave ferrite devices, including isolators, circulators, attenuators, and yig filters.

10. Have knowledge of the operational characteristics and bias requirements of yig-tuned and varactor-tuned active microwave devices.

KLYSTRONS

As noted in the introduction to this chapter, klystrons were the early popular microwave-tube type, and they remain in widespread use in modern microwave equipment. There are two categories of klystron tubes: the *multicavity* power amplifier klystron and the single-cavity *reflex* klystron. The tubes operate on a principle called *velocity modulation,* wherein the velocity of the electron beam is varied as it moves within the tube. The velocity of electrons increases (accelerates) as they move from cathode to anode, just as a falling rock accelerates from the force of gravity, with the velocity and acceleration depending on the force of the anode voltage on the cathode. Of course, the electron velocity cannot reach the speed of light, according to our physicists, because electrons have mass and the resulting energy requirements and expenditures are infinite. The electrons that are emitted from the cathode of the klystron tube are attracted by the high positive voltage of the anode, forming an electron beam as they accelerate. Some klystron tube designs require an external magnetic field to force the electrons into a narrow beam from cathode to anode. Figure 13-1a shows a simplified version of a two-cavity amplifier klystron.

Velocity Modulation

Referring to Fig. 13-1a you should observe that an input RF signal applied to the first cavity will affect a circulating current in its equivalent resonant network. The resonant network will, in turn, alternately charge grids 1 and 2 at the frequency and magnitude of the input signal. As the electron beam passes this first cavity, the alternating voltage on grid 1 will slightly acceler-

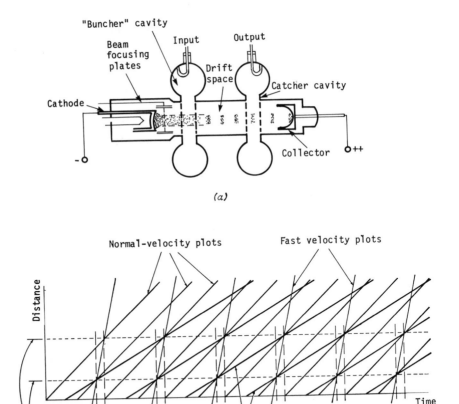

(a)

Bunching occurs at intersections of plotted lines, at distances indicated by --- lines, and at time intervals indicated by the vertical − − − lines.

(b)

FIGURE 13-1

ate and decelerate the electrons, to an extent depending on the magnitude and polarity of the voltage. With critical spacing of grids 1 and 2, in accord with the velocity of the electron beam, the accelerated or decelerated ''bunch'' of electrons will reach grid 2 in the same time that the input signal and resonant-network voltage change phase. The voltage on grid 2 will thus enforce the *bunching* of the electron beam. When the bunched electrons reach the second and third resonant circuits (cavities), they set up a circulating current caused by the variation in charge that the bunches transfer to the grids. Since the electrons are accelerating toward the anode, the velocity of the bunched electron beam is greater when it passes each successive cavity. This higher velocity results in greater energy in the bunched electrons, which thus transfer a higher signal level to the third (output) cavity, resulting in amplification.

Figure 13-1*b* shows the *Applegate diagram* used by tube designers to determine the spacing of the grids based on the velocity and acceleration of the electron beam. The slope of the solid lines represents the normal velocity of the electron beam; the dashed lines represent the decelerated velocity, and the dotted lines represent the accelerated velocity. The intersections of these lines indicate the distance between bunches on the horizontal axis. The graph dictates the spacing of the grids at one-half the distance, and the frequency of operation as the inverse of the time interval. If, for example, the distance between bunches is 12 cm and the time interval is 0.8 ns, as indicated on the diagram of Fig. 13-1*b*, the required spacing of the grids would be 12 cm/2 = 6 cm; and the frequency of operation would be $\frac{1}{8} \times 10^{-10} = 1.25 \times 10^9 = 1250$ MHz. From this relationship you should conclude that tuning of the klystron requires adjustment to the cavities and accelerating voltages in order to adjust the slopes for optimum performance.

Power Amplifiers

Figure 13-2 shows a practical two-cavity power amplifier klystron. Note the terminology used to identify each segment of the tube. The heater, cathode, and associated focusing and accelerating elements that form an electron gun are called the *beam source*. The signal input circuit and electrodes are identified as the *buncher cavity* and *grids;* the signal output circuit and electrodes are identified as the *catcher cavity* and *grids*. The space between the cavities is called the *drift tube,* and the anode is called the *collector.* The particular tube shown achieves beam focusing electrostatically, much the same as electrostatically focused cathode-ray tubes. Each cavity in the klystron must have a high positive voltage in order to attract all the high-velocity electrons and prevent secondary emission (emission from the anode). Often the high collector voltages result in the emission of X rays, thus requiring that the tube structure be shielded to prevent radiation. Such high voltages also require insulation between elements of high-power klystrons. These tubes

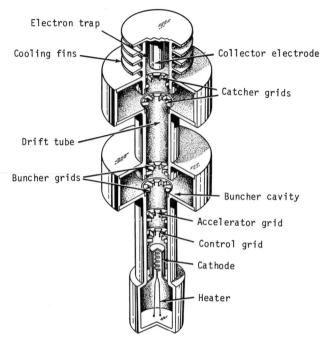

Electron trap

Cooling fins

Collector electrode

Catcher grids

Drift tube

Buncher grids

Buncher cavity

Accelerator grid

Control grid

Cathode

Heater

FIGURE 13-2

are designed and constructed using metal, glass, and ceramic parts, making
the fabrication technology quite complex. The power and gain capabilities of
these tubes, however, more than justify their complexity. Table 13-2 compares the specifications of a three-cavity tube with those of a five-cavity,
magnetically focused device. You should note that the gain of the five-cavity

TABLE 13-2.   THREE- AND FIVE-CAVITY KLYSTRONS

Parameter	Type of Klystron	
	QKK1036 Three-Cavity	QKK1053 Five-Cavity
Frequency, MHz	960–1215	1053–1235
Peak power output, kW	37	300
Average power output, kW	1.2	10
Gain, dB	30	50
Beam voltage, kV	24	45
Beam current, A	6.2	3
Heater power, W	160	300
Weight, lb	40	100
Length, in.	21.5	55

tube is 50 dB, with a CW power output of 10 kW, thus requiring only 100 mW of input signal power.

Reflex Klystrons

In contrast to the high-power-amplifier (HPA) klystron, the popular reflex klystron oscillator is a relatively low-power device, typically operating in the 10 mW to 1 W range. It also operates on the principle of velocity modulation, but does not contain a collector. The reflex klystron achieves oscillation through positive feedback of a bunched electron beam, which is internally obtained with an electrode called a *repeller* (sometimes *reflector*). Unlike the collector, the repeller is a negatively charged electrode. With this structure the device requires only one resonant cavity for operation. Figure 13-3 shows the functional operation of a reflex klystron oscillator. In operation the electron beam from the cathode is accelerated by the positive voltage on the cavity grids. As the initial accelerated electrons pass the cavity, they charge the grids, producing a small oscillatory current in the resonant network. The velocity of the accelerated electrons is sufficiently high to move them beyond the cavity, however, and into the field of the negatively charged repeller. The bunching of the mass of electrons occurs in this field, with the higher and lower velocities being equalized to bunch the electrons and return them, at low velocity, to the cavity where they are collected in the form of cavity current.

You might visualize the effect of the repeller by thinking of a large sponge being bombarded with pellets of varying velocities. The slower pellets will not compress the sponge as much as the faster moving pellets, so both will be *grouped* as they bounce off the sponge. The feedback necessary to sustain

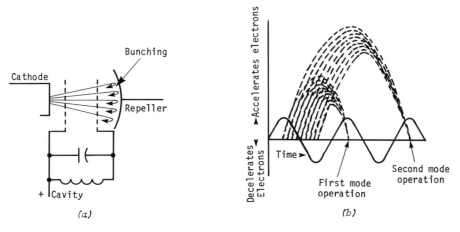

FIGURE 13-3

oscillation in the cavity of the reflex klystron comes from the bunched electrons returning from the repeller to the cavity in phase with the initial oscillatory current.

The graph of Fig. 13-3*b* shows the sinusoidal waveform of the cavity as
compared to the return time of the bunched electrons from the repeller of the
reflex klystron. You should note that the bunched electrons must return to
the cavity in phase during the negative swing (³/₄-cycle point) of the oscillatory signal in order to sustain oscillation. The magnitude of the negative
repeller voltage determines this return time of the bunched electrons. The
critical adjustment of this voltage will return the bunch at the ³/₄-cycle peak
of the oscillatory waveform to assure maximum power output from the device. The *transit time* is the period required for the electron bunches to
return to the cavity after initially accelerating past it into the repeller field. In
practice the bunches may return to the cavity after 1 + ¾ cycle, 2 + ¾ cycle,
3 + ¾ cycle, 4 + ¾ cycle, or 5 + ¾ cycle, or later, beyond the time they
initially passed the cavity. These return-time periods define the reflex klystron's *modes* of operation. As you would expect, the earlier (shorter time)
modes yield higher output power, Q, and efficiency, but require much higher
repeller voltages. Often, broader bandwidth signals are desired, however, so
later, lower repeller-voltage modes are selected for operation.

The reflex klystron can be electronically tuned, to an extent, by slight
variations of the repeller voltage. A slight increase in repeller voltage, above
its optimum X + ¾-mode value, will cause the electron bunches to arrive at
the cavity early, thus reducing the resonant period and increasing the operating frequency. A slight decrease in the repeller voltage, below its optimum
mode value, will conversely cause the electron bunches to arrive at the
cavity late, thereby increasing the resonant period and decreasing the operating frequency. A feedback signal applied to the repeller will thus conveniently achieve AFC (automatic frequency control) action in the tube;
whereas an ac modulating signal can be applied to the repeller and yield
direct FM of the microwave output signal. Table 13-3 compares the specifica-

TABLE 13-3.   50 mW AND 1 W REFLEX KLYSTRONS

Parameter	Type of Reflex Klystron	
	QKK753 Low-Power (used as LO)	QKK759 High-Power (used as transmitter)
Frequency, MHz	7750–8400	7750–8400
Mode	3¾	2¾
Power output, mW	50	1200
Resonator voltage, V	300	800
Cathode current, mA	25	80
Reflector voltage, V	−60 to −150	−220 to −450
Electronic tuning, MHz	36	24

tions of a 50 mW reflex klystron intended for use as an LO with those of a 1 W device used for FM communications service.

MAGNETRONS

Whereas the klystron is one of the most popular high-power microwave amplifiers, the magnetron is the most popular high-power microwave oscillator. Like the klystron, the magnetron operates on the principle of velocity modulation. Instead of a linear (straight line) beam, however, the magnetron takes advantage of magnetic deflection of the electron beam to develop high-power microwave signals. The magnetron tube is structured, as shown in Fig. 13-4, with a cathode that is coaxially centered within a surrounding multicavity cylindrical anode. An intense magnetic field is applied, with its north-south orientation transverse (at right angles) to the cathode-to-anode path. You may recall that electrons are deflected at right angles to an applied

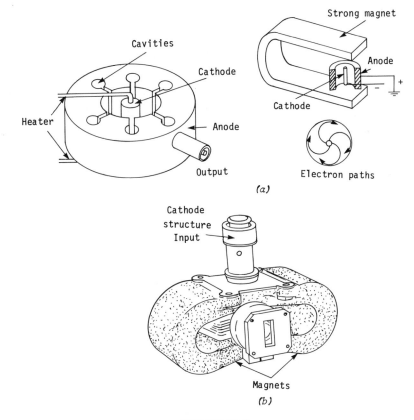

FIGURE 13-4

magnetic field; so those electrons that leave the magnetron's cathode are both attracted by the positive anode voltage and deflected by the applied magnetic field. The effect of these two forces yields an electron beam that follows a spiral path from cathode to anode.

In operation, electrons are emitted from the entire surface of the cathode. As these disbursed electrons spiral past the cavities in the anode structure, they excite circulating currents within each of the resonant networks. These currents in turn alternately accelerate and decelerate the electrons, effecting bunching, as with the linear-beam klystron. This bunching is intensified as the electrons spirally accelerate past each cavity, transferring higher energy levels and increasing the cavities' circulating current with each successive pass. The microwave output signal is obtained by coupling from one of the cavities within the anode structure.

Pulse Magnetrons

With no electron-gun structure and no beam focusing as required for the klystron, and with its 360° cathode emission, the magnetron is a most practical and low-cost source of high-power microwave signals. This simplicity in design and operation is responsible for the popularity of low-cost 2450 MHz microwave ovens as home appliances. Since the magnetron is functionally a diode, the only power supply component required for oven applications is a transformer. The tube both rectifies the applied high voltage (typically 2 to 4 kV) and generates microwave power (typically 500 to 1000 W).

In their most popular radar applications, magnetrons are pulsed for the generation of very high power, short-duration microwave signals. For such applications peak pulse power levels above a megawatt are not uncommon. A typical MW pulse-radar magnetron may operate with a 50 kV, 100 A pulse, yielding a 1 to 3 μs burst of microwave energy. Magnetrons specified for such operation have short *duty cycles* (the ratio of on time to off time) of 1% or less. Since these intense power levels do generate tremendous heat and require massive power supplies, pulse-radar generators are not in any way comparable to the simple microwave ovens. The magnetic fields for the magnetrons are most often provided by heavy permanent magnets that become a part of the total tube assembly. Additional cost savings are realized in microwave ovens by using the core of the power transformer as a solenoid-type electromagnet. Table 13-4 compares the characteristics of typical CW and pulse tubes.

Since the output signal of the magnetron is obtained through coupling from one of its resonant cavities, the tube is both frequency and power sensitive to impedance mismatches which may reflect signals from its output back into the cavity. Although a matched load (VSWR = 1 : 1) yields minimum frequency *pulling,* maximum power from the magnetron is obtained from a slightly mismatched load that offers a *conjugate match* to the cavity. When

TABLE 13-4. CW AND PULSE MAGNETRON COMPARISON

	Type of Magnetron		
Parameter	QK904 Low-Power CW	RK6249 High-Power Pulse	QKH790 Low-Power Pulse
Frequency, MHz	2450	8500–9600	8900–9400
Power output, kW	1.25	200	1
Anode current, A	3	25	1
Anode voltage, kV	4.6	28	5
Duty cycle	CW	0.001	0.001

magnetrons are put into service, they are accompanied by a *Rieke* performance diagram that describes their frequency and power performance with varying load impedances. Figure 13-5 shows this popular magnetron chart. Referring to the Rieke diagram, you should observe that the concentric circles represent constant VSWR values, while the shallow arcs (solid lines) represent pulling frequency values above and below the center-tuned frequency of the magnetron. The crossing arcs (dashed lines) represent the output power of the magnetron compared to different load VSWR values. The optimum operating conditions for the particular tube specified by the diagram of Fig. 13-5 are a VSWR of 1.2 and a 45° angle-of-reflection

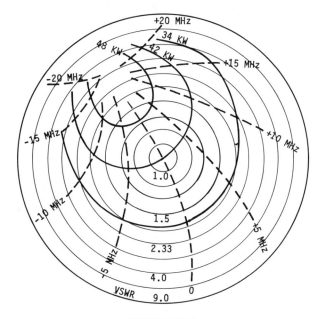

FIGURE 13-5

coefficient, yielding a 100 kW power output with zero frequency pulling. In practice, such operation is obtained with tuning stubs where the magnetron output connects to the transmission line.

Crossed-Field Amplifiers

The magnetron tube can be called a *crossed-field* oscillator, because it operates with *crossed* (at right angles) electric and magnetic fields. The crossed-field amplifier (CFA) operates like the magnetron, but has both input buncher *and* output cavities like the linear-beam klystron amplifier. With the exception of the two transmission-line connections, CFA's and magnetrons are almost identical in appearance. The internal structure of the CFA is also similar to that of the magnetron, but the operating characteristics of the tube differ. The magnetron bunches the electrons as they move past the stationary RF fields of the cavities in the anode, but the CFA achieves bunching at the signal input cavity. These electron bunches are then deflected in the CFA by its applied magnetic field, and caused to rotate (revolve) past each of the anode's cavities—from input to output. As the bunched electrons pass each cavity, they excite resonances, thus converting the energy of acceleration to signal amplitude, as with the linear-beam klystron. The path of rotation within the tube is called the *slow-wave structure,* because the velocity of rotation is actually less than that of the electrons spiraling from cathode to anode. It is the bunching pattern that rotates—not the individual electrons. Table 13-5 identifies the typical characteristics of a popular CFA tube type.

TRAVELING-WAVE TUBES

The traveling-wave tube (TWT) is a most popular linear-beam microwave tube that uses a slow-wave structure. Although some TWT's are built with coupled cavities as though they are "stretched-out" CFA's, mechanically simpler TWT's with helix-type slow-wave structures are the popular choice

TABLE 13-5.   TYPICAL SPECIFICATIONS
OF CROSSED-FIELD AMPLIFIER

Parameter	QKS1319
Frequency, MHz	1250–1350
Power output (peak), kW	100
Power output (average), kW	3
Gain, dB	14
Cathode voltage, kV	−10
Cathode current, A	20
Efficiency	0.5

for lower power applications. If you have a good understanding of how the helix functions to bunch and amplify microwave signals, you should be able to apply the same concepts to the operation of coupled-cavity TWT's and CFA's. Figure 13-6*a* shows a typical helix-type TWT, where the helix acts as a transmission line for the input signal. Even though the electrical input signal travels along the transmission line at virtually the speed of light, its true velocity along the *length* of the tube is effectively reduced by coiling the line to form the helix. The actual signal velocity is thus determined by the pitch of the helical windings. An electron-gun structure is located at the input end of the TWT, and a *collector* anode is located at the output end. The entire tube is placed coaxially within a magnetic field to obtain a narrow cathode-anode electron beam that will not touch the helix. Moreover, a high positive voltage is applied to the helix of the TWT, in conjunction with a higher collector voltage, to assure continuous acceleration of the electron beam from cathode to anode.

Helix Slow-Wave Function

In operation, the signal waveform along the length of the TWT's helix has a velocity that is slightly less than that of the emitted electron beam. As the electron beam encounters the input signal, the electrons are alternately accelerated and decelerated, resulting in bunching, as with the klystron amplifier. The bunched electrons then move along the length of the tube, with increasing energy levels caused by the forces of acceleration. The helix

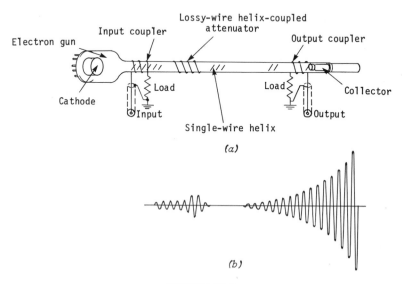

(a)

(b)

FIGURE 13-6

signal and electron bunches actually move in phase along the length of the tube, so that the increased energy levels of the electron bunches *interact* with the signal along the helix. Since the fixed-velocity helix signal reacts *against* the acceleration of the electron bunches, energy transfers to the helix with increasing intensity as the signal and electron beam progress toward the anode, thus resulting in amplification. Figure 13-6*b* shows the exponential amplification achieved within the tube. In many cases the high amplification factors result in some signal reflecting back along the helix transmission line and causing the tube to oscillate. The required isolation is achieved with an attenuator section placed along the length of the helix (most often at the center as shown in Fig. 13-6), with the tradeoff a slight loss of gain.

Coupled-Cavity TWT's

The coupled-cavity TWT is used where high-power or high-frequency requirements make the helix impractical. Functionally, successive cavities along the tube's length behave like a helix, guiding the input signal to the output. The slow-wave structure is actually a series of low-Q waveguide cavities that are excited as the input signal is sequentially coupled from one cavity to the next. The overall signal phase shift along the length of the tube is thus effectively reduced by the series of resonant networks, just as phase velocity of lower frequency signals is reduced with delay lines. As with the helix-type TWT's, coupled-cavity tubes require an intense magnetic field to focus the electron beam, preventing its actual contact with the cavity slow-wave structure; and they require operation at high positive dc voltages to assure continuous acceleration of the electron beam from cathode to anode.

The coupled-cavity TWT differs from the amplifier klystron in that the high-Q klystron cavities and drift tubes enforce bunching, but do not shift the phase of the input signal in step with the electron bunches from cathode to anode, as is achieved with the TWT and CFA slow-wave structures. The coupled-cavity TWT and CFA also differ from the helix-type TWT in frequency response, since the operating frequency range of the tubes is a function of their low-Q cavity bandwidth, whereas the nonresonant helix tubes operate over exceptionally wide frequencies. Table 13-6 compares the specifications of typical helix and coupled-cavity TWT's.

Backward-Wave Oscillators

Oscillator tubes that operate with slow-wave structures similar to those of the TWT and CFA are called *backward-wave oscillators* (BWO's). You may recall that the attenuators placed in the center of the TWT helix suppressed the possibility of oscillatory feedback in that amplifier tube. Oscillation of the tube could thus be assured with a large signal reflecting back along the

TABLE 13-6.   HELIX AND COUPLED-CAVITY TRAVELING-WAVE TUBES

Parameter	Type of Traveling-Wave Tube		
	QKW1049 Medium Power Helix CW	L5089 High Power Helix Pulsed	L5519 High Power Coupled-Cavity
Frequency, MHz	7100–8500	7000–11,000	9500–10,000
Power output, W	10	1000	20,000
Gain, dB	37	35	57
Duty cycle	—	0.02	0.015
Helix voltage, V	2800–3200	0	—
Collector current, mA	50	—	—
Collector voltage, V	2800–3200	—	−10,500
Anode voltage, V	2850–3250	—	—
Cathode voltage, V	—	−10,500	−23,000
Cathode current, A	—	1.6	4.5
Noise figure, dB	20	—	—
Pulsed grid, V	—	—	−400 to +250

helix transmission line and with the elimination of the isolating attenuator. This basic BWO design is achieved by terminating the helix in an *open* (no output connection) at the collector end of the tube, assuring a large reflected signal, and by developing standing waves along the slow-wave structure, which in turn bunch the electron beam. Since the acceleration of the electron beam is a function of the helix voltage, the operating frequency of the BWO can be varied by simply changing the level of positive dc voltage that is applied to the helix. The output signal is obtained from the end of the helix that is closest to the cathode—from the reflected waves moving back along the helix. The TWT-related linear-beam BWO just described is designated as the O-type. As you can conclude, BWO's can also be designed along the lines of the CFA, with magnetically deflected rotating electron bunches. The CFA-related BWO's are designated as the M-type. Table 13-7 compares the specifications of two typical tubes.

TABLE 13-7.   O-TYPE AND M-TYPE BWO's

Parameter	Type of Backward-Wave Oscillator	
	QKB916 O-Type Low Power	QKA857 M-Type Medium Power
Frequency, GHz	8–12.4	8.5–11
Power output, W	0.020	200
Helix voltage, V	325–1340	2000–5000
Anode voltage, V	50–130	800–2100
Sole voltage, V	—	−900 to −3500

S-PARAMETERS

Microwave tubes are specially characterized and usually sold as complete structures, with specified input and output requirements, but with general isolation from external circuitry. Most cavity tubes, for example, have their resonant circuits fabricated within the tube structure to assure practicality in installation and application. Many solid-state microwave devices are also available as complete structures, but they are *also* abundantly available as components: diodes, transistors, etc. Although few electronics engineers and technicians need to know how to build tubes, most of them need to know how to build solid-state amplifiers and oscillators using diodes and transistors.

As you know from your study of Chapters 11 and 12, microwave circuit analysis requires an understanding of both lumped and distributed constants and the application of transmission-line principles. Transistors and other active devices must likewise be analyzed for application in microwave circuits. For microwave applications the familiar low-frequency transistor specifications (*h*-parameters, etc.) are not useful, however, because they do not account for distributed constants. You may recall that the Smith chart is *the* tool used by engineers and technicians to quickly obtain passive microwave circuit information. The Smith chart can also be used to easily characterize transistors and other active microwave devices. The microwave transistor specifications that use the Smith chart are called *S-parameters* (for *scattering* parameters).

The *h*-parameters characterize transistors based on open- and short-circuit conditions that are impractical to achieve with microwave circuits. The *S*-parameters characterize transistors based on standard 50 Ω transmission-line equivalents that *are* practical and convenient to achieve in microwave circuits. Just as there are four *h*-parameter values, there are also four *S*-parameter values: input and output reflection coefficients, and forward and reverse transmission coefficients (insertion gains). These parameters are summarized in Table 13-8; the equivalent and practical test circuits are shown in Fig. 13-7. The reflection-coefficient characteristics, S_{11} and S_{22}, basically provide a Smith chart plot of the input and output impedances of

TABLE 13-8.   *S*-PARAMETER DEFINITIONS

Parameter	Definition
S_{11}	Device input reflection coefficient
S_{22}	Device output reflection coefficient
S_{21}	Forward transfer voltage (insertion) gain
$|S_{21}|^2$	Forward transfer power gain
S_{12}	Reverse transfer voltage gain (leakage)
$|S_{12}|^2$	Reverse transfer power gain

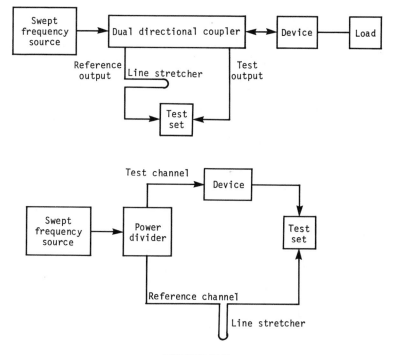

FIGURE 13-7

the transistor. These measurements can be made with a slotted line, reflectometer, or specialized network analyzer. The transmission-coefficient characteristics, S_{21} and S_{12}, specify the forward voltage gain at a given phase angle, and the voltage and phase angle of an output signal that appears back at the transistor input (reverse-direction voltage gain), respectively. These measurements can be made with a vector voltmeter or network analyzer. Since the S-parameters specify matched resistive input and output impedances, the power gain of a microwave transistor can be simply computed as the *square* of the *real* part of S_{21}.

Specifications

Many UHF and microwave transistors are specified by manufacturers with typical S-parameter performance characteristics in order to make it convenient to apply and operate the devices in high-frequency circuits. The type of mass production typical of low-frequency circuits is impossible to achieve with transistors at microwave frequencies where circuit elements must be critically adjusted to achieve optimum performance with each unit produced. Since microwave transistors are neither "perfect"—in the sense of a theoretical model—nor identical from one device to the next, S-parameter measurements are often made on each device used in critical applications. The ideal operating characteristics of a transistor amplifier include a zero-value reflection coefficient at its input and output terminals, a reasonable forward transmission coefficient with a coincident 180° phase shift, and an insignificant reverse transmission coefficient. These characteristics would translate to S-parameter values of $S_{11} = 0\angle 0°$, $S_{22} = 0\angle 0°$, $S_{21} = 20\angle 180°$, and $S_{12} = 0\angle 180°$. These characteristics and S-parameter values are, of course, practically impossible to obtain in a given transistor, but they can be approached with the addition of circuit components external to the transistor.

Figure 13-8 shows the S-parameters for a 2N3478 bipolar transistor operating from 100 MHz to 1700 MHz. S_{11} and S_{22} are plotted on the Smith chart; S_{21} and S_{12} are plotted as magnitude and phase on a linear graph. The data in Table 13-9 was obtained from the plots of Fig. 13-8 at frequencies of 100 MHz, 900 MHz, and 1700 MHz. You should note that the angles of reflection coefficient are emphasized on the periphery of the Smith chart, while the magnitudes are plotted as equivalent VSWR values. At 100 MHz, for example, the S_{11} plot crosses the 1.4 : 1 VSWR circle at an angle of $-45°$. The equivalent magnitude of the reflection coefficient (ρ) is computed from this VSWR value as:

$$\rho = (\text{VSWR} - 1)/(\text{VSWR} + 1) = (1.4 - 1)/(1.4 + 1)$$
$$= 0.4/2.4 = 0.167\angle -45°$$

The S_{22} plot crosses the 8 : 1 VSWR circle at an angle of 0°; and its equivalent magnitude of reflection coefficient is computed as:

$$\rho = (\text{VSWR} - 1)/(\text{VSWR} + 1) = 7/9 = 0.778\angle 0°$$

The S_{21} plot on the linear graph indicates a voltage gain of 20 dB (10 : 1) at a phase angle of 160° at 100 MHz; the S_{12} plot indicates a voltage gain (loss for isolation) of -44 dB (0.006) at a phase angle of $-160°$ at 100 MHz. You

TABLE 13-9. 2N3478 CHARACTERISTICS

Characteristic	Frequency		
	100 MHz	900 MHz	1700 MHz
Z_{in}	$1.4 - j1.1$	$0.2 - j0.5$	$0.42 + j0.35$
Z_{out}	$8 - j0$	$1 - j0.36$	$0.4 - j0.2$
VSWR_{in}	2.6	6	2.62
VSWR_{out}	8	1.45	5.8
S_{11}	$0.444\angle -46°$	$0.713\angle -126°$	$0.449\angle -133°$
S_{22}	$0.778\angle -0°$	$0.183\angle -80°$	$0.708\angle -153°$
S_{21}	20 dB @ $\angle 160°$	4 dB @ $\angle 70°$	2 dB @ $\angle 20°$
S_{12}	44 dB @ $\angle -160°$	34 dB @ $\angle -140°$	25 dB @ $\angle -135°$

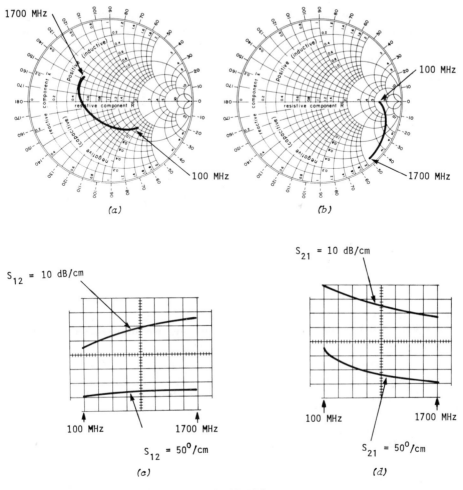

FIGURE 13-8

should observe that the gain (S_{21}) and isolation (S_{12}) are typically reduced, along with their associated phase angles, as their frequency applied to the transistor increases.

Impedance Matching

The S_{11} and S_{22} parameters may be directly translated to impedance values by reading the R and X coordinates which they intersect on the Smith chart. With this information, impedance-matching techniques can be applied to cancel mismatches, as with similar problem examples in Chapters 11 and 12. Either lumped-constant or distributed-constant reactive elements can be used external to the transistor to effect impedance matching, depending on wavelength and practicality. Input (S_{11}) matching for the 2N3478, for example, is relatively simple for single-frequency operation: at 1700 MHz a series inductance can be used to bring the $+150°$ phase angle to $180°$, and a quarter-wave transformer section can be used to match the resulting 0.4 R value at the transistor input; or a lossless match can be achieved by rotating the VSWR circle to intersect the $R = 1.0$ circle and applying a $-j0.95$ reactance at that point (0.12λ) from the transistor input.

Gain and Isolation

The parameters S_{21} and S_{12} determine the gain and isolation of the transistor's input and output terminals. The application of these parameters is least complicated when S_{11} and S_{22} are first matched to 50 Ω as in the previous paragraph. Under such ideal conditions the power gain of the transistor is simply $|S_{21}|^2$, reduced by the feedback $|S_{12}|^2$. For the 2N3478 of our previous example the power gain can be computed at 900 MHz as:

$$G = |S_{21}|^2(1 - |S_{12}|^2) = 1.6^2(1 - 0.025^2) = 2.55$$

You should note that the gain reduction from isolation is relatively insignificant for S_{12} values of less than 0.2. Although there are bipolar transistors with higher S_{21} values than the 2N3478 of this example, the imperfections of this device were chosen to illustrate that poor, low-gain performance usually results from matched input and output terminals—a power gain of 2.55 for a transistor circuit is more or less insignificant. Transistors yield higher gain factors and they operate with lower noise figures when their input and output impedances are purposely mismatched. But here is a problem for engineers and technicians: Purposely mismatching transistors for optimum gain, stability, and noise performance requires the use of S-parameters in a series of complicated mathematical operations that account for the combined vector quantities of S_{11}, S_{22}, S_{21}, and S_{12}. The fundamentals of applying S-parameters for *optimum* performance of bipolar and field-effect transistor circuits will be reviewed on the following pages. As you proceed through the material, however, please understand that most optimization calculations are done with the aid of computers; thus only the fundamentals are presented in this text.

Sample Problem 1

Given: A 2N3478 bipolar transistor with characteristics as shown in Fig. 13-8.

Determine: The S-parameter values, complex input and output impedances, simple impedance-matching networks, and matched-impedance power gain, for operation at 500 MHz.

(a) The complex input impedance and S_{11} are obtained from the Smith chart plot of Figure 13-8: $Z = 0.8 - j0.5$, $\angle\theta = -96°$, and VSWR $= 1.8$. The equivalent magnitude of the input reflection coefficient is computed as:

$$\rho = (\text{VSWR} - 1)/(\text{VSWR} + 1) = 0.8/2.8 = 0.286$$

thus $\quad S_{11} = 0.286\angle -96°$

(b) The complex output impedance and S_{22} are also obtained from the Smith chart plot of Fig. 13-8: $Z = 1.0 - j4.2$, $\angle\theta = -26°$, and VSWR $= 19:1$. The equivalent magnitude of the output reflection coefficient is computed as:

$$\rho = (\text{VSWR} - 1)/(\text{VSWR} + 1) = 18/20 = 0.900$$

thus $\quad S_{22} = 0.900\angle -26°$

(c) If the complex impedance of S_{11} cannot be *practically* transformed to cross the $1.0 \pm jX$ circle on the Smith chart (the wavelength is too long at 500 MHz), the simplest match can be obtained by canceling the input reactance of $-j0.5$ with an inductor of $+j0.5$, leaving the input impedance at $0.8 \pm j0.0$:

$$+jX = Z_0(jX_{\text{normalized}}) = 50 \times 0.5 = 25 \ \Omega$$
$$L = X/\omega = 25/(6.28 \times 5 \times 10^8) = 7.96 \ \text{nH}$$
$$\text{New VSWR} = Z_0/R_{\text{in}} = 1.0/0.8 = 50/40 = 1.25:1$$

If the $1.25:1$ VSWR is too high, a quarter-wave transformation can be employed:

$$Z' = \sqrt{Z_0 Z_L} = \sqrt{50 \times 40} = 44.7 \ \Omega$$
$$\lambda = v_0/f_0 = 1.18 \times 10^{10}/(5 \times 10^8) = 23.6 \ \text{in.} = 60 \ \text{cm}$$
$$\lambda/4 = 23.6/4 = 5.9 \ \text{in.} = 15 \ \text{cm}$$

These lengths would be reduced by the velocity factor of the transmission structure on which the transistor is mounted, but further shortening can be achieved by folding the line-structure routing.

(d) The complex impedance of S_{22} can be directly matched with a series inductor of $+j4.2$ (normalized), leaving the input impedance perfectly matched at $1.0 \pm j0.0$:

$$+jX = Z_0(jX_{\text{normalized}}) = 50 \times 4.2 = 210 \ \Omega$$
$$L = X/\omega = 210/(6.28 \times 5 \times 10^8) = 66.9 \ \text{nH}$$

(e) S_{12} is obtained from the linear (log) graph as -37 dB $\angle -160°$. The feedback voltage ratio and S_{12} are computed as:

$$V_1/V_2 = \log^{-1}(\text{dB}/20) = \log^{-1}(37/20) = 70.8:1$$
$$S_{12} = 1/(V_1/V_2) = 1/70.8 = 0.014\angle -160°$$

Since S_{12} is substantially less than 0.2, it can be neglected in the gain calculation below.

(*f*) S_{21} is obtained from the linear (log) graph as 9.6 dB $\angle 90°$. The forward voltage ratio, S_{21}, and the power gain factor are computed as:

$$V_1/V_2 = \log^{-1}(dB/20) = \log^{-1}(9.6/20) = 3.02 : 1$$
$$S_{21} = 3.02\angle 90°$$
$$G = \quad G = S_{21}{}^2 = (3.02)^2 = 9.12$$

Check: $G_{dB} = 10 \log G = 10 \log 9.12 = 9.6$ dB

MICROWAVE BIPOLAR AND FIELD-EFFECT TRANSISTORS

You should have some basic understanding of microwave transistors through the S-parameter characteristics reviewed in the previous section. The prevailing technology clearly divides transistor types by frequency range: inexpensive bipolars are mostly used below 4 GHz; whereas the expensive FET's are mostly used from 4 to 22 GHz. The transistors are applied as *small-signal* and *power* microwave amplifiers and oscillators. Like their lower frequency counterparts, optimum performance of the microwave transistors requires proper biasing of the devices before they can function as signal processing elements. Unlike low-frequency transistors, however, optimum microwave performance additionally depends on the transistor package and the physical structure in which the devices are placed. The popular microwave transistor package styles include coaxial and stripline mounting structures, with many transistors available in *chip* form for direct bonding onto thick-film circuits.

Bipolar Transistors

Microwave transistors are available in both *npn* and *pnp* types. They may be connected in either common-emitter or common-base configurations; and they may be operated class A, B, or C depending on power and efficiency requirements. The lower power, small-signal transistors are typically connected as class A common-emitter amplifiers. The familiar emitter resistor/bypass capacitor bias-stability circuit arrangement, commonly used at low frequencies, is impractical for microwave circuits, primarily because of capacitor limitations. Microwave bias stability for small-signal CE amplifiers is more practically achieved with the voltage-feedback/constant base-current circuit arrangement of Fig. 13-9*a*. This circuit uses low resistance values that are practical to achieve at microwave frequencies, and it does not require signal-bypass capacitors. For class A operation the transistor *and* bias circuit consume significant supply power, limiting this class of operation to low RF-power applications.

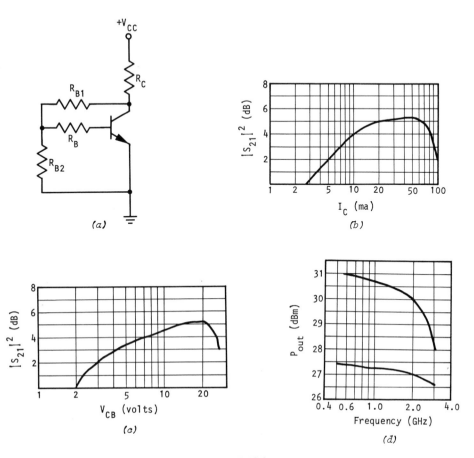

FIGURE 13-9

Figure 13-9*b* and *c* graphs the small-signal gain (S_{21}) of a typical micro-wave transistor against I_C and V_{CE} values, indicating optimum performance at 20 to 60 mA and at 12 to 20 V dc. Figure 13-9*d* graphs the *saturated* and *1 dB compression* output power (P_{out}) against frequency. If you select the 2 GHz operating point with a 1 dB compression P_{out} of 500 mW (27 dBm), and compare this to a nominal dc input of 50 mA at 20 V (equal to 1 W), 50% efficiency appears to be excellent. With the addition of the bias circuit, however, the efficiency drops to less than 25%, as will be shown in the following sample problem.

Sample Problem 2

Given: The bipolar transistor of Fig. 13-9 to be used as a small-signal CE amplifier (class A).

Determine bias circuit values and power supply requirements for the following conditions: $V_{CE} = 20$ V, $I_C = 50$ mA, $\beta(h_{FE}) = 64$, $V_{CC} = 2V_{CE}$, $I_{BB} = I_C/\sqrt{\beta}$, $V_{BB} = V_{BE} + 2$ V (for constant current), and $V_{BE} = 0.7$ V (silicon).

(*a*) The circuit voltages are determined for the operating (Q) point as specified:

$$
\begin{aligned}
V_{CC} &= 2V_{CE} = 2 \times 20 = 40 \text{ V dc} \\
V_{RC} &= V_{CC} - V_{CE} = 40 - 20 = 20 \text{ V dc} \\
V_{BB} &= V_{BE} + 2 = 0.7 + 2 = 2.7 \text{ V dc} \\
V_{RB} &= V_{BB} - V_{BE} = 2.7 - 0.7 = 2 \text{ V dc} \\
V_{RB1} &= V_{CE} - V_{BB} = 20 - 2.7 = 17.3 \text{ V dc} \\
V_{RB2} &= V_{BB} = 2.7 \text{ V dc}
\end{aligned}
$$

(*b*) The circuit currents are determined for the operating point as specified:

$$
\begin{aligned}
I_B &= I_C/\beta = 50/64 = 0.78 \text{ mA} \\
I_{BB} &= I_C/\sqrt{\beta} = 50/8 = 6.25 \text{ mA} \\
I_T &= I_C + I_B + I_{BB} = 50 + 0.78 + 6.25 = 57.03 \text{ mA}
\end{aligned}
$$

(*c*) The power requirements are determined as follows:

$$
P_T = V_{CC}I_T = 40 \times 57.03 \text{ mA} = 2.28 \text{ W}
$$

(*d*) The resistance values are computed as follows:

$$
\begin{aligned}
R_C &= V_{RC}/I_T = 20/57.03 = 350 \ \Omega \\
R_B &= V_{RB}/I_B = 2/0.78 = 2564 \ \Omega \\
R_{B2} &= V_{BB}/I_{BB} = 2.7/6.25 = 432 \ \Omega \\
R_{B1} &= V_{RB1}/(I_B + I_{BB}) = 17.3/(0.78 + 6.25) = 2460 \ \Omega
\end{aligned}
$$

The bias chokes in this circuit configuration eliminate circuit loading; and the series R_B can be trimmed for critical bias control.

Power Transistors

Because of device inefficiencies and power supply requirements, microwave *power* transistors are usually biased for class C operation in the CB (common-base) configuration. With this form of operation there is only one power supply connection (V_{CC}) to the collector (V_{CB}). As you may recall from your study of Chapter 5 of this text, class C amplifiers are biased *off*, with conduction during only a portion of the applied input signal. Under these operating conditions the instantaneous value of the input signal must exceed 0.7 V (silicon V_{BE}) for the required signal conduction period. The optimum efficiency of operation is thus achieved by adjusting the amplitude of the input signal to exceed the V_{BE} cutoff voltage.

Figure 13-10a shows the bias and signal processing circuitry of a typical microwave power amplifier. The transistor is configured for class C, CB operation on a micro-stripline structure; with C_1-L_1 and C_4-L_2 providing input and output matching, respectively. The power supply is isolated and de-coupled with the RFC's and feedthrough capacitors, C_2 and C_3. The operating (Q) point of the transistor is adjusted for an excess input signal amplitude by the selection of R_1 values to obtain the desired conduction angle. Figure 13-10b and relates class C output power and efficiency to input power and V_{CC}, respectively, for a typical bipolar microwave power transistor. You may observe that the efficiency levels off above 55%, indicating the desired operation of the device, with a corresponding 12 W P_{out} at a V_{CC} of 30 V dc.

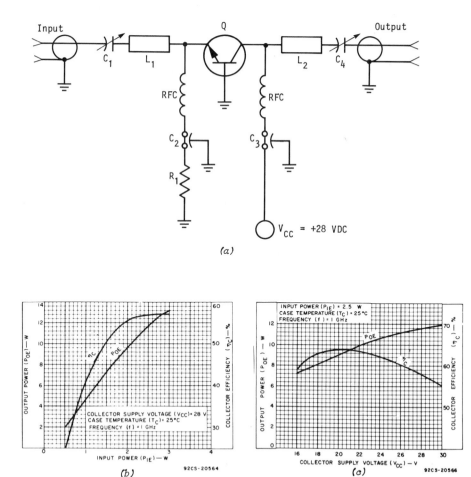

FIGURE 13-10

Since the supply voltage (V_{CC}) is applied directly across the transistor (V_{CB}) in class C operation, the circuit efficiency is equal to the transistor efficiency, without the power supply losses that are incurred with class A operation.

Field-Effect Transistors

Field-effect transistors are fast becoming important microwave devices. GaAs FET's, often identified as *MESFET's* (for metal-epitaxial semiconductor field-effect transistors), are widely used in microwave small-signal and power amplifier and oscillator applications. The fabrication technology has not yet matured, however, so there are reliability problems with the devices that raise their cost far above that of bipolar devices. As this technology improves, we can look to the FET as one of the most popular microwave devices because, unlike the bipolar transistor, its operation is not limited to the low microwave frequencies. FET oscillators at 80 GHz are now practical. Among the tradeoff problems with microwave FET's are their fragile structure and susceptibility to burnout from power supply or signal transients. The devices thus require critical biasing and signal processing circuits in order to assure reliable performance when they are placed in operation.

Microwave GaAs FET's are exclusively N-channel devices, operating with a positive-polarity drain power supply and negative-polarity gate-to-source voltage. Although some circuits are more conveniently biased with two power supplies, it is possible to use a single drain supply, developing V_{GS} through a bypassed series source resistor. This approach is not appropriate for bipolar transistors because the bypass capacitance is transferred to the input S_{11} through the low-impedance, forward-biased base-emitter junction. The bypassed source resistor approach is most appropriate for microwave FET's, however, as the input gate is isolated from the source bypass capacitor.

Figure 13-11a shows the dual-supply bias circuit for a typical GaAs FET amplifier. You should observe that zener diodes and filter capacitors are connected to the power supply input terminals to remove any transients from the line. Experience with these microwave devices has shown that even short-duration transients caused from turning on the supply or connecting the lead will destroy GaAs FET's. Figure 13-11b graphs the gain vs I_{DS}/I_{DSS} for a typical FET; and Fig. 13-11c graphs the familiar *family of curves* for the device. You should note that S_{21} is optimum for I_{DS}/I_{DSS} ratios of 0.3 to 0.8 as shown in Fig. 13-11b, indicating I_{DS} values of 20 mA to 50 mA, and corresponding V_{GS} values of -2.2 V and -0.8 V as in Fig. 13-11c.

S-Parameter Characterization

The typical S-parameter characteristics of small-signal CE bipolar transistors are indicated in Fig. 13-12. The S-parameters typical of bipolar power

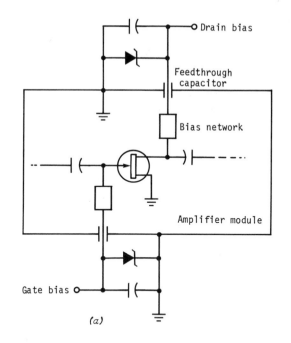

(a)

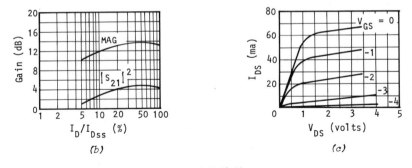

(b)   (c)

FIGURE 13-11

transistors, as applied in CB circuits, are shown in Fig. 13-12a. All four of the S-parameter characteristics are combined for convenience on a Smith chart, with the S_{21} and S_{12} values plotted on the wedge of a circular graph overlay.

If this transistor is biased for class C operation, as in Fig. 13-10, only the input and output gain/impedance-matching networks remain to complete the circuit. With S_{11} and S_{22} as low values, input and output impedance transformation is required for optimum performance of the device. This matching is achieved with the microstrip networks of Fig. 13-12b. Input transformation is achieved with LC networks, and taper transformers,

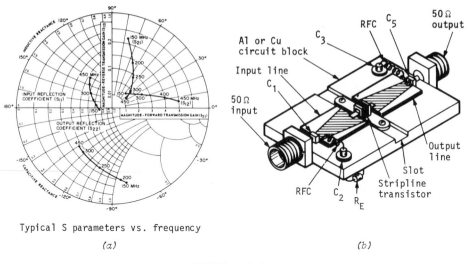

Typical S parameters vs. frequency

(a)   *(b)*

FIGURE 13-12

allowing broader frequency response than a quarter-wave transformer. Unlike bipolar devices, microwave FET's have higher S_{11} characteristics, while the other three S-parameters are similar to those of bipolars.

Bipolar and FET devices can be efficiently applied as microwave oscillators, with the lower-frequency range below 4 GHz appropriate for bipolars. Although there are a number of diode-type, two-terminal devices that produce microwave signals, transistors are, in many cases, a better choice for oscillator applications. In particular, higher efficiency transistor oscillators have low FM noise and good frequency stability, and they can easily be voltage tuned and phase locked. The most practical configuration for bipolar transistors is the CB connection, where the oscillator circuits have the highest gain, stability, and efficiency of operation. The FET equivalent of the bipolar CB connection is the CG (common-gate) configuration, which is also the more practical choice for oscillator applications. There are some oscillator circuits, however, that employ CE or CC, or CS or CD (common-drain), connections successfully.

MICROWAVE DIODES

The important microwave diode devices and their functions were initially identified in the introductory section of this chapter. You may recall that these devices function as microwave oscillators, amplifiers, modulators, mixers, switches, attenuators, detectors, and reactive network elements.

The diode devices are generally categorized by type/function as defined in *Learning Objectives* 5 through 8 of this chapter: *harmonic sources* and *amplifiers; mixers* and *detectors; bulk-effect oscillators* and *amplifiers;* and *attenuators* and *switches.* The detailed operational characteristics of these devices and functions will be reviewed in the following paragraphs.

Varactors

The use of voltage variable capacitance diodes, called *varactor diodes* (for *var*iable re*actor*), to generate and amplify signals was one of the earlier microwave applications of solid-state electronics. Low-noise amplification of microwave signals with a varactor diode was achieved in a subsystem component called a *parametric amplifier.* The early parametric amplifiers used a klystron to generate a microwave signal at twice the frequency of the one to be amplified. This high-frequency source was called a *pump,* as its output was applied to *pump* the varactor diode that was in the transmission path cavity (coax or waveguide) of an incoming microwave signal. The pump signal caused the varactor diode to change its capacitance (and reactance) at twice the frequency of the incoming microwave signal, releasing energy to the signal peaks and thus achieving amplification. Figure 13-13a shows the operation of a modern parametric amplifier along with its functional equivalent circuit, as used in applications requiring very low noise, high-gain amplification of microwave signals.

You may rightly conclude that obtaining the exact pump frequency and phase are critical factors in achieving parametric amplification as described in the above paragraph, since the varactor must transfer its energy to the *peaks* of the input signal. One solution to this frequency and phase problem is to operate the pump at some frequency other than twice that of the incoming signal. Of course, the pump and input signal then applied to the nonlinear varactor diode will yield mixing-product frequencies. Adding an additional resonant circuit at the difference frequency (or sum frequency) will maintain the resultant energy in phase with that of the input signal. This additional resonant circuit, called the *idler,* is placed in parallel with the input signal cavity to assure in-phase amplification of the input signal. Since the idler is resonant, amplified power can be extracted from it at the difference (or sum) frequency, instead of from the input signal path. Used in this way the parametric amplifier can provide very low noise upconversion or downconversion of microwave input signals. Figure 13-13b shows these additional parametric amplifier functions. Table 13-10 compares typical parametric amplifier specifications with those of other microwave devices.

Another important application of varactor diodes is for tuning modern microwave solid-state signal sources. Since the varactor diode is a voltage-variable capacitor, it can be used to provide electronic tuning, FM, and sweep functions to a microwave resonant circuit or cavity. In this application

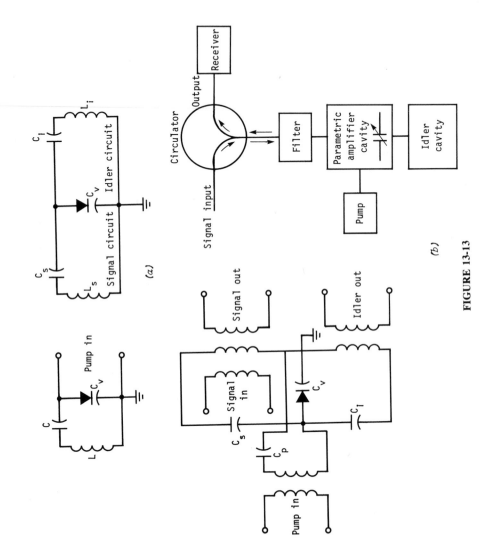

(a)

(b)

FIGURE 13-13

TABLE 13-10. TYPICAL PARAMETRIC AMPLIFIER
SPECIFICATIONS COMPARED WITH THOSE OF OTHER
ACTIVE MICROWAVE DEVICES

Device	Gain (dB)	Bandwidth (MHz)	Efficiency	Noise Figure (dB)
Parametric amplifier	20	50	Low	0.2
Klystron	40	20	Low	20
TWT	30	Octave	Moderate	6
Bipolar transistor	30	Octave	High	4
FET	30	Octave	High	3

the device performs like the varicap described in Chapter 7, but with small capacitance values. The typical junction capacitance of a diode so used in a 10 GHz varactor-tuned solid-state oscillator, for example, would be only 2 pF, with a 4 : 1 change in junction capacitance. This diode yields only a small (6 pF) change in tuning capacitance in proportion to the applied bias voltage. The approximate tuning range of these varactor-tuned signal sources may be computed, based on diode characteristics, as:

$$\Delta f = (C_{j\max} - C_{jo})V^2 \pi f^2 / Q_L P_T$$

C_{jo} is the zero-bias diode capacitance and $C_{j\max}$ is the maximum diode capacitance before breakdown, V is the applied bias voltage, f is the operating frequency, Q_L is the loaded Q of the resonant network, and P_T is the total power dissipated by *all* of the circuit elements.

As a nonlinear device, varactor diodes produce harmonic products from signals applied across their terminals. This property is used to advantage in the generation of microwave signals from lower frequency sources. The approach uses a high-power bipolar oscillator, usually in the 200 to 800 MHz range, to drive the varactor, thus producing harmonics at the microwave frequency to which an output cavity is tuned. Some varactor multipliers produce the full range of harmonic signals. They are called *comb generators* because of the comb-like display they produce on the face of a spectrum analyzer. In practice, higher harmonic-number multiplication factors are more practical to achieve in steps—using a doubler and tripler, for example, rather than a single ×6 multiplier. Varactor diodes are available for operation at microwave frequencies above 100 GHz, and they are capable of handling over 10 W at the lower microwave frequencies. The devices are most reliable and relatively inexpensive, making their use most practical for solid-state microwave signal sources.

Step-Recovery Diodes

Another popular device used for harmonic generation of microwave signals is the step-recovery diode (SRD). Unlike the variable-reactance varactor, the SRD produces harmonics from an extremely short-duration pulse at its output. In operation an RF signal is applied to the diode, which stores a charge in its junction capacitance during its conductivity period (forward bias). When the polarity of the applied RF signal reverses to turn the diode *off,* its stored charge is rapidly released in the form of a short-duration pulse. The typical pulse time of an SRD intended for 1 to 3 GHz operation is about 300 ps (picoseconds); and the typical pulse time of an SRD intended for 10 to 20 GHz operation is about 50 ps. As with the varactor diodes, the SRD's produce a *comb* spectrum when terminated in a resistive load, and *lumped* energy at some harmonic frequency when terminated in a high-Q resonant circuit (or filter). SRD's are more efficient than varactors, with only 10 dB conversion loss for a $\times 10$ multiplication, but their frequency limits fall short of the varactors at about 20 GHz. SRD's are available to handle above 10 W at the lower microwave frequencies, and about 1 W at 18 GHz.

Tunnel Diodes

Along with the varactor, the *tunnel* diode was another early successful device used in solid-state microwave amplifiers and signal sources. Owing to a number of factors, including the development of other more stable, higher power solid-state devices, the tunnel diode is fast becoming obsolete. The tunnel diode has unique properties, including a region of *negative resistance,* which allows this two-terminal device to amplify signals. Figure 13-14 compares the forward conductivity curve of the tunnel diode with conventional *pn* junction diodes. Because of its structure, electrons can easily cross the very thin diode barrier with low forward voltages across the *pn* junction. This effect is called *tunneling.* As the voltage increases, however, the barrier becomes thicker and the number of electrons decreases from the *peak* current level to the *valley* current level of Fig. 13-14. Increasing the junction voltage above this valley current will force increasing current through the barrier, just as with conventional diodes.

  The tunnel diode achieves its amplifier/oscillator functions when biased in the peak-to-valley current transition area, called the *negative resistance region,* because the response of the device exhibits *decreasing* current with *increasing* voltage. The tunnel diode may be operated as an amplifier, sinewave oscillator, relaxation oscillator, or high-speed switch by adjusting the loading and bias conditions on the device. As you may observe from its equivalent circuit, the tunnel diode has L and C properties which make it self-resonant at microwave frequencies. In its operation as an oscillator, the device must be loosely coupled to a cavity through an impedance-matching

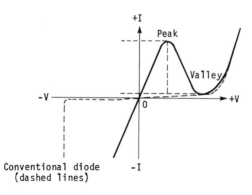

FIGURE 13-14

network that cancels the device's internal capacitance or inductance, depending on its operation above or below self-resonance. In its operation as an amplifier, isolation must be provided between signal input and output terminals in order to maintain stability. Such isolation is typically obtained with quarter-wave transformer coupling or a ferrite circulator. (Circulators and other ferrite devices are covered in a later section of this chapter.) Although the tunnel diodes are low-power devices, they have some advantages as microwave amplifiers, including a broad operating bandwidth and high microwave frequency operation to 50 GHz. Their simplicity and minimum power requirements make them an attractive choice for satellite and other space applications.

Gunn Diodes

Negative-resistance properties common to tunnel diodes were more recently discovered by J. B. Gunn to exist in certain doped GaAs semiconductor materials. This discovery and development yielded yet another solid-state oscillator/amplifier, appropriately called the *Gunn diode*. The Gunn diode is not really a *diode,* as it has only one semiconductor material with no *pn* junction; but since the convention has been established to identify all two-terminal devices having anode and cathode connections as diodes, "Gunn diode" is acceptable. Gunn diodes are also often identified as *transferred electron devices* (TED's) or *oscillators* (TEO's), *bulk-effect devices,* or *limited space-charge accumulation devices* (LSA's). As with the tunnel diode, the Gunn diode exhibits a negative-resistance region when a suitable bias voltage is applied across its terminals. Since its operation does not depend on a fragile, thin *pn* barrier, however, the device is stable and capable of higher power and higher frequency operation than the tunnel diode. Gunn devices are available with power ratings above 10 W at the lower microwave frequencies; with lower power ratings, they are available for operating through 100 GHz.

The negative-resistance region of the Gunn diode's operation is achieved by increasing the applied bias voltage until electron mobility in the GaAs is reduced. This physical phenomenon occurs when the applied voltage increases the electron velocity until they shift from one state in the GaAs structure to another. Figure 13-15a shows the conductivity curve of the Gunn diode. You should observe that the material behaves as a resistor, with linear *V-I* response, until the *threshold* value is reached, after which increasing voltage across the device results in decreasing current, or "negative resistance." Increasing the applied voltage across the diode will eventually overcome the negative resistance and increase the current above the device's *valley* level. Valley voltage generally occurs at about 4 or 5 times the threshold value, depending on the particular diode; the optimum operating point for the device occurs at about 3 to 3.5 times threshold voltage. You may observe that the Gunn diode has similar characteristics with either bias polarity, since it is not a junction device. In fact, some Gunn oscillators are designed for optional-polarity bias operation; others specify bias polarity and grounding based on their case and cavity structure.

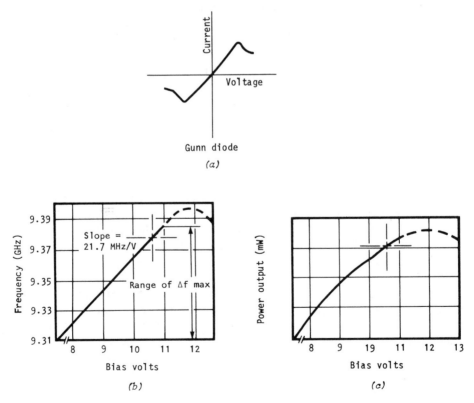

FIGURE 13-15

The more popular applications of Gunn diodes are in microwave oscillators. The devices will actually oscillate on their own, with the application of appropriate bias voltages, but their operating frequency will be determined by the transit time of the electrons across the GaAs chip. In practice, controlled *modes* of operation are preferred, however, so the frequency of oscillation can be controlled by external high-Q LC resonant networks. There are presently three popular modes of operation for the Gunn diode: *delayed domain, quenched domain,* and LSA (limited space-charge accumulation). The delayed-domain mode identifies the most popular operation of the diode below its transit-time frequency, where the resonant-network signal swing drives the device below its threshold voltage. In the quenched-domain mode a very high Q network develops a sufficiently large signal swing to cut off the device. This less popular quenched-domain mode is used for Gunn-diode operation above its transit-time frequency. The LSA mode of operation of Gunn diodes yields the highest power and frequency of operation, but at the date of this writing the successful production of LSA-mode diodes is limited, making them impractically expensive. As you may conclude, the power-supply bias voltage and LC network (or cavity) Q are critical for proper operation of Gunn diodes in any particular mode. Figures 13-15*b* and *c* graph frequency pulling (electronic tuning) and power output vs bias supply voltage for a typical Gunn-diode oscillator operating in the delayed-domain mode, with a fixed-tuned resonant-cavity LC network.

Impatt/Trapatt Diodes

In addition to the specific negative-resistance properties of tunnel and Gunn devices, a third type of negative-resistance effect was discovered to exist within avalanche diodes, making them possible sources of microwave signals. Coincident with the development of the Gunn diode in the mid-1960's, two special avalanche diodes were developed for application as microwave oscillators. These diodes are identified by their function (using technical language, of course) as the *impact avalanche transit-time* diode and the *trapped-plasma avalanche triggered-transit* diode. Fortunately, both devices are more easily identified by their acronyms: impatt (pronounced "im-pat") and trapatt (pronounced "trap-at"). The impatt is the more popular device, and is used extensively as a source of low-power microwave signals from 3 GHz to above 100 GHz, with higher power operation to 10 W at the lower microwave frequencies. The trapatt is a higher power device, practical as a power oscillator at the lower microwave frequencies—to 12 GHz. The trapatt also produces harmonics, however, making it a practical harmonic generator for the higher microwave frequencies.

The negative resistance effect in the impatt is achieved for the generated ac signal—its dc response shows no negative resistance. You should observe that the response curve of the impatt is like that of a zener breakdown diode, as shown in Fig. 13-16*a*. When a sufficiently high reverse-bias voltage is applied to the device's junction, it breaks down, releasing electrons to flow through the device. These released high-velocity electrons impact other electrons (or holes) in the semiconductor material, causing the avalanche process to take place. If a constant-current power supply is used to bias the impatt, the avalanche process will greatly increase current while loading the power supply, thus instantaneously reducing the applied voltage, achieving negative resistance due to phase delay, and generating a signal as a relaxation oscillator. When the impatt is coupled to a high-Q LC resonant network (or cavity), the 180° *V-I* phase relationship sustains sinusoidal oscillation at the resonant frequency. In contrast to the Gunn diodes, which require a constant-voltage power supply in the 7 to 15 V dc range, the impatts require a constant-current power supply typically in the range of 30 to 120 V dc, depending on the power rating and operating frequency of the device. Figure 13-16*b* and *c* graph the frequency pulling (electronic tuning) and power output vs bias supply current for a typical 10 GHz, low-power impatt diode oscillator in a fixed-tuned resonant-cavity LC network.

Although its function is similar to the impatt, the operation of the trapatt diode differs slightly from the former device. When a high reverse-bias voltage breaks down the trapatt junction causing it to avalanche, the semiconductor material is ionized and a *plasma* of electrons (and holes) is created, which in turn develops a voltage pulse that "bucks" the applied supply voltage while the avalanche current is increasing. The negative-resistance property is thus created by the plasma, reducing voltage while current is increasing. The current in the form of a *group* of released electrons (and holes) passes through the trapatt, "trapped" behind the voltage created by the plasma, and thus at a slower drift velocity than in the impatt diode. By the time the electron (or hole) *group* of current arrives at the trapatt's terminals, the phase of the signal voltage across the device will have shifted toward zero, yielding lower power dissipation and higher efficiency perfor-

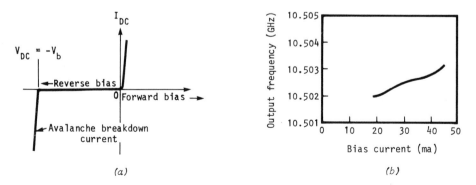

(a)

(b)

FIGURE 13-16

mance than can be achieved with impatt operation. With these properties the trapatt diode can be pulsed for high-power operation in excess of 100 W at the lower microwave frequencies.

Modulators and Switches

In addition to their operation as amplifiers/oscillators, semiconductor diodes are available to perform signal processing functions as subsystem components. Among their more popular functions are microwave attenuation and switching. A conventional *pn* junction silicon diode can be used as a microwave switch, for example, by placing it in series or in shunt with a transmission circuit. Such a switch can be turned *on* by forward biasing and *off* by reverse biasing. A simple form of pulse modulation of microwave signals can be achieved if high-speed switching signals are applied to the diodes in this configuration. A crude form of attenuation can be achieved by controlled forward bias of the diode along its conductivity "knee." Amplitude modulation (AM) of microwave signals can be achieved by applying the modulation signals along the bias knee of the diode's response curve. Whereas these four functions can be achieved with questionable results using conventional small-signal diodes, they can be achieved with excellent results using diodes specifically designed for application as transmission-line elements. Such diodes have a very low capacitance *pn* or *pin* junction structure, and are placed in low-capacitance packages for coaxial, microstrip, or waveguide applications. With typical junction capacitance values of less than 1 pF, these devices can operate with low forward losses through the higher microwave frequencies.

Microwave *pn* junction diodes provide fast switching capabilities and are typically used in low-power applications. Their junction reverse-bias capacitance values usually fall in the 0.1 to 0.3 pF range, with switching times from 2 to 10 ns. The forward resistance of these diodes to microwave signals is typically 1 Ω or less, while the reverse resistance is high. The switching reverse-bias isolation cannot be achieved with these diodes simply from high reverse resistance, however, because the high resistance is shunted by junction capacitance which determines the *actual* reverse impedance of the diode. If we use 0.2 pF as a typical junction capacitance value for a diode operating at 16 GHz, for example, its reactance is only about 50 Ω. If the package capacitance is now added to the junction value, the reactance would be reduced even further. Placing such a diode in a transmission line to achieve a series switching function would result in poor isolation, making switching impractical. Good switching isolation is achieved in practice by cancellation of the junction and package capacitances with an inductance, raising the reverse impedance to that of a parallel-resonant network ($Q \cdot X$). Practical small-signal *pn* diode switches of this design can easily achieve isolation factors of 40 to 60 dB.

If conventional *pn* diodes were to be used for higher power switching applications, their increased size would also increase their capacitance, making them inoperable as microwave switches. The *pin* diode was developed for microwave applications because, among other favorable characteristics, it has properties that keep its capacitance low with higher power capabilities. It is a three-layer diode junction with an *intrinsic* (*i* of the *pin*) layer of semiconductor material separating the *p* and *n* layers. The capacitance of the intrinsic layer, rather than the junction capacitance, determines the high-frequency performance of the device. Typical values of this intrinsic capacitance fall in the range of 0.05 to 0.2 pF; and this capacitance value will not change with forward and reverse bias, as with *pn* junction diodes. In practice, the shunt mode of operation is preferred for microstrip and waveguide components; the series mode is more practical for application along the center conductor of coax lines.

Diode switches are specified by insertion loss (*on*) and isolation (*off*) characteristics, based simply on their electrical properties at their frequency of operation. If the diodes are to be applied for narrowband microwave operation, their reverse-bias reactances are first canceled, then their *on/off* resistance characteristics are converted to equivalent insertion loss and isolation values. For series operation the insertion loss occurs when the diode is forward biased, whereas isolation is obtained through reverse biasing the diode. The isolation is a result of the mismatch loss that occurs when the diode is biased for high impedance operation. If very high coax-line (or other TEM-line) diode-switch isolation is required, more than one diode can be placed in series along the center conductor of the line. For shunt operation the insertion loss occurs when the diode is reverse biased across the transmission line, where a high impedance will hardly change the dielectric properties of the insulating material between the conductors or within the waveguide structure. The shunt diode-switch isolation occurs when the device is forward biased across the line, thus effecting a high mismatch loss. You should conclude from the above description that diode switches achieve *isolation* through reflection of signals—not dissipation. The diodes are not completely reactive, however, so they do dissipate some power. The shunt mode of operation is thus preferable for diode switching of high-power signals, because the I^2R dissipation is lower in both *on* and *off* conditions, as compared to the series-mode switch operation. For very high power switching, a number of diodes can be placed in shunt with the transmission line for parallel operation. The mode tradeoffs, of course, are limited isolation in shunt operation and limited power dissipation in series operation.

PIN Modulators

In addition to its operation as a switch, the *pin* diode has unique properties that make its use as an active attenuator/AM modulator most practical.

Although the device has conventional diode properties at lower frequencies, the presence of the *intrinsic* semiconductor layer makes the *pin* diode operate like a resistor at the high radio (and microwave) frequencies. Figure 13-17a graphs the *pin* diode's RF resistance vs applied bias current. You should observe that the device does not display the typical diode response, but is quite linear between 1 Ω and 5 kΩ, with the application of 1 mA to 100 mA of

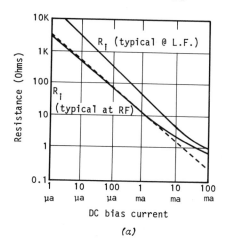

(a)

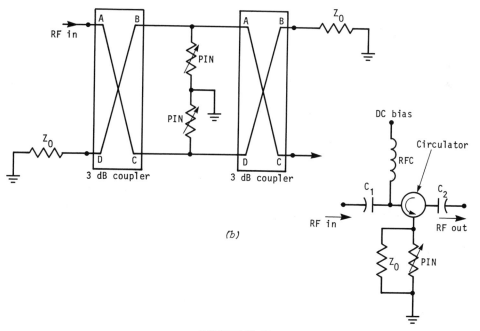

(b)

FIGURE 13-17

bias current. If such a *pin* diode is placed in series or shunt with a transmission line, the resulting mismatch loss would achieve electronic current-controlled attenuation of the line signal. One prominent application of *pin* diodes is in *leveling* the RF output of microwave sweep generators. Since both tube-type and solid-state microwave signal sources have a wide variation in their RF output levels over broad frequency ranges, precise "flat" response requires some form of "automatic" attenuation, most practically achieved with a *pin* diode attenuator as shown in Fig. 13-17*b*. Amplitude or pulse modulation of microwave signal sources is yet another popular application of *pin* diodes, where such modulation applied directly to the signal source would defeat its frequency stability. Such modulation with the *pin* diode can be achieved independently without affecting the performance of the signal source.

Mixers and Detectors

As you may recall from the introductory section of this chapter, the point-contact diode was the earliest popular semiconductor device used in microwave detector and mixer applications. This basic diode type, using even the same 1N21/23 designations, is still in widespread use in modern microwave equipment. The diode structure employs a fine catwhisker to contact a pellet of *n*-type silicon material. This extremely low capacitance contact junction allows the device to function as a mixer and detector at frequencies well above 100 GHz. Figure 13-18 shows the point-contact structure and graphs its response. As you may conclude from the device's structure, it is an extremely fragile, low-power diode since the small point-contact area is its only path of conductivity. Experience teaches that checking point-contact diodes with an ohmmeter burns them out! The diode's response curve in Figure 13-18 shows a gradual turn-on above 10 to 20 mV, providing *square-law* response without the necessity of external biasing. It is advantageous to use square-law response diodes at microwave frequencies, as they allow direct conversion of their output current in proportion to the applied RF power.

A variety of diode types are now used for microwave detection and mixing functions in addition to the point-contact version. Figure 13-18 also shows the comparative response of a "hot-carrier" or Schottky-barrier diode which provides nearly equal low-capacitance and square-law performance of the point-contact device, but without its power and reliability problems. The apparent tradeoff is, of course, a 150 mV required turn-on bias for the Schottky device. Incidentally, *Schottky barrier* refers to the metal-to-semiconductor junction separated by a planar contact surface. The technique, named after its developer, is applied to many semiconductor devices, but the name *Schottky-barrier diode* almost always refers to this special microwave detector/mixer. A recently developed *low-barrier, hot-carrier diode*

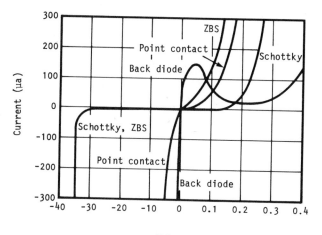

FIGURE 13-18

(LBHCD), or *zero-bias Schottky* (ZBS) *diode*, uses thin-film technology to achieve conventional Schottky-diode performance, without biasing and at even lower turn-on voltages than possible with the point-contact device. The response of this ZBS square-law response diode is shown in Figure 13-18, with turn-on voltage in the 1 to 2 mV region.

Another technique used to obtain low barrier response is with the use of a germanium tunnel diode in its region of square-law conductivity, as shown in Fig. 13-18. In this detector/mixer application the tunnel diodes are often identified as *back diodes,* in that they are selected for their reverse-conduction square-law characteristics. Although they have limited power ratings, often below 10 dBm, they are easily impedance matched to transmission lines and circuit loads.

The important characteristics of detector diodes include the following:

1. Frequency response
2. Tangential sensitivity
3. Power rating
4. VSWR

5. Square-law response
6. Loading (video) characteristics
7. General temperature, package, and mechanical characteristics

The general frequency response of diodes is a function of junction and package capacitances and of internal and package inductances. The point-contact diode still has higher frequency response than other microwave diodes.

Tangential sensitivity (TSS) is specified as the minimum input signal power that produces an 8 dB signal-to-noise ratio at the output of a diode/amplifier combination. The Schottky diodes exhibit lowest noise sensitivity, with the back diodes' response almost as good. The TSS of point-contact diodes can be improved with biasing in applications requiring high sensitivity at high microwave frequencies.

The Schottky diode has the best power-handling capabilities with good performance to +20 dBm. The point-contact diodes operate best below 0 dBm, and the back diodes respond best to signals below −10 dBm. All diodes can, of course, be *padded* to extend their power-handling capability, with a tradeoff loss in sensitivity.

The VSWR of a microwave detector takes into account the diode and package performance at power levels in the device's square-law region. Back diodes have the lowest VSWR values in the range of 1.5 to 2.5, but point-contact and Schottky devices must be padded down to achieve impedance matching, thus reducing their sensitivity.

The point-contact diode has the widest range of square-law response, with the back and Schottky diodes meeting acceptable performance specifications.

Video resistance refers to the value of detector *load* resistance, which lowers the diode's output voltage to half its open-circuit value for maximum power transfer. Since most detector applications recommend low-impedance operation, the back-diode detector performs best in these situations, with its video resistance in the 50 to 200 Ω range. Point-contact diodes have video resistance values in the 5 kΩ to 15 kΩ range; the video resistance of Schottky devices varies depending on bias current. Again, both higher impedance detectors can be padded down for improved video resistance, with a coincident loss of sensitivity.

The important characteristics of diodes used in mixer applications include those factors noted in the previous paragraph, plus noise figure, conversion loss, and compression and distortion levels. Noise figure principles and the general characteristics of RF mixers are reviewed in Chapter 8. The operation of microwave mixer diodes is similar to that of their RF counterparts, except for the high-frequency limitations of the devices. As you may conclude, the point-contact diode is still the best choice for high microwave frequency mixing; the back diode is the best choice for low-noise mixing of low-level signals; and the Schottky diode is the best choice for mixing higher power-level microwave signals. The most important characteristic of mixer diodes is their noise figure specification, which defines the lowest detectable signal and *sensitivity* of the receiving system in which the diodes are employed. In microwave practice the mixer is often the first component that receives the antenna signal, and thus contributes substantially to the *total* noise figure of the receiving system. The *minimum detectable signal* (MDS) specification of mixer diodes is obtained from a measured $S/N + N$ level of 3 dB, approximately 4 to 5 dB below the diode's TSS specification for a 10 MHz bandwidth. This MDS specification is computed, based in the noise voltage formulas of Chapter 8, as MDS $= \sqrt{4kTBR}$, where R in this instance is the equivalent noise resistance of the diode.

The *conversion loss* of an ideal diode mixer is 6 dB, based on mixing/modulation theory, where the 100% modulation sideband signal power is one-fourth the carrier power. In mixer applications the LO can be compared to the carrier, while the RF can be considered as an upper or lower sideband signal; thus the LO level into the mixer should be established at 6 dB *above* the highest anticipated RF input level. The *1 dB compression point* of the mixer specifies the RF input level and the LO level that yield 7 dB conversion loss (1 dB above the ideal 6 dB) caused by harmonic distortion from overdriving the diode(s). A projection of this 1 dB distortion level to an imaginary level where the magnitude of the third harmonic is equal to that of the desired output signal is specified as the *third-order intercept point*. Mixers are often specified by this intercept point, from which compression-point values can be obtained. Table 13-11 lists the general characteristics of popular mixer and detector diodes.

MICROWAVE FERRITE COMPONENTS

There is a family of specialized microwave devices that rely on the unique magnetic properties of ferrite materials at microwave frequencies. These devices function as isolators, circulators, attenuators, and filters, and operate in transmission lines and networks. The ferrite materials are compounds of magnetite (iron oxide) and zinc, nickel, cobalt, manganese, yttrium, and aluminum, etc. Three more popular ferrite compounds are manganese ($MnFe_2O_3$), zinc ($ZnFe_2O_3$), and yttrium-iron-garnet [$Y_3Fe_2(FeO_4)_3$], called *yig*. Ferrites are dielectric materials, so they will support and propagate the

TABLE 13-11. TYPICAL MIXER/DETECTOR DIODE CHARACTERISTICS

A. Mixers

Number/Type	Frequency (GHz)	Noise Figure (dB)	VSWR	Impedance (Ω)
1N21F/point contact	4	6	1.3	350–450
1N23F/point contact	12	7	1.3	330–460
P4210/Schottky	4	5	1.3	150–350
P4280/Schottky	12	5.5	1.3	150–350

B. Detectors

Number/Type	Frequency (GHz)	Tangential Sensitivity (dBm)	Video Resistance (kΩ)
1N830/point contact	3	−40	4–22
1N31/point contact	12	−45	6–23
1N1611/point contact	12	−53	6–8
PK1954/Schottky	9	−50	5–7
PK0214/Schottky	12	−58	4.5–18

electromagnetic energy of microwave signals. When they are thus placed in waveguide or coaxial circuits, they can be either transparent, absorptive, or reflective to the applied microwave signals, depending on their magnetic characteristics.

Ferrite dielectric materials are magnetic, so they respond to the high-frequency magnetization effects of applied microwave signals, as well as externally applied magnetizing forces. In practice the external magnetic *bias* is used to control the electrical propagation through the ferrite. This control is achieved through the deflection of spinning electrons within the ferrite material caused by a property called *gyromagnetic resonance.* Figure 13-19*a* shows a spinning electron within the ferrite material in relationship to a fixed magnetic bias and to a combination of fixed and RF-induced magnetism. The spinning electron does, in fact, behave like a gyroscope or spinning top. The combination of magnetic forces on the spinning electrons in the ferrite causes them to wobble like an unbalanced spinning top. This wobbling effect, called *precession,* occurs at a rate (frequency) that is proportional to the combined magnetic forces acting upon each of the electrons in the ferrite material. Since the applied (external) magnetic bias is substantially more intense than

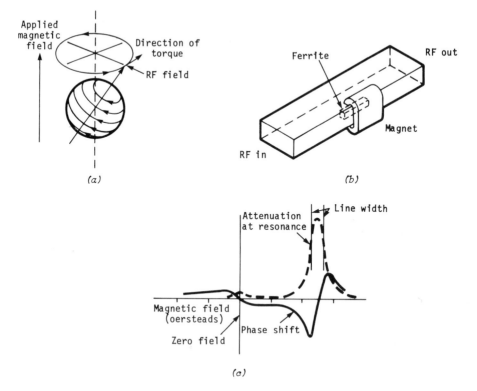

FIGURE 13-19

the magnetic effect of a propagating signal, the gyromagnetic resonant frequency can be adjusted to equal some microwave (or RF) frequency by varying this bias intensity. The approximate gyromagnetic resonant frequency may be computed as:

$$f_r = \gamma H \quad \text{or} \quad H = f_r/\gamma$$

H is the magnetizing force, in oersteds, of the applied bias, and γ is the *mass-charge ratio* of an electron, called *gyromagnetic ratio,* which has a value of 2.8 MHz/oersted. If, for example, the ferrite is to be biased for 8 GHz operation, the required bias can be computed as:

$$H = f_r/\gamma = 8 \times 10^9/(2.8 \times 10^6) = 2857 \text{ oersteds}$$

Attenuation and Isolation

When the gyromagnetic resonant frequency equals that of the propagating microwave signal, two conditions of interaction may exist between the spinning electrons and the smaller magnetic forces from the propagating signal:

1. If the magnetic effect from the propagating signal *opposes* the direction of precession, the microwave energy will be delivered into, and extracted from, the electron with no effect on the ferrite material or the signal; thus the ferrite is transparent to the signal.

2. If the magnetic effect from the propagating signal is in the *same* direction as the precessing force, the microwave energy will be delivered to the electron, enforcing its precession and causing the signal to be absorbed by the ferrite and dissipated in heat.

These two conditions define two popular ferrite device applications as *attenuators* and *isolators*. The ferrite attenuator is composed of a ferrite material with magnetic bias oriented in the direction of the propagating signal, and it provides attenuation in proportion to the length of the ferrite section. The ferrite isolator is a similarly designed two-way device that has low forward loss, but high attenuation to signals traveling in a reverse direction.

Ferrite isolation can also be achieved with lower bias levels—where the gyromagnetic resonant frequency is below that of the propagating microwave signal—by a technique called *Faraday rotation.* In this application the biased ferrite polarizes the propagating microwave signal in waveguide to the orientation of the precessing electrons. This orientation is usually at an angle of 45°. A 45° twist in the input and output flanges of the waveguide will yield correct *E-H* polarization for the signals passing through the waveguide in one direction, while rotating the polarization of the reverse signals by 90°, to below the waveguide cutoff frequency in the short (height) dimension of the waveguide. Figure 13-19*b* and *c* show a waveguide isolator and a graph of the typical magnetic response of the ferrite material.

A unique, and most useful, application of microwave ferrites is the *circulator*. This multiterminal device provides clockwise circular rotation of a microwave signal from one terminal to an adjacent terminal, excluding the signal from all other terminals. The more popular circulators have three or four terminals, arranged in a circular pattern in waveguide, coax, and stripline structures. They are applied to achieve input-output isolation in tunnel diode, step-recovery diode, and parametric amplifiers, as well as receiver/transmitter duplexer isolation to and from a single antenna. Figure 13-20 shows waveguide and coaxial circulators. The circulators may operate as low magnetic-bias Faraday rotation devices; they may operate at high magnetic bias for gyromagnetic resonance; or they may rely on phase shift. Each technique yields a different physical form and obvious tradeoff advantages and disadvantages for manufacture and application. The resonance-type isolators handle the highest power levels, and the rotation-type isolators have broad frequency response within waveguide limits. For coax and stripline applications, gyromagnetic resonance is achieved with a pill-shaped magnet above the line's ferrite *Y* junction. Ferrite switching can be achieved in circulator structures by using electromagnetic rather than permanent-magnet bias. By reversing the applied current to the electromagnet, a reversal in the direction of the signal rotation will be achieved.

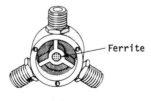

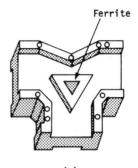

(a)

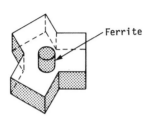

(b)

(o)

FIGURE 13-20

Filters and Resonators

The magnetic response graph of Fig. 13-19 shows the *ferrimagnetic resonant line width,* representing the operating bandwidth for the typical ferrites used in isolators and circulators. Whereas the 3 dB response for these applications requires a low Q for broadband operation, the ferrites used in *filter* applications must be compounded of high-Q materials that have a very narrow *line width.* The yig ferrite compound meets this requirement with high Q factors in the microwave region. Yig filters thus provide the capabilities of resonant cavities with the advantages of small size, light weight, electronic (magnetic) tunability, and stacking options for multisection response. Figure 13-21a shows the polished yig sphere placed between the pole pieces of a biasing electromagnet. The transverse loops around the sphere provide input and output coupling of the microwave signal. With zero magnetic bias the sphere is not resonant and there is no coupling from the input-to-output loops, since they are at right angles to each other. With the application of magnetic bias, for high-Q ferrimagnetic resonance at the signal frequency, the precession of the electrons to a 45° angle provides a high coupling coefficient from input-to-output loop, thus passing the signal through the device. Multistage filters can be achieved with two or more adjacent yig spheres within a single unit, to obtain higher off-band isolation. Typical yig filters of this type provide up to double-octave tuning with 70 dB or greater off-channel isolation and 3 to 8 dB insertion loss. Table 13-12 shows the specifications of typical yig filters.

TABLE 13-12. SPECIFICATIONS OF TYPICAL YIG FILTERS

Number of Stages	Frequency Range (GHz)	−3 dB Bandwidth (MHz)	Insertion Loss (dB)	Off-Resonance Isolation (dB)	Magnetic Circuit Tuning (MHz/mA)
2	0.5–1	20	5	55	6
3	↓	20	6	75	6
4	↓	20	7	85	6
2	2–4	30	3	55	20
3	↓	30	4	75	20
4	↓	30	6	85	20
2	4–8	30	2	55	20
3	↓	35	3	75	20
4	↓	40	4	80	20
2	8.2–12.4	30	2	55	21
3	↓	35	3	75	21
4	↓	40	4	80	21
2	12.4–18	40	3	45	25
3	↓	45	4	65	25
4	↓	50	5	75	25

   One of the most useful applications of yig filters is in electronically controlled solid-state microwave signal sources. With unloaded Q values from 10 k to 50 k, these ferrites can be used in conjunction with Gunn and impatt/trapatt diodes to obtain stable, low-noise microwave signals that can be swept over an octave frequency range. The yig filters are also used with step-recovery and varactor diodes in fixed-tuned harmonic generator applications for lower cost sources of microwave signals. Figure 13-21b shows the structure of a typical yig/Gunn-diode source. You should note its similarity to the filter unit represented in Fig. 13-21a. Yig-tuned signal sources are available for wide-range operation to 40 GHz, with power ratings to 500 mW at the lower microwave frequencies.

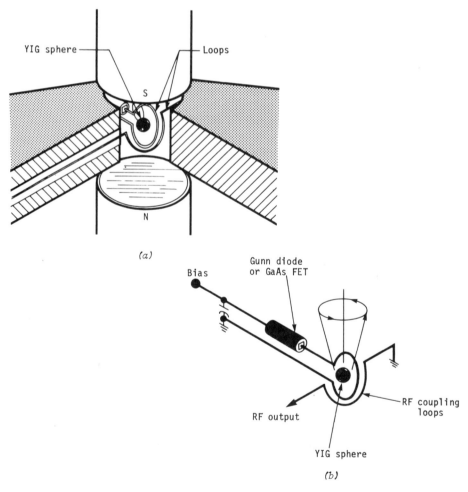

FIGURE 13-21

CONCLUSION

Chapter 12 reviewed the principles and operation of *passive* microwave devices; this chapter has been concerned with those *active* devices used to generate, amplify, and process microwave signals. As with Chapter 11, this chapter merely introduced you to a variety of popular devices, without intensive detail. Engineers and technicians who specialize in the microwave field identify most "new technology" to be in this area of solid-state devices, making the topics of this chapter somewhat transitional. Many of the present frequency and power limitations cited for the solid-state devices specified should be greatly improved or eliminated within the next few years. It is therefore recommended that you enhance your interest in this area by your regular review of microwave-oriented technical magazines and journals. The problems that follow are intended to test your general knowledge of active microwave devices, and determine your ability to correctly deploy and bias the popular solid-state and vacuum-tube units.

Problems

1. Given: A requirement to recommend the most appropriate microwave tube types for specific applications, including typical examples and their related power supply requirements.

Identify and specify:
(*a*) A 20 mW local oscillator for fixed-frequency operation
(*b*) A 20 kW oscillator for fixed-frequency operation
(*c*) A 1 W electronically tunable broadband oscillator
(*d*) A 1 kW electronically tunable broadband oscillator
(*e*) A 10 W broadband amplifier
(*f*) A 20 mW broadband low-noise amplifier
(*g*) A 20 kW fixed-frequency pulsed amplifier
(*h*) A 20 kW broadband pulsed amplifier

2. Given: A microwave transistor with input and output impedances of $0.3 +j0.2$ and $0.5 -j0.4$, respectively, and a power gain of 20 with a 160° phase shift of input-to-output signal voltage across 50 Ω source and load resistances.

Determine:
(*a*) S_{11}
(*b*) S_{22}
(*c*) S_{21} (approximate, with S_{12} insignificant)

3. Given: A microwave transistor operating at 4 GHz with the following S-parameter characteristics:

$$S_{11} = 0.2\angle175° \qquad S_{21} = 4.5\angle110°$$
$$S_{22} = 0.8\angle-53° \qquad S_{12} = 0.02\angle80°$$

Determine:
(a) The input impedance and VSWR
(b) The output impedance and VSWR
(c) The power gain in dB
(d) Specifications for a stub transformer to match the input impedance
(e) Specifications for a quarter-wave transformer to match the output impedance

4. Given: The bipolar transistor of Fig. 13-9 to be used as a small-signal CE amplifier (class A), biased for the following conditions:

$$V_{CE} = 12 \text{ V dc} \qquad\qquad I_{BB} = I_C/5$$
$$I_C = 30 \text{ mA} \qquad\qquad V_{BB} = V_{BE} + 2 \text{ V}$$
$$h_{FE} = 50 \qquad\qquad V_{BE} = 0.65 \text{ V}$$
$$V_{CC} = 2V_{CE}$$

Determine:
(a) The dc circuit bias voltages for the operating point specified above
(b) The dc circuit bias currents for the operating point specified above
(c) The power supply requirements
(d) The resistance values of the bias circuit components

5. Given: The GaAs FET of Fig. 13-11 to be used as an amplifier of small (low-level) microwave signals at 2 GHz (refer also to Fig. 13-12).

Determine:
(a) The dc circuit bias voltages and currents for an I_D of 40 mA
(b) Input and output impedance and VSWR values
(c) Specifications for input impedance matching with an LC network
(d) Specifications for output impedance matching with a taper transformer
(e) The power gain for the matched and biased FET amplifier circuit

6. Given: A step-recovery diode to be used as a harmonic generator of microwave signals in the 3 GHz region in a $\times10$ resonant circuit.

Determine:
(a) The value of series inductance in the 300 MHz section of the circuit if the capacitance of the step-recovery diode is 5 pF
(b) The required Q of the $\times10$ output network specified as $\pi n/2$, where n is the multiplication factor
(c) The length of the output quarter-wave resonant transformer
(d) The LC characteristics of a resonant output network given a 50 Ω output load resistance and the value of Q from b above

7. Given: A *pin* diode with characteristics as shown in Fig. 13-17.

Prepare and determine:
(*a*) A circuit drawing and computed bias values for use of the *pin* diode in a coax line as a shunt attenuator and a shunt switch
(*b*) A circuit drawing and computed bias values for use of the *pin* diode within a stripline structure as a series attenuator and a series switch
(*c*) The minimum and maximum attenuation, in dB, and the bias limits for the *pin* diode in the shunt mode
(*d*) The minimum and maximum attenuation, in dB, and the bias limits for the *pin* diode in the series mode

8. Given: A matched pair of Schottky-barrier diodes with the following characteristics to be used in the balanced mixer circuit of Fig. 13-22:

Noise figure = 6 dB
IF impedance = 250 Ω
VSWR = 1.5
Tangential sensitivity = -54 dBm
Voltage sensitivity = 6 mV/μW
Video resistance = 1400 Ω
Balanced mixer conversion loss = 6.5 dB
Balanced mixer 3rd-order intercept point = +8 dBm

Determine:
(*a*) The minimum detectable signal based on the TSS specification (approximate)
(*b*) The voltage across a 50 Ω line at the MDS level
(*c*) The +1 dB compression level based on the 3rd-order intercept point level

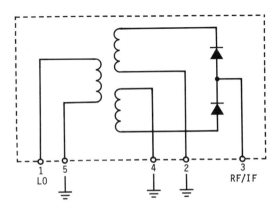

FIGURE 13-22

(*d*) The maximum LO and RF input levels at the +1 dB compression point

(*e*) The output level of the IF signal at the 1 dB compression point

(*f*) The output level of the IF signal with an RF input signal of −20 dBm

9. Given: A yig-tuned low-power Gunn oscillator, similar to that shown in Fig. 13-21, to operate in the 4 to 8 GHz frequency range with resonator characteristics as follows:

$Q = 150$
Off-resonance isolation = 50 dB
Tuning sensitivity = 20 MHz/mA
Coil resistance = 8 Ω

Determine:

(*a*) The initial approximate voltage and current settings for the Gunn oscillator based on the characteristics shown in Fig. 13-15

(*b*) The required change in magnet current and voltage to tune the oscillator over its operating frequency range

10. Given: Gunn and impatt diode signal sources with identical fixed-frequency operation and power output characteristics.

Determine or specify:

(*a*) The distinction in general power supply requirements for the two devices

(*b*) The comparative performance of the two devices in terms of stability, ruggedness, and supply and temperature tolerances

(*c*) Your recommendations for particular applications best suited for one particular device over another

CHAPTER 14

ANTENNAS AND RADIO WAVE PROPAGATION

The two main topics of this chapter are regarded as among the most complex in radioelectronic communications. The launching, transmitting (*propagating*), and receiving of RF signals with nothing but air space between what often looks like simple metal rods, must seem miraculous to anyone who thinks that electrical signals can only travel from place to place through wires. Many of us take for granted the apparent simplicity of radio communications and antenna systems and do not consider the physical science involved in understanding and applying antenna and propagation theory to practice. These topics are, in fact, most complicated and warrant thorough treatment. Because of space limitations, however, you are referred for greater detail to textbooks that cover these subjects.

Some years back the author fabricated some UHF TV antennas from metal coathangers. Their performance was excellent! People who examined the wire-form antennas exclaimed, ''How simple!'' They did not know that some 36 hours had been spent in making calibration measurements with costly precision test equipment, and that the data from those measurements and related calculations filled 29 pages of a technical notebook. The measurements relating to the ''coathanger antennas'' included impedance matching and frequency response; radiation and loss resistances; noise figure; polar and azimuth gain/directivity plots referenced to a standard dipole; and field intensity, ground effects, and path losses. With such characterization the simple coathanger antennas became ''complex electromagnetic structures''; but they hid the fact very well—they still looked like bent coathangers, even though it took more than 15 minutes to bend the wire. So here is the dilemma with antennas: their physical appearance does not convey the tremendous amount of engineering development and propagation measurements required to deploy them properly. Although the study of this chapter may not prepare you as an antenna and propagation expert, you will at least be aware of the techniques, practices, and capabilities of achieving successful radioelectronic communications with antennas. As an additional objective, it is hoped that you will be able to justify 36 hours of work on an antenna to someone who thinks it can be done in 15 minutes.

Antennas are electrical and optical devices. From an electrical viewpoint, an antenna terminates a transmission line and should appear as a matched resistive load impedance over its operating frequency range. To achieve this matched condition with minimum heat loss, the antenna must, in effect, act as a *transformer* from the characteristic impedance of its feed transmission line to the characteristic (wave) impedance of air, free space, or whatever medium the signal is propagating. (The characteristic impedance of air and free space is approximately 377 Ω.) From an optical viewpoint, the antenna must efficiently *beam* or *disperse* a signal in the form of electromagnetic radiation through whatever medium the signal is propagating. Since the behavior of radiated signals varies considerably with frequency, medium, and boundary conditions (ground, water, mountains, buildings, atmospheric variations, etc.), the applied physical science of propagation becomes a most significant aspect of antenna deployment. This chapter covers both the applied electrical and optical aspects of antenna operation, and the media through which radio signals are propagated.

In its simplest form an antenna can be represented by a rod, as shown in Fig. 14-1*a*, driven from an RF signal generator operating at a high frequency. Since one lead of the unbalanced generator output is grounded, the ground acts as one pole (polarity) of a dipole (two-pole) antenna. The "hot" lead of the generator drives the rod as though it is one leg of a transmission line. If the rod is one-half wavelength long, the applied current will flow up the length of the rod with maximum amplitude at the center of its length, reducing to zero current by the time the wave reaches the rod's open end. The flow of current along the rod develops transverse waves of magnetic energy that are attracted from the rod by their opposite-polarity magnetism in the ground plane.

This simplified example of antenna operation illustrates how energy radiates from what might otherwise be considered an open circuit. It should be mentioned here that antennas, like many other electrical components, are reciprocal—they operate the same in a transmitting mode as they do for receiving radiated signals. Such antenna systems as described can be found in use where high-frequency, *omnidirectional* (all directions, one plane) performance is required. The most abundant examples of these *vertical* antennas are CB (U.S. Citizens Band) base stations. The radials extending from the bottom of these antennas serve to provide an *artificial ground plane* for rooftop mounting. Although the shorter mobile CB antennas are made less than one-half wavelength for practicality, half-wavelength antennas, operating in the 450 MHz UHF band, can be found on taxicabs and tow trucks, etc., that use the metal roof of the vehicles as a ground plane. Figure 14-1*b* shows typical vertical antennas that operate like the simplified example of the previous paragraph.

The electromagnetic performance of antennas can be simply related to your visual experience and judgment about the behavior of light. The concept of an-

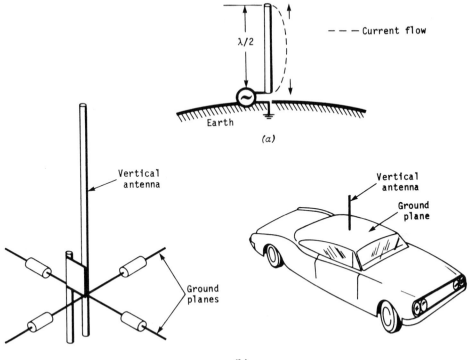

Earth

— — —Current flow

λ/2

(a)

Vertical
antenna

Vertical
antenna

Ground
plane

Ground
planes

(b)

FIGURE 14-1

tenna *gain,* for example, is actually achieved by reflection and focusing of radio signals in the same sense that light appears brighter by concentration of its energy through reflection and focusing. Propagated radio waves can likewise be deflected, reflected, defracted, guided, and absorbed just as is visually observed with light waves. The properties of the materials (media) through which the radio waves are propagated determine the particular effects imposed upon the waves. The ideal medium is, of course, transparent to radio waves, so that radio energy is directed *line of sight* from one antenna to another. Since the earth is round, however, line-of-sight transmission limits the range of radio communications around its curvature. Long distance radio communications takes advantage of the optical effects of propagation to achieve bending and reflection of waves for the transmission and reception of signals "over the horizon." The details of this important physical science of propagation will be covered, in conjunction with antenna applications, to complete the content of this chapter.

LEARNING OBJECTIVES

Upon completing this chapter, including the problem set, you should be able to:

1. Understand the principles of electromagnetic radiation as applied to antennas and the media through which radio waves propagate.

2. Understand the basic *electrical* properties of antennas, including radiation and loss resistances, and gain and propagation patterns.

3. Identify the general characteristics of various popular antenna types and the applications in which they are deployed.

4. Compute the dimensions and matching/loading networks required for the efficient operation of popular antennas.

5. Specify and interpret the significant electrical measurements that are applied to evaluate antenna performance.

6. Understand the relationship of the frequency of propagated signals to optical performance, or "wave behavior," in various media.

7. Recognize the atmospheric, geographic, and environmental boundary conditions imposed on the propagation of radio waves.

8. Understand and specify sight-path and field-intensity propagation pattern measurements, and predict conditions for successful radioelectronic communications.

ELECTROMAGNETIC RADIATION

The introductory section of this chapter described a simplified ground-plane dipole-type antenna as a basis for launching and receiving waves of magnetic energy. An even simpler example of the phenomenon on which radio transmission and reception operate is the RF transformer. Current passing through the primary windings of a transformer creates a magnetic field of force lines. The lines of magnetic force then encounter and "cut" the secondary windings of the transformer, inducing a voltage across these windings. The voltage that appears across the secondary windings is a function of the primary current and coupling coefficient ($V_{sec} = X_m I_{pri}$), as you may recall from the *RF Transformers* section in Chapter 3. The relationship of RF transmitting and receiving antennas is, in fact, the same as *very loosely coupled* RF transformers, with the small mutual reactance (X_m) value inversely proportional to the distance between the antennas. If the antennas happen to be resonant, the voltage on the receiving antenna will increase by their Q factors, just as with the operation of undercoupled transformers.

The concept of electromagnetic radiation—radio transmission and reception—follows that of the loosely coupled RF transformer, but with a rather complicated "core" structure. In this comparison the media between the antennas can be considered as the core material and structure that exist between the primary and secondary windings of an RF transformer. If the *permeability* of the core is high, there will be greater transfer of energy from primary to secondary, provided that the core material is not *lossy* (conductive). A conductive core material will completely "shield" the transfer of energy through absorption or reflection. As you may conclude, the earth, the oceans, and the many man-made structures act in the large scale to influence the transfer of electromagnetic radiation, just as various core materials and structures influence magnetic induction in the smaller scale RF transformer.

Field Forces

Comparison with RF transformers does not completely describe the phenomena of radiation, because transformer windings are usually separated by less than 0.1 wavelength, whereas transmitting and receiving antennas are separated by many wavelengths. The description of what happens to an alternating magnetic field force in a medium is again in the realm of physics. The accurate relationships of this wave behavior are credited again to the 19th-century mathematician, J. C. Maxwell. They serve to define wave propagation in transmission lines, in waveguides (see Chapter 11), and of course between antenna pairs. A verbal translation of *Maxwell's equations* is more useful here than any attempt to apply their complicated mathematical derivations:

1. "A changing magnetic field in a given medium will produce an electric field in that medium."
2. "A changing electric field in a given medium will produce a magnetic field in that medium."

An expanded description of these relationships may thus proceed as follows:

An alternating current flowing in a wire will produce an alternating magnetic field in the space surrounding it. The alternating magnetic field will produce an alternating electric field in the space surrounding *it*. The magnetically induced electric field will in turn produce a second magnetic field in the space surrounding *it*. This electrically induced alternating magnetic field will produce yet another alternating electric field; and so on, with each field alternately being reproduced by its complementary field force.

The electric and magnetic fields are transverse (perpendicular) both to the direction of propagation and to each other. The polarization of these fields depends on the orientation and design of the antenna that launches the electromagnetic wave. The polarization is identified by the direction of the *E* (electric) field; e.g., a vertical antenna produces vertically polarized radio waves. Although you have observed many examples of vertically and horizontally polarized antennas, you should be aware that varied forms of polarization are possible, including even circularly polarized radio waves. Figure 14-2 shows representational diagrams of the field forces and polarization as defined by Maxwell. *Maxwell's equations* are thoroughly covered in many theoretical textbooks on *wave propagation*.

Characteristic Wave Impedance

The magnitudes of the two field forces depend on the magnetic and electric properties of the medium through which the electromagnetic waves propagate. The relationship can be specified as a characteristic *wave* impedance (Z_0) based on the absolute permeability and *permittivity* of the medium. You may recall the use of *relative (normalized)* permeability factors as applied to the inductance of coils and transformers in Chapter 3. The *absolute* permeability, as applied to wave propagation, is rated in *henrys per meter*. The related *absolute* permittivity factor, also applied to wave propagation, is rated in *farads per meter*. The *relative* form of permittivity used to define capacitance is, of course, the familiar *dielectric constant*. The characteristic impedance of a lossless medium may thus be computed based on *L* and *C*, just as with transmission lines:

$$Z_0 = \sqrt{L/C} \quad \text{and} \quad Z_0 = \sqrt{\mu/\epsilon}$$

μ is the absolute permeability of the wave medium, and ϵ is the absolute permittivity of the wave medium. In practical application the values for free space and air are approximately equal to $\mu = 4\pi \times 10^{-7}$ H/m and $\epsilon = 1/(36\pi \times 10^9)$ F/m. Based on these factors the characteristic impedance of air and free space can be computed as:

$$Z_0 = \sqrt{\mu/\epsilon} = \sqrt{4\pi \times 10^{-7}/[1/(36\pi \times 10^9)]} = 120\pi = 377 \ \Omega$$

The characteristic wave impedance of media other than air and free space can be found by applying the particular absolute permeability and permittivity values of the media to the above formula. The *relative* permeability and dielectric constant values, if known, can be factored into the free-space constants in the above formulas for calculations involving other media. If, for example, an antenna is to launch a wave to propagate a plastic material with a relative permeability of 1.1 and a dielectric constant of 2.25, the wave impedance of the plastic material can be computed as:

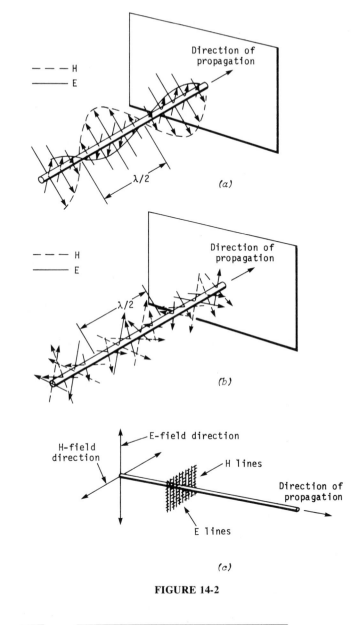

$$Z_0 = \sqrt{\mu/\epsilon} = \sqrt{1.1 \times 4\pi \times 10^{-7} \times 36\pi \times 10^9/2.25}$$
$$= 84\pi = 246 \ \Omega$$

As you may recall from Chapter 11, the velocity of propagation in transmission lines is a function of the inductance and capacitance of a given line.

Likewise, the velocity of electromagnetic wave propagation in air, free space, or other media is a function of the corresponding effects in the fields surrounding the antenna, that is, permeability and permittivity. The velocity of propagation can thus be computed, using the same absolute values used to determine wave impedance, as:

$$v = 1/\sqrt{\mu\epsilon} = 1/\sqrt{4\pi \times 10^{-7}/(36\pi \times 10^9)} = 3 \times 10^8 \text{ m/s}$$
$$= 9.84 \times 10^9 \text{ ft/s}$$

Just as with the wave impedance values, the velocity of propagation in other media can be computed based on their absolute or factored permeability and permittivity/dielectric constant values.

Sample Problem 1

Given: A requirement to propagate a 60 MHz electromagnetic signal in a body of water that is determined to have a relative permeability of 1.2 and a relative dielectric constant of 25.

Determine: The absolute permeability of the medium, the absolute permittivity of the medium, the characteristic wave impedance, the velocity of propagation, and the wavelength of the signal.

(a) First, the absolute permeability (μ') is computed, based on the relative permeability (μ) and the physical constant (μ_o), as:

$$\mu' = \mu\mu_o = 1.2 \times 4\pi \times 10^{-7} = 1.5 \times 10^{-6}$$
$$= 1.5 \ \mu\text{H/m}$$

(b) Then, the absolute permittivity (ϵ') is computed, based on the relative permittivity value of the dielectric constant (ϵ) and the physical constant (ϵ_o), as:

$$\epsilon' = \epsilon\epsilon_o = 25/(36\pi \times 10^9) = 2.21 \times 10^{-10}$$
$$= 221 \ \text{pF/m}$$

(c) The characteristic wave impedance is then computed as:

$$Z_0 = \sqrt{\mu'/\epsilon'} = \sqrt{1.5 \times 10^{-6}/(2.21 \times 10^{-10})} = 82.7 \ \Omega$$

(d) The velocity of propagation (v) is computed as:

$$v = 1/\sqrt{\mu'\epsilon'} = 1/\sqrt{1.5 \times 10^{-6} \times 2.21 \times 10^{-10}}$$
$$= 5.46 \times 10^7 \text{ m/s}$$

(e) The wavelength of the signal is finally computed as:

$$\lambda = v/f = 5.46 \times 10^7/(6 \times 10^7) = 0.91 \text{ m}$$

ANTENNA PRINCIPLES

To launch an electromagnetic wave an antenna must maintain an alternating current along its radiating conductor in such a way that it efficiently develops intense magnetic fields in the space surrounding it. From an electrical viewpoint, in order to achieve optimum power transfer these high currents must be obtained with minimum impedance mismatch from the circuits and line driving the antenna. In this function the antenna must behave like a transformer, matching the characteristic impedance of the transmission line that feeds its input, to the wave impedance of the medium into which the antenna propagates electromagnetic waves. In most applications the output impedance of the antenna must thus match the 377 Ω value of air and free space. From this point of view the antenna can be considered as an extension of the transmission line that feeds it.

The Dipole

Figure 14-3 shows how a simple antenna is evolved from an open transmission line. If the ends of the line are separated over one-quarter wavelength to form a simple antenna as shown in Fig. 14-3b, the antenna (line) actually operates as a quarter-wave taper transformer. The antenna impedance at the end of the taper section is *high* owing to the increased separation of the conductors. If the quarter-wave "arms" are moved to their maximum separation, as shown in Fig. 14-3c, the impedance at their ends will be at the highest value since the separation is maximum. You should not confuse this quarter-wave separation with that of open quarter-wave *stubs,* as reviewed in the preceding chapters. Stubs maintain transmission-line spacing over their quarter wavelength, whereas in this example the quarter-wave arms are at a maximum spacing at their ends. This type of antenna is called an *elementary doublet* or *dipole.* (A doublet is a dipole with insignificant length or conductor diameter.)

Although the dipole achieves impedance matching by quarter-wave transformation, it also provides a structure for coupling electromagnetic waves into the field that surrounds it. You should easily visualize the electrical response of the dipole as shown in Fig. 14-3d, where voltage and current standing waves develop the field forces for radiation. The end-to-end distance of the dipole is one-half wavelength and the phase of the driving signal is changing, so the center-to-end transit time of the current flow just equals the time required for the signal to change phase by 180°. The ends of the dipole are like open transmission-line elements, as they do create standing waves where the current drops to zero and the voltage rises to twice its applied value. Away from the ends of the dipole the voltage decreases, while the current increases just as with any standing wave pattern. At the center of the

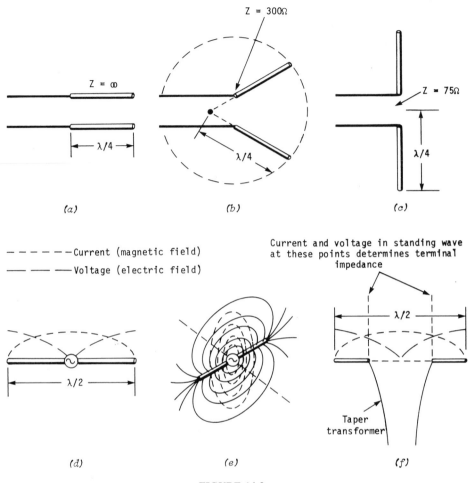

FIGURE 14-3

dipole the voltage passes through zero and changes polarity, unlike a conventional standing-wave pattern, however, because each arm of the dipole is driven with an opposite-polarity signal from its transmission-line feed. This is the reason that the *antenna* can have standing waves, but still match the driving line impedance and reflect no signals back into the line.

The instantaneous voltage and current values along the length of the dipole vary with the distance from the center of the antenna toward its ends. The points at which this antenna can be fed, dictating the spacing of the arms and conductors of the feed transmission line, will thus depend on the voltage-to-current ratio (instantaneous impedance) along the length of the dipole. The optimum terminal impedance for a given spacing with the *elementary half-wave dipole* shown in Fig. 14-3c is about 75 Ω. The terminal impedance of the tapered antenna of Fig. 14-3*b* is, by contrast, about 300 Ω owing to the increased, steeper slope of the current and voltage along its arms. As you may conclude, a wide range of terminal impedances can be achieved with the half-wave antennas of this example. An extreme case is shown in Fig. 14-3*f*, where the half-wave dipole is driven at higher impedance levels along its length. It should be obvious that this choice will actually yield poor radiation efficiency, because the arms are too short to develop an intense magnetic field. The choice of antenna operation and driving impedance thus depends on obtaining optimum radiation efficiency rather than a wide range of terminal-impedance options.

Radiation Efficiency

The *ideal* optimum radiation efficiency of an antenna would be 100%, providing a conversion of all the applied electrical energy to electromagnetic radiation. The radiation efficiency of antennas can be compared with the operation of an *LC* resonant network, with the related *Q* factor defining the ratio of energy converted to radiation, to that which is lost in resistive dissipation. The odds for 100% efficiency in an antenna are about as good as for infinitely high *Q* values in a resonant network. The practicality of *Q* values in the range of 2 to 25 can be applied to antenna efficiency by inverting these *Q* factors for percentage values of 50% to 96% as:

Percent efficiency $= 1 - (1/Q) \times 100$

Most antenna efficiencies are rated in less direct terms that are more likely to confuse you, however. These terms are *radiation resistance* and *loss resistance*. If you think in terms of efficiency, the *total* input impedance indicates the value of applied power, while the *radiation* and *loss* resistances indicate the proportion of *radiated* and *dissipated* power, respectively.

The antenna can be considered as an equivalent circuit of *both* radiation and loss resistance values in series, connected to the driving transmission line. A 70 Ω dipole, for example, may be operating at 90% efficiency, so the radiation resistance would be 63 Ω and the loss resistance would be 7 Ω. The simple relationships may be computed using the following formulas:

$$Z_{in} = R_r + R_1 \qquad R_r = \eta Z_{in}$$
$$\eta = R_r/(R_r + R_1) \qquad R_1 = (1 - \eta)Z_{in}$$

Z_{in} is the terminal impedance into the antenna, η is the decimal (not %) value of the efficiency, R_r is the radiation resistance, and R_1 is the loss resistance.

Sample Problem 2

Given: A 70 Ω transmitting antenna with low VSWR, fed with 10 W of RF power, with a measured 8 W of radiated RF.

Determine: The efficiency, radiation resistance, and loss resistance of the antenna.

(*a*)   The efficiency of the antenna is determined by the ratio of input to output power:

$$\eta = P_{out}/P_{in} = 8/10 = 0.8 \quad (80\%)$$

(*b*)   The radiation and loss resistances are computed based on the efficiency and input impedance:

$$R_r = \eta Z_{in} = 0.8 \times 70 = 56 \ \Omega$$
$$R_1 = (1 - \eta)Z_{in} = (1 - 0.8) \times 70 = 14 \ \Omega$$

Sample Problem 3

Given: A 300 Ω TV receiving antenna that converts only 60% of the radiated power applied to its input to RF signals at its transmission-line terminals.

Determine: The radiation resistance and loss resistance for the antenna. The radiation and loss resistances are computed based on the efficiency and input impedance:

$$R_r = \eta Z_{in} = 0.6 \times 300 = 180 \ \Omega$$
$$R_1 = Z_{in} - R_r = 300 - 180 = 120 \ \Omega$$

Further comparison of the simple dipole antenna with the *LC* resonant network can be made since the antenna is electrically one-half wavelength long and maintains out-of-phase currents and voltages, much like the circulating current and terminal voltage of a parallel-resonant network. The actual electrical half-wavelength of the antenna is less than one-half free-space wavelength, because the antenna's capacitance and inductance affect the velocity of propagation, as they would within a transmission line. In practice, the tedious calculations required to obtain an accurate measured length are discarded in favor of an approximate 2% to 5% reduction in the antenna length based on free-space wavelength computation. If, for example, a half-wave dipole is to be "cut" for operation at 600 MHz, its shortest length would be computed as:

$$l = 0.95(v_o/f_o)/2 = 0.95 \times (3 \times 10^8/6 \times 10^8)/2$$
$$= 0.2375 \text{ m} = 23.75 \text{ cm}$$

This distance represents the overall length (*l*) of the antenna, including the "cutout" spacing for the transmission-line feed at its center.

PROPAGATION (RADIATION) PATTERNS

Antennas radiate electromagnetic energy in patterns of varying intensity, much like a light bulb, flashlight, or other source of light energy. Some antennas confine their output to a very narrow beam; others spread their output, broadly distributing the radiation. The theoretical ideal source of radiant energy is the *point source*, like the sun in our solar system, distributing equally intense radiation in all directions. Although such point sources can be approximated with light, the longer wavelengths of radio energy prevent such even distribution. The simple dipole antenna of our previous examples provides the most uniform radiation pattern, but not in all directions. Since it is the best source, however, it is considered as a standard against which other antennas are compared.

Figure 14-4a shows the radiation pattern of the standard resonant half-wave dipole. As mentioned earlier in this chapter, these optical radiation patterns are best interpreted visually since the descriptive mathematics is rather abstract. A convenient visual/mechanical analog can probably be obtained with the example of an imaginary, perfectly round balloon. If you inflate the balloon to a given size, the force at its surface represents the energy you expended in its inflation. The balloon, so inflated, will be used as our example of a theoretical point source of radiant energy at its exact center, radiating equal values of energy in all directions, and with a magnitude represented by the force at the surface of the balloon. If you now poke the balloon from both sides, bringing the centers together, the distribution of forces on the balloon will be distorted into the doughnut-shaped pattern shown in Fig. 14-4c. This represents the radiation pattern of the dipole, with increased force in a direction perpendicular to its elements and virtually zero force at the ends of its elements.

If you compare the relative diameters of the two patterns in Fig. 14-4, you should see that the doughnut shape has an increased force in its larger dimension as compared to the spherical-shaped balloon. This increased force can be translated to a special kind of *antenna gain*, which obtains increased directional radiation through displacement of its energy pattern, as compared

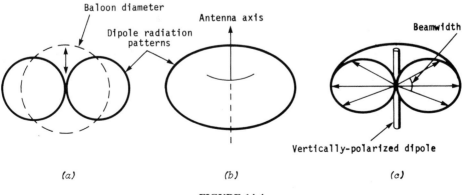

FIGURE 14-4

to the perfect sphere. The technical term for the point source of radiant energy that produces a spherical energy pattern is *isotropic radiator* (from the Latin *isotropic:* equal, in all directions). The *gain* of the dipole, in its large-force dimension, over the isotropic radiator is a factor of 1.64 : 1, as shown in Fig. 14-4, which calculates to 2.15 dB. The radiation patterns and gain factors of all antennas thus describe their electromagnetic performance, based on requirements of energy distribution. Antennas with narrow radiation patterns have high *directivity* and high gain; those with broad radiation patterns have low directivity and low gain.

Gain

Practical antennas are specified in terms of gain, referenced to an isotropic radiator or standard dipole, and *beamwidth*. From a sales standpoint, gain specifications in reference to an isotropic radiator always appear 2.15 dB better than in comparison with a standard dipole. In practice, most antenna measurements are made in reference to an operating standard dipole, however, and they should thus be appropriately specified. Measurements of beamwidth are made by rotating the antenna to both sides of its maximum radiation intensity point until the power drops off by -3 dB. The included angle between the adjacent -3 dB points is specified as the antenna's beamwidth. The gain and beamwidth of most antennas are measured for both vertical and horizontal response, with the side-view pattern called the *azimuth* plot, and the top-view pattern called the *polar* plot.

The approximate gain of an antenna can be computed with reference to an isotropic radiator based on its beamwidth occupying a portion of the surface area of a sphere, as shown in Fig. 14-4c. Since the surface area of a sphere is directly proportional to its radius, the concentration of energy at this distance will provide an increased intensity, equal to the gain of the antenna. The relationship may be computed as follows:

$$A_P = \eta/\sqrt{(\text{VBW}/360)(\text{HBW}/360)}$$

A_P is the power gain, η is the antenna efficiency, VBW is the vertical beamwidth, and HBW is the horizontal beamwidth of the antenna. If, for example, an antenna had a 45° beamwidth in both horizontal and vertical directions, with 80% efficiency, the approximate gain would be computed as:

$$A_P = \eta/\sqrt{(\text{VBW}/360)(\text{HBW}/360)} = 0.8/(45/360)$$
$$= 6.4 = 8.06 \text{ dB}$$

The approximate gain with reference to a standard dipole would be:

$$A_{P(\text{dipole})} = A_{P(\text{isotropic})}/1.64 = 8.06 \text{ dB} - 2.15 \text{ dB}$$
$$= 5.91 \text{ dB} = 3.9$$

Harmonic and Nonresonant Operation

The dipole antenna and its characteristics described thus far in this chapter were for operation within resonant, half-wavelength conditions only. Antennas are operated under a wide range of resonant and nonresonant conditions, however, and that is why the subject is so complicated. In the case of the dipole, it can be operated over a wide range of frequencies, with resonant operation at 0.5, 1.0, 1.5, 2.0, etc., wavelengths of the applied signal. Since these wavelengths occur at harmonically related frequencies, operation of the antenna above the half-wavelength frequency is called *harmonic* operation. With harmonic operation the standing-wave patterns along the arms of the antenna vary through more than one half-cycle, so the radiation pattern from the antenna changes from the basic doughnut shape of our earlier example. Figure 14-5 shows a variety of propagation patterns and related waveforms for harmonic operation of the dipole. You should note the terminology applied to define these waveforms and patterns. The term used to describe the balloon shapes of the propagation pattern is *lobe;* the term used to describe a zero, or zero-crossing, waveform is *node.* (The less often used term for the *peak* in the waveform is *loop.*) The 1.5 wavelength pattern in the figure thus has four *nodes* in the waveform, and four *major lobes* and two *minor lobes* in its propagation pattern.

Under certain conditions it may be desirable to operate a dipole slightly off fundamental or harmonic resonance. At the higher frequencies the diameter of the antenna arm rods becomes a significant part of the electrical size of the antenna. The ends of these rods can thus radiate as a result of capacitive coupling into the field. If the antenna is operated on the "inductive side" of series resonance, the coupling capacitance will be canceled, thus curtailing the side radiation. The antenna may also be made inductive by altering its

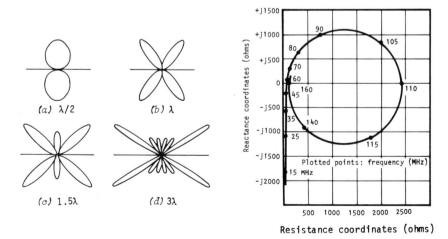

(a) λ/2

(b) λ

(c) 1.5λ

(d) 3λ

(e)

FIGURE 14-5

feed network for cancellation of unwanted lobes. There are *some* circumstances, however, where end lobes are wanted; so rather than suppress the capacitive coupling, measures are taken to increase capacitance at the ends of the arm rods. There may also be certain other requirements for a given antenna performance that make nonresonant operation desirable, including a requirement for altering lobe patterns, and especially for broadband (wider frequency) operation. Figure 14-5e shows the input impedance of the basic dipole antenna operated over a wide range of frequencies. With careful impedance matching the antenna can be made to operate over a wide range of frequencies with varying efficiencies. Resonant operation is shown in the figure where the locus of response crosses the zero line. You should observe that broadband operation is not practical with this antenna, but *wide* frequency operation *is* possible over the range of 10 to 60 MHz by canceling the capacitive reactance at its input.

Sample Problem 4

Given: The 100 inch (254 cm) dipole antenna with the impedance locus of Fig. 14-5e.

Determine: The input impedance and frequency for resonant operation at 0.5, 1.0, and 1.5 wavelength modes; the rectangular-form input impedance at 20, 30, 40, and 50 MHz; and the range of variable inductance required to tune the antenna for operation from 20 to 50 MHz.

(a) The input impedance for resonant operation is obtained from the locus plot of Fig. 14-5e:

   0.5 wavelength $= 65 \; \Omega$
   1.0 wavelength $= 2400 \; \Omega$
   1.5 wavelength $= 75 \; \Omega$

(b) The frequency for resonant operation is computed based on the wavelength of the signal:

   $1.0 \; \lambda = 2.54$ m
   $f = v/\lambda = 3 \times 10^8/2.54 = 118$ MHz
   $0.5 \; \lambda = 59$ MHz   and $1.5 \; \lambda = 177$ MHz

(c) The rectangular-form input impedance is obtained from the locus plot of Fig. 14-5e:

   Z_{in} at 20 MHz $= 45 - j1300 \; \Omega$
   Z_{in} at 30 MHz $= 50 - j800 \; \Omega$
   Z_{in} at 40 MHz $= 55 - j500 \; \Omega$
   Z_{in} at 50 MHz $= 60 - j200 \; \Omega$

(*d*)   The range of variable inductance can be computed at 20 MHz and 50 MHz to cancel the capacitive reactance at those frequencies:

L for $+j1300$ at 20 MHz $= X/\omega = 1300/(6.28 \times 20 \times 10^6) = 10.4 \; \mu\text{H}$
L for $+j200$ at 50 MHz $= X/\omega = 200/(6.28 \times 50 \times 10^6) = 0.64 \; \mu\text{H}$

So the variable inductor should range from about 0.6 to 12 μH.

If you consider the wide range of operating possibilities for this *basic* dipole antenna, and then extrapolate the variables to the great variety of antenna types in use, you should appreciate the extreme complexity of antenna design and application. It is much more involved than just throwing a length of wire out of a window or bending a coathanger. The mathematics required to describe the varied antenna and field characteristics is abstract and voluminous. The people who work in antenna technology often say that antennas are built with a combination of art, science, and a little "black magic." The following section of this chapter will review the operation of a variety of popular antenna types used for radioelectronic communications. The antennas, like the dipole, are characterized for electrical response (radiation and loss resistances) and for optical response (gain and propagation patterns).

PRACTICAL DIRECTIONAL ANTENNAS

The wide variety of antennas used in communications practice fall into *omnidirectional* and *directional* categories. The omnidirectional antennas, like the dipole, radiate signals in all directions but in only one plane. They achieve directional gain by confining their radiation to either vertical or horizontal directions, with the horizontal application being predominant. Omnidirectional antennas are used in applications that require communications from many directions in one plane. Vehicular radio communication, for example, requires transmission and reception regardless of the direction the vehicle is moving or its location. Centrally located radio broadcast stations must likewise propagate signals in all directions in order to obtain the widest coverage of radio receivers surrounding the transmit antennas.

When radio propagation in all directions is not necessary or desirable in particular applications, directional antennas are deployed to advantage in achieving higher efficiency radio communications. These antennas serve to "compress" rather than "waste" radio energy in directing the propagation pattern only in a given direction. Such directional antennas are used for point-to-point communications, where the locations of the transmit and receive stations are fixed and known. The directional gain of these antennas provides for successful communications, whereas omnidirectional antennas will fail to do so. These antennas are often rotor mounted to obtain some benefits of the "omnis" without loss of gain. The familiar roof-mounted TV antennas serve as excellent examples for your reference identification of directional antennas.

Omnidirectional propagation patterns are achieved with antennas similar to the simple dipole described in the previous sections of this chapter. Directional propagation can be achieved with a *variety* of techniques with particular antennas, many of which rely on principles of phasing. Such effects, again, fall into the category of the antenna's *optical* response. If two resonant dipole antennas were placed adjacent to each other, their propagated signals would interact with each other in the field surrounding them. This effect is called *interference,* and the resulting energy field is called the *interference pattern.* By placing the antennas exactly one-half wavelength apart and driving them with identically phased signals, the radiation fields will be in phase and additive in one direction, *X-Y;* but they will be out of phase and will cancel in the perpendicular direction, *A-B.* The directional gain achieved by this technique will simply be 3 dB (twice power) over that of the single dipole because of the in-phase addition of the two field forces. The polar-pattern response, shown in the figure, confirms the tradeoff gain vs coverage achieved by this phasing technique.

Directional gain can also be achieved by placing the two resonant dipole antennas adjacent to each other, but with *quarter-wave* spacing, and driven from the transmission-line feed at one end of the pair. Such an arrangement yields in-phase field forces that are additive in the direction that the signal is moving along the feed line. The field forces that combine in the direction opposite to that of the signal motion cancel, however, because they are 180° out of phase. Since the electrical signal is moving along the transmission line, the second dipole is encountered at 90° ($\lambda/4$) after the first dipole. The field developed by the first dipole reaches the field of the second dipole with the same 90° delay, so the interference yields in-phase addition of the two forces. The reverse-direction field force, however, encounters the field at the first dipole as developed by the electrical signal that is 90° ahead of that at the second dipole. Since the field from the second dipole is delayed by 90° in moving back to the first, the two fields are 180° out of phase, and the interference results in cancellation of the field forces.

Driven Arrays

Antennas that are driven to achieve directional gain by phasing dipoles are categorized as *driven arrays.* They can be driven at one end as in the example of the previous paragraph, or they can be center driven with phase shifting. These antennas are called *end-fire driven arrays,* because the radiated signal propagates off the end of the antenna. Increased gain can be achieved with the end-fire array by adding dipole sections spaced at quarter-wave increments along the feed line as shown in Fig. 14-6*a*, with each section

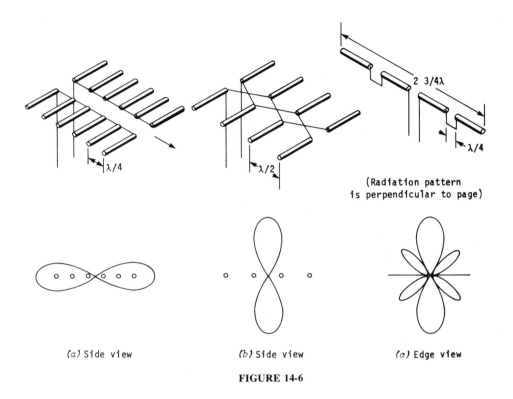

(a) Side view   (b) Side view   (c) Edge view

FIGURE 14-6

increasing the gain to an exponential maximum of about 11 dB (practically achieved with a six-section antenna).

A second popular driven-array antenna is called the *broadside,* because the radiated signal propagates off the side of the antenna rather than its end as in the previous example. Broadside arrays are simply a number of dipoles driven in parallel and in phase. The dipoles of the broadside array are most often spaced at half-wavelengths in order to achieve the in-phase interference that yields directional gain. Three-quarter wavelength spacing of the dipoles is sometimes used, however, to achieve higher directional gain with some loss in efficiency. The multielement broadside antenna and its two-direction radiation pattern are compared in Fig. 14-6b with the radiation pattern from a single dipole. A typical two-element broadside array has a directional gain of about 4 dB, and a six-element array may be optimized for a gain of about 10 dB.

A third popular driven antenna is called a *collinear array.* As its name describes, the antenna elements are in line (linear) and driven together (center fed). Figure 14-6c shows a simplified collinear array that connects four half-wave dipoles in series rather than in parallel, as with the broadside. The polar response of the collinear array is omnidirectional, as with the dipole, but the azimuth response shows increased gain resulting from the in-phase interference effects of the field forces. This antenna design is most useful as a vertically polarized omnidirectional antenna, providing about 2 dB gain with two driven elements, and a maximum gain of about 5 dB with four elements.

Many variations of *driven* antennas are used for differing application requirements. The more popular of these include the *Zepp, Lazy H, flat-top,* and a variety of *curtain* antennas (so named because they look like a "curtain" of wires). The Zepp antenna is of collinear design with the dipole elements at 0.64 λ, instead of 0.5 λ, and with greater than normal spacing between the elements. This Zepp design achieves about 1 dB additional gain over the collinear array. The Lazy H array is basically two collinear arrays connected to be driven in parallel, achieving a gain of about 6 dB with two parallel pairs of half-wave dipole elements. The *Sterba curtain* antenna is an example of a closed-circuit antenna requiring no insulators or unterminated arms. This antenna is simply two collinear arrays with phasing achieved by crisscrossing the driven elements. An optimum gain of 3 dB over a collinear, or a maximum gain of about 7 dB, is possible with this design. The flat-top antenna is a variation of the end-fire array, with the driven dipoles spaced at less than a quarter-wavelength. The antenna is center fed, but the elements are phased to achieve radiation in two directions—off both *ends* of the array. Since this design radiates in two directions, however, its gain is 3 dB less than the optimum of the end-fire array, or about 6 dB maximum.

Parasitic Arrays

The concept of electromagnetic induction, used to describe basic antenna principles in the earlier sections of this chapter, is also applicable to a popular family of directional antennas called *parasitic arrays*. Parasitic arrays achieve directional gain by driving only one dipole element, but inductively coupling energy into additional elements which, in turn, radiate in varying phase relationships to obtain interference effects. If a *parasitic element*—a rod or wire—is placed at a fraction of a wavelength away from the driven element, and is of approximate resonant length, it will intercept and reradiate the energy from the adjacent driven element.

If the parasitic element of the antenna is longer than the dipole, and is placed one-quarter wavelength from that driven element, it will reradiate a field that cancels the energy from the dipole in one direction and enforces the energy from the dipole in the other direction. This element is called a *reflector* since its direction of in-phase interference is back *toward* the driven dipole: its optical behavior *reflects* the energy back in the direction that it originated. If the parasitic element is shorter than the dipole, and it is placed one-quarter wavelength from that driven element, it will also reradiate a field that cancels the energy in one direction and enforces it in the other direction. Unlike the reflector, however, this shorter element develops in-phase interference in a direction *away* from the driven dipole: its optical behavior *directs* the energy ahead of the dipole like a lens. The shorter element is thus appropriately called a *director*. Parasitic arrays may employ more than one director to achieve greater directional gain.

Figure 14-7*a* and *b* shows the polar and azimuth radiation patterns that result from the addition of a reflector and a director to a single driven dipole, respectively. Figure 14-7*c* shows the radiation patterns that result when both the reflector and the director are connected for operation with the dipole. In addition to the spacing of its elements, the gain of the parasitic array depends on the electrical length of the reflector and the director with respect to the driven element. As the length of the parasitic reflector is increased to greater than one-half wavelength of the radiated signal, the element becomes inductive. The total electrical properties of the antenna are thus influenced by these parasitic elements, with optimum resonant response achieved when the inductive reactance of the longer reflector is canceled by the capacitive reactance of the shorter director.

Given the variables of element spacing and element length, you may conclude that the simple three-element parasitic array can achieve a large number of structural combinations to yield varying directional gain and propagation patterns. If additional parasitic director elements are added to the antenna, its directive gain will increase with a coincident reduction in the beamwidth of the radiation pattern. If additional reflectors are added, the directive gain will be increased; but often more important, the reverse-direction isolation will increase. This reverse-direction isolation in the radiation pattern is called the antenna's *front-to-back ratio;* it is a most

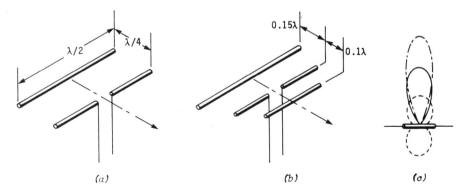

(a)   (b)   (c)

FIGURE 14-7

important requirement for applications that require transmission or reception in one direction only. Typical multielement parasitic antennas are called *beams,* and they are further defined by the number of elements used, such as "four-element beam." The directional gain of these multielement beams may be optimized to greater than 12 dB, with greater than 30 dB front-to-back ratio as achieved by the parasitic reflector.

A popular parasitic antenna with a single reflector and a number of more closely spaced director elements is called the *Yagi* or *Yagi-Uda* after its inventors. This antenna design is characterized by high directive gains and front-to-back ratios. Table 14-1 lists the optimum spacing of elements of the Yagi. You may note that the gain becomes linear with the number of elements, so the graphs can be extended beyond the 16 elements shown. The practical limitation on directional gain for the Yagi is the practicality of its physical length. Obviously, high-gain Yagis are more practical for high-frequency applications where short wavelengths allow many elements to occupy a reasonable length boom.

Parasitic antennas can be connected for parallel operation just as with broadside or collinear arrays. This technique is called *stacking.* The antennas can be stacked horizontally (side by side) or vertically (one above the other) to achieve either greater directional gain or a broader radiation pattern. Stacking two identical antennas generally achieves an increased gain of 3 dB. A *four-bay* stack will typically achieve a 6 dB increase in gain over a single antenna. The effective vertical stacking of the antennas requires that they be spaced in excess of three-quarters wavelength to achieve increased gain.

TABLE 14-1.   ELEMENT SPACING FOR YAGI ARRAY
(All units in wavelength, λ)

Element Spacing	Number of Elements							More than
	2	3	4	5	6	7	8	8, each
Reflector from driven element	0.19	0.19	0.19	0.18	0.18	0.18	0.18	0.18
Director 1 from driven element	.09	.17	.16	.16	.16	.16	.15	.15
Director 2 from director 1	....	....	.16	.18	.20	.21	.22	.22
Director 3 from director 2	....	....	....	.20	.25	.30	.30	.30
Director 4 from director 3	....	....	....	....	.28	.28	.29	.29
Director 5 from director 4	....	....	....	....	....	.30	.30	.30
Director 6 from director 5	....	....	....	....	....	....	.35	.39
Director 7 from director 6	....	....	....	....	....	....	....	.39

Such vertical stacking results in a quasi-broadside array that will increase the azimuth response, but have little effect on the polar radiation pattern.

The horizontal stacking of antennas requires that they be spaced as close as possible, just as with the collinear array. They are typically spaced from one-half to one wavelength, center to center, and are phased to achieve in-phase interference of the radiated signals. The horizontal stacking will increase the polar radiation response, but have little effect on the azimuth response. If an increase in the directional gain is required in *both* the polar and azimuth radiation patterns, the antennas can be stacked vertically and horizontally to achieve about 3 dB gain in each pattern. Figure 14-8 shows various stacking arrangements and their related radiation patterns.

OTHER DIRECTIONAL ANTENNAS

In addition to those antennas that use some form of half-wave dipole as the driven element(s), there are other antenna types that use loops. One popular antenna of this design is the *quad*. This antenna has a square-shaped loop as its driven element, with the length of each side of the square equal to a quarter wavelength of the driving signal. Like the Yagi, the quad has parasitic elements to reflect and direct the signals. These parasitic elements are of the same square-loop shape and size as the driven element, but they are tuned with tuning stubs to achieve the required inductive and capacitive reactance for reflector and director functions, respectively. With element spacing similar to that of the Yagi (see Table 14-1), the quad antenna can achieve directional gains of about 2 dB greater than the Yagi. Figure 14-9 shows two- and four-element quad antennas along with their respective radiation patterns.

The Rhombic

Another loop-type antenna, called a *rhombic* because of its rhomboid (diamond) shape, is used in applications requiring extremely high directive gain with a narrow radiation pattern. Unlike the quad antenna, the rhombic may be nonresonant, and it employs no parasitic elements. It consists of a four-sided rhomboid loop, with all sides of equal length, driven from one end, and with the opposite end terminated in an open circuit or a load resistor. The open-circuit resonant V-beam is shown in Fig. 14-10. Two sections of transmission line with each leg at least two wavelengths long make up the resonant rhombic antenna of Fig. 14-10*b* that radiates from both its feed and open ends. Its directional gain varies with its length. Figure 14-10*c* graphs the gain of the V-beam and resonant rhombic antennas against the length of each of their legs.

Horizontal patterns   Stacking arrangements   Vertical patterns.

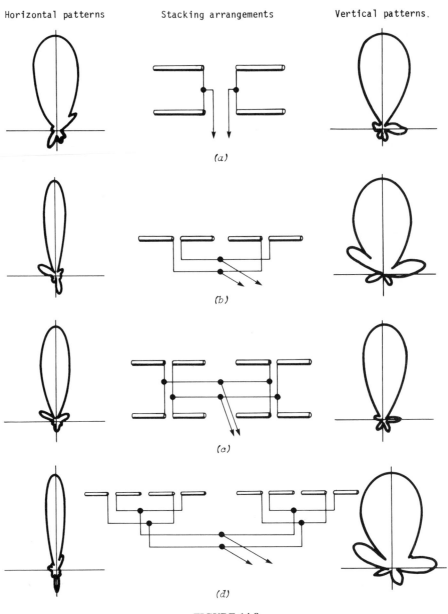

(a)

(b)

(c)

(d)

FIGURE 14-8

The nonresonant rhombic antenna design is shown in Fig. 14-10*d*. With this antenna terminated in a resistive load equal to its characteristic imped-ance of about 800 Ω, it radiates in one direction only. The field created at the feed-end "V" of this rhombic combines in phase with that at the load end

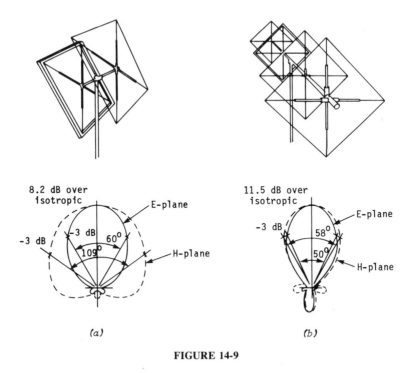

8.2 dB over isotropic

11.5 dB over isotropic

FIGURE 14-9

to achieve increased gain and high front-to-back response. The directional gain of this antenna varies with the length of its legs and its *tilt angle*. Figure 14-10*e* graphs the optimum gain of the nonresonant rhombic against the length of each of its legs. The obvious disadvantage of the rhombic design is its size; but like the Yagi, high directional gains are easily achieved at higher frequencies where wavelengths are short. The writer recalls building an effective rhombic antenna for "deep-fringe area" TV reception by running wires supported by power poles across a 2-acre field.

The Log-Periodic

The *log-periodic dipole array* (LPDA) is not *nonresonant* in the sense of the rhombic; but it is also not *resonant* in the sense of the Yagi. This popular antenna can best be described as a true broadband antenna that is practical in size. Figure 14-11 shows the design of the LPDA simply as a series of dipoles cut to a range of frequencies, increasing in size from its feed to its large end. The unique property of this antenna is the way in which each of the *nonresonant* dipoles develop interference fields at various frequencies to provide in-phase directional gain to the *resonant* dipole. The antenna functionally achieves the effects of a driven dipole, with reflector and director, over a 2:1 (octave) frequency range by the combination of the resonant properties it exhibits. The LPDA's name is derived from its electrical properties (Z_0,

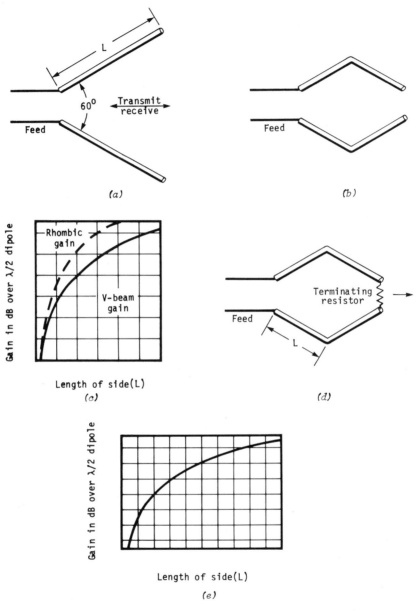

(a)

(b)

Gain in dB over λ/2 dipole

—Rhombic
gain

V-beam
gain

Length of side(L)
(c)

Feed

Terminating
resistor

L

(d)

Gain in dB over λ/2 dipole

Length of side(L)
(e)

FIGURE 14-10

radiation resistance, etc.), as they vary periodically (in a repetitive response) with the logarithm of the frequency variation within its operating range.

If you examine Fig. 14-11 closely, you should note that each of the dipole elements is driven at 180° out of phase by the alternate arrangement of

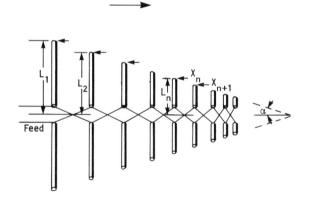

FIGURE 14-11

element connections to the feed line. One practical technique employed for fabrication of the antenna uses two vertically oriented parallel lines, with dipole sections alternately connected to the top and bottom lines, respectively. This structure should aid your understanding of the line's function in driving the dipole elements. You may recall the phasing of the fields due to the quarter-wave spacing of the dipole elements in the *end-fire* antenna, and that aiding and canceling of the fields was due to the interaction of the phase shift in the traveling electrical signal on the line, as compared to the phase shift in the radiating signal. The spacing of the resonant elements in the log-periodic antenna is one-half wavelength; but since they are driven out of phase, the basis of aiding and canceling interference also results from the interaction of the traveling electrical signal and radiated signals. The phase shift in the LPDA design, however, yields maximum radiation toward the feed end of the antenna, rather than away from it, as occurs with the end-fire array.

To achieve broadband response with the LPDA, both the spacing and the length of the dipole elements must increase from its feed end to its back end. For uniform antenna response the resonant frequency of each adjacent dipole should not exceed a ratio of 1.25 : 1. Smaller increments of resonant frequencies for each dipole will obviously require a greater number of dipoles to cover a given bandwidth, with the tradeoff advantage of increased directional gain. If, for example, a four-element LPDA were designed with a frequency increment of 1.25 between adjacent pairs of half-wave dipoles, the total resonant-frequency bandwidth would be $1.25^4 = 2.44 : 1$. The same resonant-frequency bandwidth could be achieved with a six-element LPDA with a frequency increment of $\sqrt[6]{2.44} = 1.12 : 1$, or with an eight-element LPDA with a frequency increment of $\sqrt[8]{2.44} = 1.06 : 1$. The -3 dB beamwidth of the four-, six-, and eight-element LPDA's would be approximately

20°, 10°, and 5°, however, with corresponding approximate directional gain factors of 6, 9, and 12 dB, respectively.

Circular and Cross Polarization

All the antennas described thus far have transverse polarization of their electric and magnetic fields, with the *electric* field identifying vertical or horizontal polarization. A horizontally oriented dipole, for example, develops its magnetic field perpendicular to the current flowing in its elements. The resulting magnetic field is thus vertically polarized, so the electric field is likewise vertically polarized; and the antenna polarization is identified as *horizontal*. With such a horizontally polarized transmitting antenna, the coincident receiving antenna must also be horizontally polarized, as no signal will be induced into a receiving antenna that is oriented perpendicular to the polarization of the radiated signal. Antennas for receiving signals from space vehicles present a special problem, as they must be polarized in *all* planes to achieve alignment coincident with a revolving or random-attitude transmitting antenna. Reception of such arbitrary polarity signals can be achieved by combining the outputs of two cross-polarized antennas, or by crossing the dipoles of conventional Yagi antennas and phasing them at 90° as shown in Fig. 14-12a. Since the radiation patterns overlap, the technique is a most satisfactory means of receiving random polarization signals. With the field polarization in two directions, however, such antenna arrangements reduce the directional gain by 3 dB below the performance of vertically or horizontally polarized antennas.

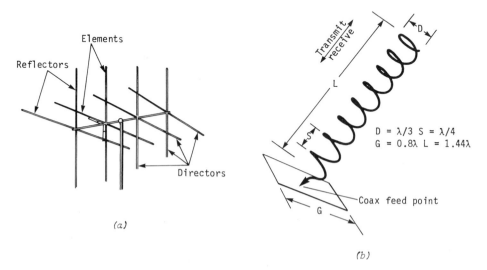

FIGURE 14-12

Another popular technique for higher frequency space communications is the deployment of *circularly polarized* radiation patterns. Such patterns are practically achieved with helical antennas that are wound in the form of clockwise or counterclockwise spirals as shown in Fig. 14-12b. The helical antenna combines the characteristics of collinear and loop antennas, with each sequential loop in its spiral structure approximately one wavelength long and spaced at approximately one-quarter wavelength. The actual design criteria for the antenna depend on the choice of the turns diameter and the pitch of the helix. The helical antenna is also relatively broadband in its response, covering approximately two-thirds of an octave (1.67 : 1). In operation the helical antennas are usually connected in a four-bay stack with two each clockwise (cw) and two counterclockwise (ccw) spirals in order to receive signals with either cw or ccw polarization. The recent applications of circularly polarized antennas to television and FM stereo have served to eliminate the distracting *ghost* or *multipath* signal distortion common to these communications services. The reflected signals change the rotation of their circular polarization by 180° (mirror effect: cw to ccw or ccw to cw), thus they are not received by a oppositely polarized antenna.

Reflectors

More efficient reflector designs can be practically employed to increase the directional gain of antennas in shorter wavelength, higher frequency applications. Among the more popular of these designs are the corner reflector and parabolic reflectors shown in Fig. 14-13. In these antennas the reflector unit functions to return in-phase interference fields to the driven elements in order to achieve a narrow beamwidth and high-gain radiation patterns. The corner reflector is usually designed with a 90° *apex angle,* but 60° and even 45° designs are practical. The driven element of this antenna is typically some form of simple half-wave dipole, placed at a minimum of a quarter-wavelength from the apex. If this distance is increased, the beamwidth of the radiation pattern also increases with a tradeoff reduction in directional gain. If the dipole is placed at about one-half wavelength from the apex, but one-quarter wavelength from the reflector surfaces, the radiation pattern divides into two major lobes. The corner reflector design is capable of achieving an optimum gain in excess of 12 dB.

The parabolic reflector has the highest gain of all antenna designs. The reflector element of this antenna can be oriented for vertical or horizontal polarization, or it can be fabricated in the form of a *dish* as shown in Fig. 14-13b, c, and d, respectively. The driven element of the parabolic antenna is usually some form of half-wave dipole, but other driven element designs are also used. For optimum response the driven element is placed at the exact focal point of the reflector, which is some odd number of quarter-wavelengths from the reflector surface. In general, parabolic reflectors are most

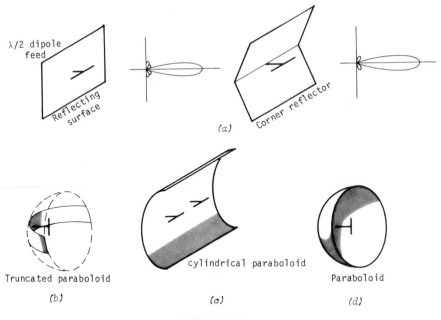

λ/2 dipole feed

Reflecting surface

Corner reflector

(a)

Truncated paraboloid

(b)

cylindrical paraboloid

(c)

Paraboloid

(d)

FIGURE 14-13

practical for short wavelength signals in the UHF frequency region and above, because the dish must be large with respect to the wavelength of the signal. The theoretical directional gain of a parabolic reflector can be computed, based on its size, as:

$$G = 4\pi A/\lambda^2$$

G is the power gain, A is the area of the parabolic dish ($A = \pi D^2/4$), and D is the diameter of the parabolic dish. If, for example, a 3-m-diameter parabolic dish were to be operated at 1000 MHz ($\lambda = 0.3$), its gain would be computed as:

$$G = \pi^2 D^2/\lambda^2 = 6.28^2 \times 3^2/0.3^2 = 3944 = 43.2 \text{ dB}$$

This calculation is theoretical and does not take into account the efficiency of the driven elements and dish, which may practically reduce the gain figure by 6 dB. You should easily conclude, however, that this antenna design with a dish diameter of 10 wavelengths provides considerably more directional gain than is possible with other antennas.

Specialized Antennas

As noted earlier, the intent of this chapter is to provide some familiarity with popular antenna types, and some general understanding of their operation and characteristics. Numerous less popular antenna types are used for special applications, but they are beyond the scope of this text. Three of these less popular antennas are so unique in their operation, however, that some mention of their function is important here. These antenna types are the *horn, slot,* and *discone.* The horn antenna is used in microwave waveguide applications as a most practical means of radiating short wavelength signals. At the end of a waveguide the horn antenna acts as a taper transition (transformer) from the wave impedance within the waveguide to the 377 Ω impedance of air. The horn structure can also be used with conventional balanced transmission lines, providing a "bow tie" structure with terminal impedances of 300 Ω to 400 Ω depending on the apex angle at the throat of the horn structure. If the sides and aperture of the bow tie or waveguide horn are just one wavelength, the directional gain of about 9 dB will be achieved. The sides and aperture of two wavelengths will yield a directional gain of about 15 dB. Figure 14-14*a* shows two variations of the horn antenna.

Like the horn antenna, the slot antenna is more useful at the UHF and microwave frequencies where wavelengths are short. You should have little difficulty in understanding the slot antenna if you look at the edge of the slot rather than the hole. If a balanced transmission line is connected to the center of two quarter-wave shorted stubs, the conditions for radiation will be met. The edges of a slot antenna are basically just shorted stubs, with the remaining metal serving only to support the structure. The slot antenna can be any multiple of half-wavelengths long in order to radiate signals, with the half-wavelength version providing a bidirectional gain equal to that of a dipole, and a terminal impedance of about 500 Ω. A unique slot array antenna consists of a section of rigid coaxial line with slots along its outer conductor as shown in Fig. 14-14*b*. This antenna can provide a wide range of radiation patterns with proper phasing of the slot pattern.

The most unusual discone antenna provides broadband omnidirectional radiation over a wide range of frequencies. As its structure shows in Fig. 14-14*c*, the antenna is fabricated of disk- and cone-shaped radiators, with dimensions of one-sixth and one-quarter wavelength, respectively, for its lowest frequency of operation. The major advantage of the discone antenna is its wide frequency response at a nominal 50 Ω impedance. Properly designed discones will provide 3 to 4 dB directional gain over an 8 : 1 (three-octave) frequency range.

ANTENNA MATCHING AND LOADING

The optimum performance of any antenna system requires efficient power transfer of electrical signals from the transmission line to the antenna's radiating elements. All losses due to impedance mismatch must thus be eliminated with the use of transformers or reactive elements in order to match the antenna to its transmission-line feed. Some antenna designs afford varied impedance options, depending on their electrical properties for a

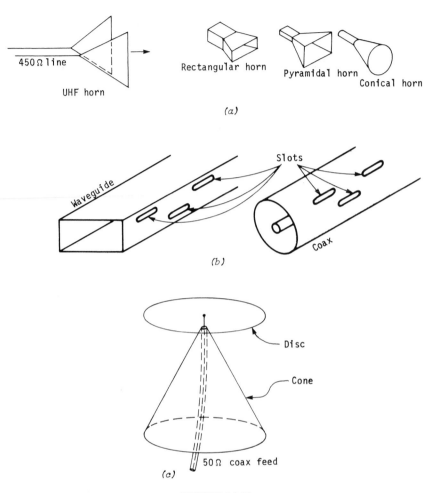

450 Ω line

UHF horn

Rectangular horn

Pyramidal horn

Conical horn

(a)

Slots

Waveguide

Coax

(b)

Disc

Cone

50 Ω coax feed

(c)

FIGURE 14-14

given mode of operation. Since there are limited standard transmission-line impedances, however, some form of matching is almost always a necessity. You may recall, for example, that the impedance locus for the dipole of Fig. 14-5, and the related Sample Problem 4, allowed for resonant operation input impedances of 65 Ω, 2400 Ω, and 75 Ω. With your knowledge of RF transmission lines, you should conclude that a 75 Ω line can drive the antenna directly; the 65 Ω operation will result in an acceptable VSWR of 1.15 : 1 with a 75 Ω line; and, if necessary, a simple matching network can be employed to match the 65 Ω operation to either a 75 Ω or a 50 Ω feed line. You should also conclude that the 2400 Ω operation of the antenna is not practical for most applications because the current will be small and the capacitive losses will be high with high-impedance operation.

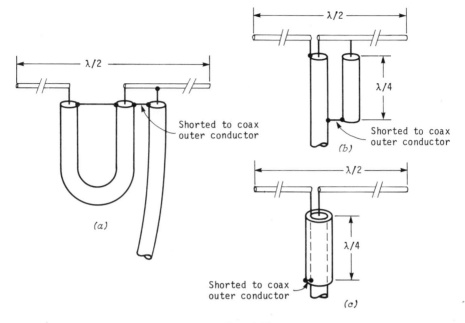

FIGURE 14-15

The Balun

In practice, transmission lines have been standardized at characteristic impedance values around 50, 75, 93, 125, 300, 450, and 600 ohms. For most antenna feed applications, the lower impedance lines are of coaxial structure to minimize radiation with the use of a grounded outer shield. The higher impedance balanced lines, such as the familiar 300 Ω TV twin-lead and open-wire antenna feeds, must be supported away from structures and conductive materials in order to prevent line radiation losses. With the exception of the 300 Ω twin-lead and higher power, lower frequency antenna feeds, the coaxial cables are the more popular choice for antenna feeds. Since most antenna designs use two *balanced* dipoles, the unbalanced coax feed presents problems that are generally corrected with the use of a special kind of transformer, called a *balun* (for *bal*anced to *un*balanced, or vice versa). If the balun is not employed, the outer shield of the coax becomes an extension of one dipole element, and will radiate along its length as well as detune the electrical length of the dipole element. So, in addition to impedance matching networks, baluns are often employed at the feed inputs of antennas in order to assure their function with coaxial cables.

Figure 14-15 shows a variety of balun designs as achieved with coax and transformer elements. The electrical equivalent of a balun is simply an RF transformer with one terminal of its unbalanced winding grounded along with the center tap of its balanced winding. Maximum broadband efficiency and power transfer are achieved with closely coupled baluns, but they may also be loosely coupled and tuned for high-Q, narrowband applications. A more popular narrowband balun may be achieved with a quarter-wave section or sleeve of coax line, where the second section of line and sleeve acts as a transformer. Since transformer action is required for the operation of a balun, it is most practical to also use this function for impedance as required for particular antenna types. In addition to the "wound" transformers, the quarter-wave transmission-line transformer technique is effectively employed at higher frequencies.

Balanced Feeds

The necessity for a balun is, of course, eliminated with higher impedance antennas that are fed with balanced transmission lines. In addition, a number of techniques can be used to raise the input impedance of antenna feed points in order to allow them to function effectively with balanced lines. One popular design employs a *folded* half-wave dipole (Fig. 14-16a) as the driven element of the antenna to raise its feed-point impedance from 72 Ω to 300 Ω. The quarter-wave transformer feed can, of course, also be used with balanced transmission lines just as effectively as with coax lines to raise the antenna feed-point impedance. Another alternative is the use of shorted dipole sections that are fed at some distance along their length, where the

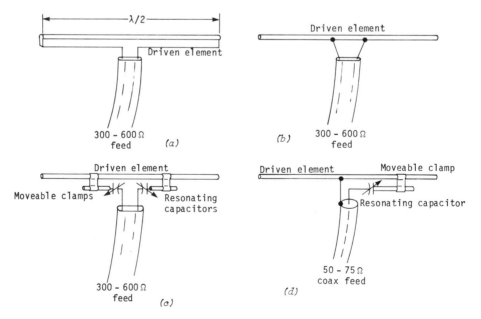

FIGURE 14-16

voltage/current ratio of the element's standing-wave pattern provides a higher impedance. This technique (Fig. 14-16b) is called *delta matching*, because the shape of the spread balanced feed line looks like the inverted Greek letter. Yet another balanced-line matching technique that combines quarter-wave transformation with delta matching is the *T-match* shown in Figure 14-16c.

Figure 14-16 also shows matching techniques that may be used to reduce the higher impedance antenna feeds in order to match lower impedance coax lines. The gamma-match (Fig. 14-16d) is the unbalanced line equivalent of the T-match technique as identified in the previous paragraph. These matching techniques and those of the previous paragraphs are appropriate for changing the resistive impedances of antennas to match those of the feed transmission lines. Although antennas ideally have resistive input impedances, few *ideal* examples exist in practice, so additional reactive matching techniques are usually required to obtain optimum performance from antenna and feed systems. The T-match and gamma-match elements, as well as the quarter-wave and wound transformers, can be made reactive with the addition of an inductance or capacitance in order to cancel reactive mismatches at antenna inputs. Shorted and open matching stubs can also be used to cancel reactive impedance mismatches in antenna systems, just as they are used with other transmission-line circuits. The actual procedures and applications of these matching techniques are detailed in Chapters 11 and 12.

HF Beam and Wire Antennas

Lower frequency operation of wire or beam-type antennas presents an obvious problem with size and space. This problem is probably most pronounced with the lower frequency mobile *whip* antennas. A quick calculation of the length of a quarter-wave vertical dipole for 27 MHz vehicular use, for example, will show that a whip in excess of 9 ft long is required for efficient radiation. Mounting a 9 ft rod on the roof or trunk of an automobile creates problems with overhead obstructions, not to mention birds and low-flying aircraft. The solution of the antenna size problem requires reduction of dipole length without creating large mismatches or poor radiation efficiencies.

You may recall from the earlier sections of this chapter that the standing-wave pattern reached a voltage peak and current null at the ends of the resonant dipole, and that shortening the dipole created capacitive effects, as with a series-resonant network operated below resonance. The simple solution, then, is to add inductance to the shortened antenna dipole in order to resonate its capacitive effects at the operating frequency. This inductance is typically referred to as a *loading coil*. Loading coils can be placed at either end or at the center of shortened dipole elements, with efficient operation available with dipole reductions to about 40% of its resonant length. Figure 14-17a shows typical loading coils as used to reduce antenna size. (The capacitor shown for reference increases antenna length.)

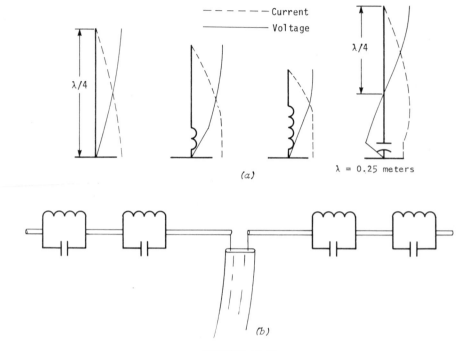

FIGURE 14-17

Traps

Another general problem with resonant antennas is their limited frequency coverage. There are many applications that require radio transmission at various frequencies, which may, in turn, require a number of beam-type antennas. In many cases the cost of extensive real estate and support towers, etc., makes multiple antenna installations costly and impractical. Under these conditions, resonant elements, called *traps,* may be applied to a single antenna structure to extend its operation to different frequency bands. Figure 14-17*b* shows a multiband beam-type antenna with traps for operation at different frequencies. The traps are parallel-resonant networks that resonate with high impedances across their *tanks.* If the antenna is operated at a high frequency, only its inner sections will resonate, with the traps resonant and simulating an open circuit, thus isolating the remainder of the antenna from the signal. At lower midband frequencies, the first trap is not resonant, so its low impedance passes the signal to resonate two sections of the dipole. The second trap is now resonant, however, and isolates the remaining section of the antenna from the midband frequency signal. Finally, the entire length of the antenna is resonant at the lowest frequency of operation, with both pairs of traps off resonance and at low impedance.

ANTENNA MEASUREMENTS

Since there are so many variables in the design and execution of antennas, measurements of their electrical and optical properties are indispensable. The basic *electrical* properties of antennas are measured using the same techniques and procedures that are applied to transmission line measurements: VSWR, characteristic impedance, power transfer, frequency response, bandwidth (Q), and losses (loss resistance). These measurements are detailed in Chapters 11 and 12. One additional important electrical measurement of receiving antennas is *noise figure,* the techniques of which are outlined in Chapter 8. The basic *optical* properties of antennas, and the characteristics of the media through which signals are to radiate, are measured with more specialized instrumentation and techniques unique to the antenna field. Among these specialized instruments are *calibrated receivers, plotters, field-strength (field-intensity) meters, and standard reference antennas.*

Field Measurements

Unlike other electronic measurements, most antenna measurements are made in the field or on the *test range*—not in the laboratory—since operating conditions are difficult to simulate artificially. Shipboard antennas designed for operation over water, for example, must be evaluated under considerably different environmental conditions than those antennas designed for overland use. The major piece of *laboratory test equipment* for antenna measurements is thus the test range itself. One of the more critical factors affecting the radiation of signals across a test range is the *ground effect.* Since the electrical properties of the earth approach that of a conductor, worst-case radiation patterns result when the transmitted signals that are reflected from the earth reach the receiving antenna out of phase with the directly radiated signals. In such cases cancellation will occur, yielding apparently inefficient antenna performance. Antenna test ranges must thus be *calibrated* in order to predetermine the ground effects and calculate them out of the measured antenna response.

Figure 14-18 shows how ground-effect reflections will change the signal strength received when the antennas are raised to different heights above the ground. The resulting plot of signal strength looks somewhat like a standing-wave pattern with a vertical rather than horizontal orientation. The plot is, in fact, a standing-wave pattern, since it is a result of incident and reflected signals alternately combining in phase and out of phase. Just as it is possible to determine the ratio of incident and reflected signals from the transmission line VSWR pattern, it is also possible to determine the ground-effect signal magnitude using these same techniques. You may thus apply the relationships of VSWR and ρ from Chapter 11 to determine the ground effects:

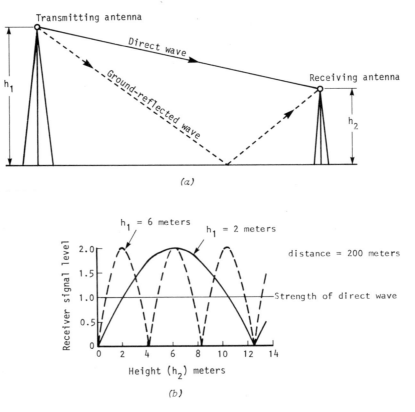

FIGURE 14-18

$$\text{VSWR} = V_{max}/V_{min} = (V_{inc} + V_{refl})/(V_{inc} - V_{refl})$$
$$\rho = \%V_{refl}/100 = V_{refl}/V_{inc} = (\text{VSWR} - 1)/(\text{VSWR} + 1)$$

If, for example, the maximum and minimum signal voltages at the terminals of a receiving antenna are 900 and 300 μV, respectively, when the transmitting and receiving antennas are raised aboveground across an antenna test range, the percent of ground-effect reflection can be computed as:

$$\text{VSWR} = V_{max}/V_{min} = 900/300 = 3:1$$
$$\rho = (\text{VSWR} - 1)/(\text{VSWR} + 1) = (3 - 1)/(3 + 1) = 0.5$$
$$\%V_{refl} = \rho \times 100 = 50\%$$

The magnitude of the incident signal can be computed as:

$$V_{inc} = V_{max}/(1 + \rho) = 900/(1 + 0.5) = 600 \ \mu\text{V}$$

The same form of standing-wave pattern as was obtained from raising the antennas aboveground can be obtained by separating the antennas in the test range. As you should observe, the ground effect can only be obtained if the antennas are separated by a minimum distance in order to achieve the effects of phasing. The establishment of this distance with respect to antenna height is a critical part of setting up the antenna tests. The computation requires that the transmitting antenna be placed at the *same height* or *above* the receiving antenna is order to maintain a uniform radiated field. The *minimum* distance between the transmitting and receiving antennas can be calculated as:

$$d = 2h_t h_r/\lambda$$

where d is the distance between antennas, h_t is the height of the transmitting antenna, h_r is the height of the receiving antenna, and λ is the free-space wavelength of the radiated signal. If, for example, transmitting and receiving antennas, operating at 100 MHz (λ = 9.84 ft or 3 m), are set at 12 ft above the ground of an antenna test range, the minimum distance required for their separation can be calculated as:

$$d = 2h_t h_r/\lambda = 2 \times 12 \times 12/9.84 = 29.27 \text{ ft} = 8.92 \text{ m}$$

Placing the antennas at any practical distance in excess of 30 ft would provide satisfactory measurement results. You should observe that the separation of the antennas increases with their height; so for a minimum distance the antennas can be placed at slightly more than one wavelength above the ground. The formula can, of course, be transposed to determine minimum height if the distances in the test range are fixed. You should keep in mind, however, that the previously established maximum/minimum signal response must be obtained by either raising the transmitting antenna or lowering the receiving antenna. The subsequent measurements are then typically performed with the antennas set to a maximum ($V_{inc} + V_{refl}$) signal level.

After establishing the test range heights and distances, the optical properties of antennas may be determined by *substitution* and *comparison* techniques. These techniques require the use of known standard-gain antennas or calibrated receivers that can compare the signals obtained from antennas under test against known signal levels. The standard antennas are of four types: the *small loop*, the *short vertical*, the *standard dipole*, and the *standard-gain horn*. The calibrated receivers may be quality spectrum analyzers, or specially designed receivers with RF and IF attenuators and metered detector outputs to indicate signal levels.

The loop and vertical antennas are typically used for measurements below 30 MHz, with their electrical length less than one-tenth the wavelength of the signal frequency. Under these conditions the induced signals and rotation patterns of the antennas can be computed based on their function as very loosely coupled transformer windings, as described in the earlier section of this chapter. When used in a receiving function, for example, the induced voltage across the small square or circular loop, and the radiated field strength can be computed using the following formulas:

$$E = \xi 2\pi NA/\lambda \quad \text{and} \quad \xi = E\lambda/2\pi NA$$

E is the induced voltage, ξ is the field strength of the radiated signal in volts per unit length, A is the cross-sectional area of the loop, and λ is the free-space wavelength of the signal. The units of the length for ξ, A, and λ are in meters or feet, etc. If, for example, a one-turn small-loop antenna with a cross-sectional area of 4 m had a terminal voltage of 220 μV when it was used to measure the field strength of a 6 MHz ($\lambda = 50$ m) radiated signal, the field strength of the signal would be computed as:

$$\xi = E\lambda/2\pi NA = 2.2 \times 10^{-4} \times 50/(6.28 \times 4) = 438 \ \mu V/m$$

Standard Dipole

The resonant half-wave standard dipole is most often used for measurements in the range of 30 MHz to 3 GHz, making it the most popular measurement antenna. Measurements above 3 GHz are more practical with directional standard-gain horn antennas. As with the loop, the induced voltage and radiated field strength of the dipole can be computed as:

$$E = \lambda\xi/\pi \quad \text{and} \quad \xi = E\pi/\lambda$$

In practice, the simple half-wave dipole is not often used, however, because it is susceptible to considerable proximity effects from obstructions above, below, and behind it; so dipoles with reflectors or directors, and known gain, are employed, with the gain values factored into the above formulas. If, for example, a 5 dB (1.78 : 1 voltage ratio) standard-gain dipole were used for a field-strength measurement at 100 MHz ($\lambda = 3$ m), with a resulting terminal voltage of 800 μV, the field strength would be computed as:

$$\xi = E\pi/G\lambda = 8 \times 10^{-4} \times 3.14/(1.78 \times 3) = 470 \ \mu V/m$$

The use of the standard-gain dipole in the antenna test range is the most common technique used to characterize the optical properties of antennas under test. The popularity of the dipole technique is also due to the prevailing greater proportion of antenna work for modern electronic communications at frequencies above 30 MHz. The techniques used for antenna measurements with the dipole are almost identical to those used with the standard-gain horn antennas at microwave frequencies, so both antennas will be here described. The incident field strength at a given distance, without considering, of course, the ground-effect signal, can be computed as follows for low-loss-resistance standard-gain half-wave dipoles and standard-gain horn antennas:

$$\xi_d = 7.02 \ \sqrt{P_t G}/d \quad \text{and} \quad \xi_h = 5.48 \ \sqrt{P_t G}/d$$

ξ_d and ξ_h are the field strength of the dipole and horn, respectively, at a given distance in meters from a transmitting antenna; P_t is the power radiated from a transmitting antenna; G is the directional gain (voltage ratio) of the antenna; and d is the distance in meters from the transmitting antenna.

If, for example, the distance between antennas, as established for the earlier test-range example (12 ft antenna height and 100 MHz operation), is 8.92 m, the field strength at that distance for the 5 dB gain dipole with a transmitted power of 10 mW can be computed (presuming negligible loss resistance) as:

$$\xi_d = 7.02 \sqrt{P_t G}/d = 7.02 \sqrt{0.01 \times 1.78}/8.92 = 105 \text{ mV/m}$$

Now if the *same* standard-gain dipole antenna were used for receiving this signal, the voltage at its output terminals can be computed, using the transposed relationship of the previous paragraph, as:

$$E = G\lambda\xi/\pi = 1.78 \times 3 \times 0.105/3.14 = 178 \text{ mV}$$

The above procedure and calculations can now be used to determine the directional gain of an antenna under test by *substitution*. Either the standard receiving or transmitting antenna can be replaced with the antenna under test, with the resulting voltage at the receiving antenna's output terminals compared to that of the standard antenna. The measurement, made with a Yagi antenna under test as the receiving antenna, might, for example, yield an output terminal voltage of 520 mV (hypothetical). The directional gain of the Yagi would then be computed as:

$$G_Y = G_S E_Y/E_S = 1.78 \times 0.52/0.178 = 5.2$$

and in dB,

$$G_Y = 20 \log 5.2 = 14.32 \text{ dB}$$

or

$$G_{YdB} = G_{SdB} + 20 \log E_Y/E_S = 5 + 20 \log 0.52/0.178 = 14.32 \text{ dB}$$

G_Y is the directional gain of the Yagi under test, G_S is the directional gain of the standard-gain dipole antenna of the hypothetical example of the previous paragraph (5 dB), E_Y is the terminal voltage at the output of the Yagi antenna under test, and E_S is the terminal voltage of the standard-gain dipole of the previous example.

Calibrated Receivers

The function of the spectrum analyzer or calibrated receiver in these measurements of antenna gain is to provide some sensitive and accurate detector reference for measuring very low signal levels. For some measurements, such as front-to-back ratio, the signals at the antenna terminals may be less than 1 μV. With high-gain antennas, however, the maximum signal levels may exceed 1 V. The requirements, then, are for a sensitive, low-noise receiver, with calibrated attenuators to prevent overloading of its RF or IF inputs, and with some meter or CRT display to indicate the relative magnitude of its outputs. The advantage of the spectrum analyzer for these measurements is its ability to simultaneously display sideband or spurious signals which may be radiated by the transmitting antenna. The spectrum analyzer will also provide a display of the broadband frequency response of antennas with the use of a nonresonant transmitting antenna driven with a swept signal source.

In operation the spectrum analyzer or receiver used for antenna measurements should provide an accurate logarithmic or linear response of the dynamic range of signal levels to be measured at the receiving antenna terminals. The polar or azimuth response of the antenna under test can be measured by rotating the antenna 360° on its mast and observing or recording the terminal voltage at the receiving antenna. The antenna *beamwidth* is simply the included angle within the maximum −3 dB response of the radiation pattern. Rather than rotating the antenna in elevation, which may complicate the ground effect and present some mechanical problems, the vertical radiation pattern of the antenna under test is most often obtained by turning the transmitting and receiving antennas on their sides (by 90°), and then repeating the 360° mast rotation as used for the polar plot. The measurement of antenna *polarization* is also generally made simultaneously with the vertical response, while only the antenna under test is on its side.

While the above procedures are in no way comprehensive, they should provide you with a solid understanding of antenna measurement techniques. The principles and generalizations may, of course, be applied to antennas other than the hypothetical dipole used in the examples. Each antenna type has its own set of theoretical characteristics against which actual performance may be compared. These characteristics are detailed in many antenna handbooks, but they are generally augmented by actual measured data from accumulated experience with building and characterizing various specific antenna designs. Even with accurate procedures and equipment, considerable *art* and experience are required for expertise with antennas where many factors beyond textbook explanations must be considered.

PROPAGATION

Successful electronic communications requires the intelligent application of antennas to specific needs. The antenna is a most critical link in a communications system, where it provides both an increase in the *effective* power of a transmitter and the *effective* sensitivity of a receiver. A dish antenna with a directive gain of 30 dB, placed at the output of a 2 W transmitter, will provide a field intensity 1000 times (30 dB) more intense at a given distance, just as though a 2000 W transmitter were distributing power equally in all directions. In the case of a receiver with an inefficient antenna of only 3 dB of directional gain, substituting a 9 dB gain antenna ($\simeq$ +6 dB) will double its voltage sensitivity. One important factor to recognize, however, is that successful electronic communications cannot be achieved, even with the highest gain antennas, if the radiated signal does not propagate the medium between transmitter and receiver—there just won't be any signal for the receiving antenna to intercept. So this final, but most important, aspect of *propagation* must be added to antenna properties and characteristics in order to complete this chapter.

Obviously, the least complicated propagation concepts have been applied in the previous section of this chapter in determining the field strength of a radiated signal. You may recall that the relationships identified the volts-per-meter at a given distance, with respect to the input power and gain of standard-gain dipole and standard-gain horn antennas. If the distance factor in these relationships is doubled, the field-strength voltage is reduced to half the initial value; whereas the *power density* of the field at twice the distance is reduced to one-fourth its initial value (Watt's law). So, as with light, the power of a radiated electromagnetic signal varies *inversely* with the distance from its source—the transmitting antenna—provided that the medium through which the signal travels is uniform (*homogeneous*). This relationship is most often generalized in practice as "6 dB of radiated signal *attenuation,*" even though this is not attenuation in the strict sense of *heat* loss.

Radiation Complications

The above *basic* relationships used to determine field strength do not take into account the many factors that can influence the *path loss* and other complications of radiation. Among these more prominent *factors* are *absorption, reflection, refraction, diffraction,* and the resulting *interference* phenomena. You may recall that interference phenomena provided the basis for directional gain in both driven and parasitic antenna arrays. Interference is strictly the result of the phase relationships of coincident field forces in the radiation pattern. The interference phenomena also provided the basis for measuring the ground effect of the antenna test range, as described in the previous section. The results of this interference yielded alternative enforcement and cancellation of the radiated signal strength as measured by varying the height or distance of the antennas. The receiving test antenna in this example, in fact, plotted a portion of the radiation pattern within the antenna test range as peak and null variations resulting from the effects of *reflection* (from the ground) and interference. The number of such interference patterns due to reflections depends, of course, on the wavelength of the radiated signal, with the phenomena being more pronounced at high frequencies where the wavelengths are short.

Reflections

Interference patterns due to reflections can, of course, create *nulls* in the radiation patterns and thus impair successful communications. Again the familiar ghosting and multipath interference identified with television and FM stereo reception are results of interference phenomena due to reflections. A more pronounced effect is also often observed when reflections from low-flying aircraft cause interference with the reception of television signals. The effect, as displayed on the TV picture tube, yields a sequence of alternate in-phase and out-of-phase variations in picture intensity, often including *Doppler-effect* changes in the rate of phase shift (flutter).

Reflections originate from radiated signals that encounter electrically conductive surfaces, including earth, mountains, trees, buildings, and other man-made structures. Signals radiated toward the sky, or from flying objects, and those in orbit obtain reflections from dense clouds, our atmosphere, other planets and moons, and other flying or orbiting objects. All these obstructions that create reflections basically represent an abrupt change in the medium through which a radiated signal is traveling. The abrupt change is simply a change in impedance, which, like mismatched transmission lines, causes reflections that, in turn, cause standing waves.

Although reflections can create undesirable effects, as noted with TV and FM broadcast, they often provide increased communications capabilities. At frequencies below 30 MHz the atmospheric layers surrounding the earth act to reflect radiated signals back toward the earth. Most of this reflected energy that reaches the earth is again reflected toward the sky, encountering the reflective atmospheric layers, and again returning to earth. This series of reflections and re-reflections "bounces" the radiated signal around the earth's circumference, providing a basis for convenient worldwide radio communications. These atmospheric layers are composed of ionized gases that surround the earth at elevations between about 40 and 250 miles above its surface. This upper region of our atmosphere is appropriately called the *ionosphere*. From an electrical viewpoint, the characteristics of the ionosphere vary with elevation and the effects of the *sun's radiation*. Figure 14-19 shows the structure of the ionosphere with its major designated distinct layers, D, E, F_1, and F_2, for typical day, night, summer, and winter elevations. The characteristics of these ionospheric layers must be exploited for successful long-distance communications at frequencies below 30 MHz.

The reflection properties of these ionospheric layers depend on their degree and density of ionization. These properties vary considerably with time of day or year and other cosmic influences, but some generalizations can be made about their effects on radiated antenna signals. The ionization of the lower density D and E layers, for example, is greatly influenced by the sun's radiation, so these layers are significant reflectors of signals below about 5 and 10 MHz, respectively, during their higher ionization levels of daytime hours. With increased altitude the ionization density of F_1 and F_2 layers is

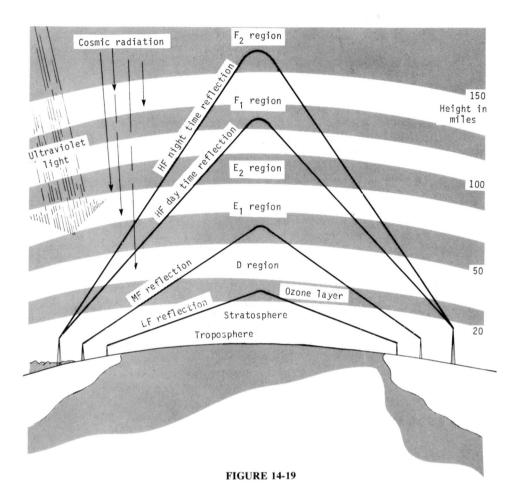

FIGURE 14-19

also greater during the daytime hours; but with the reduced high-altitude atmosphere, these layers combine to maintain a single dense ionized layer through the night. The F_2 layer is the major reflector of signals from about 10 to 30 MHz during the day, and the combined F layers act as the significant single reflective ionospheric layer at night.

Absorption

A second electrical property of the ionospheric layers is absorption. During the daylight hours a 20 MHz signal, for example, must pass through three layers—D, E, and F_1—before encountering sufficient ionization density for reflection. As this signal passes through each ionized layer, it sets the free electrons in motion, losing energy with each of their collisions. After reflection from the F_2 layer, the signal again encounters the three preceding layers, losing energy through absorption on its way back to the earth's surface. Under these conditions, only the most intense radiated antenna signals will achieve successful communications through reflection during daytime hours at the higher frequency portion of the HF region. Successful communications can be achieved, however, by using lower radio frequencies that will reflect off the daylight D or E layers of the ionosphere. Since these layers are at lower elevations, they are only useful for short-distance "hops." Figure 14-20a shows the reflection and absorption of signals from the various ionospheric layers.

Ionospheric Effects

A more detailed analysis of the behavior of radiated antenna signals in the ionosphere requires application of the basic propagation principles introduced at the beginning of this chapter and in Chapter 11. You may recall that the wave impedance of the media through which the energy propagates depends on the permeability and permittivity of the media. When the radiated electromagnetic energy from the antenna signals encounters the free electrons and ions in the ionospheric layer, the field forces cause these particles to oscillate, abruptly reducing the permittivity of the medium. This permittivity characteristic approaches zero, and thus lowers the wave impedance to appear like a short circuit terminating a transmission line. A complicating factor in this process is the relative density and degree of ionization of the particular layer. With the sparse distribution of ions in the D and E layers, for example, only a small degree of increased permittivity results, and the higher frequency signals pass through the layers with only the effects of attenuation.

When the frequency of the radiating signal is low enough to cause oscillations with a particular density of free electrons within a given ionospheric layer, the permittivity of the medium will rapidly fall toward zero, creating the conditions for reflection. This frequency, called the *critical frequency* of ionospheric propagation, varies directly with the square root of the electron density within the D, E, F_1, and F_2 layers. Although this frequency is calculable, there is little value in the exercise since so many additional variables influence the effects of these ionospheric layers on radiated signals. In practice, U.S. engineers and technicians rely upon regularly published data from the government laboratories that specify the ever-changing ionospheric conditions and characteristics. For purposes of generality the typical critical frequencies are approximately as follows: 300 kHz for the D layer, 4 MHz for the E layer, 5 MHz for the F_1 layer, 8 MHz for the daylight F_2 layer, and 6 MHz for the nighttime combined F_1/F_2 layer.

There may seem to be a contradiction in the relationship of the above critical frequencies when they are compared with earlier generalizations about 5, 10, and 30 MHz as the frequencies of "significant" reflections for

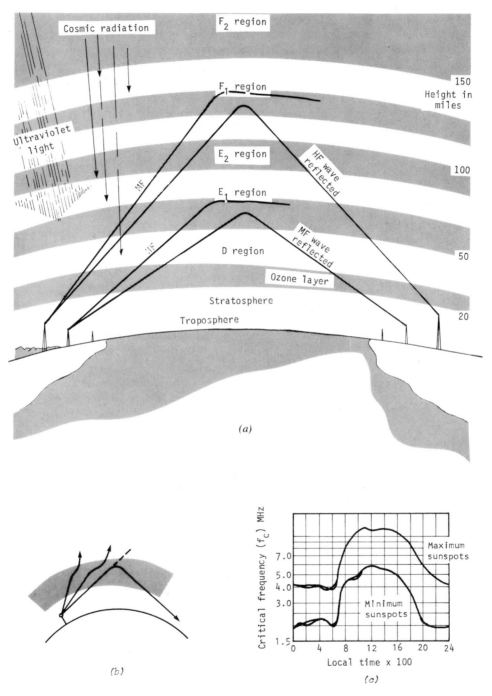

FIGURE 14-20

the *D*, *E*, and *F* layers, respectively. Nevertheless, both groups of figures are conditionally valid, but further definition is apparently here required. The *critical frequency* refers to the highest frequency of nearly zero permittivity, with a radiating signal that is *perpendicularly* incident upon the "shortest thickness" of the ionospheric layer. Such a signal would radiate vertically from the earth and reflect back to its exact point of origin—a practice useful perhaps in determining the altitude of the reflective layer, but not much use in bouncing a signal to some other point on the earth.

Figure 14-20*b* shows the more practical use of the reflective ionosphere for communications, where the transmitted signal from the earth radiates at some angle, thus encountering the reflective layer obliquely where its cross section is considerably thicker. The resulting frequency at which *this* reflection will occur is called the *maximum usable frequency* (muf), which is always higher than the critical frequency.

Refraction

The oblique entry of the radiated signal into the ionospheric layer results in another effect on its propagation. Since the broad wavefront does not encounter the ionosphere simultaneously across its beamwidth, *refraction* (bending) occurs along with the reflection process. This refraction of the wavefront actually extends the density of the ionospheric layer and the resulting muf. If the critical frequency and the angle of radiation are known from current reports and measurements, the muf for reflection can be calculated as:

$$\text{muf} = f_c/\cos \angle \phi$$

f_c is the critical frequency, and $\angle \phi$ is the deviation from perpendicular entry of the radiated signal into the ionospheric layer. If, for example, the reported f_c for the *E* layer is 4 MHz, and the radiation pattern of the antenna transmits a wave that is at 60° from vertical (30° above horizontal), the muf can be computed as:

$$\text{muf} = f_c/\cos \angle \phi = 4/\cos 60° = 8 \text{ MHz}$$

Figure 14-20*c* graphs an example of critical frequency for a typical 24-hour period in summer and winter. You should observe that sunspot radiation increases the midday winter level; whereas the nighttime ionization of the F_2 layer in winter is of low density, reducing the critical frequency.

In practice, the frequency of operation for reliable reflection is limited to about 85% of muf, making the practical frequency limit of the example in the previous paragraph about 6.8 MHz instead of 8 MHz. If radiated signals at frequencies above muf obliquely enter the ionospheric layers, they may also encounter refraction effects, bending their paths as they pass through the layers into space. Since frequencies considerably below the muf do not require the greater ionospheric density to effect reflection as compared to the higher frequencies, reflection occurs at reduced penetration of the ionospheric layer. The penetration of the ionospheric layer at various angles and frequencies is compared in Fig. 14-20*a*, which also shows the effects of refraction on the oblique incident wave. The distance at which reflected signals return to earth after ionospheric reflection is called the *skip distance*. This skip distance is measured along the earth's surface. In order to obtain full radio coverage of a wider area along the earth's surface, multilobe *sky-wave* radiation patterns may be employed to enter the ionosphere at different angles.

Diffraction

In addition to the sky-wave radiation at frequencies generally below 30 MHz, lower frequency propagation can use the electrical properties of the earth's surface to support propagation. This electromagnetic/optical phenomenon is called *diffraction;* and the radiated signals that follow the curvature of the earth's surface are appropriately called *surface waves*. Your experience with AM and FM broadcast radio at about 1 MHz and 100 MHz, respectively, should provide a good example of how the diffraction effects of the earth's surface bend the 1 MHz waves around mountains and obstacles, while blocking the line-of-sight propagation of the 100 MHz signals. The limitations of such surface-wave transmission are a result of the *same* characteristics of the earth that initially created the diffraction phenomenon. Since the conductivity and permittivity of the earth's surface are greater than the air above it, energy is lost in the earth as the velocity constant diffracts the waves along its surface. The resulting attenuation from conductivity limits the practical deployment of this technique to frequencies below about 2 MHz and for distances of less than about 200 miles. The combined effects of the sky and surface waves are, for example, used for broad coverage of AM broadcast radio.

Higher Frequency Propagation

For frequencies above 30 MHz the complications of propagation become greatly simplified. As mentioned earlier, the effects of refraction will bend, but not reflect, the waves as they encounter varying atmospheric layers at oblique angles. In line-of-sight, point-to-point transmission above the earth's surface, only a slight degree of diffraction results from the surface effects. The effects obviously decrease with signal frequency as the wavelengths and wavefronts are decreased, but the tradeoff losses in atmospheric attenuation and reflections from obstacles add other complications to meet the prevailing criterion of electronics and other fields: "There is no such thing as a free lunch!" The problems thus associated with the VHF, UHF, SHF, and EHF regions of the radio spectrum are treated differently than those of frequencies below 30 MHz.

One significant limitation on the transmission of higher frequencies along the earth's surface is imposed by the curvature of the earth. Unlike the surface waves of lower frequency signals, the higher frequency signals encounter only a slight degree of diffraction from the earth's surface, and only a slight degree of refraction from the variations in the atmosphere above the earth. The primary limitation for line-of-sight communications is thus the curvature of the earth. The *radio horizon* is actually greater than the optical horizon, because of the slight bending of the radiated signals; and it can be approximated by the following formulas:

$$d = 7245(\sqrt{h_t} + \sqrt{h_r}) \quad \text{for units in feet}$$
$$d = 4000(\sqrt{h_t} + \sqrt{h_r}) \quad \text{for units in meters}$$

where d is the distance between antennas (the radio horizon), h_t is the height of the transmitting antenna, and h_r is the height of the receiving antenna. You should observe that the height of the respective antennas plays a significant part in assuring increased distance of successful communications. The significance of this relationship can be generally observed in the location of broadcast antennas. AM broadcast antennas, operating at frequencies about 1 MHz, for example, are typically located in a low, flat area; while VHF FM broadcast antennas, operating at frequencies about 100 MHz, are elevated, or located on mountaintops. The radio horizon for such a VHF FM broadcast antenna placed on a hilltop with 900 ft elevation can achieve line-of-sight communications with a rooftop receiving antenna with 16 ft elevation, at a distance computed with the above formula as:

$$d = 7245(\sqrt{h_t} + \sqrt{h_r}) = 7245(\sqrt{900} + \sqrt{16})$$
$$= 2.46 \times 10^5 \text{ ft} = 46.65 \text{ miles} = 75.07 \text{ km}$$

Although long-distance point-to-point, line-of-sight communications have been achieved over the earth's surface by placing both transmitting and receiving antennas on mountaintops, most of these "shots" are limited to practical distances of about 100 miles or less.

Those line-of-sight characteristics that limit the radio horizon of higher frequency signals over the earth's surface serve to extend our communications capabilities in high atmospheric and space applications. Although the higher frequency signals experience some refraction and significant attenuation in the atmosphere, they encounter no such limitations in space. If a UHF or SHF antenna is directed vertically to the sky, for example, atmospheric influences are significant for only the first 5 to 10 miles, after which the only reduction in signal intensity is from the inverse-square-of-the-distance relationship—a loss that becomes insignificant at great distances. With the use of high-gain dish antennas, which are practical at these higher frequencies, it is possible to transmit a signal to a satellite and back to earth again with much less loss than would occur with a 200-mile repeater shot over the earth's surface.

The effects of the higher frequency signals vary, of course, with frequency and wavelength, with specific characteristics most suitable for particular communications applications. One such application is called *tropospheric scatter*. This technique uses the properties of the troposphere—the upper regions of the earth's atmosphere within about 10 miles of the earth's surface—as a partial reflector of UHF signals, somewhat like the way that the ionosphere reflects lower frequency signals. Unfortunately, the troposphere is not the zero-permittivity, abrupt reflector of the UHF signals incident to it; rather, it acts as only a partial reflector, *scattering* the energy just as dust and smoke scatter the light beam of a motion-picture projector. The tradeoff advantage of the troposphere is its general uniformity and reliability: its elevation and characteristics remain constant. Since the troposphere is only a partial reflector, however, intense signal levels and high-directivity antennas are required to achieve communications. The technique does allow over-the-horizon transmission and reflection of UHF signals without the use of satellites or repeater stations.

Another special application of higher frequency signals is called *super-refraction*. Whereas troposcatter is more appropriately suited to the UHF region, superrefraction (also called *ducting*) is more practical with microwave signals in the SHF region of the spectrum. You should recall the effects of refractive bending of lower frequency signals in the ionosphere as a result of wavefronts obliquely entering a medium of changed characteristic impedance and refractive index. When the shorter wave microwave signals obliquely encounter even the warm-air layer of a *temperature inversion* at a high angle of incidence, the change in refractive index is great enough to bend the waves back toward the earth. When the signals return to the earth, they reflect back up to the inversion layer where they are again bent back toward the earth's surface. The effects of this repeated superrefraction and surface reflection thus achieve over-the-horizon transmission of microwave energy without the use of oscillators or repeaters.

Although such *ducting* depends on the presence and elevation of a temperature inversion for successful communications, there are many places in the world where such inversions are typical. Such inversions almost always exist over large bodies of water in tropical and subtropical climatic zones, for example, where microwave communications over hundreds of miles is practical.

Satellite Communications

Satellite communications is perhaps the most effective and reliable means of transmitting and receiving radio signals at different points on the earth's surface. Satellite communications techniques also provide for a wide range of extraterrestrial radio capabilities. As mentioned earlier in this chapter, radio signals are not *affected* beyond the earth's atmosphere and ionosphere; so even with the use of higher frequencies for space communications, it is the atmospheric limitations that are of major concern. The lower frequency limitation for successful transmission and reception through our atmosphere is about 100 MHz, primarily to refractive bending of signals below this frequency that pass through the ionosphere. The higher frequency limitation for radiation through our atmosphere falls between 12 and 20 GHz, where the atmospheric absorption develops high levels of attenuation in the signal.

There are some specialized microwave applications that operate at the higher SHF and EHF regions of the spectrum, but attenuation losses at these frequencies are significant. An additional factor, called *Faraday rotation,* affects signals in the frequency range from 100 to 1000 MHz. This effect actually rotates the polarization of the radiated signal as it passes through the ionosphere, and so receiving antennas used for satellite applications at these frequencies must be circularly polarized.

CONCLUSION

The content of this chapter is concerned with *antennas* and *propagation,* which are among the most complex topics of electronic communications. In addition to applying a number of specialized concepts, antenna theory relies heavily on basic transmission-line principles, as reviewed in Chapter 11. As noted in the introductory section of this chapter, the coverage here is in no way complete; but it should serve to provide you with a general perspective for successful further study in the field. Since this subject field is so complex, there is always a demand for the employment of people who are competent antenna engineers and technicians; so your interest in the subject is encouraged. The following problems are structured to apply the basic aspects of antenna topics that were defined in the *Learning Objectives* section of this chapter. Should you encounter difficulty with any of the problems or concepts, you may do well to review Chapter 11 in order to better prepare yourself for a review of this chapter.

Problems

1.   Given: A "rabbit ears" indoor TV antenna, with a length of balanced twin-lead line, to be used for 100 MHz FM broadcast radio reception.

Determine:
(a)   The length of each of the rabbit-ear arms and the angle of their separation for operation as a theoretically accurate 300 Ω (radiation resistant) dipole

(b)   The loss resistance and actual input impedance values if the antenna is 75% efficient, and the resulting VSWR with the 300 Ω input of the twin-lead and FM receiver

(c)   Your predicted recommendations for cutting the antenna to remove the capacitive and inductive effects and to reduce its radiation resistance for minimum VSWR

(d)   The actual absolute permeability and permittivity of the twin-lead line if the velocity of the signal propagating along its length is 0.8 that of free space

2.   Given: A half-wavelength antenna with 85% efficiency and measured 3 dB azimuth and elevation beamwidths of 90° and 60°, respectively.

Determine:

(a)   The antenna gain, in dB, in reference to an isotropic radiator

(b)   The antenna gain, in dB, in reference to a standard dipole

3.   Given: A 20 ft (6.1 meter) dipole antenna used for higher frequency short-wave radio communications, operating according to the impedance locus of Fig. 14-5.

Determine:

(a)   The resistive input impedances and frequencies for resonant operation of the antenna at 0.5, 1.0, and 1.5 wavelength modes

(b)   The rectangular-form input impedances at 30 and 40 MHz

(c)   The range of variable inductance required to tune the antenna for operation from 25 to 50 MHz

4.   Given: A requirement to categorize and define popular driven antenna types for specific applications.

Sketch and dimension:

(a)   An end-fire array to be used at 150 MHz for directional radiotelephone service

(b)   A vertically oriented collinear sleeve array to be used for omnidirectional radiotelephone service at 450 MHz

(c)   A Sterba curtain antenna to be used for broadcast radio service at 1 MHz

5.   Given: A requirement to categorize and define popular parasitic antenna types for specific applications.

Sketch and dimension:

(a)   A three-element beam to be used for directional short-wave radio broadcast service in the 19-meter band (*b*)   A seven-element Yagi-Uda array to be used for television reception at U.S. VHF TV channel 11 (199.25 MHz)

(*c*)   A three-element quad array to be used for radio communications at 50 MHz

6.   Given: A log-periodic antenna for broadband operation.

Sketch and dimension:

(*a*)   An LPDA to cover the U.S. VHF TV low band and FM broadcast band (54 to 108 MHz)

(*b*)   An LPDA to cover the U.S. VHF TV high band (174 to 216 MHz)

(*c*)   An LPDA to cover the U.S. UHF TV band (470 to 890 MHz)

7.   Given: An earth-station satellite dish antenna 8 meters in diameter, and with 6 dB loss in gain compared to a theoretical model.

Determine:

(*a*)   The approximate gain for the 4 GHz downlink signal

(*b*)   The approximate gain for the 6 GHz uplink signal

8.   Given: A requirement for a rooftop mobile-mount vertical "whip" CB (U.S. Citizens Band) antenna for operation at 27 MHz that uses the metal body of the vehicle as a ground plane.

Determine:

(*a*)   The length, in inches, of a quarter-wave (half of a half-wave, as the other half of the dipole is the vehicle body) dipole section whip

(*b*)   The length of the half-dipole section to bring the input impedance of the whip to 50 Ω

(*c*)   The complex input impedance of a whip half-dipole section 60 inches in length

(*d*)   The required inductance in the loading coil that will best match the 60-inch antenna to a 50 Ω line impedance

9.   Given: A requirement to establish an antenna test range for evaluating antennas in the UHF region above 300 MHz.

Determine:

(*a*)   The minimum distance between antennas if the transmitting antenna is fixed at 20 ft above the ground, and the receive antenna under test is at 10 ft above the ground

(*b*)   The percent ground-effect reflection when raising and lowering the receiving antenna that yields 800 μV maximum and 200 μV minimum signal levels at the antenna terminals

10. Given: A standard, 1-meter-square, 10-turn small loop antenna, used with a calibrated receiver to make field strength measurements of 1.5 MHz radiated signals at a distance of 20 km from a transmitting station.

Determine:
- (*a*) The field strength if the loop terminal voltage measures 200 μV
- (*b*) The approximate computed field strength at 15 and 50 km from the station
- (*c*) The approximate directional gain of the loop if its azimuth and elevation beamwidths are 120° and 360°, respectively, and its efficiency is 0.95 in reference to an isotropic radiator
- (*d*) The approximate directional gain of a substitute antenna under test at the 20 km distance if its terminal voltage measures 700 μV

11. Given: A requirement for recommending operating frequencies for specific broadcast communications services based on the effects of the earth and its atmosphere.

Identify and specify:
- (*a*) The conditions and frequency range for nighttime transmission of signals halfway around the earth by reflection from the ionosphere
- (*b*) The conditions and frequency range for daytime transmission of signals halfway around the earth by reflection from the ionosphere
- (*c*) The conditions and frequency range for daytime short hop reflections off the D and E layers of the ionosphere
- (*d*) The conditions and frequency range for overall ground-wave coverage of the area within 150 miles of a radio transmitting station

12. Given: The critical frequency reported for the combined F layers of the ionosphere at night to be 6 MHz.

Determine:
- (*a*) The highest muf with a transmitted signal entering the layer at 45°
- (*b*) The required angle of the radiation signal above the earth's surface to achieve ionospheric reflection of an 18 MHz signal
- (*c*) The estimated skip distance for the 18 MHz signal around the earth's surface

13. Given: A television transmitting antenna located on a mountaintop 2000 ft above the relatively flat plain of land to the east of the antenna.

Determine:
- (*a*) The radio horizon, in miles, for the signal to a rooftop receiving antenna at an elevation of 20 ft
- (*b*) The required height of a receiving antenna 100 miles distant from the transmitting station

INDEX